ANAEROBIC BACTERIA:
Role In Disease

Publication Number 940

AMERICAN LECTURE SERIES®

A Monograph in

The BANNERSTONE DIVISION *of*
AMERICAN LECTURES IN CLINICAL MICROBIOLOGY

Editor

ALBERT BALOWS, Ph.D.

Bacteriology Division
Bureau of Laboratories
Center for Disease Control
Public Health Service
U.S. Department of Health, Education, and Welfare
Atlanta, Georgia

$27.50 90-794

The British School of Osteopathy

* 2 8 7 2 *

This book is to be returned on or before
the last date stamped below.

BALOWS, A.

90-794

...OOL OF OSTEOPATHY,
...ST., LONDON. SW1Y 4HG
... 01 930 9254-8

THE BRITISH SCHOOL OF OSTEOPATHY
1-4 SUFFOLK ST., LONDON. SW1Y 4HG
TEL. 01 - 930 9254-8

ANAEROBIC BACTERIA:

Role In Disease

Edited by

ALBERT BALOWS, Ph.D.

Bacteriology Division
Bureau of Laboratories
Center for Disease Control
Public Health Service
U.S. Department of Health, Education, and Welfare
Atlanta, Georgia

RAYMOND M. DEHAAN, M.D.

Clinical Infectious Diseases Research
The Upjohn Company
Department of Infectious Diseases
and Clinical Pharmacology
Bronson Methodist Hospital
Kalamazoo, Michigan

V.R. DOWELL, JR., Ph.D.

Enterobacteriology Branch
Bacteriology Division
Bureau of Laboratories
Center for Disease Control
Public Health Service
U.S. Department of Health,
Education, and Welfare
Atlanta, Georgia

LUCIEN B. GUZE, M.D.

Research and Education
Veterans Administration Hospital
(Wadsworth) Los Angeles, California
Infectious Diseases Division
Harbor General Hospital
Torrance, California
Department of Medicine, UCLA
Los Angeles, California

CHARLES C THOMAS • PUBLISHER
Springfield • Illinois • U.S.A.

Published and Distributed Throughout the World by
CHARLES C THOMAS • PUBLISHER
Bannerstone House
301-327 East Lawrence Avenue, Springfield, Illinois, U.S.A.

This book is protected by copyright. No part of it
may be reproduced in any manner without written
permission from the publisher.

© *1974, by* CHARLES C THOMAS • PUBLISHER

ISBN 0-398-03074-X
Library of Congress Catalog Card Number: 74–1009

With THOMAS BOOKS *careful attention is given to all details of manufacturing and design. It is the Publisher's desire to present books that are satisfactory as to their physical qualities and artistic possibilities and appropriate for their particular use.* THOMAS BOOKS *will be true to those laws of quality that assure a good name and good will.*

Printed in the United States of America

BB-14

Library of Congress Cataloging in Publication Data

International Conference on Anaerobic Bacteria, Center
for Disease Control, 1972.
Anaerobic bacteria: role in disease

(American lecture series, publication no. 940. A
monograph in the Bannerstone division of American
lectures in clinical microbiology)
1. Bacteria, Anaerobic—Congresses. 2. Bacteria,
Pathogenic—Congresses. 3. Bacterial diseases—
Congresses. I. Balows, Albert, ed. II. Title.
[DNLM: 1. Bacterial infections—Congresses. WC200
I58a 1972]
QR89.5.I56 1972 616.01′4 74–1009
ISBN 0–398–03074–X

CONTRIBUTORS

WILLIAM A. ALTEMEIER, M.D.

Department of Surgery
University of Cincinnati
College of Medicine
Cincinnati, Ohio

HOWARD R. ATTEBERY, D.D.S.

Anaerobic Bacteriology Laboratory
Wadsworth Hospital Center, Veterans Administration
Department of Pediatric Dentistry
UCLA School of Dentistry
Los Angeles, California

ALBERT BALOWS, Ph.D.

Bacteriology Division
Bureau of Laboratories
Center for Disease Control
Atlanta, Georgia

JOHN G. BARTLETT, M.D.

Department of Medicine
UCLA School of Medicine
Infectious Disease Section
Veterans Administration Hospital
Sepulveda, California

HENRI BEERENS, Ph.D.

Laboratorie des Anaerobis
Institute Pasteur de Lille
Lille, France

HARVEY R. BERNARD, M.D.

Department of Surgery
The Albany Medical College of Union University
Albany, New York

KENNETH S. BRICKNELL

Anaerobic Bacteriology Laboratory
Wădsworth Hospital Center
Veterans Administration
Los Angeles, California

W.H. BRUMMELKAMP, M.D., Ph.D.

Universiteitsklinek voor Heelkunde
Binnengasthuis, Amsterdam
The Netherlands

ELIZABETH P. CATO, M.S.

Anaerobe Laboratory
Virginia Polytechnic Institute and State University
Blacksburg, Virginia

MICHAEL CHERINGTON, M.D.

Division of Neurology
University of Colorado Medical Center
Denver, Colorado

ANTHONY W. CHOW, M.D.

Infectious Diseases Division
Harbor General Hospital
Torrance, California
Department of Medicine, UCLA
Los Angeles, California

DAN DANIELSSON, M.D., Ph.D.

Department of Clinical
Bacteriology and Immunology
Regional Hospital
Orebro S-701 85
Sweden

RAYMOND M. DeHAAN, M.D.

Clinical Infectious Diseases Research
The Upjohn Company
Kalamazoo, Michigan

V.R. DOWELL, Jr., Ph.D.

Bacteriology Division
Bureau of Laboratories
Center for Disease Control
Atlanta, Georgia

B.S. DRASAR, Ph.D.

Bacteriology Department,
Wright-Fleming Institute
St. Mary's Hospital Medical School
London W2 1PG, England

ANN FALLON

Infectious Disease Section, Medical Service
Wadsworth Hospital Center
Veterans Administration and
UCLA School of Medicine
Los Angeles, California

JOEL FELNER, M.D.

Department of Medicine
Emory University School of Medicine
Atlanta, Georgia

SYDNEY M. FINEGOLD, M.D.

Infectious Disease Section
Wadsworth Hospital Center,
Veterans Administration
Department of Medicine
UCLA Medical Center
Los Angeles, California

LUCILLE K. GEORG, Ph.D.

Mycology Branch, Laboratory Division
Center for Disease Control
Atlanta, Georgia

RONALD J. GIBBONS, Ph.D.

Forsyth Dental Center
Boston, Massachusetts

C. GOMPERTZ, Ph.D.

Royal Post Graduate Medical School
Henfield Hospital
London W12, England

JAY S. GOODMAN, M.D.

Mercy Hospital, Inc.
Department of Medicine
University of Maryland School of Medicine
Baltimore, Maryland

SHERWOOD L. GORBACH, M.D.

Infectious Disease Section
Veterans Administration Hospital
Sepulveda, California

LUCIEN B. GUZE, M.D.

Research and Education
Veterans Administration
Hospital (Wadsworth)
Los Angeles, California
Infectious Diseases Division
Harbor General Hospital
Torrance, California
Department of Medicine, UCLA
Los Angeles, California

A.H.W. HAUSCHILD, Ph.D.

Food Research Laboratories
Department of National Health and Welfare
Tunney's Pasture
Ottawa, Ontario, Canada

PAUL E. HERMANS, M.D.

Department of Laboratory Medicine
Mayo Clinic
Rochester, Minnesota

M.J. HILL, Ph.D.

St. Mary's Hospital Medical School
Wright-Fleming Institute
Paddington
London W2, England

TOR HOFSTAD, M.D.

Department of Microbiology
The Gade Institute
University of Bergen
Schools of Medicine and Dentistry
500 Bergen, Norway

LILLIAN V. HOLDEMAN, Ph.D.

Anaerobe Laboratory
Virginia Polytechnic Institute and State University
Blacksburg, Virginia

EDWARD W. HOOK, M.D.

Department of Internal Medicine
University of Virginia School of Medicine
Charlottesville, Virginia

YUNG-YUAN KWOK

Anaerobic Bacteriology Laboratory
Wadsworth Hospital Center,
Veterans Administration
Los Angeles, California

ADOLF W. KARCHMER, M.D.

Department of Medicine
Harvard Medical School
Infectious Disease Unit
Massachusetts General Hospital
Boston, Massachusetts

DWIGHT W. LAMBE, Jr., Ph.D.

Department of Pathology
Emory University Hospital
Atlanta, Georgia

PHILLIP I. LERNER, M.D.

Infectious Disease Section
Veterans Administration Hospital
Cleveland, Ohio

WALTER J. LOESCHE, D.M.D., Ph.D.

Department of Oral Biology
University of Michigan School of Dentistry
Ann Arbor, Michigan

WILLIAM J. MARTIN, Ph.D.

Section of Clinical Microbiology
Department of Laboratory Medicine
Mayo Clinic and Mayo Foundation
Rochester, Minnesota

T.C. MICHAELSON, M.D.

Section of Infectious Diseases
Temple University Health Sciences Center
Philadelphia, Pennsylvania

W.E.C. MOORE, Ph.D.

Anaerobe Laboratory
Virginia Polytechnic Institute and State University
Blacksburg, Virginia

JEANETTE NORSEN, M.A.

Infectious Disease Service
Cook County Hospital
Chicago, Illinois

SVEN PERSSON

Department of Clinical Bacteriology and Immunology
Regional Hospital
Orebro S-701 85
Sweden

S. MADLI PUHVEL, Ph.D.

Department of Medicine
Division of Dermatology
UCLA School of Medicine
Los Angeles, California

RONALD M. REISNER, M.D.

Department of Medicine
Division of Dermatology
UCLA School of Medicine
Los Angeles, California

CHARLES W. RIETZ, M.D.

Department of Pathology
Emory University Hospital
Atlanta, Georgia

JON E. ROSENBLATT, M.D.

Infectious Disease Section
Veterans Administration and UCLA School of Medicine
Los Angeles, California

EDWARD B. ROTHERAM, Jr., M.D.

Division of Infectious Diseases
Allegheny General Hospital
Pittsburgh, Pennsylvania

H. SCHJÖNSBY, M.D.

Royal Post Graduate Medical School
Henfield Hospital
London W12, England

JOSEPH W. SEGURA, M.D.

Department of Urology
Mayo Clinic and Mayo Foundation
Rochester, Minnesota

JOHN C. SHERRIS, M.D.

Department of Microbiology
University of Washington School of Medicine
Seattle, Washington

LOUIS DS. SMITH, Ph.D.

Anaerobe Laboratory
Virginia Polytechnic Institute and State University
Blacksburg, Virginia

ALEX C. SONNENWIRTH, Ph.D.

Departments of Microbiology and Pathology
Washington University School of Medicine
The Jewish Hospital of St. Louis
St. Louis, Missouri

EARLE H. SPAULDING, Ph.D.

Department of Microbiology and Immunology
Temple University Health Sciences Center
Philadelphia, Pennsylvania

STUART E. STARR, M.D.

Department of Pediatrics
Emory University School of Medicine
Atlanta, Georgia

PAUL T. SUGIHARA

Anaerobic Bacteriology Laboratory
Wadsworth Hospital Center
Veterans Administration
Los Angeles, California

VERA L. SUTTER, Ph.D.

Anaerobic Bacteriology Laboratory
Wadsworth Hospital Center
Veterans Administration
Los Angeles, California

MORTON N. SWARTZ, M.D.

Department of Medicine
Harvard Medical School
Infectious Disease Unit
Massachusetts General Hospital
Boston, Massachusetts

ROBERT M. SWENSON, M.D.

Section of Infectious Diseases
Temple University Health Sciences Center
Philadelphia, Pennsylvania

SOAD TABAQCHALI, M.D.

Department of Medical Bacteriology
St. Bartholomew's Hospital
London, EC1A 7BE, England

HARAGOPAL THADEPALLI, M.D.

Infectious Disease Section
Veterans Administration Hospital
Sepulveda, California

CLYDE THORNSBERRY, Ph.D.

Bacteriology Branch
Laboratory Division
Center for Disease Control
Atlanta, Georgia

KAZUE UENO, D.V.M., Ph.D.

Department of Bacteriology
Gifu University School of Medicine
Gifu City, 500, Japan

VALERIE VARGO, M.S.

Department of Microbiology and Immunology
Temple University Health Sciences Center
Philadelphia, Pennsylvania

DAVID H. VROON, M.D.

Department of Pathology
Emory University Hospital
Atlanta, Georgia

JOHN A. WASHINGTON, II, M.D.

Section of Clinical Microbiology
Department of Laboratory Medicine
Mayo Clinic and Mayo Foundation
Rochester, Minnesota

ARNOLD N. WEINBERG, M.D.

Department of Medicine
The Cambridge Hospital
Harvard Medical School
Cambridge, Massachusetts

TRACY D. WILKINS, Ph.D.

Anaerobe Laboratory
Virginia Polytechnic Institute and State University
Blacksburg, Virginia

A.T. WILLIS, M.D., Ph.D.

Public Health Laboratory
Luton and Dunstable Hospital
Luton, LU4 ODZ, England

FOREWORD

The genesis of this series, *The American Lecture Series in Clinical Microbiology,* stems from the concerted efforts of the Editor and the Publisher to provide a forum from which well-qualified and distinguished authors may present, either as a book or monograph, their views on any aspect of clinical microbiology. Our definition of clinical microbiology is conceived to encompass the broadest aspects of medical microbiology not only as it is applied to the clinical laboratory but equally to the research laboratory and to theoretical considerations. In the clinical microbiology laboratory we are concerned with differences in morphology, biochemical behavior and antigenic patterns as a means of microbial identification. In the research laboratory or when we employ microorganisms as a model in theoretical biology, our interest is often focused not so much on the above differences but rather on the similarities between microorganisms. However, it must be appreciated that even though there are many similarities between cells, there are important differences between major types of cells which set very definite limits on the cellular behavior. Unless this is understood it is impossible to discern common denominators.

We are also concerned with the relationships between microorganism and disease—any microorganism and any disease. Implicit in these relations is the role of the host which forms the third arm of the triangle: microorganism, disease and host. In this series we plan to explore each of these: singly where possible for factual information and in combination for an understanding of the myriad of interrelationships that exist. This necessitates the application of basic principles of biology and may, at times, require the emergence of new theoretical concepts which will create new principles or modify existing ones. Above all, our aim is to present well-documented books which will be informative, instructive and useful, creating a sense of satisfaction to both the reader and the author.

Closely intertwined with the above *raison d'etre* is our desire to produce a series which will be read not only for the pleasure of knowledge but which will also enhance the reader's professional skill and extend his technical ability. *The American Lecture Series in Clinical Microbiology* is dedicated to biologists—be they physicians, scientists or teachers—in the hope that this series will foster better appreciation of mutual problems and help close the gap between theoretical and applied microbiology.

This book represents the published proceedings of an International Conference on Anaerobic Bacteria. It includes major aspects of the bacteriological and clinical considerations of the organisms and disease entities of what constitutes a most important segment of infectious diseases today. The Conference was held at the Center for Disease Control in Atlanta, Georgia, in November 1972, and was jointly sponsored by Emory University, The Upjohn Company, and the Center for Disease Control. The stimulus for the Conference came from the increasing isolation of anaerobic bacteria from clinical specimens and a mounting awareness of their association with significant numbers of infectious and other diseases. The Conference provided a meaningful interchange of information on the bacteriology and clinical relevance of anaerobes with a clear indication that this interchange paved the way for a better understanding of the microorganisms and the diseases they cause. There was no question but what the proceedings needed to be published in a permanent form so that the information could be widely disseminated and shared with clinicians and microbiologists throughout the world. It seemed only proper that *The American Lecture Series in Clinical Microbiology* serve as the vehicle for this purpose. One need only to glance at the Contents to realize the extent of coverage of the subject and the expertise of the authors in their presentations dealing with this most intriguing area of infectious disease. This book is designed not only to share the data and information that were presented, but also to provide direction for additional research leading to improved diagnoses and therapy. If these objectives are met (and all indications are that they will be), then this book will have well served its purpose.

Albert Balows, Ph.D.
Editor

PREFACE

The decision to publish the proceedings of the International Conference on Anaerobic Bacteria was made once it became apparent that the nature and scope of the material presented by the distinguished group of speakers would be in large demand, not only by those who attended the Conference, but also by clinicians and microbiologists who daily face the problem of anaerobic bacterial infections. The Editors soon recognized that their responsibilities were many and diversified. The publication of these proceedings which represents the culmination of the Conference, was accomplished with the excellent support and collaborative efforts of many individuals. There are three who are particularly deserving of mention because, without their "behind the scenes" efforts, the Conference and this published account of that Conference would have been most difficult to achieve and assuredly would not have had the measure of success that it has been accorded.

To Mrs. Ruby W. Caplan of the Center for Disease Control we express our sincere thanks and gratitude for the excellent manner in which she managed the entire Conference. To Mrs. Diana Schellenberg of The Upjohn Company who served as the Editors' editor with her mastery and skill in handling the manuscripts we are most grateful and thankful. We are indebted to Dr. Dwight W. Lambe, Jr., of Emory University for his capable assistance in the planning of the Conference and in handling the fiscal aspects of the Conference.

ALBERT BALOWS
RAYMOND M. DEHAAN
V.R. DOWELL, JR.
LUCIEN B. GUZE

CONTENTS

PART II
INTESTINAL FLORA AND ASSOCIATION WITH DISEASE
Moderators: R.M. DeHaan and W. McCabe

Chapter *Page*

PART III
ANAEROBIC INFECTIONS—GENERAL CONSIDERATIONS
Moderators: H.D. Isenberg and L.B. Guze

PART IV
ANAEROBIC INFECTIONS—DISEASE SYNDROMES
Moderators: E. Hook and L. Sabath

PART V
DISEASE SYNDROMES (CONTINUED) AND *IN VITRO* ANTIBIOTIC SUSCEPTIBILITY TESTING
Moderators: G. Jackson and W.M. Kirby

PART VI
CLINICAL RESULTS OF TREATMENT AND SUMMARIES
Moderators: L.B. Guze and D. Kaye

ANAEROBIC BACTERIA:
Role In Disease

CHAPTER I

ANAEROBIC BACTERIA PERSPECTIVES

ALBERT BALOWS

IT IS MY TASK to introduce the subject of this Conference, anaerobic bacteria, by placing it in proper perspective. This can best be done by paraphrasing a comment in a recent editorial on anaerobic infections: medicine and microbiology must synergize lest, once again, all the world become anaerobic (Medeiros, 1972).

Not too many years ago, we had no great problems in recognizing anaerobic diseases: botulism, tetanus and gas gangrene were distinct clinical entities, each caused by specific sporeforming clostridia. Today, the situation has taken a 180-degree turn. Anaerobic bacteriology is literally in a renaissance period of development, and with it comes a new approach to the clinical relevance of infections caused by anaerobes. Improvements in technology during the past 10 years have permitted more definitive bacteriologic and clinical studies which continue to emphasize the importance of anaerobes in human and animal disease. These studies have dispelled certain misconceptions—so prominent in the literature—in relation to the disease potential of these anaerobes that are indigenous to man. When we recall that the microbiota of man is heavily favored for the anaerobes (by factors of 10 : 1 in the vagina, oral cavity, and skin to as much as 1,000 : 1 in the large intestine), it is no wonder that life-threatening diseases caused by one or more of the endogenous anaerobes are more common than those due to anaerobic bacteria of exogenous origin.

Technological achievements make it possible to cultivate virtually all anaerobes that may be present in a clinical specimen. This is accomplished by using improved methods of specimen collection, coupled with one or more anaerobic systems and improved isolation media. A wide variety of differential media and sophisticated end-product determinations facilitate the classification of these anaerobes. More recently, several laboratories have directed their efforts toward the development of *in vitro* antimicrobic susceptibility techniques to aid in the selection of appropriate therapy.

At the outset of this Conference, we should note that some indigenous anaerobes appear to be frequently associated with human disease, whereas others, although present in large numbers in the normal microflora, seldom, if ever, are involved in human infections (Table I-I). For example, all

TABLE I-I

GENERA OF ANAEROBIC BACTERIA

Frequently encountered in significant infections	*Seldom, if ever, encountered in significant human infections*
Actinomyces	*Acidaminococcus*
Arachnia	*Borrelia*
Bacteroides	*Butyrivibrio*
Bifidobacterium	*Lachnospira*
Clostridium	*Lactobacillus*
Eubacterium	*Leptotrichia*
Fusobacterium	*Ruminococcus*
Peptococcus	*Succinimonas*
Peptostreptococcus	*Succinivibrio*
Propionibacterium	
Treponema	
Veillonella	

five subspecies of the *Bacteroides fragilis* group may be present in large numbers in normal gut flora. We rarely, if ever, isolate subspecies *ovatus* from clinical specimens associated with human infection, but the remaining four subspecies are isolated with almost regular frequency. To emphasize this point, let me show you a list (Table I-II) of anaerobic bacteria most frequently submitted to the CDC laboratories for identification (Dowell and Hawkins, 1972). These isolates are received from laboratories across the country, so this list is representative of those anaerobes most frequently isolated in the *average* hospital laboratory.

TABLE I-II

ANAEROBIC BACTERIA FROM HUMAN INFECTIONS MOST FREQUENTLY SUBMITTED TO THE CDC ANAEROBE LABORATORY: 1962–1972

1. Clostridia
   C. bifermentans
   C. butyricum
   C. cadaveris (*C. capitovale*)*
   C. innocuum
   C. limosum (*Clostridium* sp CDC group P-1)
   C. perfringens
   C. ramosum (*Catenabacterium filamentosum, Bacteroides terebrans*)
   C. septicum
   C. sordellii
   C. sporogenes
   C. subterminale
   C. tertium
2. Nonsporeforming gram-positive bacilli
   Actinomyces israelii
   Actinomyces odontolyticus
   Actinomyces naeslundii
   Arachnia propionica (*Actinomyces propionicus*)
   Bifidobacterium eriksonii (*Actinomyces eriksonii*)
   Eubacterium alactolyticum (*Ramibacterium* species)
   Eubacterium lentum (*Corynebacterium diphtheroides*)
   Eubacterium limosum
   Propionibacterium acnes (*Corynebacterium acnes*)
   Propionibacterium granulosum (*Corynebacterium granulosum*)
3. Nonsporeforming, gram-negative bacilli
   Bacteroides fragilis ss. *fragilis* (*B. fragilis*)
   Bacteroides fragilis ss. *thetaiotaomicron* (*B. variabilis*)
   Bacteroides fragilis ss. *vulgatus* (*B. incommunis*)
   Bacteroides melaninogenicus ss. *asaccharolyticus*
   Bacteroides melaninogenicus ss. *intermedius*
   Fusobacterium mortiferum (*Sphaerophorus ridiculosis*)
   Fusobacterium necrophorum (*Sphaerophorus necrophorus*)
   Fusobacterium nucleatum (*Fusobacterium fusiforme*)
4. Anaerobic cocci
   Peptococcus sp. CDC group 2
   Peptostreptococcus sp. CDC group 1
   Peptostreptococcus sp. CDC group 2
   Peptostreptococcus sp. CDC group 3
   Veillonella alcalescens
   Veillonella parvula

* Former name.

From the bacteriologic viewpoint, we have made considerable progress; the mystique of the anaerobic bacteria is not nearly so overwhelming as it was a dozen years ago. We also have developed a clearer understanding of the clinical importance of these anaerobes. Bacteremia caused by *Bacteroides fragilis* is an established clinical entity, with a consequence of which most clinicians are well aware. A striking association of *Clostridium septicum* with certain malignancies, such as leukemia and various types of carcinoma, has been established. *Propionibacterium acnes,* a well recognized member of the normal cutaneous flora and frequently discarded by the clinical laboratory as a "diphtheroid," has been definitely incriminated as an etiologic agent of subacute bacterial endocarditis and may also be involved in some cases of chronic meningitis and chronic actinomycotic-like illnesses.

This Conference stems from the desire to reappraise our observations and data so that we can better assess the role of anaerobic bacteria in disease and the directions future investigations should take.

REFERENCES

Dowell, V.R., Jr., and Hawkins, T.M.: *Laboratory Methods in Anaerobic Bacteriology. CDC Laboratory Manual.* Atlanta, Center for Disease Control, HEW, PHS, 1972.

Medeiros, A.E.: Once all the world was anaerobic. *N Engl J Med, 287:*1041, 1972.

PART I

NOMENCLATURE, TAXONOMY AND GENERAL METHODOLOGY

CHAPTER II

COLLECTION OF CLINICAL SPECIMENS AND PRIMARY ISOLATION OF ANAEROBIC BACTERIA

V.R. Dowell, Jr.

ABSTRACT: *Techniques for collection of clinical specimens and primary isolation of anaerobic bacteria are described briefly. The aspects emphasized are (1) proper selection of specimens to avoid contamination with normal flora and consequent erroneous results; (2) collection of specimens under anaerobic conditions; (3) use of fresh or prereduced primary isolation media; and (4) selective isolation procedures.*

Anaerobic bacteria associated with human infections are widely distributed in nature. Their habitats include soil, water, and the oral cavity, gastrointestinal tract, genitourinary tract, and skin of man and lower animals (Rosebury, 1962; Smith and Holdeman, 1968; Willis, 1969). Although there are a number of diseases involving anaerobic bacteria from exogenous sources (Table II-I), endogenous infections involving these microorganisms (Table II-II) are much more common. Factors commonly predisposing to endogenous anaerobic infections include surgery, malignancy, diabetes

TABLE II-I

DISEASES INVOLVING ANAEROBIC BACTERIA OF EXOGENOUS ORIGIN

Foodborne illnesses
Botulism
Clostridium perfringens gastroenteritis
Wound Infections
Tetanus
Myonecrosis ("gas gangrene")
Crepitant cellulitis
Benign superficial infections
Infection involving human or animal bite
Botulism
Septic abortion (contaminated instruments)

TABLE II-II

ANAEROBIC INFECTIONS OF ENDOGENOUS ORIGIN

Central Nervous System:
Brain abscess, meningitis
Dental, ENT, Respiratory:
Peridontal infection, otitis media, pharyngitis, tonsillitis, sinusitis, pulmonary abscess, pneumonia, empyema, "actinomycosis"
Intra-abdominal:
Appendicitis, diverticulitis, colitis, malabsorption disease; post-surgical infection including cellulitis, myonecrosis, and tetanus; abscess of any organ, peritonitis
Genitourinary:
Endometritis, salpingitis, ovarian abscess, infection of Bartholin's gland, urethritis, nephritis, abscess of kidney
Other:
Bacteremia, subacute bacterial endocarditis, osteomyelitis, perirectal abscess, decubitus ulcer, etc.

mellitus, arteriosclerosis, alcoholism and antibiotic, immunosuppressant, corticosteroid or X-irradiation therapy (Bornstein *et al.*, 1964; Felner and Dowell, 1971).

As discussed in previous publications, proper selection and collection of specimens; culture of the material as soon as possible after collection; use of fresh, properly reduced media; and provision of adequate anaerobic conditions are some of the more important considerations in the isolation of anaerobic bacteria (Dowell and Hawkins, 1968; Dowell, 1970).

COLLECTION OF SPECIMENS

Proper selection and collection of specimens are extremely important for laboratory confirmation of anaerobic infections. Otherwise, culture results may be misleading or meaningless. Since various anaerobic bacteria are present in large numbers in the normal flora of the oral cavity and gastrointestinal tract and some inhabit the genitourinary tract and skin (Dowell, 1970; Finegold, 1970), specimens likely to be contaminated with these microorganisms should not be cultured. These include the following:

1. Throat or nasopharyngeal swabs.
2. Sputum or bronchoscopic specimens.
3. Feces or rectal swabs.
4. Voided or catheterized urine.

5. Vaginal or cervical swabs (not collected by visualization via a speculum).
6. Material from superficial wounds or abscesses not collected properly to exclude surface contaminants.
7. Material from abdominal wounds obviously contaminated with feces, e.g. open fistula. On the other hand, *all* body fluids and tissues from sites not contaminated with normal flora should be cultured for anaerobic bacteria (Sutter *et al.*, 1972). Examples of acceptable clinical specimens for laboratory diagnosis of anaerobic infections are listed in Table II-III.

Exposure of clinical specimens to oxygen is probably one of the major reasons some laboratories have little success in cultivation of anaerobes. The most suitable samples for isolation of anaerobic bacteria are aspirated materials and tissue biopsies; swab samples are much less satisfactory (Dowell and Hawkins, 1968). When a sample is collected with a needle and syringe, the syringe should be cleared of air and the fluid injected immediately into a sterile *gassed out* tube or vial (Holdeman and Moore, 1972; Sutter *et al.*, 1972) to minimize exposure to oxygen. If it is not possible to culture tissue samples immediately, these should also be placed in an anaerobic environment until processed. A miniature anaerobic jar utilizing a 35-mm film container and steel-

TABLE II-III

EXAMPLES OF ACCEPTABLE CLINICAL SPECIMENS FOR LABORATORY DIAGNOSIS OF ANAEROBIC INFECTIONS

CNS:
Cerebrospinal fluid, abscess material, tissue biopsy
Dental, ENT:
Carefully aspirated material from abscesses and biopsied tissue
Pulmonary:
Transtracheal aspirate, tissue biopsy, direct lung aspirate, pleural fluid, "sulfur granules" from draining fistula
Intra-abdominal:
Aspirate from loculated abscess, ascitic fluid, tissue biopsy
Genitourinary:
Urine (suprapubic aspirate), aspirate from loculated abscess, tissue biopsy from normally sterile site, cervical material collected by direct visualization
Other:
Blood, bone marrow, bile, aspirated "joint" fluid, muscle biopsy from suspected gas gangrene, biopsied tissue from any normally sterile site

wool immersed briefly in an acidified copper sulfate solution to absorb oxygen has been recommended by Attebury and Finegold for this purpose (Sutter *et al.*, 1972).

If it is absolutely necessary to obtain material with swabs, at least three swabs should be used to provide sufficient material for microscopic examination and culture. These should be processed immediately or placed in prereduced anaerobically sterilized Carey-Blair medium with a *head* of oxygen-free CO_2 as recommended by Sutter *et al.* (1972) or in *gassed out* tubes containing oxygen-free CO_2 (Holdeman and Moore, 1972; Sutter *et al.*, 1972). Anaerobic transport tubes and vials are now available commercially (Hyland Laboratories, Scott Laboratories). For maximum recovery of anaerobes in the same relative proportions as present in the infected tissue, all clinical specimens except blood samples should be held at room temperature for no longer than 2 hours before processing. The specimens should not be refrigerated, as chilling is detrimental to some anaerobes (Dowell, 1970).

Because of the complexity of the subject, anaerobic blood culture techniques will not be discussed in detail in this report. However, the following points regarding blood cultures should be kept in mind:

1. Before performing the venipuncture the skin must be carefully cleansed and disinfected to avoid contamination with normal skin inhabitants such as *Propionibacterium acnes* (Dowell, 1970; Sutter *et al.*, 1972).

2. Precautions must be taken to expel air from the syringe or collection device to prevent introduction of air into the blood culture system.

3. An adequate volume of blood should be cultured to allow detection of microorganisms in small numbers (at least 5 ml, preferably 10 ml) and the volume of culture medium should be at least 9 to 10 times that of the blood sample.

4. The medium must be nutritionally adequate for the anaerobic bacteria and as fresh as possible. Since some strains of *Bacteroides melaninogenicus* require vitamin K compounds for growth, addition of menadione (vitamin K_1) is recommended (Dowell and Hawkins, 1968; Holdeman and Moore, 1972; Sutter *et al.*, 1972).

5. Use of a medium containing polyanetholsulfonate (Liquoid) is advantageous but it should be kept in mind that some anaerobic cocci and *B. melaninogenicus* may be inhibited (Sutter *et al.*, 1972).

6. The medium must be properly reduced and as free of molecular oxygen as possible. Therefore, the medium should contain a reducing agent such as cysteine to lower the oxidation-reduction potential and air should be excluded during storage. This can be accomplished most expediently by preparation of blood culture bottles containing prereduced anaerobically sterilized (PRAS) medium (Holdeman and Moore, 1972; Sutter *et al.*, 1972).

7. PRAS blood culture media are now available commercially (Hyland Laboratories, Scott Laboratories).

Laboratory Confirmation of Botulism

The most effective means for confirming a diagnosis of botulism is to demonstrate the presence of botulinal toxin in the serum of the patient (CDC, 1970). It is also useful to test stomach contents, feces and suspect foods for toxin and to culture foods, feces, stomach contents and excised tissue (wound botulism) for *Clostridium botulinum*. Specimens to be tested for botulinal toxins and/or *C. botulinum* should be collected and handled as follows:

BLOOD. Collect sufficient blood, before the patient is treated with botulinal antitoxin, to provide 15 to 20 ml of serum. Refrigerate serum at 4°C until examined.

FOOD. Leave unopened containers sealed. Collect others in sterile, unbreakable containers. Refrigerate at 4°C until examined.

GASTRIC CONTENTS, FECES. Collect in sterile, unbreakable containers and refrigerate at 4°C until examined.

EXCISED TISSUE. Collect in sterile *gassed out* tubes if possible and hold under anaerobic conditions at ambient temperature until examined.

If it is necessary to ship the materials to a distant laboratory, place specimens in a leakproof container, wrap with a cushioning material, and pack with ice or refrigerant in a second leakproof, insulated shipping container and ship by the most rapid means available.

MICROSCOPIC EXAMINATION OF CLINICAL MATERIALS

Numerous investigators have emphasized the importance of microscopic examination of clinical specimens when anaerobic infections are suspected (Dowell and Hawkins, 1968; Dowell, 1970; Finegold, 1970; Holdeman and Moore, 1972; Smith and Holdeman, 1968; Sutter *et al.*, 1972; Willis, 1969). Microscopic examinations can give immediate presumptive evidence of the presence of anaerobes and aid the physician in his choice of therapy. A Gram-stained direct smear should be examined from all types of clinical materials except blood. In addition to providing information on the cellular character of the specimen and the types of microorganisms present, the Gram-stained smear also aids the microbiologist in his choice of selective media for isolation of anaerobes from polymicrobic infections (Sutter *et al.*, 1972). Microscopic examination of unstained clinical material by regular light, dark-field and phase contrast microscopy is also useful, particularly when spirochetes are suspected (Holdeman and Moore, 1972) and in the examination of "sulfur granules" from patients with suspected actinomycosis. Acid-fast and Giemsa-stained direct smears can also provide useful information (Dowell, 1970).

PRIMARY ISOLATION

General Considerations

Media may become inhibitory for anaerobic bacteria if stored in the presence of oxygen (Aranki *et al.*, 1969; Dowell, 1970; Finegold, 1970; Hobbs *et al.*, 1971; Holdeman and Moore, 1972; Killgore *et al.*, 1973; Martin, 1971; Smith and Holdeman, 1968). For this reason, all media used for cultivation of anaerobes should be at optimum freshness; this is particularly true for plating media used for primary isolation.

Also, overheating of media during or after preparation should be avoided (Dowell and Hawkins, 1968; Smith and Holdeman, 1968). Plating media for primary isolation should be prepared on the day of use or freshly prepared media can be stored under anaerobic conditions for a period no longer than two weeks. The media can be stored in an anaerobic jar (Smith and Holdeman, 1968), an anaerobic glove box (Aranki *et al.*, 1969; Killgore *et al.*, 1973), or in an airtight cabinet with a constant flow of carbon dioxide as described by Martin (1971). Liquid media not prepared by the PRAS technique (Holdeman and Moore, 1972) should be stored in tightly capped tubes in the dark at room temperature for no longer than two weeks.

Provided clinical specimens are collected properly and the materials are cultured with fresh, properly reduced media, successful cultivation of anaerobes can be obtained by use of the GasPak anaerobe jar (BBL) or an anaerobic jar with a gas replacement method (Collee *et al.*, 1972; Killgore *et al.*, 1973). The anaerobic glove box system (Aranki *et al.*, 1969; Killgore *et al.*, 1973), and the roll-streak tube system (Holdeman and Moore, 1972) using PRAS media also give excellent results. Since most clinical laboratories do not use an anaerobic glove box or roll-streak systems at present the following isolation procedures are designed for use with anaerobic jars.

Media

All clinical specimens except blood should be cultured by both direct plating and enrichment procedures, and selective media should be employed if warranted. At a minimum, the following should be inoculated with each specimen:

One tube of chopped meat-glucose medium enriched with hemin-menadione solution (Dowell and Hawkins, 1968);

One tube of thioglycollate broth (BBL 135C or equivalent) enriched with 10 percent V/V sterile rabbit serum and hemin-menadione solution;

Two plates of blood agar containing menadione (Dowell and Hawkins, 1968)—incubate one plate anaerobically and the other in a candle jar or CO_2 incubator.

Unless prereduced, the liquid media (chopped meat-glucose and thioglycollate broth) should be heated in a boiling water bath for 10 minutes and cooled. The rabbit serum and hemin-menadione solution are added after the medium has been heated and cooled, taking care to avoid aeration.

If microscopic examination of the direct smear reveals a mixture of microorganisms, also inoculate appropriate selective media. Some of the more useful selective media and their use are listed in Table II-IV.

TABLE II-IV

SELECTIVE MEDIA* FOR ISOLATION OF ANAEROBIC BACTERIA

Medium	*Use*
1. Phenethyl alcohol blood agar† + menadione, 0.5 mcg/ml medium	Isolation of gram-negative and gram-positive anaerobes. Inhibits facultative gram-negative bacteria, including *Proteus* species.
2. Kanamycin blood agar§ + menadione, 0.5 mcg/ml medium	Isolation of gram-negative and gram-positive anaerobes. Inhibits most facultative gram-negative bacteria and some anaerobes.
3. Kanamycin-vancomycin blood agar§ + menadione, 0.5 mcg/ml medium	Isolation of *Bacteroides* species. Inhibits most facultative bacteria and other anaerobes.
4. Paromomycin-vancomycin† blood agar + menadione, 0.5 mcg/ml medium	Isolation of *Bacteroides* and *Fusobacterium* species.
5. Egg yolk agar#	Isolation of clostridia. Allows detection of lecithinase, lipase and proteolytic activities.
6. Egg yolk agar§ + neomycin, 100 mcg/ml	Isolation of clostridia. Inhibits some facultative and anaerobic bacteria.
7. Egg yolk agar ½ antitoxin (*C. perfringens* type A) plates††	Identification of *C. perfringens.*
8. Stiff blood agar (blood agar containing 4 to 6% agar)§	Inhibits swarming of clostridia.

* Never use selective media without nonselective media to serve as controls.
† See Dowell *et al.*, 1964.
§ See Dowell, 1970.
# See Dowell and Hawkins, 1968.
†† See Dowell, 1970; Hobbs *et al.*, 1971.

Spore Selection Techniques

In addition to selective media it is also useful at times to employ a spore selection technique for isolation of clostridia from mixed populations. Two techniques are commonly used, heat treatment and alcohol treatment (Dowell, 1970; Hobbs *et al.*, 1971). The following procedure for isolation of *Clostridium tetani* employs a heat treatment technique:

1. Macerate the excised tissue with sterile scissors and inoculate 2 tubes of freshly boiled (10 minutes) and cooled chopped meat-glucose medium or 2 tubes of PRAS chopped meat-glucose medium.
2. Heat one tube at 80°C for 10 minutes to destroy vegetative cells; cool.
3. Incubate both tubes under anaerobic conditions for 48 hours at 35 to 37°C.
4. After incubation examine Gram-stained smears of the heated and unheated cultures.
5. If slim gram-positive or gram-variable rods with or without terminal spores are present, subculture to plain blood agar (inoculate one streak across the plate to take advantage of the tendency of *C. tetani* to spread) and stiff blood agar (4 to 6% agar). Streak the stiff agar to obtain isolated colonies.
6. Incubate the plates of each medium in an anaerobic system for 24 to 48 hours at 35 to 37°C.
7. Look for the typical thin film of spreading growth of *C. tetani* on the plain blood agar and small translucent colonies on the stiff agar. Prepare Gram-stained smears of colonies on each medium and pick colonies of slim gram-positive rods with or without spores to tubes of chopped meat-glucose medium which have been prereduced as described previously. Incubate in an anaerobic system until good growth is obtained (usually 24 hours).
8. Check purity, inoculate differential media, and perform toxin neutralization tests to confirm identity.

The alcohol treatment is performed by mixing an aliquot of a mixed culture containing spores of clostridia with an equal volume of absolute ethanol and incubating at room temperature for one hour to kill vegetative cells (Dowell, 1970). Both a heat treatment procedure and alcohol treatment are routinely used for selective isolation of *Clostridium botulinum* (CDC, 1970; Dowell and Hawkins, 1968; Hobbs *et al.*, 1971).

Handling of Cultures after Inoculation

Inoculated media, particularly plating media, should be placed in an anaerobic system immediately to avoid undue exposure to air. Failure to do so will result in decreased recovery of anaerobes. One of the major advantages of the roll-streak tube system (Holdeman and Moore, 1972), the original roll tube procedure developed by Hungate (1950), Bryant (1963) and others (Holdeman and Moore, 1972) and the anaerobic glove box system (Aranki *et al.*, 1969; Killgore *et al.*, 1973) is that specimens are not exposed to aerobic conditions during culture procedures and the cultures can be incubated, inspected, and subcultured at any time without exposing the bacteria to air.

Recently Martin (1971) described a practical procedure to decrease exposure of cultures to air after inoculation; this shows promise. In his laboratory, as soon as a plating medium is inoculated the plate is immediately placed in a Brewer anaerobe jar receiving a constant stream of O_2-free CO_2. The lid assembly rests in the jar without fastening so that it can be quickly lifted and another plate inserted. After the jar is filled with plates of inoculated media, a GasPak Hydrogen-CO_2 envelope is placed in the jar a GasPak lid is secured as in the regular GasPak procedure. Another jar then serves as a container for gassing additional cultures with O_2-free CO_2 as they are inoculated.

Incubation, Colonial Observations and Subculture

The most suitable temperature range for primary isolation of clinically important anaerobes is 35 to 37°C with few exceptions (Dowell, 1970). The cultures should be incubated under anaerobic conditions for at least 48 hours (preferably 3–5 days) before jars are opened; otherwise slow growing anaerobes may not be detected. After the anaerobe jars are opened, inspect the colonies with a hand lens and a dissecting microscope and immediately subculture each colony type to appropriate prereduced media and place in an anaerobic system. In addition, examine a Gram-stained smear of each. After 24 to 48 hours incubation, confirm the identity of pure culture isolates with appropriate differential tests (Dowell and Hawkins, 1970; Holdeman and Moore, 1972).

REFERENCES

Aranki, A., Syed, S.A., Kenney, E.B., and Freter, R.: Isolation of anaerobic bacteria from human gingiva and mouse cecum by means of a simplified glove box procedure. *Appl Microbiol, 17:*568, 1969.

Bornstein, D.L., Weinberg, A.N., Swartz, M.N., and Kunz, L.J.: Anaerobic infections—review of current experience. *Medicine, 43:*207, 1964.

Bryant, M.P.: Symposium on microbial digestion in ruminants: identification of groups of anaerobic bacteria active in the rumen. *J Anim Sci, 22:*801, 1963.

Center for Disease Control: *Botulism in the United States and Review of Cases 1899–1969* and *Handbook for Epidemiologists, Clinicians, and Laboratory Workers.* Atlanta, DHEW, PHS, 1970.

Collee, J.G., Watt, B., Fowler, E.B., and Brown, R.: An evaluation of the GasPak system in the culture of anaerobic bacteria. *J Appl Bacteriol, 35:*71, 1972.

Dowell, V.R., Jr., Hill, E.O., and Altemeier, W.A.: Use of phenylethyl alcohol in media for isolation of anaerobic bacteria. *J Bacteriol, 88:*1811, 1964.

Dowell, V.R., Jr., and Hawkins, T.M.: *Laboratory Methods in Anaerobic Bacteriology.* Atlanta, National Center for Disease Control, PHS No. 1803, 1968.

Dowell, V.R., Jr.: Anaerobic infections. In: *Diagnostic Procedures for Bacterial, Mycotic and Parasitic Infections.* New York, American Public Health Assoc., Ch. 15, 1970.

Felner, J.M., and Dowell, V.R., Jr.: "Bacteroides" bacteremia. *Am J Med, 50:*787, 1971.

Finegold, S.M.: Isolation of anaerobic bacteria. In: Blair, J.E., Lennett, E.H., and Truant, J.P. (Eds.): *Manual of Clinical Microbiology.* Bethesda, American Soc. for Microbiology, Ch. 32, 1970.

Hobbs, G., Williams, K., and Willis, A.T.: Basic methods for the isolation of clostridia. In: Shapton, D.A. and Board, R.G. (Eds.): *Isolation of Anaerobes.* London, Acad Pr, 1971.

Holdeman, L.V., and Moore, W.E.C.: *Anaerobe Laboratory Manual.* Blacksburg, Va., Virginia Polytechnic Institute and State University, 1972.

Hungate, R.E.: The anaerobic mesophilic cellulolytic bacteria. *Bacteriol Rev, 14:*1, 1950.

Killgore, G.E., Starr, S.E., Del Bene, V.E., Whaley, D.N., and Dowell, V.R., Jr.: Comparison of three anaerobic systems for ioslation of anaerobic bacteria from clinical specimens. *Am J Clin Pathol, 59:*552, 1973.

Martin, W.J.: Practical method for isolation of anaerobic bacteria in the clinical laboratory. *Appl Microbiol, 22:*1168, 1971.

Rosebury, T.: *Microorganisms Indigenous to Man.* New York, McGraw, 1962.

Smith, L.DS., and Holdeman, L.V.: *The Pathogenic Anaerobic Bacteria.* Springfield, Thomas, 1968.

Sutter, V.L., Attebery, H.R., Rosenblatt, J.E., Bricknell, K.S., and Finegold, S.M.: *Anaerobic Bacteriology Manual.* Los Angeles, Dept. of Continuing Education in Health Sciences, Univ. Ext. and School of Medicine, UCLA, 1972.

Willis, A.T.: *Clostridia of Wound Infection.* London, Butterworth, 1969.

CHAPTER III

COMPARISON OF METHODS FOR ISOLATION OF ANAEROBIC BACTERIA

Jon E. Rosenblatt, Ann Fallon, and Sydney M. Finegold

Abstract: *Twenty-three clinical specimens were collected with precautions to avoid contamination by normal flora, transported under anaerobic conditions, and cultured, using liquid media, anaerobic jars, roll tubes and the anaerobic chamber, with both selective and nonselective media. Under these conditions the recovery of anaerobes using a GasPak jar or evacuation-replacement jar was as good as that with prereduced anaerobically sterilized (PRAS) media in roll tubes or on plates in the anaerobic chamber. Use of liquid media as a "backup" method was disappointing. Although overall recovery was consistently poorer with selective media than with nonselective media, the selective media were useful, particularly in isolation of anaerobes in the presence of heavy growth of facultative organisms.*

There has recently been an increased awareness of the role of anaerobic bacteria in clinical infections. Unfortunately, good data are not available on the frequency with which anaerobes are found in various infections. Older studies (Bornstein *et al.*, 1964; Goldsand and Braude, 1966; Stokes, 1958) provided incidence figures of 2 to 10 percent whereas a recent report (Moore *et al.*, 1969) indicated recovery of anaerobes from 85 percent of clinical specimens. These studies have all had serious deficiencies including inadequate methods of collection, transport, culture and identification and/or a lack of clinical correlation with the culture results.

Expansion of our knowledge about the role of anaerobes in clinical infections has been limited by the lack of adequate anaerobic bacteriology in most hospital laboratories. Although sophisticated methods for the cultivation of anaerobes have been developed in recent years, their use has been primarily restricted to research facilities. Many clinical microbiologists consider the anaerobic chamber and roll tube methods too complex and expensive for routine hospital laboratory use. Other workers consider the simpler anaerobic jar and liquid media methods inadequate. The study reported here was designed to determine which methods of anaerobic culture are adequate for the recovery of clinically significant anaerobic bacteria. Each clinical specimen was cultured using liquid media, anaerobic jars, roll tubes, and the anaerobic chamber utilizing both selective and nonselective media. The recovery of anaerobes by these different methods was then compared. The results of this comparative study, however, have validity only if certain other variables were rigidly controlled. Thus, specimens were carefully obtained, avoiding contamination with the body's normal bacterial flora, which abounds with anaerobes. Since anaerobic bacteria are variably susceptible to the toxic effects of oxygen, specimens were protected from oxygen during transit to the laboratory by placing them in anaerobic transport tubes immediately after collection. In addition, the clinical significance of each of the specimens reported here (with one exception noted in the discussion section) was verified by close observation of the patients' clinical course by physicians from the Infectious Disease Section of Wadsworth Hospital Center. This study, then, is a comparison of anaerobic culture methods using specimens from clinically significant infections, which were properly collected and anaerobically transported. The latter three factors are at least as important as the method of culture. Each is a vital link in the chain of isolation of anaerobic bacteria.

MATERIAL AND METHODS

SPECIMENS. Specimens were obtained avoiding contamination with normal flora. Transtracheal aspiration (TTA) and thoracentesis were used for pulmonary specimens. Closed abscesses were aspirated using a needle and syringe. Peritoneal fluid was aspirated by transabdominal percutaneous needle puncture and tissue specimens were obtained using aseptic surgical techniques. As soon as the specimens were obtained they were injected into a double stoppered transport tube containing oxygen-free CO_2 or N_2 (Attebery and Finegold, 1969) or placed in the anaerobic "mini-jar" (Attebery and Finegold, 1970).

Twenty-three specimens were included in the study. Seven pulmonary specimens were obtained by TTA. Material was obtained from two subphrenic abscesses, one liver abscess, an abscess adjacent to the bowel, a brain abscess, an abscess of the retroperitoneal space, an abscess of the chest wall, and a lung abscess (surgically drained). Three specimens of peritoneal fluid were obtained by transabdominal needle puncture. Five other specimens included aspirate from an osteomyelitic sinus tract, gallbladder aspirate, aspirate from cellulitis of a gangrenous toe, empyema fluid, and a tissue specimen obtained at surgery from a soft tissue infection and osteomyelitis of the foot. Because our hospital serves primarily male veterans, we processed no specimens from the female pelvis, which is a not uncommon site of anaerobic infection.

PROCESSING OF SPECIMENS. On arrival in the laboratory, the transport tube was placed in the anaerobic chamber and the specimen was transferred to a second rubber stoppered tube (appropriate for later use with a gassing cannula). The chamber-plated media were then inoculated using a standard .01-ml stainless steel loop. This same inoculum was used for all plates and tubes. Plates were placed inside an anaerobic jar which was sealed and removed from the chamber. Roll tubes were then inoculated according to procedures described by Cato *et al.* (1970). Plates for the GasPak jar (BBL) and an evacuation-replacement jar were then inoculated on the open bench. The GasPak jars were set up in the prescribed manner and evacuation-replacement jars were sealed and then a vacuum of 27 to 29 inches Hg was drawn. This vacuum was filled with oxygen-free N_2 and the procedure was repeated four times. A fifth vacuum was filled with a gas mixture containing 80 percent N_2, 10 percent H_2, and 10 percent CO_2. Each time an anaerobic jar was set up, a "rejuvenated" catalyst packet was used. These palladium coated alumina pellets were "rejuvenated" by heating to 160°C for 30 minutes. A methylene blue saturated paper strip indicator was enclosed in each jar.

The liquid media were the last to be inoculated. Each tube of thioglycollate media was boiled for five minutes before using. Plates for isolation of facultative organisms were inoculated with the remaining specimen and all containers were incubated at 37°C.

MEDIA. The following set of four plates was inoculated for both the GasPak and the evacuation-replacement anaerobic jars: A Brucella blood agar plate containing menadione (BMB); a neomycin blood agar plate (NEO); and a laked blood agar plate containing vancomycin and kanamycin (LKV). The preparation of these media has been described by Finegold (1970). A fourth plate consisted of the standard BMB medium to which 0.1% dithiothreitol (DTT) and 0.1% cysteine were added. Moore (1968) has shown that inclusion of DTT in blood agar plates facilitates recovery of the fastidious anaerobe *Clostridium novyi* type B in the anaerobic jar. The DTT plate was included in this study to determine whether this agent might enhance the recovery of other anaerobes as well. Media were stored by wrapping in Mylar bags and refrigerating at 4°C, and were generally used within one week of preparation.

Three kinds of liquid media were utilized. Commercial (BBL 11260) "regular" fluid thioglycollate medium (R-thio) was prepared according to the manufacturer's directions. The "concentrated thioglycollate" medium (C-thio) contained 29.5 g of the commercial fluid thioglycollate medium and 1 ml of a hemin solution in 750 ml of distilled water as well as 10% sterile ascitic fluid and 0.1% menadione. Prereduced anaerobically sterilized (PRAS) chopped meat glucose (CMG) medium was either purchased commercially or prepared in our laboratory. The PRAS CMG was not gassed out with CO_2 during the time the stopper was removed for inoculation. In spite of this less than optimum procedure the resazurin indicator did not become oxidized.

PRAS media were prepared according to methods in the V.P.I. *Outline* (Cato *et al.*, 1970) for CMG, roll tubes and plating media used in the anaerobic chamber. The composition of PRAS BMB, LKV, and NEO was similar to that of the non-PRAS media, except that sodium thioglycollate and cysteine HCl were included as reducing agents. Roll tubes contained PRAS brain heart infusion agar (BHIA), BMB (utilizing laked blood), LKV̂, and NEO (also utilizing laked blood) media. Plates containing PRAS media for use in the anaerobic chamber (BMB, LKV, and NEO) were poured in the chamber.

IDENTIFICATION OF ANAEROBES. Identification of the nonsporeforming gram-negative rods was by methods outlined by Sutter and Finegold (1971). Other anaerobes were identified according to the methods in the V.P.I. *Outline* (Cato *et al.*, 1970), including the gas chromatographic analysis of end-products of glucose fermentation. Most organisms were identified to genus and species, although a number could only be identified to the genus level.

EVALUATION OF RECOVERY OF ANAEROBES. A semi-quantitative scheme was devised to evaluate the growth of the anerobes based on scores of 1+ to 4+ for streaked plates and roll tubes. Any growth in liquid media was considered *positive* and the degree of growth was not graded. Because this is only a semi-quantitative method and the grading was necessarily different for plates and roll tubes, a difference in growth of at least two grades, i.e. from 1+ to 3+ or from 2+ to 4+, was required before the difference in recovery of organisms was considered significant. Plates and tubes were examined at 24, 48 and 72 hours and presence or absence of growth and quantitation of growth recorded. Subcultures for isolation and identification were made at 72 hours. The original plates and tubes were further observed at irregular intervals over the next two weeks for the appearance of new anaerobic growth.

RESULTS

ANAEROBIC CULTURE DATA. A single anaerobe was recovered in pure culture from four specimens, and multiple anaerobes were found in a single specimen. Four specimens yielded a single anaerobe mixed with one or more facultative organisms, while 14 yielded multiple anaerobes and facultative organisms. Thus, five of the 23 specimens contained only anaerobes and 18 contained mixtures of anaerobes and facultative organisms; 13 of these latter 18 contained multiple facultative organisms. Table III-I lists the 51 different isolates of anaerobic bacteria recovered from the 23 specimens. Twenty-seven were nonsporeforming gram-negative rods. The eleven anaerobic cocci included one gram-positive species that could be identified no further than "anaerobic coccus," and two so-called "microaerophilic streptococci." The latter two organisms would not grow on plates incubated in room air, but grew on anaerobic plates and in the presence of 2 to 5 percent CO_2 in air. In addition to the 51 isolates included in the present study, we have recovered the following organisms from clinical specimens using the GasPak method: *Bacteroides oralis*, *Bacteroides corrodens*, *Peptococcus prevotii*, and unspeciated members of the genera *Eubacterium* and *Propionibacterium*.

RESULTS OF COMPARATIVE STUDIES. Table III-II shows the comparative recovery of the anaerobes by the various isolation methods. The three different liquid media are presented as

TABLE III-I

ANAEROBES ISOLATED BY ALL METHODS

Organism	*No. Isolates*
Bacteroides fragilis	12
Bacteroides melaninogenicus	10
Fusobacterium nucleatum	4
Fusobacterium necrophorum	1
Clostridium perfringens	1
Clostridium ramosum	1
Clostridium subterminale	1
Clostridium species	1
Propionibacterium acnes	3
Propionibacterium species	3
Eubacterium species	3
Peptococcus asaccharolyticus	3
Peptostreptococcus magnus	2
Peptostreptococcus intermedius	1
Peptostreptococcus anaerobius	1
"Anaerobic coccus"	1
Veillonella species	1
"Microaerophilic streptococcus"	2
	51

TABLE III-II

COMPARATIVE RECOVERY OF ANAEROBIC BACTERIA

Organism	*Total No. of Isolates*	*GasPak Jar*	*Evacuation-replacement Jar*	*Roll Tubes*	*Chamber*	*Liquid Media**		
						CMG	*R-thio*	*C-thio*
B. fragilis	12	10 (83) †	11 (92)	10 (83)	10 (83)	10 (83)	9 (75)	9 (75)
B. melaninogenicus	10	9 (90)	9 (90)	8 (80)	7/9§ (78)	8/9§ (89)	7 (70)	6 (60)
F. nucleatum	4	4 (100)	4 (100)	4 (100)	4 (100)	4 (100)	3 (75)	3 (75)
F. necrophorum	1	1 (100)	1 (100)	1 (100)	1 (100)	1 (100)	1 (100)	1 (100)
Propionibacterium species	6	6 (100)	6 (100)	6 (100)	5/5§ (100)	4/5#§ (80)	5# (83)	5# (83)
Eubacterium species	3	3 (100)	3 (100)	3 (100)	3 (100)	3 (100)	3 (100)	3 (100)
Clostridium species	4	4 (100)	4 (100)	2 (50)	3/3§ (100)	3# (75)	3# (75)	3# (75)
Peptostreptococcus species	4	4 (100)	4 (100)	2 (50)	4 (100)	1/3#§ (33)	2# (50)	2# (50)
Peptococcus species	3	3 (100)	3 (100)	2 (67)	3 (100)	2# (67)	2# (67)	2# (67)
"Anaerobic coccus"	1	1	1	1	1	1	1	1
Veillonella species	1	1	1	1	1	1	1	1
"Microaerophilic Streptococcus"	2	2	2	2	2	2	2	2

* CMG = chopped meat glucose, R-thio = "regular" commercial thioglycollate, C-thio = "concentrated thioglycollate".
† Number of isolates recovered by each method (% recovery).
§ Indicates this method not used with one specimen.
# Same isolate not grown by each of the liquid media.

separate *methods*. Growth in any of the media used in a method was considered evidence of recovery by that method.

In regard to the gram-negative nonsporeforming rods, the regular thioglycollate (R-thio) and concentrated thioglycollate (C-thio) liquid media yielded consistently poorer recovery of *B. fragilis*, *B. melaninogenicus* and *F. nucleatum* than did the other methods, including the chopped-meat glucose (CMG) liquid medium. Recovery of these isolates was only 60 to 75 percent in R-thio or C-thio, compared to 78 to 100 percent using other methods. Using the semi-quantitative grading system, the roll tubes and plates showed no consistent differences in yield of organisms.

The liquid media failed to grow one isolate of *P. acnes* and one isolate of *C. ramosum*, while the roll tubes failed to grow one isolate of *C. perfringens* and one isolate of *C. ramosum*. Otherwise, recovery rates for the gram-positive rods were 100 percent. Semi-quantitative grading of yield showed no consistent differences between the methods.

Recovery of the anaerobic cocci was 100 percent by each method, with the following exceptions: the liquid media and roll tubes failed to grow two isolates of *Peptostreptococcus* (one *P. intermedius* and one *P. anaerobius*) and one isolate of *Peptococcus asaccharolyticus*. The semi-quantitative grading system revealed no differences in yield by the different methods.

Overall, liquid media were disappointing as a "backup" method. There were only two instances in which isolates not recovered by other methods grew in liquid media. In contrast, there were eight isolates recovered by at least one of the other methods which were not isolated from liquid media.

Table III-III summarizes recovery of anaerobic isolates by different media. Data from the GasPak jar and the evacuation-replacement jar are grouped together under "Anaerobic Jars," since the same media were used in both. Recovery of nonsporeforming gram-negative rods, of gram-positive rods, and of anaerobic cocci was consistently poorer on selective media than on nonselective media. Nevertheless, the selective

TABLE III-III

COMPARATIVE % RECOVERY OF ANAEROBES ON SELECTIVE AND NONSELECTIVE MEDIA

Organisms	*Anaerobic Jars*		*Roll Tubes (PRAS)**		*Chamber (PRAS)*	
	Nonselective BMB-DTT†	*Selective LKV-NEO§*	*Nonselective BHIA-BMB#*	*Selective LKV-NEO*	*Nonselective BMB*	*Selective LKV-NEO*
Nonsporeforming gram-negative rods	85%	65%	80%	63%	85%	70%
Gram-positive rods	96%	29%	83%	21%	100%	23%
Anaerobic cocci	100%	55%	67%	45%	100%	64%

* Prereduced, anaerobically sterilized media.

† Combined recovery on Brucella blood agar (with menadione) plates and Brucella blood agar plates containing dithiothreitol ("nonselective").

§ Combined recovery on Brucella blood agar plates containing neomycin and laked blood agar plates containing kanamycin and vancomycin ("selective").

Combined recovery on Brucella blood agar (containing menadione) roll tubes and on tubes containing brain-heart infusion agar ("nonselective").

media were considered very useful. In particular, specimens containing heavy growth of facultative organisms were more easily evaluated using selective media which inhibited the growth of most of these organisms. The inclusion of laked blood in the LKV medium greatly enhanced the appearance of the brown pigment of *B. melaninogenicus*, making colonies of this organism easily recognizable.

Colony size was not noticeably larger, growth was not superior by semi-quantitative grading, and colonies did not appear earlier on one type of media or in one specific method compared to any of the others.

Roll tubes were technically more difficult to work with and more time consuming than the other methods. Colonies were often not well separated and it was difficult to pick them cleanly. Specimens containing heavy growth of facultative bacteria were especially troublesome, necessitating extensive subculturing for isolation of individual colony types. Selective media were less helpful in roll tubes than with the other methods. Inclusion of blood (even laked blood) in roll tube media made colony types less recognizable. Different colony types were more difficult to recognize and less distinctive, even with the use of a dissecting microscope. The characteristic colonial morphology described for many anaerobes was not apparent on examination of the roll tubes. The larger diameter (25-mm) tubes were used in the latter part of the study and were preferable in terms of colony separation and recognition, but recovery rates did not improve.

DISCUSSION

Older methods of anaerobic culture, in particular the anaerobic jar and liquid media containing reducing agents have been criticized as being inadequate for the isolation of many important anaerobes. Moore *et al.* (1970) reported the recovery of 144 anaerobic isolates from 81 clinical specimens using the roll tube method and stated that in their experience over 85 percent of clinical specimens have contained obligate anaerobes. This is certainly a remarkable figure compared to older incidence figures of 2 to 10 percent. However, their report provided no assurances that their specimens came from patients with clinically significant infections or were collected avoiding contamination with normal flora. Since this was not a comparative study, they did not definitely demonstrate the superiority of roll tubes for the isolation of clinically significant anaerobes.

Vargo *et al.* (1971) did perform a comparative study and showed significantly better recovery of anaerobes in an anaerobic research laboratory using PRAS media and the anaerobic chamber or jars than in a routine clinical laboratory. Of course, in their study there were a great many variables in addition to culture methods, such as delays in specimen processing and differences in the experience of the technologists with anaerobic bacteriology.

Although McMinn and Crawford (1970) reported that the combined use of thioglycollate and a GasPak jar recovered only 28.6 percent of anaerobes isolated by the roll tube method, their data suggest that many of their specimens (from the oral cavity, sputum, ulcers, drain sites and urine) were contaminated with normal flora. Furthermore the authors state that the specimens were transported and stored under aerobic conditions. Spears and Freter (1958), Drasar (1967) and Aranki *et al.* (1969) all demonstrated the superiority of PRAS media in roll tubes or the anaerobic chamber over recovery of anaerobes by conventional jar methods. However, these were research studies dealing with normal intestinal flora, not clinical specimens.

The data in the present study indicate that when clinical specimens are obtained avoiding contamination with normal flora, transported in an anaerobic container, and immediately placed under anaerobic conditions, the recovery of anaerobes with the jar method is as good as with PRAS media in roll tubes or on plates in the anaerobic chamber. A corollary of this conclusion is that the isolation of clinically significant anaerobes depends more on proper collection and transportation of specimens than on the use of culture methods more sophisticated than the anaerobic jar. The jar method has been criticized because of excessive exposure of the specimen to oxygen during inoculation and the oxidized state of the media, and also because the methylene blue indicator in the GasPak jar can take as long as 1½ to 2 hours to become reduced. The reduction of the indicator does not occur until the oxygen concentration has been lowered to 0.1% (Eh, +11mv). However, the jar atmosphere has been found to contain less than 3 percent oxygen within 25 minutes after the jar was set up (J. Finegold, personal communication). Loesche (1969) has shown that anaerobes similar to those isolated in this study ("moderate anaerobes") grow maximally in oxygen tensions of 3 percent or less. Moreover, Loesche's "moderate anaerobes" had only a 20 percent reduction in colony counts after 80 to 100 minutes exposure to room air. In view of this data our good recovery rates with the anaerobic jar should not be too surprising. Attebery *et al.* (1971) have reported the isolation of so-called extremely oxygen sensitive (EOS) anaerobes from the intestinal flora which will die if exposed to room air for more than five minutes. However, since none of these EOS organisms was recovered in the present study (which included chamber and roll tube methods) they may not be significant causes of clinical infections and would therefore not be "missed" by the use of jar techniques alone.

The evacuation-replacement jar might have a theoretical advantage over the GasPak jar in the more rapid establishment of a reduced atmosphere in the former. There is no wait for generation of hydrogen and subsequent conversion of large amounts of oxygen to water as occurs in the GasPak system. However, the present study did not show any quantitative or qualitative differences between the two jar methods in recovery of anaerobes. The GasPak jar might be made more efficient by several means, including pump evacuation of air after the jar has been sealed and the storage of media and inoculated plates in jars constantly flushed with oxygen-free CO_2 as described by Martin (1971).

The importance of freshly prepared media in anaerobic bacteriology has been stressed by some workers. In the present study, media which were stored (refrigerated) in Mylar bags for approximately one week performed as well as PRAS media which were stored anaerobically. In addition, recent observations in our laboratory suggest that media prepared in room air may be reduced by overnight storage in the anaerobic chamber and will function as well as PRAS media even in intestinal flora studies.

Anaerobic jar methods are undoubtedly not suitable for studies of the normal intestinal flora where many fastidious, oxygen susceptible anaerobes are found. Moreover, some clinical laboratories processing large numbers of anaerobic cultures may find it more efficient to use the anaerobic chamber or roll tubes. Proponents of roll tubes have cited the ability to inspect each individual tube without disturbing the anaerobic atmosphere as an important advantage. The necessity to open an anaerobic jar and expose all plates therein to room air, in order to inspect one or several specific plates, is certainly a disadvantage. In this regard, newer developments should stress the need for smaller, individual anaerobic systems utilizing plates rather than larger jars containing increased numbers of plates.

Although CMG performed better than the other liquid media in this study, the overall recovery of anaerobes in liquid media was disappointing. There were eight instances in which anaerobes which did not grow in liquid media were recovered by other methods, but only two instances in which the reverse was true. Suboptimal use of CMG may have contributed to this poor record. CMG is a PRAS medium, but we removed the stopper for inoculation without gassing the tube with oxygen-free CO_2 or N_2. We were attempting to simulate conditions in the *average* hospital laboratory, which would not have a gassing apparatus. Even brief exposure of PRAS media to oxygen can cause some oxidation of the reducing agents, leading to a poorer anaerobic environment. However, we must note that our indicators did not become oxidized. The availability of PRAS CMG in anaerobic vials with rubber diaphragm stoppers would eliminate this problem. The specimen could be inoculated into the CMG vial with a syringe and needle without exposure of either the specimen or medium to oxygen.

The use of selective media (LKV and NEO) in this study was of great benefit in the isolation of anaerobes in the presence of heavy growth of facultative organisms. Considerable time and effort was saved. Since the majority of specimens from patients with anaerobic infections will contain mixtures of anaerobic and facultative bacteria, the use of selective media is strongly recommended for the clinical laboratory. However, since there is a decrease in recovery of most anaerobes on selective media, a suitably enriched nonselective medium should always be inoculated along with the selective medium.

Only clinically significant anaerobic isolates were included in this study, with one exception. This was a pure culture of a *Propionibacterium* species recovered from a TTA. This patient's course suggested his lung disease had a noninfectious etiology. Propionibacteria were also recovered from five other specimens, but in mixed culture with other anaerobes and (in four instances) facultative organisms. These organisms are part of the normal skin flora and their role in causing disease has not been adequately determined. However, several authors (Betts *et al.*, 1971; Moore *et al.*, 1969; Smith and Holdeman, 1968) have reported experiences suggesting a pathogenic role for propionibacteria. The data from our study do not allow us to evaluate the specific contribution of the propionibacteria isolates to the infections with which they were associated.

The results of this study should have considerable importance for the clinical laboratory. Adequate anaerobic work using conventional plates and the anaerobic jar can be performed as long as proper collection and transport of specimens is carried out. Strong efforts should be directed toward overcoming deficiencies in the latter two areas. Anaerobic transporters can be easily prepared in the laboratory or can be purchased commercially. A description of anaerobic transport methods can be found in the *Anaerobic Bacteriology Manual* prepared by the staff of the Wadsworth (V.A.) Hospital Center Anaerobic Bacteriology Laboratory (Sutter *et al.*, 1972). Suitable collection procedures are also of vital importance to avoid the culturing of anaerobic normal flora contaminants, leading to meaningless culture results, improper antimicrobial therapy and accumulation of erroneous information about clinical anaerobic infections. Finegold *et al.* (1972) have described proper methods of collection of specimens for anaerobic culture. In a recent study, Bartlett *et al.* (1972) concluded that TTA is a suitable collection method for reliably establishing the bacteriologic diagnosis in anaerobic pulmonary infections.

REFERENCES

Aranki, A., Syed, S.A., Kenney, E.B., and Freter, R.: Isolation of anaerobic bacteria from human gingiva and mouse cecum by means of a simplified glove box procedure. *Appl Microbiol, 17:*568, 1969.

Attebery, H.R., and Finegold, S.M.: Combined screw-cap and rubber-stopper closure for Hungate tubes (pre-reduced anaerobically sterilized roll tubes and liquid media). *Appl Microbiol, 18:*558, 1969.

Attebery, H.R., and Finegold, S.M.: A miniature anaerobic jar for tissue transport or for cultivation of anaerobes. *Am J Clin Pathol, 53:*383, 1970.

Attebery, H.R., Nastro, L.J., and Finegold, S.M.: An extremely oxygen-sensitive *Peptococcus* from normal human feces. In: *Bacteriol. Proc.—1971.* Abstr. #M265, p. 108. Bethesda, Am Soc for Microbiol, 1972.

Betts, R.F., Short, H., and Dowell, V.R.: *Propionibacterium acnes:* a clinical pathogen. Abstr. #141, p. 71. Paper presented at 11th Interscience Conf. on Antimicrob. Agents and Chemotherapy. Atlantic City, Oct. 19–22, 1971.

Bartlett, J., Finegold, S.M., and Rosenblatt, J.E.: Transtracheal aspiration (TTA) in diagnosis of anaerobic pulmonary infection. Abstr. #132, p. 70. Paper presented at 12th Interscience Conf. on Antimicrob. Agents and Chemotherapy. Atlantic City, Sept. 26–29, 1972.

Bornstein, D.L., Weinberg, A.N., Swartz, M.N., and Kunz, L.J.: Anaerobic infections—review of current experience. *Medicine, 43:*207, 1964.

Cato, E.P., Cummins, C.S., Holdeman, L.V., Johnson, J.L., Moore, W.E.C., Smibert, R.M., and Smith, L.DS.: *Outline of Clinical Methods in Anaerobic Bacteriology.* Blacksburg, Virginia Polytechnic Institute and State University, 1970.

Drasar, B.S.: Cultivation of anaerobic intestinal bacteria. *J Pathol Bacteriol, 94:*417, 1967.

Finegold, S.M.: Isolation of anaerobic bacteria. In: Blair, J.E., Lennette, E.H., and Truant, J.P. (Eds.): *Manual of Clinical Microbiology.* Bethesda, Am Soc for Microbiology, 1970, pp. 265–269.

Finegold, S.M., Rosenblatt, J.E., Sutter, V.L., and Attebery, H.R.: Collection of specimens. In: *Scope Monograph on Anaerobic Infections.* Kalamazoo, Upjohn, 1972, pp. 52–55.

Goldsand, C., and Braude, A.I.: Anaerobic infections. *DM,* Nov. 1966.

Loesche, W.J.: Oxygen sensitivity of various anaerobic bacteria. *Appl Microbiol, 18:*723, 1969.

Martin, W.J.: Practical method for isolation of anaerobic bacteria in the clinical laboratory. *Appl Microbiol, 22:*1168, 1971.

McMinn, M.T., and Crawford, J.J.: Recovery of anaerobic microorganisms from clinical specimens in pre-reduced media versus recovery by routine clinical laboratory methods. *Appl Microbiol, 19:*207, 1970.

Moore, W.B.: Solidified media suitable for the cultivation of *Clostridium novyi* type B. *J Gen Microbiol, 19:*207, 1970.

Moore, W.E.C., Cato, E.P., and Holdeman, L.V.: Anaerobic bacteria of the gastrointestinal flora and their occurrence in clinical infections. *J Infect Dis, 119:*641, 1969.

Smith, L.DS., and Holdeman, L.V.: Propionibacteria and anaerobic corynebacteria. In: *The Pathogenic Anaerobic Bacteria.* Springfield, Thomas, 1968, pp. 138–146.

Spears, R., and Freter, R.: Improved isolation of anaerobic bacteria from the mouse cecum by maintaining continuous strict anaerobiosis. *Proc Soc Exp Biol Med, 124:*903, 1967.

Stokes, E.J.: Anaerobes in routine diagnostic cultures. *Lancet, 1:*668, 1967.

Sutter, V.L., Attebery, H.R., Rosenblatt, J.E., Bricknell, K.S., and Finegold, S.M.: *Anaerobic Bacteriology Manual.* Los Angeles, Univ. of Calif., LA Ext. Div., 1972.

Sutter, V.L., and Finegold, S.M.: Antibiotic disc susceptibility tests for rapid presumptive identification of gram-negative anaerobic bacilli. *Appl Microbiol, 21:*13, 1971.

Vargo, V., Michaelson, T.C., Spaulding, E.H., Vitagliano, R., Swenson, R.M., and Forsch, E.: Comparison of a pre-reduced anaerobic method and GasPak for isolating anaerobic bacteria. In: *Bacteriol. Proc.—1971.* Abstr. #M269, p 109. Bethesda, Am Soc for Microbiol, 1972.

CHAPTER IV

A COMPARISON OF TWO PROCEDURES FOR ISOLATING ANAEROBIC BACTERIA FROM CLINICAL SPECIMENS

E.H. Spaulding, Valerie Vargo, T.C. Michaelson and R.M. Swenson

Abstract: *A series of 700 paired clinical specimens were compared by two procedures. One was a collection, culture and GasPak procedure being used routinely in a clinical bacteriology laboratory in 1970. The other was an anaerobic collection tube—continuously anaerobic procedure using prestored media and an anaerobic chamber (AC-PS-AC). The latter produced nearly 3 times as many positive specimens and 5.7 times as many anaerobic isolates.*

The results of several other smaller studies indicated strongly that these differences can be attributed as much, or more, to inadequacies in the way the routine procedure was carried out as to the superiority of the AS-PS-AC procedure.

A recommended routine GasPak procedure, which corrects the recognized deficiencies of the earlier methods, is outlined.

In 1970 we initiated a comparative study of two procedures for isolating anaerobic bacteria from clinical specimens. The comparison was between: (1) the methods in routine use in June 1970 in the Temple University Hospital and Clinical Diagnostic Bacteriology Laboratory and (2) an anaerobic-collection, prestored media-anaerobic chamber (glove box) procedure (AC-PS-AC) which had just been adopted in our newly established Anaerobe Laboratory. The major differences in these two procedures are listed in Table IV-I.

The primary objectives of this study were to (1) determine whether a substantially greater number of anaerobes would be isolated in our Anaerobe Laboratory than in our Diagnostic

This investigation was supported by a grant from The Upjohn Company, Kalamazoo, Michigan, and by funds from the Temple University Health Sciences Center.

We acknowledge with gratitude the invaluable participation in this study of Rose Vitagliano, R.N., and Delores Dolan, R.N.; and also the interest and cooperation of Dr. Kenneth R. Cundy and Evalyn Bernhardt; and especially the technical assistance of Elaine Forsch, Maria Carolina Martinez and Myroslava Korzeniowski.

Laboratory and (2) decide whether our Diagnostic Laboratory should consider adoption of an AC-PS-AC type of procedure for routine culturing of clinical specimens.

The results presented in this report were obtained with 700 sets of paired specimens, each pair collected simultaneously by the same person. One of the specimens was processed by the Diagnostic Laboratory, the other by the Anaerobe Laboratory.

METHODS

COLLECTION AND TRANSPORT. Most of the specimens were obtained under the direct supervision of one of the two specially trained infection control nurses who are members of our Anaerobe Group, or of one of the physician-authors of this report (T.C.M. and R.M.S.).

Most of the specimens destined for the Diagnostic Laboratory were obtained with a swab or were fluids or exudates. The collection tubes were aerobic. Transport to the Diagnostic Laboratory was via the regular delivery system; the average transit time is estimated as having been 1 to 2 hours.

In contrast, great care was taken to exclude air from the Anaerobe Laboratory specimens. We followed closely the anaerobic collection method of Attebery and Finegold (1969). Material obtained by needle aspiration was injected into a screw-capped, double-closure collection tube. This is a very satisfactory collection system, but we also made available, in addition to the usual empty gassed-out (O_2-free CO_2) collection tube, a second tube containing 1 ml of prereduced anaerobically sterilized (PRAS) brain-heart infusion broth-supplemented (BHI-S) as described in the VPI *Anaerobe Laboratory Manual* (Holdeman and Moore, 1972). The latter tube is needed when the infected site is dry. About 0.5 ml, which has been withdrawn with a needle and syringe, is used to flush the site (usually a wound) and a portion returned to the tube. This broth tube was also useful for flushing the needle when the quantity of specimen obtained was scanty.

DIAGNOSTIC LABORATORY CULTURE PROCEDURE. Upon arrival in the laboratory specimens were inoculated promptly in duplicate to commercially prepared trypticase-soy 5 percent sheep blood agar plates which were used 3 to 7 days after delivery by the manufacturer. The freshly inoculated aerobic plates were placed in the incubator without delay; the other ones (anaerobic) were placed in jars but sometimes remained exposed to air as long as 2 hours before the jars were GasPaked. The packets of palladium GasPak catalyst were not always reactivated by heat before reuse. After 24 hours incubation all jars were opened and the culture plates examined for growth. A minority of anaerobic cultures were discarded at this time; the others remained in contact with air for an estimated average of 1 to 2 hours before being reGasPaked for a second incubation of 24 to 48 hours.

It is our impression that the procedure just described is fairly representative of that being used in 1970—and even today—in many, if not the majority, of American and Canadian clinical bacteriology laboratories which do anaerobic cultures.

ANAEROBE LABORATORY CULTURE PROCEDURE. Anaerobic collection tube specimens were processed under continuously anaerobic conditions and cultured on reduced enriched media in general accordance with the principles outlined by Hungate (1950) for the isolation of highly O_2-sensitive members of the rumen microflora. Whereas Hungate employed roll tubes, we found this method inconvenient for our application. Instead, we adopted the anaerobic chamber (glove box) method of Aranki *et al.* (1969) which enabled us to carry out primary isolations, as well as subcultures, on agar plates. Our prereduced media, except for the BHI-S blood agar plate medium used for primary culture, were also anaerobically sterilized (PRAS) as described by Holdeman and Moore (1972). We refer to the procedure about to be described as AC-PS-AC (*a*naerobic *c*ollection-*p*re*s*tored *a*naerobic *c*hamber).

With very few exceptions the specimens were collected in such a manner as to avoid as much as possible contamination with normal microbial flora, including use of such procedures as transtracheal aspiration, culdocentesis, etc.

As noted in Table IV-I, these specimens were transported

TABLE IV-I

COMPARISON OF ANAEROBIC ISOLATION PROCEDURES IN 1970–1972 STUDY

	GasPak *Diagnostic Laboratory*	*AC-PS-AC** *Anaerobe Laboratory*
Collection	Aerobic tube	Gassed-out stoppered tube
Transport	Routine	Immediate
Medium	Trypticase-soy sheep blood agar plate (aerobic)	Brain-heart infusion agar supplemented with 5% sheep blood (Holdeman and Moore, 1972)
Incubation	GasPak 24 hr (+24 hr)	Anaerobic chamber 48 or 72 hr
Identification	Gram stain and aerobic subculture	Speciation (Holdeman and Moore, 1972)

* Anaerobic chamber—prestored media—anaerobic chamber

to the Anaerobe Laboratory without delay, generally by messenger who was often on hand at the time of collection. Upon arrival the collection tube was passed through the lock of the anaerobic chamber (AC), and a portion of the specimen aspirated by needle and syringe inoculated to the surface of one plate of BHI-S agar base medium to which had been added after sterilization vitamin K (menadione) and 5 percent sheep blood (Cato *et al.*, 1970). Plates of this medium were prepared aerobically two or more days in advance (Aranki *et al.*, 1969). As soon as the agar was solidified, the plates were passed into the AC where they were stored until used. In the O_2-free gas mixture of 85% N_2, 10% H_2 and 5% CO_2 which fills the chamber, the oxidation-reduction control plate (same base without blood but with resazurin as an Eh indicator) became colorless in 24 to 36 hours, indicating that the corresponding set of blood agar plates were reduced and ready for use.

After inoculation of the primary culture plate, the collection tube was passed out of the AC and used to inoculate one aerobic trypticase-soy blood agar plate and one tube of PRAS-BHI-S broth as a back-up for the plate culture. A direct smear slide was also prepared at this time.

Primary culture plates could be observed as often as desired during incubation in the AC without interrupting anaerobic incubation and subcultures of rapid growers could be made at 24 hours. But the more thorough examination for different colony types was generally postponed for 48 to 72 hours when a subculture was made to another anaerobic plate for purity. If the subculture growth appeared to be pure, one colony was picked to a tube of PRAS-BHI-S brain-heart broth. This tube was brought out of the AC and incubated. Subsequent growth was used to check for the inability to grow on agar aerobically or in a candle jar. Strains which failed to grow on either plate (obligate anaerobes) were speciated by VPI methods (Holdeman and Moore, 1972). Microaerophilic streptococci were included (growth in candle jar but not aerobically) only during the last one-third of this study.

Many blood cultures were collected during the early part of the study; 5 ml of blood was added to 50 ml of PRAS-BHI-S broth in metal-capped, perforable-stoppered bottles. This method was replaced about midway by use of the Supplemented Peptone Broth tube (Becton-Dickinson) which had been adopted by the Diagnostic Laboratory. From this time on, no more blood cultures were included in the comparative study.

When Diagnostic Laboratory specimens were unduly delayed or lost, both of the paired specimens were excluded from this study.

RESULTS

A total of 700 paired specimens collected between June 1970 and May 1972 were considered comparable in terms of collection and transport and these provided the results in Table IV-II. It can be seen that the Anaerobe Laboratory recovered nearly three times more anaerobic bacteria than the Diagnostic Laboratory (268 *vs* 95). Furthermore, the total number of anaerobic isolates from these specimens was only 103 in the Diagnostic Laboratory but 584 in the Anaerobe Laboratory, a ratio of 5.7.

The average number of anaerobic isolates per positive specimen was 2.2 (584 in 268), a figure which supports the widely-held impression that a mixture of anaerobes is frequently present in anaerobic infections. The largest number

TABLE IV-II

ANAEROBIC BACTERIA FROM CLINICAL SPECIMENS, JULY 1970–MAY 1972

	GasPak Diagnostic Lab.	*Ratio*	*AC-PS-AC Anaerobe Lab.*
Total: 700 specimens			
No. of specimens:			
No growth			200
Neg. for anaerobes			232
Pos. for anaerobes	95*	2.8	268†
% positive	14%		38%
No. of anaerobic isolates	103	5.7	584

* Includes 5 which were positive only in Diagnostic Lab.
† Includes 18 reported as "No growth" by Diagnostic Lab.

found in a single specimen was seven (endometrium). The largest number per specimen reported by the Diagnostic Laboratory was two and this occurred only eight times. It is noteworthy that the comparable specimens for 18 of the Anaerobic Laboratory positives were reported by the Diagnostic Laboratory as "No growth"; yet the Anaerobe Laboratory recovered a total of 35 isolates from them, three of which contained as many as four anaerobes.

It appears that diagnostic laboratories which perform similar aerobic collection-GasPak procedures will fail to detect the presence of anaerobic bacteria nearly two thirds of the time. And only one time in five or six will they discover in clinical specimens that a mixture of anaerobes is present.

These results provided a clearcut answer to the question raised as the first objective of this study, for the AC-PS-AC procedure had indeed grown many more anaerobes from clinical specimens than the Diagnostic Laboratory GasPak. But they in turn raised another question, *i.e.* were these differences due largely to the better collection method, the richer culture medium, anaerobic chamber *vs* GasPak incubation, or some other factor? Each of these possibilities was looked at in a series of small studies.

COMPARISON OF TECHNICAL FACTORS. When both of 11 sets of paired specimens, swab and collection tube, were brought promptly to the Anaerobe Laboratory and processed in the identical manner (AC-PS-AC), the collection tube produced about one-third more isolates than the swab (26 *vs* 19).

Similarly, a few specimens, some as swabs and others in a collection tube, were inoculated in duplicate to a trypticase-soy blood agar plate and a BHI-S blood agar plate; the latter produced about 50 percent more anaerobic isolates following the GasPak procedure outlined in Table IV-III.

TABLE IV-III

RECOMMENDED ROUTINE PROCEDURE FOR PRIMARY ISOLATION OF ANAEROBES FROM CLINICAL SPECIMENS

1. Collect specimens anaerobically (gassed-out tubes).
2. Use enriched medium for primary culture.
3. Store freshly inoculated plates in flush-jar to protect them from exposure to air until ready for GasPak.
4. Use freshly-reactivated catalyst in GasPak procedure.
5. Incubate GasPak jars 2 or 3 days without opening.

GasPak, flush jar and anaerobic chamber incubations were also compared in yet another small series; there was no clear superiority of one method over the others. Our results agree, therefore, with those recently reported by Killgore *et al.* (1972) and by Rosenblatt *et al.* (1972), both of which indicate that GasPak incubation may be as effective as an anaerobic chamber for the isolation of anaerobes from clinical specimens not contaminated with normal flora.

Because clinicians so often need culture reports in 24 hours, clinical laboratories often examine and report out anaerobic cultures after only 24 hours' incubation. Therefore, we decided to compare the effect of 24 *vs* 48 hour primary GasPak incubations upon the number of anaerobes which could be isolated. Some specimens were swabs, others were in anaerobic collection tubes. They were streaked to duplicate plates of commercially prepared BHI-S sheep blood agar plates and promptly placed in separate GasPak jars containing freshly-reactivated catalyst. One jar (A) was opened at 24 hours and the colonies were examined and subcultured for aerobic growth and then speciated by VPI methods. The other jar (B) was not opened for 48 hours; after examination and subculture these plate cultures were reincubated for a total of 5 to 7 days. Jar A produced only 15 anaerobe-positive cultures and 25 different isolates, whereas 71 anaerobes (2.8x) were recovered from 25 specimens after 48 hours' incubation (Jar B). The total number of isolates in B jars increased to 78 after 3 days and to 88 (mostly *Bacteroides melaninogenicus*) after 5 to 7 days' incubation.

Thus, no single one of the differences between the GasPak (Diagnostic Laboratory) and the AC-PS-AC (Anaerobe Laboratory) procedures (Table IV-I) accounts for all of the greatly increased number of anaerobes isolated by the latter procedure in the 1970 to 1972 comparative study. Rather, it appears that each step in the latter may be an improvement over its counterpart: anaerobe collection tube *vs* aerobic swabs, use of a richer medium for primary isolation, careful avoidance of exposure to air of freshly inoculated plates, *etc.* Our studies indicate, however, that the most important single factor is the way the GasPak procedure is carried out. The simple step of continuing primary GasPak incubation for 48 instead of 24 hours increased the number of anaerobe-positive specimens by two-thirds and the number of isolates 2.5- to 3-fold.

RECOMMENDED PROCEDURE

Since May 1972 when the 1970 to 1972 comparative study was stopped, the procedure shown in Table IV-III has been used routinely in the Temple University Clinical Microbiology Laboratories. The enriched medium is commercially prepared BHI-S blood agar. In Step Number 3 the freshly-inoculated plates are placed at once in a jar fitted with a double-vented lid which can be connected to a tank of O_2-free gas (in our case 85% N_2, 10% H_2, 5% CO_2) which is run slowly through the jar for 5 to 10 minutes. The purpose of storage in a flushed-out jar is the same as that of the method described by Martin (1971), who employs a continuous stream of CO_2. It avoids death of the more O_2-sensitive anaerobes from prolonged exposure to air. When the jar is ready for GasPak the vented lid is replaced by the usual type. The importance of heat reactivation of catalyst *each time* it is used seems not to be generally recognized. Retained moisture and especially absorption of H_2S on the surfaces of the palladium-coated pellets can reduce considerably the catalytic activity. This activity can be restored with dry heat at 160° C for 1½ hours, as recently emphasized by Sutter *et al.* (1972). The importance of Step Number 5 has already been discussed.

We have not compared the Recommended Procedure and the AC-PS-AC procedure, but we suspect that the former may be capable of recovering almost as many anaerobes from clinical specimens as the latter. Thus, the second objective of this study has also been fulfilled. We are of the opinion that our Diagnostic Laboratory, as well as clinical bacteriology laboratories in general, does not need to use an anaerobic chamber for routine anaerobic culturing of clinical specimens. On the other hand, this study made it abundantly clear that routine "anaerobic" culturing in clinical laboratories may be accomplishing very little. What is needed is widespread realization of the pitfalls and a conscientious effort to utilize the full capability of the GasPak procedure. One way to do this is to follow closely the Recommended Procedure in Table IV-III.

DISCUSSION

The apparent superiority of the AC-PS-AC procedure is due not only to technological advances but also—and probably much more so—to the human factor. Persistent efforts must be made to persuade clinicians to use properly one of several types of anaerobe collection tubes and bottles which are commercially available. Within the laboratory the principles underlying modern anaerobic culture techniques must be understood. In addition, there is the problem of convincing the bench-level personnel that full exploitation of these techniques includes the painstaking and time-consuming task of examining carefully primary cultures for different colony types, preparing and examining many stained smears and making both aerobic and CO_2 subcultures to determine whether the organism is an obligate anaerobe, a microaerophile or a facultative species. Unless the comparable aerobic culture is negative, colonies suspected of containing anaerobes should not be reported out as such until these subcultures have been incubated.

Readers will find the recently published manual by Sutter *et al.* (1972) very useful as a guide to good anaerobic culture methods in the clinical laboratory.

REFERENCES

Aranki, A., Syed, S.A., Kenney, E.B., and Freter, R.: Isolation of anaerobic bacteria from human gingiva and mouse cecum by means of a simplified glove box procedure. *Appl Microbiol, 17:*568, 1969.

Attebery, H.R., and Finegold, S.M.: Combined screw-cap and rubber-stopper closure for Hungate tubes (prereduced anaerobically sterilized roll tubes and liquid media). *Appl Microbiol, 18:*558, 1969.

Cato, E.P., Cummins, C.S., Holdeman, L.V., Johnson, J.L., Moore, W.E.C., Smibert, R.M., and Smith, L.DS.: *Outline of Clinical Methods in Anaerobic Bacteriology.* Blacksburg, Virginia Polytechnic Institute and State University, 1970.

Holdeman, L.V., and Moore, W.E.C.: *Anaerobe Laboratory Manual.* Blacksburg, Virginia Polytechnic Institute and State University, 1972.

Hungate, R.E.: The anaerobic mesophilic cellulolytic bacteria. *Bact Rev 14:*1, 1950.

Killgore, G.E., Starr, S.E., Del Bene, V.E., Whaley, D.N., and Dowell, V.R., Jr.: Comparison of three anaerobic systems for the isolation of anaerobic bacteria from clinical specimens. Abstract #M86, p. 94. Paper presented at 72nd Ann. Mtg. American Society for Microbiology. Philadelphia, April 23–28, 1972.

Martin, W.J.: Practical method for isolation of anaerobic bacteria in the clinical laboratory. *Appl Microbiol, 22:*1168, 1971.

Moore, W.E.C.: Techniques for routine culture of fastidious anaerobes. *Int J Syst Bact, 16:*173, 1966.

Rosenblatt, J.E., Fallon, A.M., and Finegold, S.M.: Recovery of anaerobes from clinical specimens. Abstract #M87, p. 94. Paper presented at 72nd Ann. Mtg. American Society for Microbiology. Philadelphia, April 23–28, 1972.

Sutter, V.L., Attebery, H.R., Rosenblatt, J.E., Bricknell, K.S. and Finegold, S.M.: *Anaerobic Bacteriology Manual.* Los Angeles, Department of Continuing Education in Health Sciences, University of California, 1972.

CHAPTER V

COMPARISON OF ISOLATION TECHNIQUES FOR ANAEROBIC BACTERIA

Stuart E. Starr

ABSTRACT: *Fifteen specimens obtained by needle aspiration of deep abscesses were cultured using the GasPak, roll-streak tube, and anaerobic chamber systems. Twenty-nine anaerobes were isolated, and recovery was comparable with the three systems. Furthermore, fresh aerobically prepared media and aerobically prepared, anaerobically stored media gave results equal to those with prereduced anaerobically sterilized media.*

Since 1861, when Louis Pasteur reported the discovery of a bacterium that lived without air and was killed by exposure to it, a number of methods for isolating anaerobic bacteria have been described. Early workers inoculated the deeper portions of deep agar tubes, where lower oxygen concentrations prevail. These tubes, introduced by Veillon and perfected by Prevot, are still in use and quite effective. Colonies are recovered by cutting the tube with a file near a colony and aspirating with a Pasteur pipet. Anaerobic jars came into use around the time of World War I, and a variety have been described. The most recent innovation is the GasPak system (BBL) with a disposable carbon dioxide-hydrogen generator (Brewer and Allgeier, 1966). The roll-streak tube system, developed by W.E.C. Moore (1966), is based on the anaerobic culture methods of Hungate. Tubes of prereduced anaerobically sterilized media with an oxygen-free gaseous phase are inoculated under a stream of oxygen-free gas. Another type of system is the anaerobic glove box, and of the various anaerobic chambers described, the flexible, clear plastic chamber described by Aranki *et al.* (1969) is probably the most practical. Specimens are introduced into the glove box through an entry lock. All procedures, including inoculation and incubation of plates, are performed inside the chamber.

COMPARISON OF SYSTEMS

About two years ago we decided to compare the various anaerobic systems mentioned. The deep agar tube technique was not included, as it had previously been shown in one study to be less effective than an anaerobic jar technique (Beerens and Castel, 1958). Using stock strains of anaerobic bacteria, we found that quantitative recovery of the strict anaerobes *Clostridium novyi* type B and *Clostridium haemolyticum* was one thousand-fold greater in the anaerobic chamber than in GasPak jars, but that recovery of less strict anaerobes was comparable with the two systems (Killgore *et al.*, 1971).

We then compared the GasPak, roll-streak tube and anaerobic chamber systems, using material for culture obtained by needle aspiration of deep abscesses in patients at Grady Memorial Hospital (Killgore *et al.*, 1973). Specimens were transported to the laboratory in tubes containing oxygen-free gas. Recommended freshly prepared or prereduced media for each of the anaerobic systems were inoculated. We found that recovery was comparable with the three systems. Of 29 anaerobes isolated from 15 specimens, 28 were isolated with the GasPak jar and anaerobic chamber systems and 23 with the roll-streak tube system. We believe that most of the organisms not isolated in roll-streak tubes could be isolated with media now recommended for this system. Results similar to ours have been reported by Rosenblatt *et al.* (1972) in their study of 23 clinical specimens.

It appears that all of the anaerobic systems tested are adequate for recovery of anaerobes from clinical specimens. There are, however, practical advantages and disadvantages for each system (Killgore *et al.*, 1973). For example, with the GasPak system one cannot examine cultures without actually opening the jars. With the other systems early and frequent observations are more feasible.

All of the anaerobes isolated in our study and in that of Rosenblatt *et al.* (1972)—organisms such as *Bacteroides fragilis, Bacteroides melaninogenicus, Fusobacterium nucleatum,* peptostreptococci, *Clostridium perfringens,* and

Propionibacterium acnes—are moderate anaerobes. "Strict" anaerobes as defined by Loesche (1969) were not encountered. This may explain why the GasPak system performed as well as the more elaborate anaerobic systems.

COMPARISON OF MEDIA

While our study was primarily of anaerobic systems, data were also obtained concerning the choice of media. It has been suggested that exposure of media to oxygen during preparation allows formation of organic peroxides which are toxic to anaerobes. Prereduced anaerobically sterilized media were developed to avoid this hazard. In our study, however, fresh aerobically prepared media and aerobically prepared, anaerobically stored media gave results equal to those with prereduced anaerobically sterilized media. Differences in recovery might have been found if stricter anaerobes had been encountered in our study.

The composition of media employed for isolation of anaerobes is another important consideration, for a number of anaerobes have been shown to have strict nutritional requirements; for example, most strains of *B. melaninogenicus* require hemin and about half require menadione. A number of organisms presently considered difficult to isolate may require other yet unknown nutritional factors.

Another important consideration is the method of collection and transport of specimens. Use of sophisticated anaerobic systems is futile if organisms are allowed to perish en route to the laboratory. Ideally, specimens should be obtained by needle aspiration or tracheal aspiration, depending on the site involved, and immediately placed in tubes containing oxygen-free gas, with or without liquid media.

REFERENCES

Aranki, A., Seyd, S.A., Kenney, E.B., and Freter, R.: Isolation of anaerobic bacteria from human gingiva and mouse cecum by means of a simplified glove box procedure. *Appl Microbiol, 17*:568, 1969.

Beerens, H., and Castel, M.M.: Procédé simplifié de culture en surface des bactéries anaérobies. Comparison avec la technique utilisant la culture en profondeur. *Ann Inst Pasteur (Lille), 10:*183, 1958.

Brewer, J.H. and Allgeier, D.L.: Safe self-contained carbon dioxide-hydrogen anaerobic system. *Appl Microbiol, 14:*985, 1966.

Killgore, G.E., Starr, S.E., Del Bene, V.E., Whaley, D.N., and Dowell, V.R., Jr.: Comparison of three anaerobic systems for the isolation of anaerobic bacteria from clinical specimens. *Am J Clin Pathol, 59:*552, 1973.

Killgore, G.E., Starr, S.E., and Whaley, D.N.: Comparison of anaerobic systems for recovery of anaerobic bacteria. In: *Bact Proc—1971.* Abstr. #M267, p. 109. Bethesda, American Society for Microbiology, 1971.

Loesche, W.J.: Oxygen sensitivity of various anaerobic bacteria. *Appl Microbiol, 18:*723, 1969.

Moore, W.E.C.: Techniques for routine culture of fastidious anaerobes. *Int J Syst Bacteriol, 16:*173, 1966.

Rosenblatt, J.E., Fallon, A.M., and Finegold, J.M.: Recovery of anaerobes from clinical specimens. Abstr. #M87, p. 94. Paper presented 72nd Ann. Mtg. American Society for Microbiology. Phila. April 23–28, 1972.

CHAPTER VI

IDENTIFICATION OF ANAEROBES FROM CLINICAL INFECTIONS

W.E.C. MOORE, E.P. CATO AND L.V. HOLDEMAN

ABSTRACT: *The clinical laboratory can provide rapid, accurate information on the kinds of anaerobes infecting tissue and the antibiotic susceptibilities of these organisms only if the physician cooperates in providing good specimens. Precautions to avoid contamination of specimens by normal flora and protection from oxygen during transport are extremely important. Genus identification of isolates is based on microscopic and chromatographic results, and can be complete in 24 to 36 hours after the specimen is received. Recent DNA homology studies have made species identification more accurate and easier. Some practical points in processing specimens efficiently and identifying anaerobes rapidly are outlined.*

When a specimen from infected tissue is sent to the clinical laboratory, the physician wants four things:

1. He wants to know that all of the kinds of organisms that are important in the infection were detected.
2. He wants accurate information concerning the antibiotic susceptibility of the kinds of organisms that are important in the infection.
3. If the information is to be useful, he needs to have it as soon as possible.
4. Usually of lowest priority on the list, he would like to know the accurate identity of the organisms present.

This last information may not greatly affect his management of the present infection, but physicians who now receive accurate and prompt information find that identification is increasingly helpful when they see new cases of infections with kinds of anaerobes that they have encountered before.

The clinical laboratory can provide good responses to these four points only if they have the cooperation of the physician in providing them with good specimens quickly.

The development of much of the information discussed here was financed by the National Institute of General Medical Sciences; GM 14604

COLLECTION, TRANSPORT AND CULTURE

The laboratory must receive a specimen that is representative of the actual infection site. Unfortunately, poor specimens are often submitted; for example, drainage that is sampled after skin contact often contains a confusing array of contaminating bacteria that require much time and expense to separate and analyze, and the analytical results are often inaccurate or misleading. The practice of sampling fresh wounds is not rewarding either. If soil was introduced into the wound, several dozen types of bacteria may be present, but many are of no consequence, and there is only a small chance that the few cells of potential significance will be detected along with the many others. Much has already been written about the need to avoid contamination of specimens with normal oral, intestinal or urogenital flora. This precaution is of extreme importance because of the many anaerobes normally present in these sites.

To allow the best possible job, samples must arrive at the laboratory with the same bacteria in the same proportion as in the infection site. To do this, the specimen should be protected from air, preferably in oxygen-free CO_2, and cultured as soon as possible. Transport medium, thioglycollate, or any other fluid should not be used, because in specimens from "mixed culture infections" important changes in the original proportions of different kinds of organisms can take place, even without nutrient. These changes make both the laboratory work and the interpretation of results more difficult. Enrichment cultures should not be used either, because the more sensitive organisms are often lost; they are overgrown by the species that grow faster in the medium used. When enrichment cultures are used, the results are often inaccurate and valuable time is lost waiting for these cultures to incubate.

If the specimen is still representative of the flora in the infection site when it is received, microscopic examination usually indicates that when there are several kinds of organisms present, they usually occur in approximately equal proportions. Direct streaking from swabs, drainage or other specimens generally yields the predominant organisms in proportions similar to those in the original material. This reduces the chance of isolating contaminants that may be present in comparatively small numbers. It also improves the accuracy of the analysis and decreases the need for selective media and extra time required to isolate important organisms that are overgrown when specimens are mishandled.

Hospital personnel may feel that proper procedures for handling specimens will be too time-consuming and expensive in their situation. Yet, when a case is of a serious nature, or is not responding to treatment, special care is frequently given the specimen. These are the specimens for which good anaerobic laboratory work can best be justified. Specimens that are placed in the refrigerator for several hours before being sent for analysis probably aren't worth the expense of analysis; the analytical results can be very misleading.

If the specimen is streaked directly in a tube of prereduced agar, it may be examined for growth at intervals without exposure to air and the colonies can be picked as soon as growth is apparent. Colonies from most specimens can be picked after overnight incubation. If colonies are picked under CO_2, the tube can be re-incubated to permit growth of slower-growing organisms that may also be present. Since only one streaked tube is handled at a time, rather than a number of different plates that must be handled when a full jar is opened, it is easier to maintain an efficient laboratory schedule.

The colonies from our clinical material are examined under a dissecting microscope, marked, then opened under CO_2 and picked to prereduced chopped meat carbohydrate broth, which is a rapid growth medium. Each colony type is picked three times. That is, the same colony is picked to the chopped meat carbohydrate broth, again to an aerobic slant to test for facultative isolates, and again to a small drop of water on a slide for a Gram's stain. This is usually possible, even with very small colonies, since thousands of cells remain at the original colony site even after the colony has been picked. The prereduced chopped meat subcultures are examined at intervals until there is good growth, usually in 12 to 24 hours.

Samples from this tube are stained to determine morphotype and to check for purity. Part of this culture is used for chromatographic analysis; part is used to inoculate the antibiotic susceptibility plates; and the remainder is used to inoculate the required media for biochemical tests.

IDENTIFICATION

The genus identification of anaerobes is based on the microscopic and chromatographic results. First, ether extracts of 2 ml of the chopped meat carbohydrate culture are chromatographed. For gram-negative rods or cocci, chromatograms are run only to the end of isovaleric acid (about 15 minutes of chromatograph time). For clostridia the chromatogram is run only until it shows the absence of the product which would be 2 carbons longer than the last product detected. For example, if acetic acid (the 2-carbon acid) is present, but butyric acid (the 4-carbon acid) is absent, then the analysis is complete; no higher molecular weight volatile acids will be present. The chromatograms for gram-positive nonsporing rods and cocci are run as for clostridia, but if (1) no acid product, (2) only acetic acid, or (3) only acetic and formic acids are detected, it is necessary to determine the presence or absence of lactic or succinic acids using a methylated sample of the culture. Methylation requires an additional milliliter of culture and 2 or 3 minutes of the technologist's time. The sample is usually ready for analysis by the time the ether extract is chromatographed. Methylated samples must be run to distinguish bifidobacteria, lactobacilli, some species of eubacteria, and several anaerobic cocci, but need not be run on other cultures.

A number of clinical laboratories now use simple gas chromatography on a routine basis to aid rapid and accurate identification. With information from chromatography and Gram's stain, genus identification of the isolates is usually complete in 24 to 36 hours after the specimen is received. Following chromatography, fewer biochemical tests are required to speciate each isolate. The required biochemical test media and the antibiotic susceptibility plates are inoculated from the same original pure culture on which the chromatographic analyses were done. The antibiotic tests and the biochemical tests can usually be read in 12 to 24 hours, so that the susceptibility pattern and the species identification can often be reported in 36 to 48 hours after the specimen is submitted. With fast growing organisms, e.g. *Clostridium perfringens*, we have completed this series of isolation, subculture, antibiotic susceptibility and species identification in 18 hours. With slow growing organisms like actinomyces, several days may be required.

As research has progressed, a number of improvements have been made which make identification more accurate and easier. Recent work with DNA homology by competition experiments has shown, for example, that weak fermentation of mannose is a distinguishing characteristic of *Fusobacterium varium* while indol production is a variable characteristic that is of no diagnostic importance in this species. In other cases, Johnson and Cummins (1971, 1972) have shown that several organisms, previously thought to be distinct, are variants of a genetically homologous group. For those groups in which DNA/DNA homology studies are complete, accurate laboratory identification has been simplified considerably. It is on the basis of the DNA work that we have been able to select the most accurate routine differential test for species identification.

To save laboratory time, improved formats for the reference information have been published. Using the new format for reference material (Holdeman and Moore, 1972) technologists can compare described reactions simply by placing the laboratory record of their results next to matching columns in the manual. This saves time and helps avoid errors. It is many such small details that help make a laboratory efficient and productive.

Although the procedures for individual analytical tests are published in detail, certain differences in work with anaerobes as compared with aerobes or facultative bacteria are worth noting:

1. In general, species of anaerobes are more pleomorphic than are aerobes or facultative bacteria.

Pleomorphism is often more pronounced in less suitable or more oxidized media.

2. The Gram reaction is sometimes difficult to interpret. In such cases, stains from very young cultures and knowledge of the fermentation products often help interpretation. For example, our present knowledge indicates that succinic acid is a major product only of *Bacteroides* species, *Actinomyces* species (if cultured under CO_2) and gram-positive cocci (which often form elongated rod-like cells in older cultures).

3. Accurate measurement of pH requires use of pH electrodes because many of the anaerobes chemically reduce colorimetric indicators.

4. Spores are not readily seen in some species of clostridia, and conversely, vacuoles produced by several nonsporeforming species may easily be mistaken for spores. Unless unmistakable spores are seen, cultures from the individual tube in which spores were suspected should be heat-tested in starch broth to verify the presence of spores, unless, of course, it is known that the isolate will not grow in starch. Except for cultures of *C. perfringens* and *C. ramosum,* it is important to demonstrate spore production to identify an isolate as a clostridium. Many nonsporing anaerobes resemble clostridia in cellular morphology.

5. Proteolytic species can usually be given a tentative identification on the basis of gelatin liquefaction, which is relatively rapid. To verify the identification, milk and meat cultures should be held until digestion is evident, or for 3 weeks before discarding as negative.

Cultures can be identified on the basis of published descriptions no matter what culture method is used as long as good growth of the isolates is obtained. Fermentation patterns may be determined by chromatography of a culture from any fermented carbohydrate as long as an uninoculated blank value for the medium is also determined. Several reference laboratories are now saving time and materials by chromatographing the referred pure culture, to identify the genus immediately after the necessary subcultures and Gram stains are made, before the biochemical tests are run.

One pitfall in laboratory identification has come to our attention. Several clinical laboratories determine only the *key* characteristic of isolates without using confirmatory tests. A number of less common species may key to a given species, e.g. *Bacteroides hypermegas,* but we have never found a verified strain of this species in clinical infections. Failure to run important confirmatory tests often results in erroneous identification and serves to decrease the value of all reports from that laboratory.

Whether or not an individual laboratory can afford to speciate anaerobic isolates is an individual decision. Certainly, thorough anaerobic work will increase the bacteriological work load by as much as 20 percent, more if the work is not restricted to those specimens where it will be of the most value. The cost of labor and materials for the local clinical specimens processed in our laboratory is currently $14.00 per isolate. This includes the antibiotic susceptibility pattern and species identification. Before we provided these analyses, the physicians we serve did not expect this type of information, but they have been pleased with the results and have used our laboratory more and more wisely. As many more laboratories are providing accurate identifications and more information is available about each species, the value of identification increases. More physicians are beginning to expect this service. With available methods and published information, an increasing number of laboratories have found that accurate identification of clinical anaerobes is practical for them. At first, anaerobic work may appear to be more difficult than work with other bacteria because there are differences in procedure. However, it is now being done routinely and efficiently at a number of locations by capable technologists who understand the work, and who interpret their data in a stepwise manner. If technologists can recognize only 8 species of anaerobes (*B. fragilis, C. perfringens, Peptococcus magnus, Eubacterium lentum, Peptostreptococcus intermedius, Eubacterium limosum, C. ramosum* and *Bacteroides corrodens*), they probably will be able to identify more than half of the anaerobic isolates from clinical material.

REFERENCES

Cummins, C.S., and Johnson, J.L.: Taxonomy of the clostridia: wall composition and DNA homologies in *Clostridium butyricum* and other butyric acid clostridia. *J Gen Microbiol, 67:*33, 1971.

Holdeman, L.V., and Moore, W.E.C. (Eds.): *Anaerobe Laboratory Manual.* Blacksburg, Virginia Polytechnic Institute Anaerobe Laboratory, 1972.

Johnson, J.L., and Cummins, C.S.: Cell wall composition and deoxyribonucleic acid similarities among the anaerobic coryneforms, classical propionibacteria, and strains of *Arachnia propionica. J Bacteriol, 109:*1047, 1972.

CHAPTER VII

METHODS AND TECHNIQUES FOR IDENTIFICATION: INVITED DISCUSSION

V.R. DOWELL, JR.

I AGREE WITH ALMOST everything Dr. Moore said in his presentation but there are a few areas where we seem to have some disagreement.

PLAIN GASSED-OUT TUBE VS GASSED-OUT TUBE CONTAINING A TRANSPORT MEDIUM FOR COLLECTION AND TRANSPORT OF CLINICAL SPECIMENS. I feel that we need more data from comparative studies in order to determine if in a transport medium there is significant growth which will cause us to misconstrue reults. The use of a transport medium in certain cases may be good, particularly for a swab specimen collected in surgery. Although the specimen may be sent to the laboratory by a courier, he may stop for a Coke on the way and it may be two or three hours before the specimen reaches the laboratory. The delay could result in detrimental drying of the specimen.

THE PREDOMINANT ORGANISM. I am not so sure that we can always judge what is the predominant organism in mixed infections on the basis of a microscopic examination. If we have a specimen containing a mixture of microorganisms, e.g. *Escherichia coli*, *Bacteroides fragilis* and a gram-positive rod, it is difficult to distinguish between *E. coli* and *B. fragilis* in a Gram-stained preparation. Therefore, it is difficult to judge the relative proportion of each organism in the specimen. Furthermore, I am not so sure that the predominant organism is always most important. In the case of a synergistic relationship an organism present in small numbers can be quite important if it is supplying essential nutrient for one or more of the other microorganisms. The work of Gibbons, Macdonald and others with mixed infections involving *Bacteroides melaninogenicus* has shown this (Socransky and Gibbons, 1965). They found that an anaerobic diphtheroid, a nonsporeforming gram-positive rod, was supplying vitamin K required by *B. melaninogenicus*, and the presence of the diphtheroid was essential for initiation of infection without the simultaneous administration of vitamin K to the experimental animal.

SELECTIVE MEDIA. In regard to the use of selective *vs* nonselective media for isolation of anaerobes, consider a hypothetical infection involving three different microorganisms A, B and C. Organism A is present in the clinical material in a concentration per gram of 10^8, B in a concentration of 10^5 and C in a concentration of 10^3. Now, in order to isolate organism B (10^5 per gram) with a nonselective medium it is necessary to select one colony out of a thousand, and organism C (10^3 per gram), one colony out of a hundred thousand. This is rather difficult to accomplish with a nonselective medium. Selective media can be extremely valuable in allowing rapid isolation of anaerobes from a mixed population such as the one described.

IDENTIFICATION. I agree that we need rapid and accurate identification of anaerobes, but, as indicated by Dr. Gorbach in his remarks, rapid presumptive identification may be more important than definitive identification to the attending physician. What concerns me now is what the hundreds or possibly thousands of hospital laboratories in this country that do not have a gas chromatograph can do to identify anaerobes.

Why identify anaerobes? Certainly they can produce a variety of life-threatening infections, and the treatment of infections with certain anaerobic organisms, e.g. *Bacteroides fragilis*, is much different from that for infections involving facultative bacteria such as *Escherichia coli*. Treatment of anaerobic infections frequently requires surgical intervention as well as the use of proper antibiotics. I would like to discuss some tests that can be used in the absence of a gas chromatograph for presumptive identification.

Microscopic examination of clinical materials can be very helpful in presumptive identification. For example, in clostridial myonecrosis ("gas gangrene") large gram-positive rods are usually observed and "pus" cells are absent. On the other hand, smears of material from other types of anaerobic infections usually show abundant pus cells. The physician will manage a pyogenic infection much differently from a case of probable myonecrosis, which usually requires surgical intervention to remove necrotic tissue.

The following are useful characteristics for preliminary identification of anaerobes:

1. Microscopic characteristics (morphology, gram reaction, presence or absence of spores, motility).
2. Characteristics of colonies on blood agar and egg yolk agar.
3. Type of growth in liquid medium.
4. Oxygen tolerance (surface growth in air, candle jar, or anaerobic conditions, type of growth in agar deeps).
5. Catalase production.
6. Growth in 20 percent bile medium.
7. Growth in the presence of penicillin (2-unit or 10-unit disc).

Microscopic characteristics are very useful for presumptive identification. However, one needs practice in order to recognize morphological differences of various species, and it is worthwhile for individuals not accustomed to working with anaerobes to practice with reference strains. This can be very helpful since some are quite characteristic once you have seen them. Reference strains are also useful for evaluation of isolation and identification procedures.

Characteristics for presumptive identification of some commonly encountered anaerobic nonsporeforming gram-negative bacilli are shown in Table VII-I. Although some strains of *B. fragilis* may be pleomorphic, the majority are not. Also, *B. fragilis* does not produce black colonies on blood agar and is not inhibited by penicillin (using a 2-unit disc) or by 20 percent bile medium. In fact, some strains of *B. fragilis* are stimulated by 20 percent bile. On the other hand, *B. melaninogenicus* isolates usually produce black colonies on blood agar and may or may not be inhibited by penicillin

TABLE VII-I

PRESUMPTIVE IDENTIFICATION OF SOME COMMONLY ISOLATED ANAEROBIC NONSPOREFORMING GRAM-NEGATIVE BACILLI

Species	*Usual Morphology*	*Black Colonies*	*Growth Inhibited by Penicillin (2-unit Disc) and 20% Bile*
Bacteroides fragilis	Small to medium rods variable in length	−	−
Bacteroides melaninogenicus	Small coccoid rods	+	variable
Fusobacterium mortiferum	Highly pleomorphic; filaments, swollen bodies, etc.	−	−
Fusobacterium necrophorum	Rods, quite variable in width and length	−	+
Fusobacterium nucleatum	Filamentous rods, uniform in width, with or without pointed ends	−	+

and bile. Most strains of *F. necrophorum* and *F. nucleatum* are inhibited by penicillin and bile.

Useful characteristics for presumptive identification of anaerobic nonsporeforming gram-positive bacilli commonly isolated from human infections are shown in Table VII-II. Notice that *P. acnes* is the only organism in this group which produces catalase. Since *Propionibacterium* species are the only anaerobic nonsporeforming gram-positive bacilli which produce catalase, the catalase test is very useful for presumptive identification. Microscopic morphology is also useful in identification of these organisms. If true branching is observed it is most likely an *Actinomyces* species or *Arachnia propionica.* Ability to ferment glucose and glycerol and the rapidity of growth are also useful characteristics for rapid identification. In addition, identification of end products can provide information which allows rapid presumptive identification if a gas chromatograph is available. For example, of the anaerobic nonsporeforming gram-positive bacilli, only members of the genera *Propionibacterium* and *Arachnia* produce propionic acid.

The purpose of presumptive identification is to provide the physician with information which he can utilize in treat-

TABLE VII-II

USEFUL CHARACTERISTICS FOR PRESUMPTIVE IDENTIFICATION OF ANAEROBIC NONSPOREFORMING GRAM-POSITIVE BACILLI COMMONLY ISOLATED FROM HUMAN INFECTIONS

Species	*Usual Morphology*	*Rapidity of Growth*	*Catalase Production*	*Glucose Fermented*	*Glycerol Fermented*	*Propionic Acid*
***Actinomyces** israelii*	Branching rods or diphtheroidal	Slow	−	+	−	−
***Arachnia** propionica*	Branching rods or diphtheroidal	Slow	−	+	−	+
***Bifidobacterium** eriksonii*	Rods with bifid ends	Rapid	−	+	−	−
***Eubacterium** alactolyticum*	Diphtheroidal rods, false branching	Slow	−	+	−	−
***Eubacterium** lentum*	Diphtheroidal rods	Moderate	−	−	−	−
***Eubacterium** limosum*	Uniform, medium size rods	Rapid	−	+	−	−
***Propionibacterium** acnes*	Diphtheroidal, false branching	Slow	+	+	+	+

ment of the patient when needed. If more accurate identification of the microorganism is required, this can be done later, or the cultures can be referred to a reference laboratory for identification. Definitive identification of anaerobes can be quite time consuming.

At present we are investigating various techniques in our laboratory to develop more rapid identification procedures. Two approaches look promising: (1) Use of a micro-method system utilizing conventional biochemical tests which is being developed by the Analytab Corporation in France and (2) use of fluorescent antibody techniques.

THE MICRO-METHOD SYSTEM. This is essentially the same as the Analytab system (Analytab Products, Inc., New York, New York) for identification of facultative gram-negative bacteria with different substrates. The micro tests are performed as follows:

1. As soon as isolated growth is obtained on an agar medium, 3 to 4 colonies are used to prepare a suspension of cells in carbohydrate free broth.
2. The 20 different substrates are then inoculated with the cell suspension, using a Pasteur pipette, and the system is incubated anaerobically at 35 to 37°C for 24 to 48 hours.
3. The reactions are recorded.

The reactions with rapidly growing anaerobes are usually complete within 24 hours. Although some of the tests will require modification and/or other tests substituted, we have found that the overall agreement with conventional differential tests is quite good. There has been approximately 91 percent agreement with conventional tests on the basis of the total number of tests performed. The majority of the work with the micro-method system has been performed within an anaerobic glove box. However, it is also possible to perform the tests outside if appropriate anaerobic conditions are provided after the substrates are inoculated. Of course, it is essential to use an adequate inoculum of viable cells in order to be successful. I agree with Dr. Moore that it is extremely important to have adequate growth when performing differential tests; otherwise the results can be very disappointing.

USE OF FLUORESCENT ANTIBODY REAGENTS. We have prepared FA reagents with a variety of anaerobic nonsporeforming gram-positive and gram-negative bacilli and so far the results of tests performed with these look promising. The specificity of the conjugates is surprisingly good; they are either species or subspecies specific. However, within a species or subspecies, we find that the conjugates prepared to date do not react with all strains of the homologous group. The number of serotypes within a species and/or subspecies is variable. Some appear to have one or two serotypes and others have a larger number. On the basis of the preliminary studies, I feel that it will be possible to develop polyvalent FA conjugates for the commonly encountered anaerobic nonsporeforming bacilli and these will be valuable for rapid identification.

REFERENCES

Socransky, S.S., and Gibbons, R.J.: A required role of *Bacteroides melaninogenicus* in mixed anaerobic infections. *J Infect Dis, 115*:247, 1965.

CHAPTER VIII

CURRENT CLASSIFICATION OF CLINICALLY IMPORTANT ANAEROBES

LILLIAN V. HOLDEMAN,
ELIZABETH P. CATO AND W.E.C. MOORE

ABSTRACT: *The majority of workers in anaerobic bacteriology are now using a standardized nomenclature, which is a marked improvement over the multiple synonyms appearing in the older literature. Because different subspecies of* Bacteroides fragilis *apparently have different pathological significance, it is important to be able to identify these organisms to the subspecies level. For example,* B. fragilis *ss.* fragilis *is the most common subspecies isolated from clinical specimens, but the least common in fecal flora of normal North Americans.*

Some differentiating characteristics are reviewed for bacteroides, clostridia, and members of other genera which are isolated less frequently from clinical specimens.

We do not presume to know all of the clinically important anaerobes. Therefore, this discussion will be limited to those anaerobes that are isolated most often from clinical infections. The kinds of anaerobes most frequently isolated from clinical infections, in decreasing frequency, are: bacteroides, clostridia, and anaerobic gram-positive cocci; followed by other gram-negative rods (called variously fusobacteria or sphaerophorus) and gram-positive rods (eubacteria, lactobacilli, propionibacteria, actinomyces); and occasionally bifidobacteria and gram-negative cocci (veillonella and acidaminococci). This listing obviously excludes *Treponema pallidum,* which has not yet been grown *in vitro.*

About 85 percent of the anaerobes now isolated from clinical infections are members of described species. The other 15 percent of the isolates belong to species not yet described, but these usually can be assigned to the appropriate genus on the basis of morphological characteristics and fermentation products. As anaerobic techniques improve, a larger variety of organisms is being isolated.

We gratefully acknowledge the support of National Institutes of General Medical Science 14604 for support of our taxonomic research.

We need to discuss nomenclature to some extent. A rose may be a rose, but a bacteroides is not always a bacteroides. Depending on the author or the bacteriologist cited, in American literature a "bacteroides" may mean:

1. any nonsporing anaerobic rod (gram-positive or gram-negative)
2. any nonsporing anaerobic gram-negative rod
3. any simple, rarely pleomorphic, anaerobic rod with rounded ends (*Bergey's Manual,* 7th ed.; Breed *et al.*, 1957)
4. any organism similar to *B. fragilis* or *B. melaninogenicus*
5. an anaerobic gram-negative rod that does not produce butyric acid as a major product of fermentation, or that also produces isobutyric and isovaleric acids when butyric acid is produced (*Bergey's Manual,* 8th ed.; Buchanan and Gibbons, In press)

Likewise, in the older literature, an *Actinomyces israelii* or *A. bovis* may, in actuality, have been a *Propionibacterium* species, any gram-positive rod that branched, or a mixture of an actinomyces and another organism.

Recently, because of the cooperative effort of members of taxonomic subcommittees working at national and international levels, uniform nomenclature is being used by the majority of workers in the field. This standardization of nomenclature is greatly improving communication, and it is this newer nomenclature that is discussed here, especially concerning the organisms most frequently isolated from clinical material.

The species which occurs with greatest frequency in clinical specimens is *Bacteroides fragilis.* There are five subspecies of *B. fragilis* (ss. *distasonis,* ss. *fragilis,* ss. *ovatus,* ss. *thetaiotaomicron,* and ss. *vulgatus*), differing in indol production and in the fermentation of a few carbohydrates. These subspecies are composed of clusters of strains with very similar reactions, and about 70 percent of all our strains of *B. fragilis* belong to one of these subspecies. However, there are strains of *B. fragilis* having characteristics intermediate between those of strains in the subspecies. These intermediate strains we call *B. fragilis* without any subspecies designation.

Several years ago we thought that there probably was little value to the physician for the clinical laboratory to differentiate between the subspecies of *B. fragilis*. We now feel differently. From clinical specimens, we have never isolated *B. fragilis* ss. *ovatus*, and we have found *B. fragilis* ss. *fragilis* 3 times to more than 10 times more often than ss. *thetaiotaomicron*, ss. *vulgatus*, ss. *distasonis*, or the intermediate strains. *B. fragilis* ss. *fragilis* was found 3 times more frequently than the other subspecies when primary isolations were done from streaks of broth enrichment cultures, and more than 10 times more often after we started doing primary isolations from direct streaks of the specimens submitted to the laboratory.

Although *B. fragilis* ss. *fragilis* is the most common subspecies found in clinical specimens, it is the least common in fecal flora of normal North Americans. From a recent study of 1136 strains of bacteria from feces of 20 individuals, 26 percent of the isolates were *B. fragilis*, and the incidence of *B. fragilis* in individuals ranged from 4 percent to 60 percent. Of these 290 strains from feces, only 0.7 percent were *B. fragilis* ss. *fragilis*. The most common were ss. *vulgatus* (47%), followed by intermediate strains (23%), ss. *thetaiotaomicron* (17%), ss. *distasonis* (9%), and ss. *ovatus* (3%). *B. fragilis* ss. *fragilis*, most frequently isolated from clinical specimens, clearly is not the most numerous in feces.

There is some correlation between serotypes and subspecies of *B. fragilis* (Beerens *et al.*, 1971; Lombard and Dowell, 1972). However, there is more than one serotype in some of the subspecies, and there are serological cross-reactions between some of the subspecies.

B. oralis, from which *B. fragilis* needs to be differentiated, is part of the normal flora of the mouth. Unlike *B. fragilis*, *B. oralis* does not grow well in 20 percent bile, does not produce gas, generally does not ferment either arabinose or xylose, and is part of the normal flora of the mouth and vagina.

Most bacteriologists recognize *B. melaninogenicus* as the gram-negative anaerobic bacillus that produces black pigment on blood agar plates. These characteristics often are considered sufficient for identification of the species. However, strains of *B. melaninogenicus* may be quite dissimilar metabolically, and on these bases have been divided into three subspecies. The subspecies of *B. melaninogenicus* most usually isolated from clinical material is *B. melaninogenicus* ss. *asaccharolyticus* (proteolytic, not fermentative). *B. melaninogenicus* ss. *intermedius* (fermentative and somewhat proteolytic) is the next most frequently encountered; and ss. *melaninogenicus* (fermentative and not proteolytic) is the least commonly encountered in our experience. Until recently there has been little attempt to characterize strains of *B. melaninogenicus*, so one usually cannot tell which metabolic group was isolated, described, or studied when one reads literature published before 1971.

Another species of *Bacteroides* isolated with a fair amount of frequency from clinical specimens is *Bacteroides corrodens*, a rather fastidious, nonfermentative organism that is part of the normal flora of the mouth. The original description of *B. corrodens* (Eiken, 1958) was of organisms that were either obligately anaerobic or facultative. In a series of studies about the "corroding bacteria", F.L. Jackson and colleagues have shown that the obligately anaerobic organisms are decidedly different from the facultative strains. They differ markedly in guanine-cytosine percent composition of the DNA as well as in serological and some biochemical properties (Jackson *et al.*, 1971). Jackson and Goodman (1972) proposed that the facultative strains (also known by Elizabeth O. King's designation of "HB-1") be called *Eikenella corrodens*. For the type strain of *E. corrodens*, they used ATCC 23834 (NCTC #10596), originally deposited as the type strain for *Bacteroides corrodens* (Henriksen, 1969). Since the specific epithet "*corrodens*" has been applied to the facultative organisms placed in *Eikenella*, it is highly probable that a new specific epithet may be proposed for the anaerobic strains. In biochemical properties, the anaerobic strains differ from the facultative strains in producing urease and H_2S and by growing in 0.02% $NaNO_3$ (Jackson *et al.*, 1971). Because the anaerobic organism is fastidious and grows very poorly, false-negative urease reactions may occur unless one is very careful to obtain adequate growth in the urease medium. Strains of *Eikenella corrodens* seem to be more common in clinical isolates than are strains of the obligate anaerobe. Therefore, one should carefully interpret the literature about "*B. corrodens*" that was published before 1972. The authors may well have been talking about only the facultative strains, only the anaerobic strains, or both.

There are about 10 other species of bacteroides that occur in clinical specimens, but they occur relatively infrequently. Care must be taken to recognize the sporeforming anaerobic rods that always stain gram-negative. These belong to *Clostridium* and should not be confused with bacteroides or fusobacteria.

The clostridia are the sporeforming anaerobic bacilli. The three clostridia we find most frequently in clinical infections are *C. perfringens*, *C. ramosum* (synonyms = *E. filamentosum*, *B. terebrans* and *B. trichoides*), and *C. innocuum*, with representatives of about 10 other species occurring with some degree of frequency, and about 15 additional species occurring relatively infrequently. *C. perfringens* is certainly the most frequently isolated clostridium, and recognition of this species is not too difficult if one has worked with a few cultures and remembers that spores are usually very difficult to demonstrate in cultures from clinical material. The growth rate of *C. perfringens* is most rapid at 45 C. This temperature can be used for rapid isolation and identification of the species. Upon occasion, by utilizing this optimum growth temperature to maximum advantage, strains of *C. perfringens* have been isolated from a mixed culture from a wound specimen, inoculated into appropriate media for differentiation and susceptibility testing, and a final report (with antibiotic susceptibility test results) made available within 18 hours of the time the culture was taken. (In these cases, the antibiotic susceptibility tests were *not* incubated at 45 C, naturally.)

The cells of *C. ramosum* and *C. innocuum* are smaller than those of "classical" clostridia. *C. ramosum* also sporulates exceedingly poorly in laboratory cultures, and may decolorize very easily when gram-stained. It was recognized for many years as either *E. filamentosum* (gram-positive nonsporing rod), or *B. trichoides* (gram-negative nonsporing rod). However, *C. ramosum* has distinctive biochemical characteristics and fermentation products that enable identification without the demonstration of spores in all strains examined. *C. septicum*, although not exceedingly frequent in human clinical specimens, should be isolated and recognized when present. Alpern and Dowell (1969) reported that in persons with *C. septicum* infections, antibiotic therapy decreased mortality from 92 percent (no treatment) to 13 percent (treatment).

Anaerobic cocci most often have been reported either as "anaerobic streptococci" or "anaerobic gram-positive cocci." The anaerobic cocci that occur in clinical specimens are easily speciated according to differences in fermentation products and a few biochemical reactions. The species most commonly isolated are *Peptococcus magnus*, *Peptostreptococcus anaerobius*, *Peptostreptococcus intermedius*, *Peptococcus asaccharolyticus* and *Peptococcus prevotii*. Of these, *Peptostreptococcus intermedius* is the only frankly fermentative organism. Although *P. intermedius* may be fastidiously anaerobic upon initial isolation, it often becomes more aerotolerant upon culture in the laboratory; some strains will grow exceedingly well in a carbon dioxide atmosphere and others may grow aerobically after several transfers in the laboratory. M. Rogosa (personal communication) has suggested that this species properly belongs in *Streptococcus*, but we do not know that it will be included in *Streptococcus* in the 8th edition of *Bergey's Manual.*

The next most commonly encountered group are the fusobacteria, which are those gram-negative rods that make butyric acid. Of the fusobacteria, *F. nucleatum* (synonym = *F. fusiforme*) is the most commonly isolated and has the cell morphology most usually associated with fusobacteria—long thin cells with tapered ends. Not all fusobacteria, however, have thin cells or cells with tapered ends. *F. necrophorum* (*Sphaerophorus necrophorus, S. funduliformis*), *F. mortiferum*, *F. naviforme*, *F. gonidiaformans* and *F. varium* are occasionally isolated from clinical specimens.

Organisms of the genus *Sphaerophorus* in the 7th edition of *Bergey's Manual* are now either members of *Bacteroides* or *Fusobacterium*, depending on the kinds of acids produced from fermentation of glucose or utilization of peptone.

Among the gram-positive rods, strains of eubacteria are the most commonly isolated, while propionibacteria, lactobacilli, actinomyces, and bifidobacteria are isolated less frequently. The eubacterium most commonly isolated is *E. lentum*. Because cells of this species are often very coccoid, *E. lentum* is apt to be misidentified as a gram-positive coccus. Many of our strains of *E. lentum* have come from specimens (often blood) from which *B. fragilis* also was isolated. In addition, some recently isolated strains of *E. lentum* frequently are resistant to tetracycline *in vitro*.

The genus *Propionibacterium* includes the nonsporing gram-positive rods that produce propionic acid, which now includes the anaerobic species formerly in the genus *Corynebacterium* (*Bergey's Manual*, 7th edition). *Propionibacterium acnes* and *P. granulosum*, both part of the normal flora of the skin, are the two propionibacteria most frequently isolated from clinical specimens. Since a *P. acnes* or a *P. granulosum* isolate may have come from either the skin of the patient, the specimen collector, or the technologist, a second specimen should be submitted for culture when either of these species is isolated. Although these organisms usually represent contamination, there are a few reports of infections caused by *P. acnes*, and isolation of the organism should not be considered contamination in every instance.

COMMENT

In the last few years, we have seen development of a standardized nomenclature. Characteristics for reliable differentiation of commonly encountered species are relatively easy, which greatly simplifies identification of isolates. Future literature reports concerning anaerobes should be easier to interpret because more and more people are using the standard nomenclature based on differentiation of species by the same methods. Once this has been accomplished, one of the major purposes of taxonomy—meaningful communication—will have been fulfilled.

REFERENCES

Alpern, R.J., and Dowell, V.R., Jr.: *Clostridium septicum* infections and malignancy. *JAMA, 209:*385, 1969.

Beerens, H., Wattre, P., Shinjo, H., and Romond, C.: Premiers resultats d'un essai de classification serologique de 131 souches de Bacteroides du groupe *fragilis (Eggerthella). Ann Inst Pasteur, 121:*187, 1971.

Breed, R.S., Murray, E.G.D., and Smith, N.R.: *Bergey's Manual of Determinative Bacteriology,* 7th ed. Baltimore, Williams and Wilkins, 1957.

Buchanan, R.E., and Gibbons, N.E.: *Bergey's Manual of Determinative Bacteriology,* 8th ed. Baltimore, Williams and Wilkins. In press.

Eiken, M.: Studies on an anaerobic, rod-shaped gram-negative microorganism: *Bacteroides corrodens* n. sp. *Acta Pathol Microbiol Scand, 43:*404, 1958.

Henriksen, S.D.: Designation of the type strain of *Bacteroides corrodens* Eiken 1958. *Int J Syst Bacteriol, 19:*165, 1969.

Jackson, F.L., and Goodman, Y.E.: Transfer of the facultatively anaerobic organism *Bacteroides corrodens* Eiken to a new genus, *Eikenella. Int J Syst Bacteriol, 22:*73, 1972.

Jackson, F.L., and Goodman, Y.E., Bel, F.R., Song, P.C., and Whitehouse, R.L.S.: Taxonomic status of facultative and strictly anaerobic "corroding bacilli" that have been classified as *Bacteroides corrodens. J Med Microbiol, 4:*171, 1971.

Lombard, G.L., and Dowell, V.R., Jr.: Preparation of fluorescent antibody reagents for identification of Bacteroides. Abstr. #M93. Paper presented at 72nd Ann. Mtg. American Society for Microbiology. Philadelphia, April 23–28, 1972.

CHAPTER IX

INVITED DISCUSSION: NOMENCLATURE, TAXONOMY AND GENERAL METHODOLOGY

Sydney M. Finegold

It is important to recognize the need for two different approaches to the nomenclature, taxonomy and general methodology of anaerobic bacteriology. On the one hand, a busy clinical laboratory must use approaches which are as simple as possible in order to provide the clinician as rapidly as possible with information needed for proper therapy of an illness. On the other hand, the research laboratory studying the normal flora or attempting to define the role of certain anaerobes in specific disease entities is always working on the frontier of new knowledge and must apply the most detailed and rigorous techniques.

With regard to the clinical laboratory, we should not insist that the Hungate technique or its modifications or the anaerobic chamber be used in processing specimens from suspected anaerobic infection unless we can establish that the particular laboratory will benefit from use of such sophisticated equipment. As you have already heard, studies done to date, although admittedly limited in scope, do not show any advantage for the roll-tube, prereduced anaerobically sterilized (PRAS) media or chamber techniques over conventional procedures in terms of recovery of anaerobic bacteria from clinical specimens. Additional studies of this type are certainly desirable to establish what percentage, if any, of clinical materials contain anaerobes which are so fastidious as to be unrecoverable by the bench technique. Studies to date suggest that this percentage would be small at most. In spite of this, certain clinical laboratories, especially in large medical centers, will find it convenient and advantageous to use one or both of the more sophisticated techniques in certain situations. The anaerobic chamber permits leisurely and detailed examination of Petri dish cultures and appropriate subculturing without concern for the need to rapidly get the subcultures back under anaerobic conditions in a new jar. Similarly, the PRAS roll-tube procedure offers the advantage of rapid inoculation of large numbers of tubes for biochemical determinations.

Techniques for identification must also be simplified for the clinical laboratory. For immediate management of a sick patient, the primary needs are gross identification of the infecting organisms, *e.g.* anaerobic gram-negative bacillus or microaerophilic streptococcus, and accurate determination of the antimicrobial susceptibility of the infecting organism. Use of selective and differential media along with nonselective media may expedite recovery and tentative identification. Eventually, simplified antibiotic disc susceptibility tests will be available to help in making recommendations for therapy.

After tentatively identifying the organism and determining its antimicrobial susceptibility pattern, the laboratory should make every reasonable effort to definitively identify the infecting organism. This is particularly important in the case of anaerobic bacteria, for gram-positive organisms may often appear gram-negative, sporeforming organisms may appear to be nonsporeformers, bacilli and cocci may be difficult to distinguish from each other, and it may even be difficult to state whether or not the organism is anaerobic. The more detailed the identification procedure, the less likely one is to categorize the organism incorrectly. Specific identification is certainly important to increase our knowledge of the role of certain species in various disease processes, the prognosis of infections due to these various organisms, and responsiveness of infections to various forms of antimicrobial therapy.

While it is clear that, at present, analysis of metabolic end-products by gas chromatography is necessary for definitive identification of certain organisms, it is equally clear that most clinical laboratories are not presently equipped to do this type of analysis, nor are they likely to be soon. Perhaps gas chromatography, or definitive identification in general, might be made available through regional reference laboratories. While it may be desirable to work toward the availability of gas chromatography for most or all clinical microbiology laboratories, another approach which may be more realistic is to develop simplified techniques which will obviate the need for a gas chromatograph and expedite rapid identification.

Those engaged in research on anaerobes must use the best techniques available and constantly search for ways to improve our ability to recover and characterize these organisms. We must resign ourselves to use of large batteries of media of various types, including selective and nonselective media and media with unusual constituents which might permit recovery of organisms with special nutritional requirements, and incubation under a variety of atmospheres and other conditions simulating the ecological niches of the various organisms we are seeking.

If we are to communicate effectively with each other, it is crucial that we use standardized nomenclature. Considerable progress has been made toward this in recent years with the *Bergey's Manual* committees, the various subcommittees of the International Committee on Systematic Bacteriology, and national subcommittees on various anaerobic bacteria. We must be certain that clinicians and bacteriologists in clinical laboratories consistently use the latest approved terminology.

With regard to collection of specimens, I stress again that there are two major considerations: (1) In obtaining clinical specimens we must be absolutely certain that there is no possibility of contamination with the indigenous flora of the body, and (2) specimens must be placed under anaerobic conditions (by any of the variety of means presented this morning) as soon as possible after collection. If these two important rules are followed, we will avoid many problems in management of patients with anaerobic infections and also much confusion and misinformation in the literature.

While a laboratory should avoid improper specimens, if

such a specimen is submitted it is important to contact the clinician. This will avoid discarding an irreplaceable specimen which might still yield some information, will help educate the physician regarding proper specimen submission, and will promote better rapport between clinician and microbiologist.

The following will assist in determining the significance of anaerobic isolates: (1) Quantitation or semi-quantitation of growth of all types of isolates (aerobic, facultative and anaerobic), and (2) correlation of culture results with direct Gram staining or other direct observation. Gram staining of clinical material is also extremely important for quality control. Direct identification by fluorescent antibody technique offers a great deal of promise. Ultimately, *both* clinical and bacteriological assessment of a case is necessary.

PART II

INTESTINAL FLORA AND ASSOCIATION WITH DISEASE

CHAPTER X

NORMAL HUMAN INTESTINAL FLORA

Howard R. Attebery,
Vera L. Sutter and Sydney M. Finegold

Abstract: *The present study points out for the first time the erratic distribution of both microorganisms and moisture content in various portions of the fecal specimen. Accordingly, it is not only necessary to correct bacterial counts for moisture content, but also to thoroughly homogenize the entire fecal specimen before withdrawing an aliquot for bacteriological study. Various diluents for microbiological studies of feces or intestinal contents were studied and several were found to be satisfactory.*

Detailed bacteriological studies of feces of seven Japanese-American individuals living on mixed western-traditional diets were conducted. The anaerobic viable bacterial count of the feces from these individuals was 2.95×10^{11}*/g dry weight. This represents 35 percent of the total bacterial count as determined by direct microscopy. Various subspecies of* Bacteroides fragilis *were the dominant members of the fecal flora, but anaerobic gram-positive bacilli of various types were also quite numerous. A large number of species of anaerobes and facultative or aerobic forms was found in the seven individuals, largely as a result of the use of a large battery of selective, nonselective and differential media.*

Although various investigators have studied the bacterial composition of feces and intestinal contents for many decades, we still have a great deal to learn about the composition of the normal intestinal flora, to say nothing of the role of such flora in various physiologic and pathophysiologic functions.

Published studies on the microbial content of the intestinal tract contain many deficiencies. Although it has been established for some time that the anaerobic flora of the intestinal tract significantly outnumber aerobic or facultative organisms, a number of studies failed to do anaerobic cultures or used unsatisfactory anaerobic techniques. Clearly, conventional anaerobic jar techniques are entirely inadequate for cultivation of certain extremely fastidious anaerobic bacteria which may be present in the intestinal tract as normal flora

Study supported in part by USPHS Contract NIHNCI-E-72–3209
We would like to acknowledge the excellent technical assistance of Palma Wideman, Walker Carter and Michel De Meo.

(Drasar, 1967). An anaerobic chamber or glove box (Drasar, 1967) or the Hungate roll-tube technique (Hungate, 1966) or modifications (Moore, 1966) are satisfactory for recovery of these fastidious anaerobes. We are quite ignorant about satisfactory procedures for storage or transport of feces or intestinal contents if there must be some delay between obtaining the specimen and processing it in the laboratory. The optimum way to handle this situation is to place the specimen in an anaerobic chamber immediately.

Virtually all published studies of the bacteriology of feces used an aliquot of 0.1 to 1.0 gram. Surprisingly, no one has published on the validity of such sampling techniques. Although it is apparent from gross observation that there is significant variation in the moisture content of fecal specimens, many workers do not routinely correct bacterial counts for moisture content of the specimen. Another widely varying factor is the diluent; investigators have recommended a number of different diluents.

Although some investigators seek to define only the numerically dominant flora in feces, using nonselective media, we feel this approach may lead us to overlook organisms present in small numbers but very important in terms of physiologic or pathophysiologic effects. Therefore, we believe it is desirable to use a wide variety of selective and differential media as well as nonselective media. Furthermore, it is desirable to incorporate various substrates to permit recovery of organisms with unique requirements. Special atmospheric and other conditions designed to pick up organisms occupying particular ecological niches should be employed. By using specialized media and procedures, Nottingham and Hungate (1968) were able to demonstrate methane bacteria in human feces for the first time, Crowther (1970) cultivated anaerobic sarcinae, and Smibert recovered anaerobic spirochetes from human feces (Holdeman and Moore, 1972).

Finally, there is a great need for studies in which an individual is sampled repeatedly over an extended period of time to determine the normal day-to-day variation in fecal or intestinal bacterial content in association with minor variations in diet and other factors. The limits of normal variation in counts must be determined before we can define the effect of administration of an antimicrobial agent or other external influence.

The purposes of the present paper are (1) to document the irregular distribution of microorganisms in fecal samples; (2) to compare various diluents for intestinal flora work; and (3) to present the results of a detailed study of the fecal microbial flora of seven adults. We have not attempted to review the literature on the bacteriology of feces or of intestinal contents at various levels in the gastrointestinal tract, as other papers are available (Donaldson, 1964; Drasar *et al.*, 1969; Finegold, 1969; Finegold *et al.*, 1970; Haenel, 1970; Moore *et al.*, 1969). Another very important aspect of intestinal microbiology not covered in this paper is the bacterial flora intimately associated with gastrointestinal mucosal epithelia (Nelson and Mata, 1970; Savage, 1970).

MATERIALS AND METHODS

Study of Homogeneity of Specimens

Six fresh stools were utilized, each studied at a different time. A total of eight samples was obtained for culture from each of the six stools, using approximately 1 g (exact weight determined later) each time. Five areas were sampled: the initial portion of the stool, the terminal portion, the middle, the surface over the initial third of the specimen (only 1 mm deep), and a deep portion from the terminal third. One of the five areas was always sampled in triplicate in order to determine the variation in count in relation to the techniques used. The eighth sample was taken at random after the entire remainder of the stool had been homogenized. All of these manipulations, including homogenization, were carried out inside an anaerobic chamber.

A series of ten-fold serial dilutions of each of these specimens was set up inside the anaerobic chamber and from these, aliquots were subcultured to the following: blood agar plates for subsequent aerobic incubation (total aerobic count); Pfizer selective enterococcus agar for aerobic incubation (counts of group D streptococci); Wadsworth Anaerobic Laboratory All-Purpose Medium (Sutter *et al.*, 1972) for anaerobic incubation (total anaerobic count); and to neomycin-egg yolk agar (Sutter *et al.*, 1972) for anaerobic incubation (counts of *Clostridium perfringens*). In addition, studies were carried out as described previously (Attebery *et al.*, 1972) to determine the percentage of extremely oxygen-sensitive (EOS) organisms in each of the eight samples. Portions of each sample were weighed before and after thorough dehydration in a drying oven to determine the moisture content of the various samples.

Comparison of Diluents

The 19 different diluents listed in Table X-I were compared, using a total of six different human fecal specimens. The pH range of the diluents was from 5.0 to 7.7, and the Eh range from +40 to −340.

TABLE X-I

DILUENTS TESTED

Diluent Number	*Diluent*	*Final pH*	*Final Eh*
1	VPI anaerobe laboratory diluent (1969)	7.4	−260
2	VPI anaerobe laboratory diluent (1972)		
3	Ueno's diluent	7.7	−280
4	Fluid thioglycollate medium	7.0	−220
5	Distilled water—chamber reduced	5.5	− 10
6	Distilled water-cysteine HCl	5.9	−160
7	Distilled water-cysteine HCl-Tween 80	6.0	−160
8	Distilled water-yeast extract	5.4	− 50
9	Winkler's diluent	5.0	− 80
10	Gall's diluent	7.0	−340
11	Osmotic diluent (14-O)	7.1	−190
12	Osmotic diluent (14-P)	7.1	−230
13	Osmotic diluent (A)	6.8	−190
14	Osmotic diluent (P)	6.7	−200
15	Charcoal water	5.8	− 10
16	Drasar's diluent	7.7	−240
17	Kreb's modified diluent	7.6	−270
18	APM diluent (Wadsworth Anaerobic Lab.)	7.4	−330
19	Distilled water—not reduced	5.2	+ 40

Normal Human Fecal Flora Studies

Of the seven Japanese-American subjects used in these studies, six, ranging in age from 71 to 83, were residents of a nursing home who were well enough to give a good dietary history and to cooperate with the procedure required for stool collection. The seventh was a 48-year-old man in good health. The average age for the whole group (four females and three males) was 74 years. None of the subjects had active gastrointestinal disease. None had received antimicrobial agents within two weeks of the sample collection date, or had required laxatives or enemas within the week prior to stool collection. The nursing home subjects all had traditional Japanese food for their supper, while the seventh subject had traditional Japanese suppers two or three times per week.

The stool specimens from the nursing home subjects were collected at the nursing home in a sterile glass jar and promptly placed into an anaerobic jar. An evacuation-replacement technique utilizing 90 percent N_2 and 10 percent H_2 was used to create anaerobic conditions. A paraffin insert was used to eliminate half of the atmosphere of the anaerobic jar so that anaerobic conditions would be achieved more quickly. These specimens were then promptly transported a distance of 15 miles to our laboratory and placed in an anaerobic chamber. The specimen from the seventh subject was voided in a toilet adjacent to our laboratory and promptly brought into the laboratory and passed into the anaerobic chamber. The stool specimens varied in weight from 12.8 to 150 g (mean, 59.3 g; median 41.5 g). Each specimen was thoroughly homogenized within the anaerobic chamber using Waring blender cups (sterile) of various sizes. Then an aliquot of approximately 1 g (exact weight determined subsequently) was obtained. Serial ten-fold dilutions were prepared using the APM diluent (diluent 18 in Table X-I) with glass beads and mechanical agitation to effect good mixing. These dilutions were kept in an ice water bath in the chamber during subsequent processing.

The media used for aerobic incubation are listed in Table X-II, and those used for anaerobic incubation, in Table X-III.

TABLE X-II

MEDIA USED IN STOOL CULTURE STUDY FOR AEROBIC INCUBATION

Blood agar plate
Blood agar plate (heated dils.)
Desoxycholate agar
Nitrogen deficient agar
Cetrimide agar
Polymyxin staphylococcus agar
Pfizer selective enterococcus agar
Molybdate agar
Sabouraud's dextrose agar with chloramphenicol
Tryptose-phosphate broth-agar with rabbit serum

TABLE X-III

MEDIA USED IN STOOL CULTURE STUDY FOR ANAEROBIC INCUBATION

Blood agar plate	LBS (Lactobacillus selective) medium
Blood agar plate (heated dils.)	Rifampin blood agar
Medium No. 10 (Bryant)	Rifampin-vancomycin blood agar
RGCA (Rumen glucose cellobiose agar)	Veillonella-neomycin agar
WAL All Purpose Medium (APM)—PRAS	Methanobacterium medium (closed roll tube—high H_2 conc.)
Egg yolk agar plate (heated dils.)	Mitis-salivarius agar (aerobic incub. subseq.)
Neomycin-Nagler egg yolk agar	Anaerobic spirochete medium (VPI)
Kanamycin-vancomycin (KV) blood agar	Anaerobic Sarcina medium (Drasar)
KV laked blood agar	Fusobacterium selective media (Ueno)
Tomato juice—Eugonagar (Bifidobacterium medium)	

Direct microscopic counts were performed on suitably diluted specimens. The techniques for inoculation and incubation of media, for determinations of pH, percent solids, and EOS anaerobes, and for examination and identification of cultures are given elsewhere (Attebery *et al.*, 1972; Sutter *et al.*, 1972).

RESULTS

Homogeneity of Specimens

Table X-IV shows the variation in total counts on anaerobic plates for each of the six areas sampled. The count in the homogenized sample was 2.83×10^{11} whereas the other five samples varied from 2.01 to 3.83×10^{11}. Counts of the triplicated samples always agreed very closely. Aerobic plate counts were quite constant. The homogenized sample had a count of 2.29×10^9; the others ranged from 2.27 to 2.77×10^9.

TABLE X-IV

HOMOGENEITY OF FECAL SPECIMENS IN ANAEROBIC PLATE COUNTS*

Area Sampled	*Count/g (dry weight)* $\times 10^{11}$
Initial portion	2.89
Terminal portion	3.16
Middle portion	2.01
Surface, initial third	2.45
Deep, terminal third	3.83
Homogenized	2.83

* Mean of 6 specimens corrected to dry weight.

Counts of group D streptococci were rather variable. The homogenized sample gave a count of 3.05×10^8 of the streptococci; the counts in the other five samples varied from 3.47 to 6.69×10^8. Furthermore, in two stools there was a marked difference in colony type noted in the initial and terminal end samples. Species differences were confirmed. One species was present in large numbers at the initial end and small numbers at the terminal end and the other species had the opposite distribution. Counts of *C. perfringens* were even more variable. The homogenized sample had a count of 6.7×10^6 whereas the other five samples ranged from 1.67 to 8.43×10^6.

The percentage of extremely oxygen-sensitive anaerobes was 5.6 in the homogenized specimen and varied from 2.3 to 6.0 in the other five samples. These EOS organisms were more numerous in the deep portion, the terminal end, and the homogenized specimen, with a marked reduction in the surface areas sampled. Microscopic examination of EOS colonies from the various sampled areas showed that anaerobic vibrios (presumably *Butyrivibrio* species) were ten times more numerous in the deep terminal third of the specimen than in the surface initial third.

The difference in distribution of water content in the six areas sampled for each of the six fecal specimens is shown in Table X-V. In specimen A, the amount of solids per 100 g weight of initial specimen ranged from 67 to 92 g in the six different portions of the specimen sampled. The other

TABLE X-V

HOMOGENEITY OF FECAL SPECIMENS IN DISTRIBUTION OF WATER CONTENT

Specimen	*Solids (g)/100 g Wet Weight (Range)**
A	67–92
B	21–32
C	16–23
D	25–40
E	21–32
F	26–38

* Six areas sampled.

stool specimens showed less variation. The initial end of each specimen tended to be more dehydrated than the remainder.

Comparison of Diluents

Five diluents (Nos. 1, 2, 3, 17 and 18) were clearly superior to the other 14. Among these five, none was clearly superior, that is, no one diluent gave the highest count with each of the six specimens studied. Diluents 1, 2 and 18 ranked first most often, and average counts with diluent 18 were slightly higher than with the other diluents.

Normal Human Fecal Flora Studies

General data on fecal flora of the seven subjects are noted in Table X-VI. Total solids in the seven specimens varied

TABLE X-VI

GENERAL CHARACTERISTICS OF NORMAL HUMAN FECAL FLORA*

	Mean Values
pH	7.24
Petroff-Hausser microscopic count (dry weight)	7.6×10^{11}
Anaerobic plate count (APM) (dry weight)	2.95×10^{11}
% of direct count recovered	35%
Ratio anaerobic/aerobic plate count	214
% extremely oxygen-sensitive (EOS) anaerobes	4%

* Seven Japanese-American subjects on mixed western-traditional diet.

from 10.7 to 41.1 percent (mean, 23.9%; median, 22.6%). Counts of EOS anaerobes varied from 7.02×10^8 to 1.6×10^{11}. EOS organisms identified included *Peptostreptococcus intermedius*, two other peptostreptococci which did not fit into definite species, a peptococcus which could not be speciated, *Bacteroides fragilis* ss. *thetaiotaomicron*, a *Bacteroides fragilis* which could not be subspeciated, *Eubacterium aerofaciens*, three other types of eubacteria which could not be speciated and *Bifidobacterium adolescentis* biotype C. Spirochetes were seen on direct examination of one of the specimens but were not recovered on culture. No sarcina or methane bacteria were recovered.

Figure X-1 shows the overall pattern of the predominant members of the normal human fecal flora in the seven subjects studied. *Bacteroides fragilis* clearly predominates over all other organisms. The medians used in Figure X-1 are based on a total of seven specimens, whether or not the given organism was isolated from a particular specimen.

Tables X-VII through X-XIII give detailed information on each of the major groupings of organisms found in the fecal flora. The number of subjects harboring each species is

TABLE X-VII

FACULTATIVE STREPTOCOCCI IN NORMAL FECAL FLORA

	Number of Subjects Harboring Organism	*Log_{10} No. Organisms per g*	
		Range	*Median*
Enterococcus Group			
S. faecalis	7	4–10	8
S. faecalis var. *liquefaciens*	3	4–8	5
S. faecalis var. *zymogenes*	2	7–10	
S. faecium	1	11	
S. durans	2	8–9	
Other Group D streptococci			
S. bovis	4	4–11	8
S. equinus	2	9–10	
Other streptococci			
S. agalactiae	1	7	
S. mitis	2	6–9	
S. salivarius	1	4	
S. sanguis	2	4–9	
Other	2	7–11	

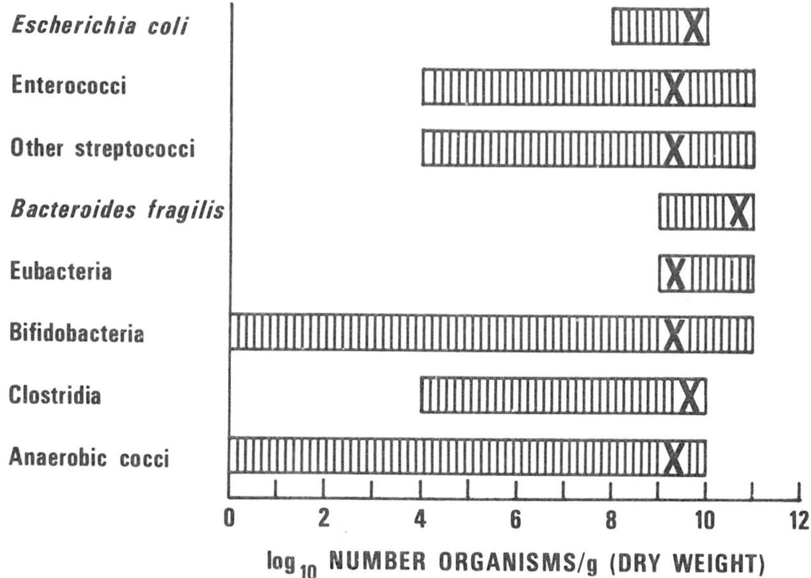

Figure X-1. Predominant normal fecal flora in 7 Japanese-American subjects on mixed western-traditional diet. The length of the blocks represents the range; the X within the block represents the median.

indicated. The range and median are given only for those specimens containing the specific organisms under consideration, and the counts (expressed as $\log_{10}$/g) are based on dry weight of feces. For example (Table X-VII), while *Streptococcus faecalis* was found in all seven individuals,

TABLE X-VIII

FACULTATIVE OR AEROBIC GRAM-NEGATIVE BACILLI IN NORMAL FECAL FLORA

	Number of Subjects Harboring Organisms	*Log₁₀ No. Organisms per g*	
		Range	*Median*
Escherichia coli	7	8–10	10
E. coli, lactose-negative	2	7	
Klebsiella pneumoniae	4	6–10	7
Klebsiella species	1	5	
Citrobacter species	1	7	
Proteus mirabilis	1	9	
Proteus morganii	1	8	
Pseudomonas aeruginosa	1	3	

S. faecalis var. *liquefaciens* was found in only three; when this latter organism was found, it was present in a range of 10^4 to 10^8/g, with a median count of 10^5/g.

Table X-VII reveals many species of streptococci, including several not in group D. As noted in Table X-VIII, *Escherichia coli* was present in all seven subjects, and *Klebsiella pneumoniae* in four. Several other facultative or aerobic gram-negative bacilli were found occasionally. Lactobacilli (Table X-IX) were found in relatively few specimens. Of interest is the recovery of *Bacillus* species in five of the seven subjects, with a median count of 10^7. Various yeasts were also found commonly.

Table X-X lists the bacteroides and fusobacteria found. One or more subspecies of *B. fragilis* was found in each individual studied and this species dominated the entire fecal flora. The two subspecies of *B. fragilis* found most commonly were ss. *thetaiotaomicron* and ss. *distasonis*. Eubacteria were found in all seven individuals studied (Table X-XI). Although seven species were recognized, the majority of eubacteria isolates could not be classified in any of the existing species. Six of the seven subjects had one or another *Bifidobacterium*

TABLE X-IX

LACTOBACILLI AND MISCELLANEOUS FACULTATIVE OR AEROBIC ORGANISMS IN NORMAL FECAL FLORA

	Number of Subjects Harboring Organisms	*Log_{10} No. Organisms per g*	
		Range	*Median*
Lactobacilli			
L. acidophilus	2	6–8	
L. fermentum	1	10	
Lactobacillus species	2	5–8	
Miscellaneous			
Sarcina lutea	1	4	
Sarcina species	1	3	
Pediococcus species	1	8	
Micrococcus species	1	8	
Staphylococcus epidermidis	1	10	
Bacillus species	5	4–11	7
Nocardia species	2	8	
Candida albicans	4	3–5	4
Candida species	1	9	
Other yeasts	4	5–6	5

TABLE X-X

BACTEROIDES AND FUSOBACTERIA IN NORMAL FECAL FLORA

	Number of Subjects Harboring Organisms	*Log_{10} No. Organisms per g*	
		Range	*Median*
B. capillosus	1	10	
B. clostridiiformis, ss. *clostridiiformis*	1	6	
B. fragilis			
ss. *distasonis*	5	9–11	10
ss. *fragilis*	3	8–11	9
ss. *ovatus*	1	10	
ss. *thetaiotaomicron*	6	9–11	10
ss. *vulgatus*	4	9–11	10
other	3	9–10	9
Bacteroides species (two types)	1	9	
Fusobacterium species	1	6	

species. In the case of clostridia (Table X-XII), although eight different species were recovered, again a number of strains would not fit into established species. The most common species were *C. perfringens* and *C. ramosum*, each isolated in four subjects. Anaerobic cocci were found in six of the

TABLE X-XI

EUBACTERIA AND BIFIDOBACTERIA IN NORMAL FECAL FLORA

	Number of Subjects Harboring Organisms	*Log_{10} No. Organisms per g*	
		Range	*Median*
Eubacteria			
E. aerofaciens	3	9–11	10
E. contortum	2	9–10	
E. lentum	4	6–8	7
E. limosum	1	6	
E. nitritogenes	1	9	
E. rectale	1	7	
E. tenue	1	7	
Eubacterium species (16 types)	7	5–11	9
Bifidobacteria			
B. adolescentis			
biotype A	2	9	
biotype C	2	9	
biotype D	1	11	
B. infantis	1	8	
Bifidobacterium species (2 types)	2	6–7	

TABLE X-XII

CLOSTRIDIA IN NORMAL FECAL FLORA

	Number of Subjects Harboring Organisms	*Log_{10} No. Organisms per g*	
		Range	*Median*
C. aminovalericum	1	6	
C. barati	1	7	
C. innocuum	2	8–10	
C. oroticum	1	8	
C. paraputrificum	2	9–10	
C. perfringens	4	5–9	8
C. ramosum	4	4–10	9
C. sporosphaeroides	1	4	
Clostridium species (11 types)	6	4–10	6

seven subjects (Table X-XIII). Of particular note are the recovery of *Acidaminococcus fermentans* in two specimens and *Ruminococcus albus* in one specimen.

DISCUSSION

It should be clear from the homogenization studies that sampling stool specimens at random for bacteriologic studies may lead to significant errors. The difference in counts is partially, but not entirely, explained by differences in dis-

TABLE X-XIII

ANAEROBIC COCCI IN NORMAL FECAL FLORA

	Number of Subjects Harboring Organisms	*Log_{10} No. Organisms per g*	
		Range	*Median*
Peptococcus			
P. asaccharolyticus	2	9–10	
Peptococcus species (3 types)	2	4–9	8
Peptostreptococcus			
P. intermedius	2	9	
P. micros	2	9–10	
Peptostreptococcus species (6 types)	3	7–10	10
Veillonella			
V. parvula	2	4–7	
Acidaminococcus fermentans	2	6–8	
Ruminococcus albus	1	10	

tribution of water content. Certainly, fecal counts should be expressed in terms of dry weight; however, it is also mandatory that the entire specimen of feces be homogenized before an aliquot is taken for analysis. The homogenization will obviously have to be done under anaerobic conditions. Further details on our homogenization study will be published elsewhere.

The survey of diluents, which will also be published in more detail elsewhere, revealed that a number of diluents were unsatisfactory and that several could be used dependably. The APM diluent was chosen for subsequent bacteriologic studies of human fecal specimens.

It is clear that diet may play a significant role in determining the bacterial content of the feces (Aries *et al.*, 1969). The results reported in the present paper are from a limited number of Japanese-American subjects on a mixed western-traditional diet. Certain of the findings would not be expected in subjects consuming the usual American diet. For example, Americans have lower counts of clostridia in their stool than people living in Japan (Akama and Otani, 1970). Ueno and his colleagues, in another paper presented at this conference, point out certain differences with regard to gram-negative anaerobic bacilli in people on the traditional Japanese diet and those on a western diet. *Fusobacterium nucleatum* is typically absent in Japanese and may be present in Americans. Other fusobacteria (formerly classified in the genus *Sphaerophorus*) are found more frequently and in higher counts in Japanese than in Americans (Ohtani, 1970; see page 140). Why more fusobacteria were not noted in the present study is unknown. It is of interest, however, that one of these subjects who had been studied on a number of other occasions (in connection with other projects) invariably had sphaerophorus in his stool in the past; however, he received therapy with clindamycin in a dosage of 1200 mg daily for one week five months prior to the time the specimen was obtained for the current study. The sphaerophorus was eliminated from his stool by the clindamycin therapy and has not returned since. An unclassifiable, very large, pleomorphic bacteroides with pointed ends may be seen at times in stools of Americans but has not been noted in Japanese (see page 142). *Bacteroides putredinis*, not recovered in the current study, typically is seen more frequently in Americans than in Japanese. Finally, the high counts of *E. coli* noted in the present study are typical for Japanese subjects (Ueno, 1972). The ratio of anaerobic to aerobic count was lower in the present study (214) than would be typical for people on an American diet (1,000 or more). Another factor accounting for the high coliform counts was the fact that these organisms were often recovered in larger numbers on anaerobic plates than on aerobic plates. These organisms, although facultative, were apparently favored by the anaerobic conditions.

The overall percentage of EOS anaerobes was relatively low in this study, but in one of the subjects these organisms accounted for 16 percent of the total anaerobic count. In general, EOS organisms, when detectable, were present in high counts.

Recovery of 35 percent of the direct microscopic count (by the Petroff-Hausser technique) is generally quite good. The recovery may actually have been better than indicated by this percentage, however. A relatively large number of fibers were noted microscopically in dilution tubes. Correction for the presence of these fibers would result in a higher recovery percentage. Furthermore, although it is clear that most bacteria seen on direct smear are alive, a certain number are undoubtedly dead. If there were a simple, convenient way to distinguish directly between live and dead organisms, correction for dead organisms would show that a greater percentage recovery was achieved.

The use of a large battery of selective, nonselective and differential media permitted recovery of a large number of genera and species of anaerobic, facultative and aerobic bacteria. Our past experience with selective media indicates that some isolates present in smaller numbers would be completely overshadowed by the larger numbers of other organisms, and therefore, these organisms present in lower counts would not be recovered at all. They may nevertheless have important metabolic activities of significance to the host.

Recovery of two strains of *Acidaminococcus fermentans* is of interest. In a previous publication (Attebery *et al.*, 1972), we reported the first isolation of this organism from humans. Members of the genus *Ruminococcus* have rarely been described in human intestinal contents. There is one report of the isolation of *Ruminococcus bromii* from human feces (Moore and Holdeman, 1972).

REFERENCES

Akama, K., and Otani, S.: *Clostridium perfringens* as the flora in the intestine of healthy persons. *Jap J Med Sci Biol, 23:*161, 1970.

Aries, V.C., Crowther, J.S., Drasar, B.S., Hill, M.J., and Williams, R.E.O.: Bacteria and the aetiology of large-bowel cancer. *Gut, 10:*334, 1969.

Attebery, H.R., Sutter, V.L., and Finegold, S.M.: Effect of a partially chemically defined diet on normal human fecal flora. *Am J Clin Nutr,* 25(12): 1391, 1972.

Crowther, J.S.: Distribution of anaerobic sarcinae in human faeces. *J Med Microbiol, 3:*ix, 1970.

Donaldson, R.M., Jr.: Normal bacterial populations of the intestine and their relation to intestinal function. *N Engl J Med, 270:*938, 994, 1050, 1964.

Drasar, B.S.: Cultivation of anaerobic intestinal bacteria. *J Pathol Bacteriol, 94:*417, 1967.

Drasar, B.S., Shiner, M., and McLeod, G.M.: Studies on the intestinal flora. I. The bacterial flora of the gastrointestinal tract in healthy and achlorhydric persons. *Gastroenterol, 56:*71, 1969.

Finegold, S.M.: Intestinal bacteria. The role they play in normal physiology, pathologic physiology, and infection. *Calif Med, 110:*455, 1969.

Finegold, S.M., Sutter, V.L., Boyle, J.D., and Shimada, K.: The normal flora of ileostomy and transverse colostomy effluents. *J Infect Dis, 122:*376, 1970.

Haenel, H.: Human normal and abnormal gastrointestinal flora. *Am J Clin Nutr, 23:*1433, 1970.

Holdeman, L.V., and Moore, W.E.C. (Eds.): *Anaerobe Laboratory Manual.* Blacksburg, Virginia Polytechnic Inst. and State Univ., 1972.

Hungate, R.E.: *The Rumen and its Microbes.* New York, Acad Pr, 1966.

Moore, W.E.C.: Techniques for routine culture of fastidious anaerobes. *Int J Sys Bacteriol, 16:*173, 1966.

Moore, W.E.C., Cato, E.P., and Holdeman, L.V.: Review. Anaerobic bacteria of the gastrointestinal flora and their occurrence in clinical infections. *J Infect Dis, 119:*641, 1969.

Moore, W.E.C., and Holdeman, L.V.: Identification of anaerobic bacteria. *Am J Clin Nutr, 25:*1306, 1972.

Nelson, D.P., and Mata, L.J.: Bacterial flora associated with the human gastrointestinal mucosa. *Gastroenterol, 58:*56, 1970.

Nottingham, P.M., and Hungate, R.E.: Isolation of methanogenic bacteria from feces of man. *J Bacteriol, 96:*2178, 1968.

Ohtani, F.: Selective media for the isolation of gram-negative anaerobic rods. Part II. Distribution of gram-negative anaerobic rods in feces of normal human beings. *Jap J Bacteriol, 25:*292, 1970.

Savage, D.C.: Associations of indigenous microorganisms with gastrointestinal mucosal epithelia. *Am J Clin Nutr, 23:*1495, 1970.

Sutter, V.L., Attebery, H.R., Rosenblatt, J.E., Bricknell, K.S., and Finegold, S.M.: *Anaerobic Bacteriology Manual.* Los Angeles, Univ. of Calif., LA Ext. Div., 1972.

Ueno, K.: Personal communication. 1972.

CHAPTER XI

ROLE OF MICROBIAL ALTERATIONS IN THE PATHOGENESIS OF INTESTINAL DISORDERS

Soad Tabaqchali, H. Schjönsby and D. Gompertz

ABSTRACT: *The small intestine of man, which usually harbors a sparse microflora consisting predominantly of gram-positive facultative microorganisms, may become colonized with fecal type microorganisms in high concentrations when the integrity of the gastrointestinal tract is deranged. This paper reviews conditions which may lead to abnormal bacterial colonization and some of the metabolic abnormalities which result. The underlying conditions include abnormalities of gastric function, conditions causing stasis, and free communications between large and small bowel. Deconjugation and dehydroxylation of conjugated bile salts by bacteria in the small intestine may contribute to steatorrhea and diarrhea. Some mechanisms are discussed for disorders in carbohydrate, protein and vitamin metabolism in conditions of bacterial overgrowth.*

Studies on the microflora of the small intestine in man have been limited. Most studies have concentrated on the upper small intestine, and it is only recent observations which have emphasized the importance of investigating the whole of the small intestine from duodenum to ileum in order to understand the relationship between the microflora and the host. Furthermore, the isolation and culture techniques used so far have been inadequate in investigating the anaerobic microflora, which has limited our knowledge of the different metabolic effects and requirements of these microorganisms in relation to man.

The normal small intestine in man usually harbors a sparse microflora, derived mainly from the oropharynx, which appears in a wave-like fashion after meals (Drasar *et al.*, 1969). The microflora consists predominantly of gram-positive facultative microorganisms in concentrations varying from 10^3 to 10^5 organisms per ml of intestinal aspirate. In the ileum fecal

From the MRC Intestinal Malabsorption Group, Royal Postgraduate Medical School, London, England microorganisms may also be present, *i.e.* enterobacteria, bacteroides and bifidobacteria, but in low concentrations up to 10^4 organisms/ml (Drasar *et al.*, 1969; Gorbach *et al.*, 1967). The controlling mechanisms which maintain this indigenous microflora in man at a definable and reproducible limit are unknown. Factors involved are gastric acidity (Smith, 1966; Gray and Shiner, 1967), intestinal peristalsis (Dack and Petran, 1934), bile (Floch *et al.*, 1970), and secretory immunoglobulins (Hersh *et al.*, 1970; Williams and Gibbons, 1972; McClelland *et al.*, 1972), and perhaps complex bacterial interactions (Donaldson, 1968) and effect of different diets (Hill *et al.*, 1971).

The normal situation is well maintained unless the integrity of the gastrointestinal tract is deranged. This may lead to abnormal colonization of the small bowel with fecal type microorganisms in high concentrations, and these may cause a variety of metabolic abnormalities. The microorganisms isolated from the small bowel in patients with intestinal disorders listed in Table XI-I differ from the normal flora; they

TABLE XI-I

CONDITIONS ASSOCIATED WITH AN ABNORMAL BACTERIAL FLORA*

1. Abnormalities of gastric function
   a) Polya partial gastrectomy—Afferent loop syndrome
   b) Malfunctioning gastro-jejunostomy
   c) Pernicious anemia
2. Conditions causing stasis
   a) Surgical blind loops
      Enteroanastomosis
   b) Strictures
      Congenital
      Crohn's disease
      Tuberculosis
   c) Adhesions
      X-ray irradiation
   d) Small intestinal diverticulosis
   e) Abnormal motility
      Scleroderma
      Diabetic neuropathy
      Vagotomy
      Ganglion blocking agents
   f) Partial biliary obstruction with cholangitis
3. Free communications between large and small bowel
   a) Gastrocolic fistula
   b) Enterocolic fistula
   c) Massive intestinal resection

* Reprinted from *Scand J. Gastroenterol, Suppl. 6*:142, 1970.

are mainly fecal in type and predominantly anaerobic. They include facultative enterococci, enterobacteria, streptococci, lactobacilli and yeast, and anaerobic microorganisms such as bacteroides, bifidobacteria, clostridia and veillonella. The concentrations and distributions are complex; they differ from patient to patient and depend on the site and extent of the causative lesion. The concentrations range from 10^5 to 10^9 organisms per ml (Gorbach and Tabaqchali, 1969; Drasar *et al.*, 1966; Tabaqchali, 1970b; Gorbach, 1971). The purpose of this paper is to list conditions which may lead to abnormal bacterial colonization and to discuss some of the metabolic effects of these microorganisms in man.

CONDITIONS ASSOCIATED WITH BACTERIAL OVERGROWTH

The conditions under which the small intestine develops an abnormal growth of bacteria are listed in Table XI-I. There are three main types of gastrointestinal abnormalities that encourage bacterial growth. The first group includes abnormalities of the stomach, such as achlorhydria and partial gastrectomy. The second group includes conditions causing stasis within the small intestine, as in strictures, whether congenital, infective, or associated with Crohn's disease (Vince *et al.*, 1972). Stagnant intestinal loops following surgical enteroanastomosis, diverticulosis of the small intestine (single or multiple), and conditions inhibiting normal intestinal motility, as in scleroderma (Kahn *et al.*, 1966; Salen *et al.*, 1966), will lead to bacterial colonization. Similarly, in tropical sprue an abnormal bacterial flora may be present in the ileum (Gorbach *et al.*, 1969). Partial biliary obstruction with cholangitis may act as a source of bacterial contamination of the small intestine (Scott and Kahn, 1968). The third group consists of conditions with free communication between large and small bowel through a fistula, as in ileocolic fistula due to Crohn's disease or diverticulitis of the colon. Massive intestinal resection may also lead to bacterial contamination of the small intestine, particularly when the ileocecal valve is resected (Gorbach and Tabaqchali, 1969; Tabaqchali, 1970a).

METABOLIC DISORDERS ASSOCIATED WITH BACTERIAL OVERGROWTH

Steatorrhea

Patients with bacterial colonization of the small intestine frequently have steatorrhea, due to malabsorption of dietary fat (Donaldson, 1965). Impaired pancreatic lipolysis has been suggested as the cause of the steatorrhea due to bacterial inactivation of lipase (Wirts and Goldstein, 1963), but experimental studies in animals (Donaldson, 1965; Kim *et al.*, 1966) and in man (Donaldson, 1965) have failed to confirm this hypothesis. It is now well established that steatorrhea in patients with the stagnant loop syndrome is associated with an altered bile salt metabolism (Dawson and Isselbacher, 1960; Donaldson, 1965; Kim *et al.*, 1966; Tabaqchali and Booth, 1966a; Tabaqchali *et al.*, 1968). This subject has been reviewed in detail elsewhere (Tabaqchali, 1970b).

The small intestine usually contains conjugated bile salts in sufficient concentrations to promote the dispersion and absorption of lipids through the formation of mixed micelles. Under abnormal conditions where there is bacterial colonization of the small intestine leading to steatorrhea, certain strains of bacteria, such as enterococci, bacteroides, bifidobacteria, veillonella and clostridia (Midvedt and Norman, 1967; Hill and Drasar, 1968), may deconjugate and further dehydroxylate the normally occurring conjugated bile salts. This may lead to a reduction of the concentration of the conjugated bile salts sufficiently to impair micelle formation, thereby causing malabsorption of fat. This is supported by direct measurement of bile salt concentrations in the lumen of patients with bacterial colonization and steatorrhea (Tabaqchali *et al.*, 1968) and by studies showing that intraluminal micelle formation was reduced in experimental animals (Kim *et al.*, 1966) and that steatorrhea, both in experimental animals and in one patient with stagnant loop syndrome, was reduced by feeding conjugated bile salts (Kim *et al.*, 1966; Tabaqchali *et al.*, 1968). Although *in vitro* experiments have suggested that the free bile acids produced by bacteria may inhibit the uptake and esterification of fatty acids by intestinal mucosa and thereby cause fat malabsorption, this has not been demonstrated *in vivo* (Cheney *et al.*, 1970). The free bile acids at the pH of intestinal contents appear to be either absorbed (Dietschy, 1968) or precipitated (Dowling and Small, 1968). However, the mere presence of abnormal bacterial growth or the presence of free bile acids in the small intestine does not necessarily lead to steatorrhea. Limited lesions of the jejunum or ileum may not be asociated with steatorrhea if there is insufficient degradation of bile salts (Gorbach and Tabaqchali, 1969; Donaldson, 1968). Recently, Ament *et al.* (1972) reported that the steatorrhea in intestinal stasis was due to absorptive cell dysfunction and not to defective intraluminal micelle formation. They described varying degrees of histological damage to the intestinal mucosa in the three patients studied.

Detection of Bile Acid Deconjugation

The presence of bile acid deconjugation in the small intestine can be detected without intestinal intubation by two methods. One is the demonstration of free bile acids in their serum. Normally only conjugated bile acids are detectable in the serum, in concentrations ranging from 1.1 to 9.6 μg/liter, whereas in patients with the stagnant loop syndrome, levels of 13 to 52 μg/liter have been reported (Lewis *et al.*, 1969; Panveliwalla *et al.*, 1969). The increase is due mainly to the presence of free bile acids; conjugated bile acids are either normal or only slightly increased. The mechanism by which the unconjugated bile acids remain in the systemic circulation and are not cleared by the liver is still not well understood. Clearance of intravenously administered ^{14}C-cholic acid showed that the half-life is longer and the pool size and absolute turnover rate greater in patients than in control subjects (Panveliwalla *et al.*, 1969). The strong avidity of free bile acids to plasma protein may play a role in maintaining the high serum levels (Burke *et al.*, 1971). The second method for detecting bile salt deconjugation in the small intestine is measuring $^{14}CO_2$ in expired air following an oral dose of ^{14}C-glycholic acid, due to bacterial metabolism of glycine, forming $^{14}CO_2$ (Fromm and Hofmann, 1971). These two tests are not specific and will give positive results in conditions where there is malabsorption of bile salts or after distal intestinal resection. Bacterial activity in the colon on the excessive bile salts forms free bile acids, which can be absorbed from the colon as well as from the small intestine. In the second method, bacterial action on the ^{14}C-glycine can occur in the colon with the formation of $^{14}CO_2$.

Diarrhea

The mechanism of diarrhea in patients with bacterial overgrowth is still not understood. Mekhjian *et al.* (1971), using colonic perfusion techniques in man, showed that increased amounts of bile acids in the colon caused colonic secretion of water and electrolytes, and suggested this as an explanation for the diarrhea that accompanies ileal disease or resection. This phenomenon has also been demonstrated in the small intestine of the rat (Harries and Sladen, 1972). Production of hydroxy fatty acids by bacteria may play a role in the pathogenesis of diarrhea. Bacteria degrade fat to long-chain hydroxy acids (Kim and Spritz, 1968); some, such as hydroxystearic acid, are chemically similar to ricinoleic acid, the major fatty acid in castor oil (Kellock *et al.*, 1969; Mekhjian *et al.*, 1971). Certain strains of clostridia, bifidobacteria and *Streptococcus faecalis* can rapidly convert oleic acid to hydroxy-stearic acid (Thomas, 1972), and bacteria can also produce unsaturated fatty acids (Schroeffer *et al.*, 1970).

Carbohydrate Metabolism

Malabsorption of D-xylose occurs in patients with bacterial overgrowth. This may be due to bacterial fermentation of xylose (Goldstein *et al.*, 1970) or to inhibition of mucosal transport. Glucose transport by jejunal mucosa was impaired in the presence of deconjugated bile salts *in vitro* (Pope *et al.*, 1966) and in experimental animals (Tabaqchali and Booth, 1966b), but in these experiments the intestinal mucosa showed marked damage histologically. Gracey *et al.* (1969, 1971) demonstrated reversible inhibition of active intestinal sugar transport by deconjugated bile salts and suggested that this may explain the temporary monosaccharide malabsorption present in infancy.

Protein Metabolism

Although hypoalbuminemia occurs in patients with malabsorption caused by bacterial overgrowth, severe protein deficiency and caloric undernutrition occur only rarely.

Protein-Losing Enteropathy

Excess protein loss from the intestine probably occurs in some patients with the stagnant loop syndrome when there is ulceration of mucosa (Jeejeebhoy and Coghill, 1961; Ament *et al.*, 1972). This is supported by studies on rats with experimental blind loops showing an increase in the excretion of ^{51}Cr from the intestine after an intravenous injection of $^{51}CrCl_2$ (Nygaard and Rootwelt, 1968).

Bacterial Breakdown of Dietary Protein

Bacterial products of dietary proteins are numerous. Bacteria metabolize various amino acids and the end products are absorbed, metabolized or excreted in the urine. Bacterial production of indol and the excretion of indoxyl sulfate (indican), indolacetic acid, volatile phenols, hippuric acid and the amines, piperidine and pyrrolidine, are significantly increased in rats with experimental blind loops as compared to control rats (Donaldson *et al.*, 1961; Miller *et al.*, 1971). Urinary indican excretion may be raised in patients with the stagnant loop syndrome and other malabsorption disorders (Greenberger *et al.*, 1968; Tabaqchali *et al.*, 1966; Hamilton *et al.*, 1970), but these findings should be interpreted with special consideration to factors which may affect indol production by bacteria, such as diet, presence of substrate, type of microorganism colonizing the bowel, pH of intestinal contents (Neale *et al.*, 1972) and presence or absence of pancreatic disease (Fordtran *et al.*, 1964).

Severe Protein Malnutrition

Disturbance of protein metabolism in the stagnant loop syndrome if associated with malabsorption may be severe enough to cause protein malnutrition, a clinical picture resembling kwashiorkor (Krickler and Schrire, 1958; Neale *et al.*, 1967). In some patients the amount of indican excreted in the urine represents 60 percent of the dietary intake of L-tryptophan. Other essential amino-acids may be degraded, thus leading to deficiency in the host. A study in a single patient showed that synthesis of albumin and fibrinogen was reduced, but urea synthesis was increased (Jones *et al.*, 1968). The production of urea suggests the bacteria were utilizing nitrogen by forming ammonia, which is diverted through the urea cycle.

Intestinal Infantilism

In children with long-continued caloric undernutrition due to the stagnant loop syndrome there may be failure of growth and sexual development leading to a clinical picture of infantilism (Bayes and Hamilton, 1969; Neale, 1968).

VITAMIN METABOLISM

Vitamin B12 Malabsorption

Vitamin B12 malabsorption is not uncommon in patients with the stagnant loop syndrome. This may lead to vitamin B12 deficiency and megaloblastic anemia. Proliferation of enteric bacteria in the small intestine has been implicated as the cause of the malabsorption, since treatment with oral broad spectrum antibiotics improves the absorption, if the ileum is intact and has not short-circuited (Mollin *et al.*, 1957). The mechanism by which bacteria interfere with vitamin B12 absorption is still not fully understood. Three possible mechanisms have been investigated.

COMPETITIVE UPTAKE OF B12 BY MICROORGANISMS. Although it is well known that bacteria can take up free vitamin B12 *in vitro*, addition of intrinsic factor (IF) appears to protect against its uptake *in vitro* but not *in vivo*. However, this protection is not complete for certain strains of enterobacteria and bacteroides, as shown in Figure XI-1. Other factors may also be involved, such as the length

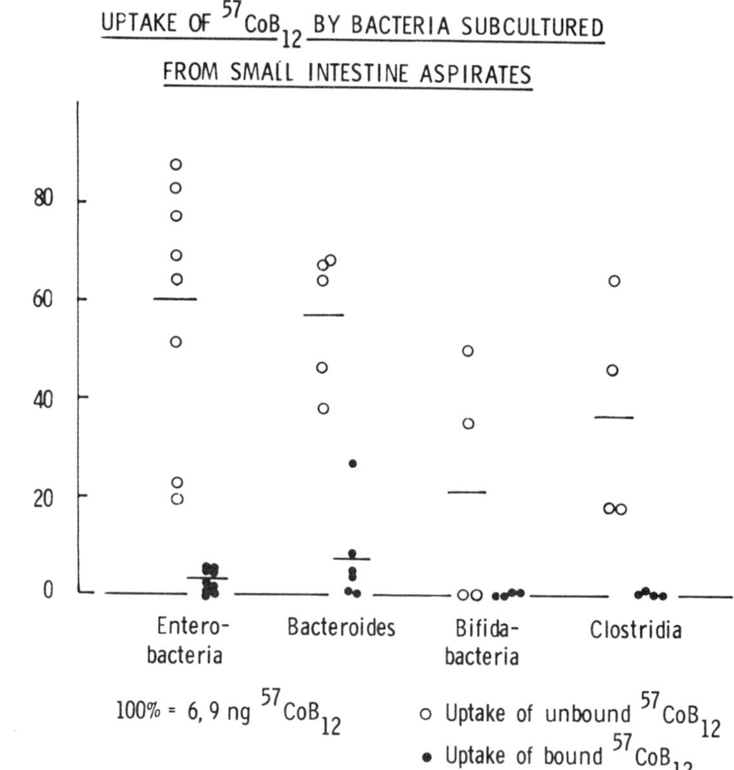

Figure XI-1. The uptake of ^{57}Co-B12 by bacteria sub-cultured from small intestinal aspirates.

of the incubation period (Donaldson, 1962); increasing the nutrient available to the microorganisms may alter the amount of uptake (Dellipiani *et al.*, 1967). *In vitro*, certain bacteria have an avidity for B12 approaching that of human IF. IF-bound B12 may interchange with B12 on the surface of the bacteria but not with the B12 within the interior of the cell (Gianella *et al.*, 1971; 1972). Studies in man support the hypothesis that bacteria can take up bound vitamin B12 *in vivo*. Patients with the stagnant loop syndrome who showed vitamin B12 malabsorption, patients with bacterial overgrowth with normal vitamin B12 absorption, and control subjects were intubated to the ileum; after an overnight fast a test meal was administered containing 1 μg ^{57}Co vitamin B12 bound to human IF. Ileal aspirates were obtained at different times for 5 hours and centrifuged. The percentage radioactivity in the deposit averaged 46 to 74 percent in patients with the stagnant loop syndrome in whom there was vitamin B12 malabsorption, whereas it was <10 percent in the other two groups (Fig. XI-2). Treatment of one of these patients with antibiotics led to a reduction in the percentages of radioactivity in the deposit; at the same time there was a decrease in the anaerobic flora colonizing the small intestine, and the B12 absorption became normal (Tabaqchali, 1970b; Schjönsby *et al.*, 1973).

Destruction of Intrinsic Factors. The second possibility investigated was the suggestion that bacteria may destroy IF activity. IF binding capacity was moderately decreased (up to 25%) after incubation of human gastric juice with microorganisms isolated from the intestine (*E. coli*, klebsiella and bacteroides) for 6 to 24 hours, but the ability of the bacterial supernatant to bind vitamin B12 and its subsequent uptake by guinea-pig brush borders was not impaired (Schjönsby and Tabaqchali, 1971b).

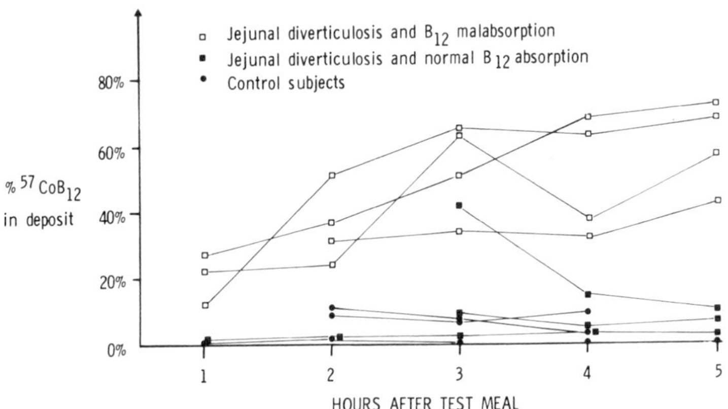

Figure XI-2. Amount of ^{57}Co-B12 in centrifuged deposits of upper ileal aspirates after feeding a test meal with 1 µg ^{57}Co-B12 bound to human gastric juice.

"Toxic" Factors. The third mechanism investigated was whether there was bacterial damage to the enterocyte receptors, *i.e.* the "toxic factor" theory. Donaldson (1962) showed normal vitamin B12 absorption by the isolated ileum of rats with experimental blind loops. Schjönsby and Tabaqchali (1971a) isolated midintestinal brush borders from control rats, and from two groups of rats with experimental blind loops (a group with normal vitamin B12 absorption and a group with abnormal vitamin B12 absorption), and measured the uptake of free and rat IF-bound vitamin B12 by the brush borders isolated from these three groups. The results showed that addition of IF enhanced the uptake of vitamin B12 by the brush borders, but there was no significant difference between the three groups (Fig. XI-3). It appears, therefore, that bacteria do not interfere with the uptake of vitamin B12 by the brush borders, but whether there is interference of vitamin B12 transport within the enterocyte has not yet been determined.

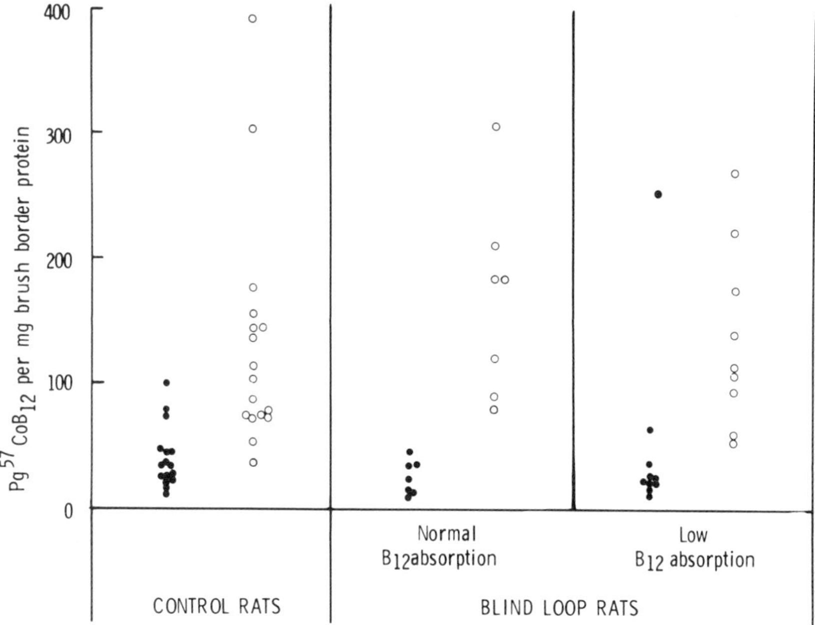

Figure XI-3. The uptake of ^{57}Co-B12 by rat mid-intestinal brush borders. • = unbound B12, o = IF-bound B12.

In summary, the evidence so far suggests that bacteria interfere with vitamin B12 absorption in the stagnant loop syndrome by competitive uptake of the vitamin within the lumen of the small intestine even when bound to IF, and that there may be minimal destruction of IF activity.

Metabolism of Other B Vitamins

There is not clear evidence of disturbance of B vitamins in patients with the stagnant loop syndrome. Signs of vitamin B deficiency states are not uncommon in malnourished patients, but these have not been related to the intestinal bacterial flora. Tabaqchali and Pallis (1970) described one patient with jejunal diverticulosis who developed an encephalopathic syndrome with EEG changes responding rapidly to treatment with a single injection of nicotinic acid (100 mg intravenously). Studies on *in vitro* incubation of fecal anaerobic microorganisms showed that the concentration of nicotinic acid in the culture medium was decreased by 5 of 14 different bacteria and in none was it increased, whereas the concentration of other B vitamins was variable (Gall, 1970).

Folic Acid Metabolism

Folic acid deficiency occurs rarely in patients with the stagnant loop syndrome (Barrett and Holt, 1966; Wakisaka, 1968). Serum folate levels may be either normal or occasionally raised. High serum folate levels have been recorded in patients with bacterial overgrowth; these were not due to vitamin B12 deficiency per se, since treatment with vitamin B12 did not reduce the serum level, and the urinary excretion of folate was also markedly raised (Hoffbrand *et al.*, 1966). Bacteria in the lumen of the intestine may synthesize folic acid, which is then absorbed, increasing the serum levels in patients with bacterial overgrowth in the upper small intestine but not in patients with distal stagnant loop syndrome, where the bacterial overgrowth is mainly in the ileum (Hoffbrand *et al.*, 1971). This may be due to preferential absorption of folic acid from the jejunum as compared to the ileum (Hepner *et al.*, 1968). The folate compound circulating in the serum of patients with stagnant loop syndrome is the same as in normal subjects, *i.e.* 5-methyl tetrahydro-folate (Hoffbrand *et al.*, 1971). These findings have been confirmed by recent work in dogs with experimental blind loops (Bernstein *et al.*, 1972).

VOLATILE FATTY ACID PRODUCTION

The production of volatile fatty acids by anaerobic metabolism of carbohydrate, fatty acids and amino acids, has been extensively studied in ruminant nutrition (Bryant, 1970). In the rat, as much as 5 percent of the animal's energy requirement is provided by production of volatile fatty acids in the cecum (Yang *et al.*, 1970). Moore, Cato and Holdeman (1969) have studied the production of volatile fatty acids by human colonic anaerobic bacteria and have used end product analysis by gas-liquid chromatography (GLC) extensively in taxonomy.

Since the production of volatile fatty acids is a primary metabolic activity of anaerobic bacteria, measurement of these acids in intestinal content may prove a good diagnostic marker in patients with bacterial overgrowth in the small intestine. Using GLC, concentrations of volatile fatty acids were estimated in intestinal aspirates in normal subjects, in control patients with malabsorption without evidence of bacterial overgrowth, and in patients with the stagnant loop syndrome (Chernov *et al.*, 1972). The levels of acetate and propionate were raised in all patients with the stagnant loop syndrome as compared to the other two groups, whereas butyrate was less markedly changed. Acetate accounted for 85 to 95 percent of the volatile fatty acids present. The concentrations of these volatile fatty acids in three patients with the stagnant loop syndrome fell to normal levels after treatment with antibiotics.

In man, the fate and metabolism of these volatile fatty acids and whether they contribute to energy requirements is not known.

CONCLUSION

There are numerous conditions which can cause abnormal bacterial growth in the small intestine and give rise to various metabolic and nutritional disorders. In considering the bacterial flora and the metabolic activity of the microorganisms that colonize the small intestine, a good analogy could be made between ruminants and patients with the stagnant loop syndrome. There has been a great deal of work on the competition between bacteria and host for nutritional substances, which leads to nutritional deficiency. In contrast, little interest has been directed toward the possibility that these microorganisms may have a beneficial role in man as they do in the ruminant, particularly since, despite the numerous and complex metabolic abnormalities that may be demonstrated, the majority of patients live in happy symbiosis with their profuse intestinal flora.

REFERENCES

Ament, M.E., Shimoda, S.S., Saunders, D.R., and Rubin, C.E.: Pathogenesis of steatorrhoea in three cases of small intestinal stasis syndrome. *Gastroenterology, 63:*728, 1972.

Barrett, C.R., Jr., and Holt, P.R.: Postgastrectomy blind loop syndrome. Megaloblastic anemia secondary to malabsorption of folic acid. *Am J Med, 41:*629, 1966.

Bayes, B.J., and Hamilton, J.R.: Blind loop syndrome in children. Malabsorption secondary to intestinal stasis. *Arch Dis Child, 44:*76, 1969.

Bernstein, L.H., Gutstein, S., Efron, G., and Wager, G.: Experimental production of elevated serum folate in dogs with intestinal blind loops. Relationship of serum levels to location of the blind loop. *Gastroenterology, 63:*815, 1972.

Bryant, M.P.: Normal flora—rumen bacteria. *Am J Clin Nutr, 23:*1440, 1970.

Burke, C.W., Lewis, B., Panveliwalla, D., and Tabaqchali, S.: The binding of cholic acid and its taurine conjugate to serum proteins. *Clin Chim Acta, 32:*207, 1971.

Cheney, F.E., Burke, V., Clarke, M.L., and Senior, J.R.: Intestinal fatty absorption and esterification from luminal micellar solutions containing deoxycholic acid. *Proc Soc Exp Biol Med, 133:*212, 1970.

Chernov, A.J., Doe, W.R., and Gompertz, D.: Intrajejunal volatile fatty acids in the stagnant loop syndrome. *Gut, 13:*103, 1972.

Dack, G.M., and Petran, E.: Bacterial activity in different levels of the intestine and in isolated segments of small and large bowel in monkeys and dogs. *J Infect Dis, 54:*204, 1934.

Dawson, A.M., and Isselbacher, K.J.: Studies on lipid metabolism in the small intestine with observations on the role of bile salts. *J Clin Invest, 39:*730, 1960.

Dellipiani, A.W., Samson, R., and Girdwood, R.H.: Factors influencing the uptake of 58 Cobalt-labelled vitamin B_{12} by *Escherichia coli.* (Abstr.) *Gut, 8:*629, 1967.

Dietschy, J.M.: Mechanisms for the intestinal absorption of bile acids. *J Lipid Res, 9:*297, 1968.

Donaldson, R.M., Jr.: Malabsorption of Co^{60} labelled cyanocobalamin in rats with intestinal diverticula. I. Evaluation of possible mechanisms. *Gastroenterology, 43:*271, 1962.

Donaldson, R.M., Jr.: Role of indigenous enteric bacteria in intestinal function and disease. In: Code, C.F., and Heidel, W., (Eds.): *Handbook of Physiology.* Vol. 5. Washington, American Physiological Society, 1968, p. 2807.

Donaldson, R.M., Jr.: Significance of small bowel bacteria. *Am J Clin Nutr, 21:*1088, 1968.

Donaldson, R.M., Jr.: Studies on the pathogenesis of steatorrhoea in the blind loop syndrome. *J Clin Invest, 44:*1815, 1965.

Donaldson, R.M., Jr., Dolcini, H.A., and Grays, J.: Urinary excretion of indolic compounds in rats with intestinal pouches. *Am J Physiol, 200:*794, 1961.

Dowling, R.H., and Small, D.M.: The effect of pH on the solubility of varying mixtures of free and conjugated bile salts in solution. *Gastroenterology, 54:*1291, 1968.

Drasar, B.S., Hill, M.J., and Shiner, M.: The deconjugation of bile salts by human intestinal bacteria. *Lancet, 1:*1237, 1966.

Drasar, B.S., Shiner, M., and McLeod, G.M.: Studies on intestinal flora. I. Effects of diet, age and periodic sampling on numbers of faecal microorganisms in man. *Gastroenterology, 53:*845, 1969.

Floch, M.H., Gershengoren, W., and Diamond, S.: Cholic acid inhibition of intestinal bacteria. *Am J Clin Nutr, 23:*81, 1970.

Fordtran, J.S., Scroggie, J.B., and Polter, D.E.: Colonic absorption of tryptophan metabolites in man. *J Lab Clin Med, 64:*125, 1964.

Fromm, H., and Hofmann, A.F.: Breath test for altered bile acid metabolism. *Lancet, 2:*621, 1971.

Gall, L.S.: Normal faecal flora of man. *Am J Clin Nutr, 23:*1457, 1970.

Gianella, R.A., Broitmann, S.A., and Zamcheck, N.: Competition between bacteria and intrinsic factor for vitamin B_{12}: implications for vitamin B_{12} malabsorption in intestinal bacterial overgrowth. *Gastroenterology, 62:*255, 1972.

Gianella, R.A., Broitmann, S.A., and Zamcheck, N.: Vitamin B_{12} uptake by intestinal micro-organisms. Mechanisms and relevance to syndromes of intestinal bacterial overgrowth. *J Clin Invest, 50:*1100, 1971.

Goldstein, F., Karakadag, S., Wirts, C.W., and Kowlessar, O.D.: Intraluminal small intestinal utilisation of D-xylose by bacteria. A limitation of the D-xylose absorption test. *Gastroenterology, 59:*380, 1970.

Gorbach, S.L.: Intestinal microflora. *Gastroenterology, 60:*1110, 1971.

Gorbach, S.L., Banwell, J.G., Mitra, R., Chatterjee, B.D., Jacobs, B., and Mazumaer, D.N.G.: Bacterial contamination of the upper small bowel in tropical sprue. *Lancet, 1:*74, 1969.

Gorbach, S.L., Plaut, A.G., Nahas, L., and Weinstein, L.: Studies of intestinal microflora. II. Micro-organisms of the small intestine and their relations to oral and faecal flora. *Gastroenterology, 53:*856, 1967.

Gorbach, S.L., and Tabaqchali, S.: Bacteria, bile and the small intestine. *Gut, 10:*963, 1969.

Gracey, M., Burke, V., and Anderson, C.M.: Association of monosaccharide malabsorption with abnormal small intestinal flora. *Lancet, 2:*384, 1969.

Gracey, M., Burke, V., and Oshin, A.: Reversible inhibition of intestinal active sugar transport by deconjugated bile salts *in vitro. Biochim Biophys Acta, 225:*308, 1971.

Gray, J.D.A., and Shiner, M.: Influence of gastric pH on gastric and jejunal flora. *Gut, 8:*574, 1967.

Greenberger, N.J., Saegh, S., and Ruppert, R.D.: Urine indican excretion in malabsorptive disorders. *Gastroenterology, 55:*204, 1968.

Hamilton, J.D., Dyer, N.H., Dawson, A.M., O'Grady, F.W., Vince, A., Fenton, J.C.B., and Mollin, D.L.: Assessment and significance of bacterial overgrowth in the small intestine. *Q J Med, 39:*265, 1970.

Harries, J.T., and Sladen, G.E.: The effects of different bile salts on the absorption of fluid, electrolytes, and monosaccharides in the small intestine of the rat *in vivo. Gut, 13:*596, 1972.

Hepner, G.W., Booth, C.C., Cowan, J., and Mollin, D.L.: Absorption of crystalline folic acid in man. *Lancet, 2:*302, 1968.

Hersh, T., Floch, M.H., Binder, H.J., Conn, H.O., Prizont, R., and Spiro, H.M.: Disturbances of the jejunal and colonic bacterial flora in immunoglobulin deficiencies. *Am J Clin Nutr, 23:*1595, 1970.

Hill, M.J., Crowther, J.S., Drasar, B.S., Hawksworth, G., Aries, U., and Williams, R.E.O.: Bacteria and aetiology of cancer of large bowel. *Lancet, 1:*95, 1971.

Hill, M.J., and Drasar, B.S.: Degradation of bile salts by human intestinal bacteria. *Gut, 9:*22, 1968.

Hoffbrand, A.V., Tabaqchali, S., Booth, C.C., and Mollin, D.L.: Small intestinal bacterial flora and folate status in gastrointestinal disease. *Gut, 12:*27, 1971.

Hoffbrand, A.V., Tabaqchali, S., and Mollin, D.L.: High serum folate levels in intestinal blind loop syndrome. *Lancet, 1:*1339, 1966.

Jeejeebhoy, K.N., and Coghill, N.F.: The measurement of a gastrointestinal protein loss by a new method. *Gut, 2:*123, 1961.

Jones, E.A., Craigie, A., Tavill, A.S., Franglen, G., and Rosenoer, V.M.: Protein metabolism in intestinal stagnant loop syndrome. *Gut, 9:*446, 1968.

Kahn, I.J., Jeffries, G.H., and Sleisenger, M.H.: Malabsorption in intestinal scleroderma. Correction by antibiotics. *N Engl J Med, 274:*1339, 1966.

Kellock, T.D., Pearson, J.R., Russell, R.I., Walker, J.R., and Wiggins, H.S.:

The incidence and clinical significance of faecal hydroxy-fatty acids. *Gut, 10:*1055, 1969.

Kim, Y.S., and Spritz, N.: Metabolism of hydroxy fatty acids in dogs with steatorrhoea secondary to experimentally produced intestinal blind loops. *J Lipid Res, 9:*487, 1968.

Kim, Y.S., Spritz, N., Blum, M., Terz, J., and Sherlock, P.: The role of the altered bile acid metabolism in the steatorrhoea of experimental blind loop. *J Clin Invest, 45:*959, 1966.

Krickler, D.M., and Schrire, V.: "Kwashiorkor" in an adult due to an intestinal blind loop. *Lancet, 1:*510, 1958.

Lewis, B., Panveliwalla, D., Tabaqchali, S., and Wootton, I.D.P.: Serum bile acids in the stagnant loop syndrome. *Lancet, 1:*219, 1969.

McClelland, D.B.L., Samson, R.R., Parkin, D.M., and Shearman, D.J.C.: Bacterial agglutination studies with secretory IgA prepared from human gastrointestinal secretions and colostrum. *Gut, 13:*450, 1972.

Mekhjian, H.S., Phillips, S.F., and Hoffmann, A.F.: Colonic secretion of water and electrolytes induced by bile acids: perfusion studies in man. *J Clin Invest, 50:*1569, 1971.

Midvedt, T., and Norman, A.: Bile acid transformations by microbial strains belonging to genera found in intestinal contents. *Acta Pathol Microbiol Scand, 71:*629, 1967.

Miller, B., Mitchison, R., Tabaqchali, S., and Neale, G.: The effects of excess bacterial proliferation on protein metabolism in rats with self-filling jejunal sacs. *Eur J Clin Invest, 2:*23, 1971.

Mollin, D.L., Booth, C.C., and Baker, S.J.: The absorption of vitamin B_{12} in control subjects, in Addisonian pernicious anaemia and in the malabsorption syndrome. *Br J Haematol, 3:*412, 1957.

Moore, W.E.C., Cato, E.P., and Holdeman, L.V.: Anaerobic bacteria of the gastrointestinal flora and their occurrence in clinical infections. *J Infect Dis, 119:*641, 1969.

Neale, G.: Protein deficiency in temperate zones. *Proc Roy Soc Med, 60:*1069, 1968.

Neale, G., Antcliff, A.C., Welbourn, R.G., Mollin, D.L., and Booth, C.C.: Protein malnutrition after partial gastrectomy. *Q J Med, 36:*469, 1967.

Neale, G., Gompertz, D., Schjönsby, H., Tabaqchali, S., and Booth, C.C.: The metabolic and nutritional consequences of bacterial overgrowth in the small intestine. *Am J Clin Nutr, 25:*1409, 1972.

Nygaard, K., and Rootwelt, K.: Intestinal protein loss in rats with blind segments on the small bowel. *Gastroenterology, 54:*52, 1968.

Panveliwalla, D., Tabaqchali, S., Wootton, I.D.P., and Lewis, B.: Bile acid metabolism in the stagnant loop syndrome. In: *Proc. 8th International Congress on Nutrition, Prague.* Amsterdam, Excerpta Medica (Int. Congress Series No. 213, 448), 1969.

Pope, J.L., Parkinson, T.M., and Olson, J.A.: Action of bile salts on the metabolism and transport of water-soluble nutrients by perfused rat jejunum *in vitro. Biochim Biophys Acta, 130:*218, 1966.

Salen, G., Goldstein, F., and Wirts, C.W.: Malabsorption in intestinal scleroderma. Relation to bacterial flora and treatment with antibiotics. *Ann Intern Med, 64:*834, 1966.

Schjönsby, H., Drasar, B.S., Tabaqchali, S., and Booth, C.C.: The mechanism of vitamin B_{12} malabsorption in patients with the stagnant loop syndrome. *Scand J Gastroenterology, 8:*41, 1973.

Schjönsby, H., and Tabaqchali, S.: Uptake of rat gastric-juice-bound vitamin B_{12} by intestinal brush borders isolated from blind loop rats. *Scand J Gastroenterol, 6:*515, 1971a.

Schjönsby, H., and Tabaqchali, S.: Effect of small intestinal bacteria on intrinsic factor and the vitamin B_{12} intrinsic factor complex. *Scand J Gastroenterol, 6:*707, 1971b.

Schroepfer, J.L., Jr., Niehaus, W.G., Jr., and McClosky, J.A.: Enzymatic conversion of linoleic acid to 10D—hydroxy—Δ_{12}—cis—octadenoic acid. *J Biol Chem, 245:*3798, 1970.

Scott, A.J., and Kahn, G.A.: Partial biliary obstruction with cholangitis producing a blind loop syndrome. *Gut, 9:*187, 1968.

Smith, H.W.: The antimicrobial activity of the stomach contents of suckling rabbits. *J Pathol Bacteriol, 91:*1, 1966.

Tabaqchali, S.: Case study of a patient with massive intestinal resection. In: *Proc. 7th International Congress Clin. Chem., Geneva/Evian 1969.* Vol 4. Basel, Kargar, pp. 119–123, 1970a.

Tabaqchali, S.: The pathophysiological role of small intestinal bacterial flora. *Scand J Gastroenterol, Suppl. 6:*139, 1970b.

Tabaqchali, S., and Booth, C.C.: Alteration of bile salt metabolism in the stagnant loop syndrome. *Gut, 7:*712, 1966a.

Tabaqchali, S., and Booth, C.C.: Jejunal bacteriology and bile-salt metabolism in patients with intestinal malabsorption. *Lancet, 2:*12, 1966b.

Tabaqchali, S., Hatziaonnou, J., and Booth, C.C.: Bile salt deconjugation and steatorrhoea in patients with the stagnant loop syndrome. *Lancet, 2:*12, 1968.

Tabaqchali, S., Okubadejo, O.A., Neale, G., and Booth, C.C.: Influence of abnormal bacterial flora on small intestinal function. *Proc Roy Soc Med, 59:*1244, 1966.

Tabaqchali, S., and Pallis, C.: Reversible nicotinamide deficiency encephalopathy in a patient with jejunal diverticulosis. *Gut, 11:*1024, 1970.

Thomas, P.J.: Identification of some enteric bacteria which convert oleic acid to hydroxy-stearic acid *in vitro. Gastroenterology, 62:*430, 1972.

Vince, A., Dyer, N.H., O'Grady, F.W., and Dawson, A.M.: Bacteriological studies in Crohn's disease. *J Med Microbiol, 5:*219, 1972.

Wakisaka, G.: Problems in post-operative patients of digestive disease. XI. Blind loop syndrome with megaloblastic anaemia. *Gastroenterology (Japan), 3:*109, 1968.

Williams, R.C., and Gibbons, R.J.: Inhibition of bacterial adherence by secretory immunoglobulin A: A mechanism of antigen disposal. *Science, 177:*697, 1972.

Wirts, C.W., and Goldstein, F.: Studies on the mechanism of postgastrectomy steatorrhoea. *Ann Intern Med*, *58:*25, 1963.

Yang, M.G., Manoharan, K., and Mickelsen, O.: Nutritional contribution of volatile fatty acids from the caecum of rats. *J Nutr*, *100:*545, 1970.

CHAPTER XII

BACTERIA AND THE ETOLOGY OF CANCER OF THE LARGE INTESTINE

M.J. Hill and B.S. Drasar

Abstract: *The incidence of colon cancer is positively correlated with the amount of dietary fat and animal protein, and the gut bacteria may play a vital role as intermediaries. This paper outlines some of the pathways by which carcinogens and/or co-carcinogens may be produced by bacteria from dietary components or the intestinal secretions produced in response to the diet. Bacteria could also play a further role by promoting the enterohepatic circulation of carcinogens (and their consequent retention within the body), and possibly by controlling hepatic detoxification mechanisms.*

There are a number of epidemiological studies showing (Fig. XII-1) that the incidence of colon cancer is much lower in Japan, East Africa and India than in Western Europe or North America (Doll 1967; Wynder *et al.*, 1967). It has been suggested that the differences might derive from differences in dietary fat (Wynder *et al.*, 1969), protein (Gregor *et al.*, 1969), refined carbohydrate (Burkitt, 1971) and fiber (Walker *et al.*, 1970). Our studies, based on WHO statistics, show the incidence of colon cancer to be strongly correlated with the amount of dietary fat and animal protein (Drasar and Irving, 1973) and not at all with dietary fiber (Table XII-I).

The search for preformed carcinogens in the diet has not revealed an adequate explanation for the correlation between diet and the incidence of colon cancer, and so we have postulated that the gut bacteria might play a role as intermediaries.

POSTULATED ROLE OF BACTERIA

We have postulated that:

1. Cancer of the colon is caused by the production of carcinogens and/or co-carcinogens by gut bacteria from dietary components or from intestinal secretions produced in response to the diet.

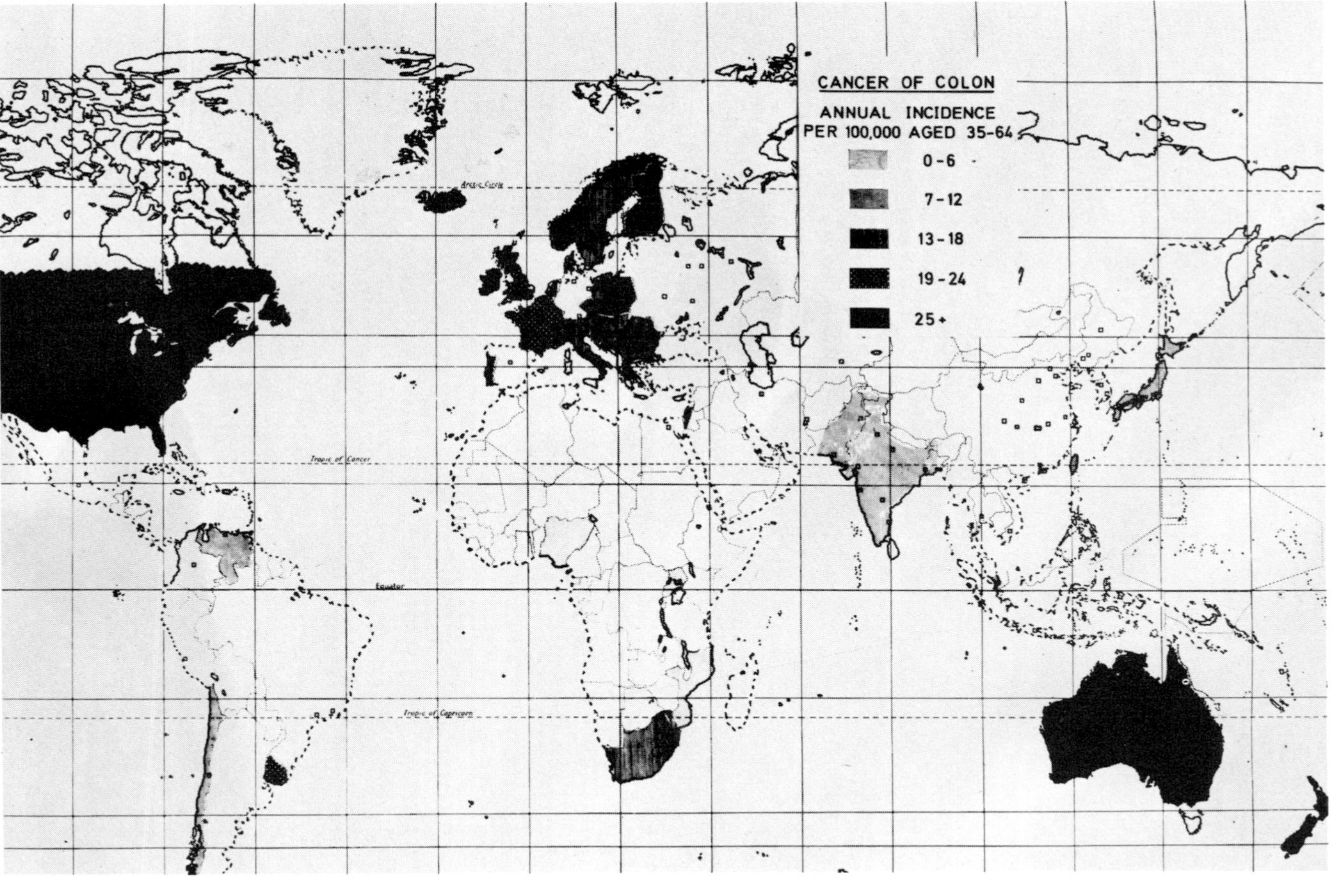

Figure XII-1. The geographical distribution of colon cancer.

TABLE XII-I

CORRELATION COEFFICIENT OF DIET AND INCIDENCE OF CANCER OF THE COLON, BREAST AND STOMACH

	Colon	*Breast*	*Stomach*
Total fat	0.81	0.80	−0.05
Animal fat	0.84	0.78	−0.08
Bound fat	0.88	0.80	−0.06
Total protein	0.70	0.59	−0.06
Animal protein	0.87	0.79	−0.02
Refined sugar	0.32	0.50	−0.04
Fiber	0.02	0.05	0.27

2. The nature of the diet affects the composition of the intestinal bacterial flora and determines the substrates available for bacterial metabolism.

3. Since the diet controls the intestinal flora, the substrates available for carcinogen production and also the physiological conditions within the gut, this would explain the correlation between diet and the incidence of colon cancer.

In our initial studies we chose fat as the dietary component most likely to be involved. The amount of dietary fat determines the concentration of steroids in feces (Hill, 1971a), as illustrated in Figure XII-2, and many acid steroids have been claimed to be carcinogenic (Fig. XII-3). Thus our working hypothesis became: (a) the amount of dietary fat determines both the concentration of bile acids and cholesterol in the large intestine and the bacterial flora acting on these steroids, and (b) bacteria can produce carcinogens and/or co-carcinogens from the biliary steroids.

To test this working hypothesis, we have examined the bacterial flora and steroid content of feces of people living in areas with both high and low risks of colon cancer (Hill *et al.*, 1971a) and our conclusions were:

1. Fecal specimens from people living in areas with a low colon cancer incidence had fewer bacteroides and more enterococci than did those from people living in high incidence areas.

2. The amount of fecal steroid, both acid and neutral, was much less in specimens from the low incidence areas.

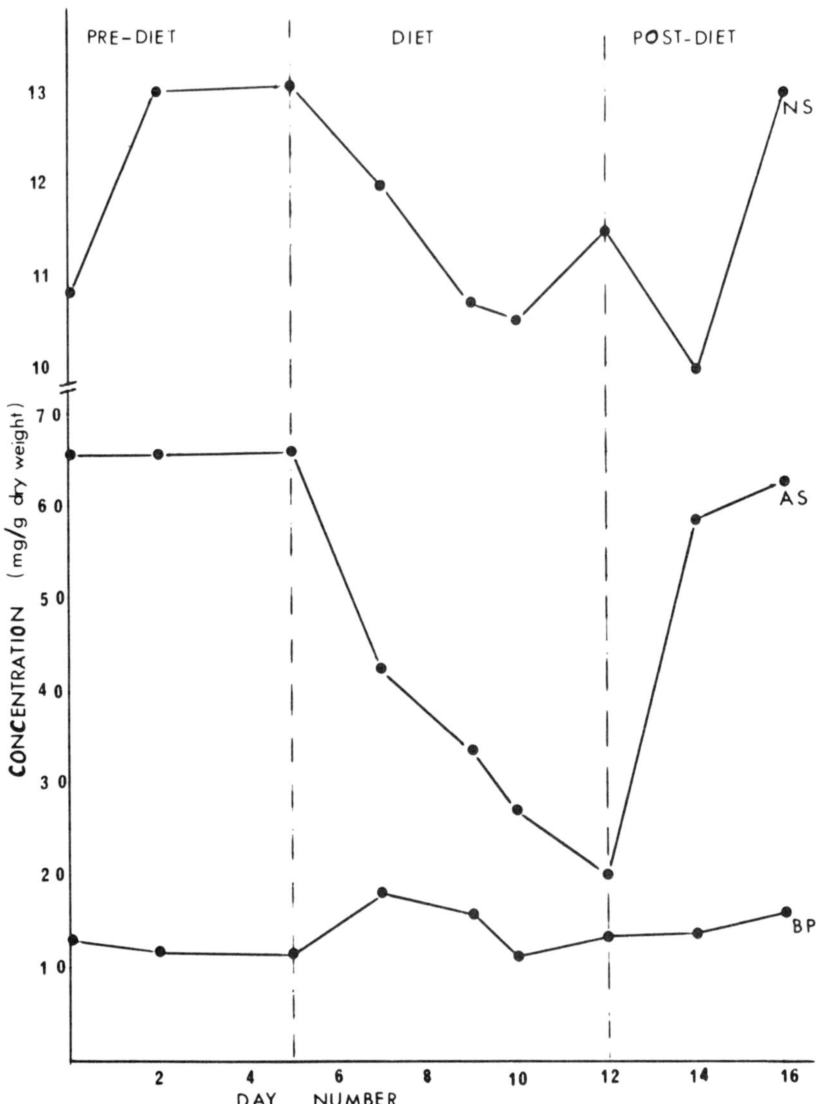

Figure XII-2. The effect of a low fat diet on the concentration of fecal acid and neutral steroids. NS=neutral steroids, AS=acid steroids, BP=bile pigment (in arbitrary units).

3. The fecal steroids from people living in low incidence areas were much less bacterially degraded than were those from high incidence areas. This finding was

BILE ACIDS WHICH HAVE BEEN SHOWN TO BE CARCINOGENS

bis nor Δ^5 cholenic acid

Lacassagne et al. Nature 1966 209 1026

apocholic acid

Lacassagne et al. Nature 1961 190 1007

deoxycholic acid

Cook et al. Nature 1940 145 627
V. Ghiron Proc. 3rd Int Canc. Con. 1939, 116

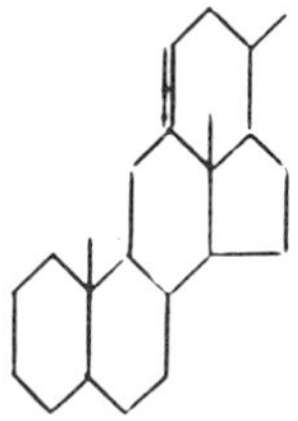

dehydronorcholene

Druckrey et al. Naturwissenschaften 1941 29 63

Figure XII-3. Some carcinogenic steroids and derivatives.

paralleled by the decreased percentage of bacterial strains able to degrade bile acids isolated from feces of people living in the low incidence countries.

Considering these results in the light of our working hypothesis the amount of presumed substrate available for carcinogen production was greater in the high risk groups and the degree of bacterial action was also greater.

PRODUCTION OF CARCINOGENS BY GUT BACTERIA

In order to support further the hypothesis we need to demonstrate that bacteria can produce a carcinogen from biliary steroids. In this section we will discuss the evidence for this. It should also be remembered that the correlation between colon cancer incidence and dietary animal protein is almost as good as that for fat (Table XII-I). We will also, therefore, discuss the possible production of carcinogens or co-carcinogens from amino acids.

PRODUCTION OF CARCINOGENS FROM BILIARY STEROIDS

The bile acids synthesized in the liver are cholic acid and chenodeoxycholic acid. Deoxycholic acid is the product of bacterial 7α-dehydroxylation of cholic acid and has been claimed to be carcinogenic (Cook *et al.*, 1940). There is an extremely good correlation between the mean fecal concentration of deoxycholic acid and the incidence of colon cancer in nine populations studied (Fig. XII-4), which would appear to support the contention that it is carcinogenic, but Bischoff (1969) in his review on carcinogenic steroids has suggested that the apparent carcinogenicity in rats was due to an artifact.

It is possible that bacteria might produce a polycyclic aromatic compound from the biliary steroids and to achieve this, four types of reaction are necessary (Hill, 1971b). These reactions are illustrated in Figure XII-5, and all have been demonstrated by us using human intestinal bacteria. The dehydration reaction (Fig. 5(b)) is carried out by *Bacteroides fragilis* and the gram-positive anaerobic bacteria; a much higher proportion of such bacteria isolated from people living in high incidence countries carried out this reaction (Table XII-II). The first and third reactions have been demonstrated with strains of *Clostridium paraputrificum* and *C. butyricum* (Aries, Goddard and Hill, 1971; Goddard and Hill, 1972) as has the fourth type of reaction (Goddard and Hill, unpublished results). Preliminary studies have isolated very few organisms capable of these reactions from feces of people living in the low incidence areas but such organisms represent a significant

>

Figure XII-4. The relationship between the fecal concentration (in mg/g dry weight of feces) of (a) deoxycholic acid, (b) total neutral steroids, and (c) total acid steroids and the incidence of colon cancer in Uganda, Japan, India, South Africa (Bantu), Hong Kong, England, South Africa (white) U.S.A., and Scotland (listed in order of increasing incidence).

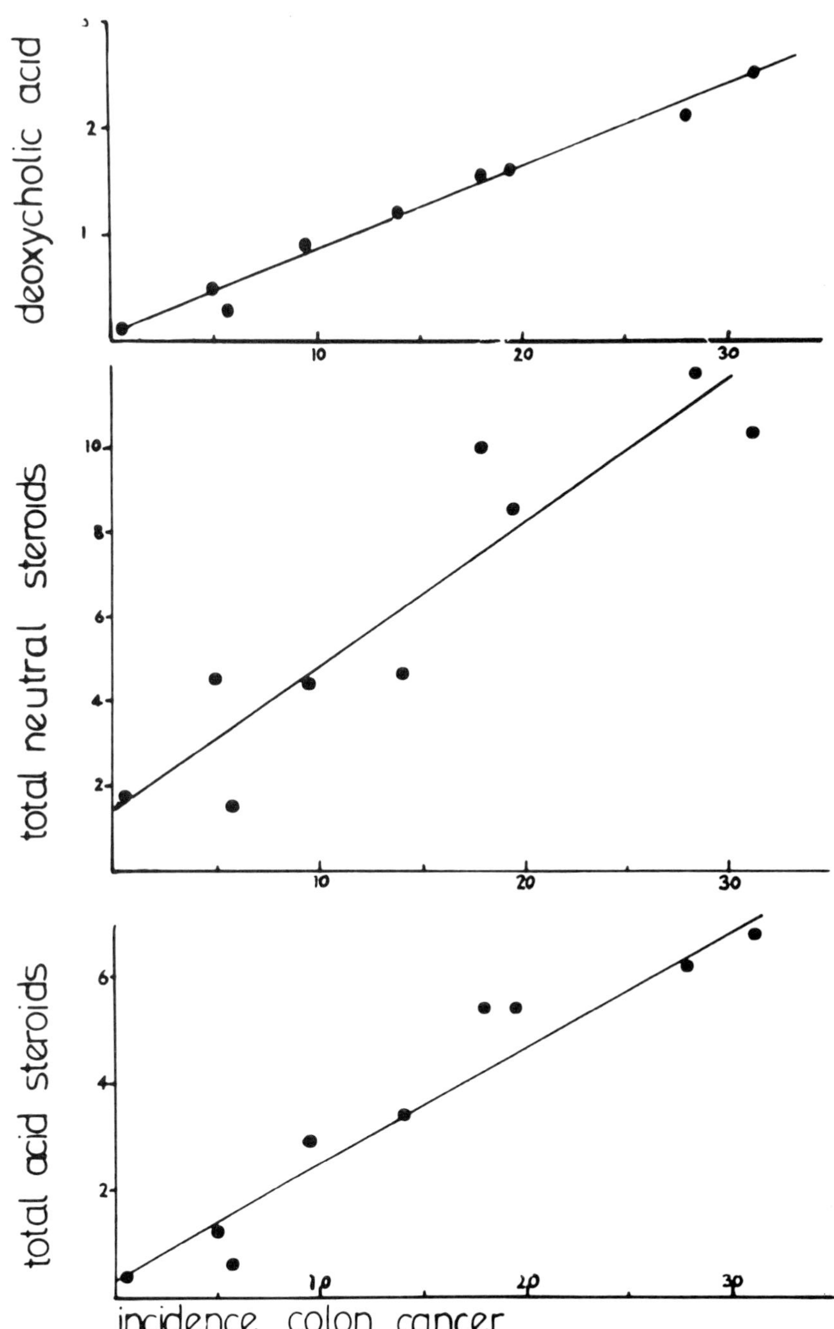
deoxycholic acid
total neutral steroids
total acid steroids
incidence colon cancer
1
2
2
4
6
8
10
2
4
6
10
20
30

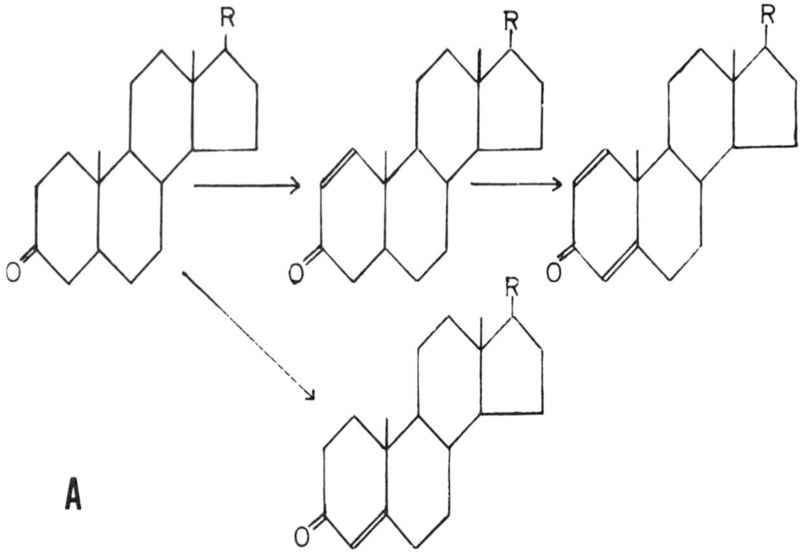
R
R
R
R
O
O
O
O
A

Figure XII-5. The mechanisms by which bacteria can dehydrogenate the steroid nucleus. (a) Dehydrogenation in conjunction with a 3-oxo group, demonstrated with clostridial strains. The enzyme is specific for 5β-steroids. (b) Dehydration demonstrated at the C6-7 bond. (c) Removal of the 10-

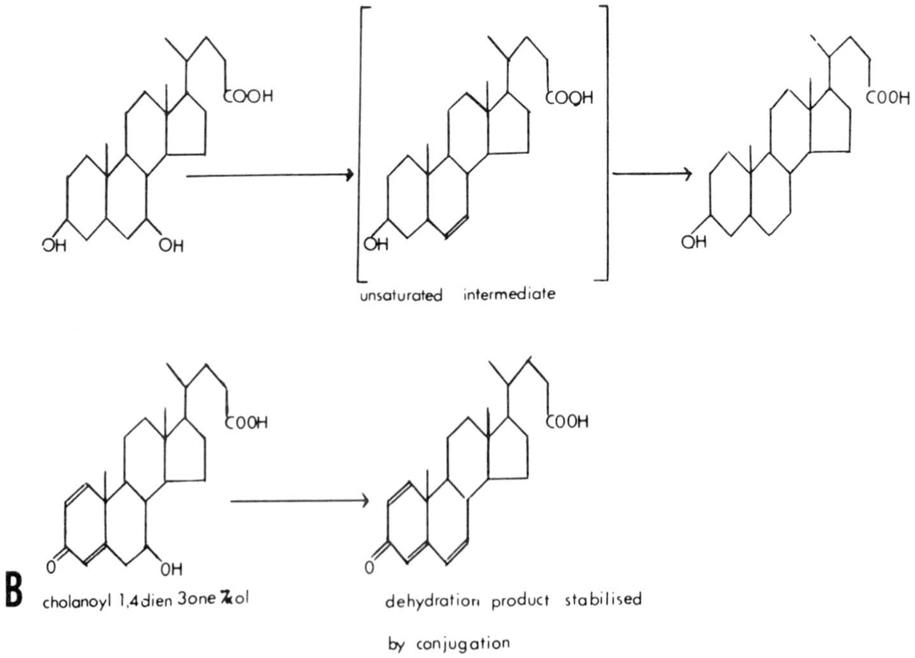
COOH
COOH
COOH
OH
OH
OH
OH
unsaturated intermediate
COOH
COOH
O
OH
O
B
cholanoyl 1,4dien 3one 7αol
dehydration product stabilised
by conjugation

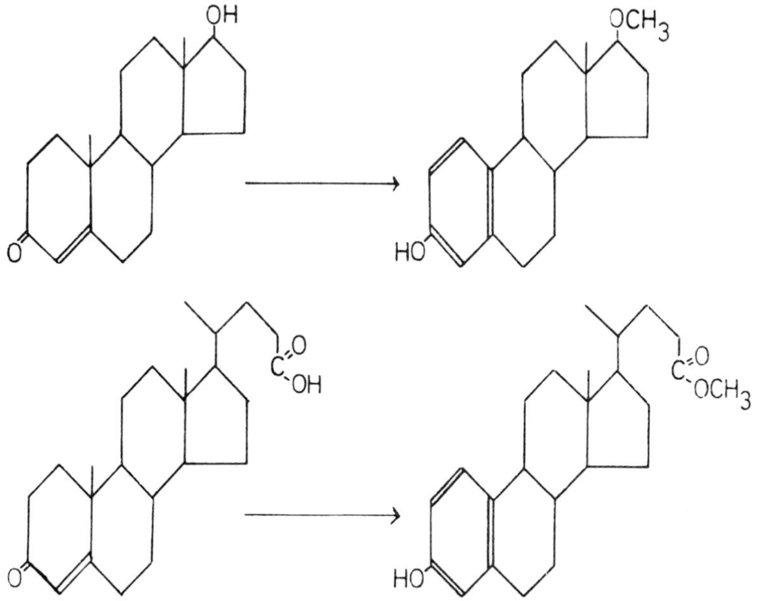

C

methyl group linked to a Cl-2 dehydrogenation demonstrated with clostridial strains. (d) Nuclear dehydrogenation in conjunction with nuclear double bonds, demonstrated with clostridia in dehydrogenation of the C8–9 to yield equilenin as illustrated.

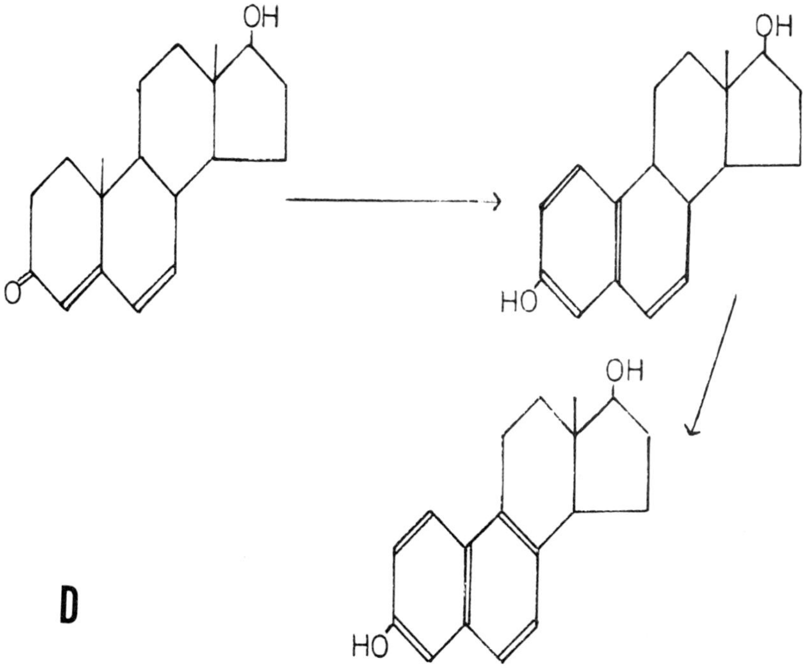

D

TABLE XII-II

PROPORTION OF BACTERIA POSSESSING 7α-DEHYDROXYLASE ISOLATED FROM FECES OF PEOPLE LIVING IN VARIOUS COUNTRIES

	England	*Scotland*	*Uganda*	*India*
Bacteroides spp.	44%	56%	33%	5%
Bifidobacterium spp.	40%	56%	4%	5%
Clostridium spp.	34%	45%	6%	0%
Enterococci	11%	40%	3%	0%
Enterobacteria	0%	0%	0%	0%

proportion of the lecithinase-negative organisms isolated from people living in the high incidence areas. One possible sequence of these four types of reactions (Fig. XII-6) yields a 17-substituted cyclopentaphenanthrene and the carcinogenicity of these has been reviewed by Coombs and Croft (1969).

One of the intermediates in this sequence would be a 17-substituted estrogen. We have demonstrated the production of estrone, estradiol and a series of unidentified steroids with a phenolic ring A from 4-cholesten-3-one; similarly 3-oxo-4-cholenic acid was aromatized to yield a bile acid with a phenolic ring A (Hill *et al.*, 1971b); the substrate for this reaction is readily produced from 3-oxo-5β-cholanic acid (Aries *et al.*, 1971), a normal fecal bile acid. Estradiol is carcinogenic in rats (Lacassagne, 1939) and hamsters (Kirkman, 1959); it would be interesting to test the carcinogenicity of the estradiol analogues produced from the bile acids and cholesterol.

The Production of N-nitrosamines

Bacteria catalyze the N-nitrosation of secondary amines at physiological pH-values (Sander, 1968; Hawksworth and Hill, 1971). Three secondary amines are produced in the large bowel, namely dimethylamine (from lecithin or choline), piperidine (from lysine) and pyrrolidine (from arginine or ornithine). All three are potentially nitrosatable by the gut flora in the presence of adequate nitrate or nitrite, but our evidence to date indicates that little, if any, nitrate or nitrite reaches the cecum and consequently the production of N-nitrosamines in the gut is unlikely. The nitrosamines are

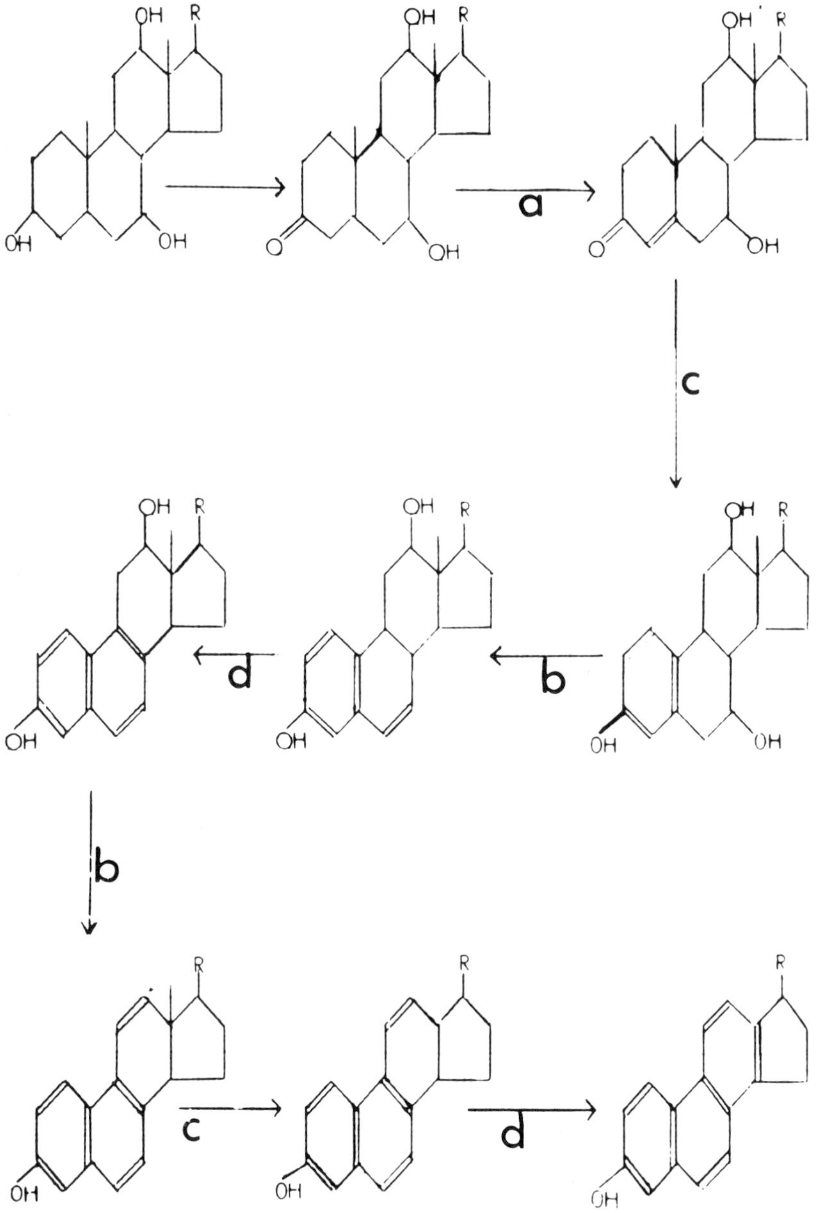

Figure XII-6. A possible sequence of the four types of nuclear dehydrogenation reactions yielding a polycyclic aromatic hydrocarbon. The letters below the arrows denote which reaction in Figure XII-5 is being utilized.

extremely potent carcinogens which rarely act at the site of production. We have hypothesized that they are produced *in vivo* in the infected urinary bladder or the stomach of achlorhydrics (Hill and Hawksworth, 1973) and that they are implicated in the etiology of human gastric cancer (Hill, 1972).

Tryptophan Metabolites

It has been noted by a number of groups of workers that there is a relation between the urinary concentration of tryptophan metabolites and the incidence of bladder cancer; this has been reviewed by Bryan (1971). All of the products of tryptophan metabolism found in human urine are produced as readily by the gut bacteria as by the liver and a few are produced only by the bacteria. Little is known of the relative contribution of the gut flora and the liver to the urinary tryptophan metabolite concentration.

Kuznezova (1969) has shown that 3-hydroxykynurenine and 3-hydroxyanthranilic acid are both mutagenic to mammalian tissue culture cells; both of these are produced by gut bacteria.

The Production of Phenols

Boutwell and Bosch (1959) demonstrated that the local application of phenol, p-cresol and 2-ethylphenol promotes the development of skin tumors in mice after a single initiating dose of dimethyl benzanthracene. The major urinary simple phenols are phenol, p-cresol and 4-ethylphenol, all of which are the products of the metabolism of dietary aromatic amino acids by the gut bacterial flora (Bakke, 1969). These simple phenols are absent from the urine of germ-free rats (Bakke and Midtvedt, 1970). These latter authors have pointed out that, although the significance of the normal production of these simple phenols has not been investigated, (a) they have been shown to have tumor promoting activity; (b) their production is decreased as the level of protein in the diet is decreased; (c) the incidence of spontaneous hepatoma is decreased in C3H mice as the amount of protein in the diet is decreased; and (d) germ-free mice are more resistant to hepatoma induction by DMAB than are conventional mice.

They might also have added that the incidence of a number of human cancers correlates well with the amount of dietary protein, and that these include such important human cancers as breast and colon cancers.

The Production of Ethionine

Ethionine, the 5-ethyl analogue of methionine, was first discovered as a synthetic carcinogen by Farber (1963) who showed that in the rat it is incorporated into protein in several tissues. Studies of bacterial cultures grown in mineral salts plus glucose medium supplemented with sulfate or methionine have shown that ethionine is produced by a number of species including *Escherichia coli* (Fisher and Mallette, 1961). The ethionine was in the free amino acid form within the cells and in the culture supernatant and was not incorporated into bacterial protein. Thus, any ethionine produced *in vivo* in the human gut would be available for carcinogenic action against the host. As yet, the production of ethionine in the gut has not been extensively studied.

Hydrolysis of Conjugated Carcinogens and Their Enterohepatic Circulation

The bacterial hydrolysis of the plant glucoside cycasin to yield the carcinogenic aglycone methylazoxymethanol is well documented (Laqueur and Spatz, 1968). To what extent this is a specific example of a generalized reaction is not known, but it is at least possible that a number of normal dietary plants contain glucosides of carcinogenic compounds.

A number of polycyclic aromatic compounds are "detoxified" by the liver as glucuronide derivatives; these glucuronides are hydrolyzed by bacterial action resulting in the enterohepatic circulation of the aglycone and a resultant failure to excrete the compound with optimal speed (see the review by Smith, 1966). It is possible that the enterohepatic circulation of carcinogens may be important in human cancer, but the postulate would be difficult to test in view of the lack of a suitable animal model (Hawksworth *et al.*, 1971).

Modification of Hepatic Detoxification Mechanisms

The mechanisms by which the liver detoxifies potentially toxic substances has been extensively studied and there is much evidence that the hydroxylation mechanism involving the P.450 microsome system is of major importance. Many compounds affect the activity of the hepatic P.450 microsome system (for example, barbiturates) and it is likely that such compounds may be produced by bacterial action on inactive substrates or alternatively that active compounds are inactivated in the gut by bacterial action. Thus the activities of the gut bacterial flora may control the detoxification mechanisms of the liver. This is virtually an unexplored subject, which has far-reaching possibilities.

REFERENCES

Aries, V.C., Goddard, P., and Hill, M.J.: Degradation of steroids by intestinal bacteria. III. 3-oxo-5β-steroid Δ^1-dehydrogenase and 3-oxo-5β-steroid Δ^4-dehydrogenase. *Biochim Biophys Acta, 248:*482, 1971.

Bakke, O.M.: Studies on the degradation of tyrosine by rat caecal contents. *Scand J Gastroenterol, 4:*603, 1969.

Bakke, O.M., and Midtvedt, T.: Influence of germ-free status on the excretion of simple phenols of possible significance in tumour promotion. *Experientia, 26:*519, 1970.

Bischoff, F.: Carcinogenic effect of steroids. *Adv Lipid Res, 7:*165, 1969.

Boutwell, R.K., and Bosch, D.K.: The tumour-promoting action of phenol and related compounds for mouse skin. *Cancer Res, 19:*413, 1959.

Bryan, G.T.: The role of urinary tryptophan metabolites in the etiology of bladder cancer. *Am J Clin Nutr, 24:*841, 1971.

Burkitt, D.P.: Epidemiology of cancer of the colon and rectum. *Cancer, 28:*3, 1971.

Cook, J.W., Kennaway, E.L., and Kennaway, N.M.: Production of tumours in mice by deoxycholic acid. *Nature, 145:*627, 1940.

Coombs, M.M., and Croft, C.J.: Carcinogenic *cyclo*-penta[a]phenanthrenes. *Progress in Tumour Res, 11:*69, 1969.

Doll, R.: Worldwide distribution of gastrointestinal cancer. *Natl Cancer Inst Monogr, 25:*173, 1967.

Drasar, B.S., and Irving, D.: Environmental factors and cancer of the colon and rectum. *Br J Cancer, 27:*167, 1973.

Farber, E.: Ethionine carcinogenesis. *Adv Cancer Res, 7:*383, 1963.

Fisher, J.F., and Mallette, M.F.: The natural occurrence of ethionine in bacteria. *J Gen Physiol, 45:*1, 1961.

Goddard, P. and Hill, M.J.: Degradation of steroids by intestinal bacteria. IV. The aromatisation of ring A. *Biochim Biophys Acta, 280:*336, 1972.

Gregor, O., Toman, R., and Prusova, F.: Gastrointestinal cancer and nutrition. *Gut, 10:*331, 1969.

Hawksworth, G., Drasar, B.S., and Hill, M.J.: Intestinal bacteria and the hydrolysis of glycosidic bonds. *J Med Microbiol, 4:*451, 1971.

Hawskworth, G., and Hill, M.J.: Bacteria and the N-nitrosation of secondary amines. *Br J Cancer, 25:*520, 1971.

Hill, M.J.: The effect of some factors on the faecal concentration of acid steroids, neutral steroids and urobilins. *J Pathol, 104:*239, 1971a.

Hill, M.J.: *Some Implications of Steroid Hormones in Cancer.* London, Heinemann, 1971b.

Hill, M.J.: Bacteria and the aetiology of cancer of the stomach. *J Med Microbiol, 5:*Pxiv, 1972.

Hill, M.J., Drasar, B.S., Aries, V.C., Crowther, J.S., Hawksworth, G., and Williams, R.E.O.: Bacteria and the aetiology of cancer of the large bowel. *Lancet, I:*95, 1971a.

Hill, M.J., Goddard, P., and Williams, R.E.O.: Gut bacteria and aetiology of cancer of the breast. *Lancet, II:*472, 1971b.

Hill, M.J., and Hawksworth, G.: *N-nitroso Compounds: Analysis and Formation. Proceedings of a Working Conference Held at the Deutsches Krebsforschungszentrum Heidelberg, Fed. Rep. of Germany, 13–15 October 1971.* Lyons, I.A.R.C., 1973.

Kirkman, H.: Estrogen-induced tumours of the kidney. III. Growth characteristics in the Syrian hamster. *Nat Cancer Inst. Monogr, 1:*1, 1959.

Kuznezova, L.E.: Mutagenic effect of 3-hydroxykynurenine and 3-hydroxyanthranilic acid. *Nature, 222:*484, 1969.

Lacassagne, A.C.: Statistique des sarcomes fusocellulaires observes chez des souris longtemps injectes avec des substances diverses. *CR Soc Biol, 132:*365, 1939.

Laqueur, G.L., and Spatz, M.: Toxicology of cycasin. *Cancer Res, 28:*2262, 1968.

Sander, von J.: Nitrosaminsynthese durch Bakterien. *Hoppe-Seyler's Z Physiol Chem, 349:*429, 1968.

Walker, A.R.P., Walker, B.F., and Richardson, B.D.: Bowel transit times in Bantu population. *Br Med J, 3:*48, 1970.

Wynder, E.L., Kajitani, T., Ishikawa, S., Dodo, H., and Takano, A.: Environmental factors of cancer of the colon and rectum. II. Japanese epidemiological data. *Cancer, 23:*1210, 1969.

Wynder, E.L., and Shigematsu, T.: Environmental factors of cancer of the colon and rectum. *Cancer, 20:*1520, 1967.

CHAPTER XIII

COMPARISON OF CHARACTERISTICS OF GRAM-NEGATIVE ANAEROBIC BACILLI ISOLATED FROM FECES OF INDIVIDUALS IN JAPAN AND THE UNITED STATES

KAZUE UENO, PAUL T. SUGIHARA,
KENNETH S. BRICKNELL, HOWARD R. ATTEBERY,
VERA L. SUTTER, AND SYDNEY M. FINEGOLD

ABSTRACT: *The distribution of Bacteroidaceae in the feces was studied in Japanese residing in Japan, persons of Japanese descent residing in the United States, and Caucasians and Negroes residing in the U.S. The influence of diet was examined in Japanese during a period of several months after they arrived in the U.S.*

In Japanese residing in Japan and Japanese in the U.S. who ate primarily Japanese foods, fusobacteria were prevalent in the feces. There were few or no fusobacteria, however, in the feces of Japanese-Americans who ate chiefly western foods. Fusobacterium varium *was not found in most Caucasian and Negro subjects. This organism was found in two persons who had been in Viet Nam or Japan for a long period of time and in one other with no history of foreign travel.* F. nucleatum *was found in three Caucasians, but not in any Japanese. Among five Japanese who were studied for five months after they came to the U.S., three who ate chiefly western food showed a marked decrease in the counts of* Fusobacterium, *while two who continued to eat Japanese foods did not.*

Prevalence of Bacteroides *species showed little difference among the groups of subjects, with two exceptions: (1) A large pleomorphic* Bacteroides, *which did not fit into any known species, was isolated only from Caucasians; and (2)* B. fragilis *ss.* fragilis *was much more common in Japanese on traditional diets than in Japanese or Caucasians and Negroes on western diets.*

One Japanese man was studied over a 10-month period after he came to the U.S. from Japan, while he switched from traditional diet to western foods and back again. There was a striking reciprocal relationship between F. varium *and* B. fragilis *ss.* vulgatus, *in that the latter would disappear as the former appeared in high count.*

This work was carried out during Dr. Ueno's tenure as a Visiting Scientist in the Wadsworth Anaerobic Bacteriology Laboratory under a fellowship sponsored by the Japan Ministry of Education.

Most of the numerous reports on the family Bacteroidaceae in human feces deal with these anaerobic gram-negative rods collectively as bacteroides (Haenel, 1958, 1961; Mitsuoka, 1965; Rosebury, 1962; Zubrzycki and Spaulding, 1962). We believe the reason for this has been the difficulty of classifying these organisms and the lack of selective media. In recent years Finegold and coworkers formulated selective media for the genera Bacteroides *and* Fusobacterium (Sphaerophorus), *making it easier to study the family Bacteroidaceae in feces (Finegold* et al., *1965, 1971; Sutter* et al., *1971, 1972). These media have been used in numerous studies (Finegold and Miller, 1968; Sabbaj* et al., *1971; Finegold* et al., *1966; Finegold, 1970; Sugihara* et al., *1972). Sugihara* et al. *(1972) found* F. varium *and* F. mortiferum *in the feces of 4 of 13 healthy Asians residing in America. The bacterial count ranged from* 1.2×10^5 *to* 3.5×10^{10} *per gram dry weight of feces. However, among 14 Caucasians and Negroes, only one had* Fusobacterium *species, and the bacterial count was low. These preliminary findings suggested that intestinal flora differs with race or diet.*

Meanwhile, Suzuki and coworkers in Japan had formulated selective media for Bacteroides *and* Fusobacterium *(Ohtani, 1970; Kozakai and Suzuki, 1968; Ninomiya* et al., *1972) and studied the distribution of the family Bacteroidaceae in the feces of healthy Japanese residing in Japan.*

The present paper reports a quantitative comparison of the distribution of the Bacteroidaceae in subjects with different racial background and dietary habits.

MATERIALS AND METHODS

The subjects in this study comprised three groups: (A) Japanese residing in Japan (these subjects, studied previously, were included for comparative purposes); (B) persons of Japanese descent residing in the United States; and (C) Caucasians and Negroes residing in the United States. The subjects were adults who did not have active gastrointestinal disease and who had not received antimicrobial agents for the preceding three months.

Enumeration of Bacteroidaceae in the Feces of Japanese Residing in Japan

Fresh fecal specimens from 10 healthy persons were collected in large sterile Petri dish bottoms (150 mm diameter) and processed within 10 minutes after voiding. A 1-g aliquot of the feces was weighed out and thoroughly mixed in 9 ml sterile diluent (KH_2PO_4, 4.5 g; Na_2PO_4, 6 g; 1-cysteine HCl, 0.3 g; Tween 80, 1 g; agar, 1 g; distilled water, 1000 ml; pH 7.3). The specimen was diluted from 10^{-1} to 10^{-10} with the diluent. The media used are shown in Table XIII-I. The anaerobic jar-steel wool method (Kozakai and Suzuki, 1968)

TABLE XIII-I

PROCEDURE FOR THE ISOLATION OF BACTEROIDACEAE IN FECAL FLORA STUDY IN JAPAN

Medium	*Dilutions Plated**	*Incubation Period (Days)*	*Purpose*
NBGT blood agar	10^{-7} 10^{-8} 10^{-9}	3	Enumeration of Bacteroidaceae
Bacteroides medium (NISSUI)	10^{-7} 10^{-8} 10^{-9}	3	Enumeration of bacteroides
Modified FM medium (NISSUI)	10^{-3} 10^{-4} 10^{-5} 10^{-6} 10^{-7}	3	Enumeration of fusobacteria
GAM agar (NISSUI)	10^{-7} 10^{-8} 10^{-9}	3	Total count and predominant flora
Liver veal blood agar (Difco)	10^{-7} 10^{-8} 10^{-9}	3	

* 0.1 ml of each dilution was plated on each medium used.

was used to maintain anaerobic conditions during incubation at 37°C for 3 to 4 days.

The Bacteroidaceae isolated were identified using the recommendations of the Subcommittee on Gram-negative Anaerobic Rods of the International Committee on Systematic Bacteriology. The monographs of Suzuki and Kozakai (1968), Buttiaux *et al.* (1966), Prévot (1957, 1967), Breed *et al.* (1957), Werner (1968), Barnes and Goldberg (1968), and Rosebury (1962) were also consulted.

Enumeration of Bacteroidaceae in the Feces of Japanese and a Filipino Residing in the U.S.

The subjects in this part of the study were as follows: one Filipino; three persons of Japanese descent (a Japanese immigrant, a second generation Japanese-American, and a third generation Japanese-American) who had resided in the U.S. for a long period of time; and two Japanese who had come from Japan and had been in the U.S. for only a short time.

Feces were processed immediately after voiding, using the methods described by Sutter *et al.* (1972) for homogenization and dilution. The diluent was the same as that used for specimens from the Japanese in Japan, described above. The media used are shown in Table XIII-II.

TABLE XIII-II

PROCEDURE FOR THE ISOLATION OF BACTEROIDACEAE FROM FECAL FLORA STUDY IN U.S.

Medium	*Dilutions Plated**	*Incubation Period (Days)*	*Purpose*
NBGT blood agar	10^{-7} 10^{-8} 10^{-9}	3	Enumeration of Bacteroidaceae
Bacteroides medium (NISSUI)	10^{-2} 10^{-3} 10^{-4} 10^{-5} 10^{-6}	3	Enumeration of bacteroides
Modified FM medium (NISSUI)	10^{-7} 10^{-8} 10^{-9}	3	Enumeration of fusobacteria
Kanamycin-Vancomycin blood agar	10^{-7} 10^{-8} 10^{-9}	3	Enumeration of bacteroides
Blood agar	10^{-7} 10^{-8} 10^{-9}	3	Total count and predominant flora
GAM agar (NISSUI)	10^{-7} 10^{-8} 10^{-9}	3	

* 0.1 ml of each dilution was plated on each medium used.

Anaerobic conditions during culture were obtained by the "evacuation-replacement" system, using an anaerobic jar with a "cold catalyst." Air was removed from a sealed jar by drawing a vacuum of 25 inches Hg; this procedure was repeated three to four times, and the jar was filled with an oxygen-free gas such as N_2 between evacuations. The final fill of the jar was made with a gas mixture containing 80 percent N_2, 10 percent CO_2, and 10 percent H_2. The plates were incubated for three to four days at 37° C.

After incubation, at least 30 colonies were taken from each medium and each strain was identified to species according to the criteria of Holdeman and Moore (1972) and Sutter *et al.* (1972).

Enumeration of Bacteroidaceae in the Feces of Caucasians and Negroes Residing in the U.S.

The feces of nine Caucasians and two Negroes were examined according to the method described above for Japanese residing in the U.S.

Examination of the Effect of Change in Diet on Bacteroidaceae in Feces

The methods described above for Japanese residing in the U.S. were applied over a period of months to study the effect of change in diet on the distribution of members of the family Bacteroidaceae in the feces of 6 subjects who came to the U.S. from Japan.

RESULTS AND DISCUSSION

Japanese Residing in Japan

A total of 191 strains were isolated and identified from the feces of 10 persons (Ohtani, 1970), as shown in Table XIII-III. The bacterial count of bacteroides in feces of these subjects was 10^8 to 10^{11} per gram wet weight of feces, and in all cases the bacteroides predominated among the gram-

TABLE XIII-III
DISTRIBUTION OF BACTEROIDACEAE IN FECES OF 10 HEALTHY ADULTS IN JAPAN*

Organism \ Specimen No.	1	2	3	4	5	6	7	8	9	10
B. convexus	++	++	++	++	++	++	++	++	+	++
B. thetaiotaomicron	+	+	+	+	+	+	+	+	++	+
B. terebrans†		+			+		+	+		
B. putidus	+				+			+		
B. melaninogenicus			+			+				
B. fragilis				+		+				
Bacteroides sp.	+								+	+
S. necrophorus	+	+	+	+	+	+	+	+	+	+
S. freundii	+	+				+	+			
S. siccus	+	+								+
S. varius					+				+	
S. pyogenes					+		+			
Sphaerophorus sp.				+			+	+	+	+

* From Ohtani (1970) + = 1∼10x10^8/g wet feces ++ = 1∼30x10^9/g wet feces

† *B. terebrans* is now classified as *Clostridium ramosum*

negative organisms. Among the *Bacteroides* species, *B. convexus* (*B. fragilis*) and *B. thetaiotaomicron* (*B. fragilis* ss. *thetaiotaomicron*) were predominant in terms of frequency and bacterial count. These two organisms were found in all subjects; *B. convexus* had a higher count, with one exception. *B. terebrans** was isolated in 4 of 10 cases, *B. putidus* in 3, *B. melaninogenicus* in 2, and *B. fragilis* in 2. Unspeciated *Bacteroides* were found in 3.

Sphaerophorus (*Fusobacterium*) species were found in all subjects, with counts of 10^8 to 10^{10} per gram wet weight. *S. necrophorus* (*F. necrophorum*) was found in all cases and was the predominant species in frequency of isolation and bacterial count among this genus. Next in order were *S. freundii*, *S. siccus*, *S. varius* and *S. pyrogenes*. Unidentified species of *Sphaerophorus* (*Fusobacterium* or *Bacteroides*) were isolated in 5.

In summary, in these Japanese subjects residing in Japan there were always many sphaerophorus, just as there were many bacteroides in the feces. Ito *et al.* (1972) found similar results (Table XIII-IV), with both sphaerophorus and bac-

TABLE XIII-IV

NUMBER OF ANAEROBES IN FECES OF 6 HEALTHY ADULTS IN JAPAN*

Organism		*Bacterial Count (average)*	*Range of Bacterial Counts*†
Gram-negative rods	*Bacteroides* sp	2.6×10^{10}	1×10^7–8×10^{10}
	Sphaerophorus sp	4.5×10^9	2×10^8–2×10^{10}
Gram-positive rods	*Corynebacterium* sp	2.6×10^8	1×10^5–2×10^9
	Eubacterium sp	$<1 \times 10^5$	
	L. acidophilus	$<1 \times 10^6$	
	Catenabacterium sp	$<1 \times 10^5$	
	Clostridium sp	1.5×10^7	1×10^5–2×10^9
Gram-neg. coccus	*Veillonella* sp	6×10^6	1×10^5–2×10^9
Gram-pos. cocci	*Peptostreptococcus* sp	$<1 \times 10^5$	
	Peptococcus sp	$<1 \times 10^5$	
Total Anaerobes		3.6×10^{10}	1×10^{10}–1×10^{11}

* 3 specimens done on each person; from Ito *et al*, 1972.
† Number per gram wet weight.

---

* *B. terebrans* is now classified as *Clostridium ramosum*, according to the CDC Manual, *Laboratory Methods in Anaerobic Bacteriology*.

teroides isolated in all of their 6 subjects. The bacterial counts per gram wet weight were 2×10^8 to 2×10^{10} for sphaerophorus and 1×10^7 to 8×10^{10} for bacteroides. The species were not identified.

Japanese and a Filipino Residing in the U.S.

As shown in Table XIII-V, members of the genus *Bacteroides* were isolated in all 6 subjects. Bacterial counts per gram wet weight of feces were as follows for the most frequently encountered subspecies of *B. fragilis* ss. *thetaiotaomicron*, 10^8 to 10^{10}; ss. *vulgatus*, 10^8 to 10^{10}; ss. *distasonis*, 10^7 to 10^{10}.

In contrast to the results in the group residing in Japan, fusobacteria were not isolated consistently, but were found in those subjects who ate Japanese foods more than 2 to 3 times a week. *F. necrophorum* was isolated in a count of 10^7 in only 2 of the 6 cases, and these subjects had been in the U.S. for a relatively short time (2 months in one case and 7 days in the other). *F. varium* was isolated in 4 cases. The 2 subjects whose feces yielded no fusobacteria were second and third generation Japanese-Americans, and their meals consisted largely of western foods; they seldom ate Japanese foods. One subject (No. 4) whose feces yielded *F. varium* had lived in the U.S. for 55 years, but ate Japanese food for breakfast, lunch and dinner. The results were similar to those obtained by Sugihara *et al.* (1972).

Caucasians and Negroes Residing in the U.S.

In the feces of the 9 Caucasians and 2 Negroes, the predominant isolates were various subspecies of *Bacteroides fragilis* (Table XIII-VI). The findings were consistent with those of Moore *et al.* (1972), using the roll-tube method. Unspeciated *Bacteroides* were found in 6 subjects. The isolates in 4 of these cases had large cells and were pleomorphic (Fig. XIII-1). Morphologically, the cells resembled those of *Fusobacterium* species, but biochemically they belonged to the genus *Bacteroides*. These *Bacteroides* were not found in persons of Japanese descent. The biochemical characteristics were not consistent with any species described to date. In terms

TABLE XIII-V

DISTRIBUTION OF BACTEROIDACEAE IN FECES OF FILIPINO AND JAPANESE SUBJECTS IN THE U.S.

Organism / Specimen No.		1	2	3	4	5	6
B. fragilis, ss	*thetaiotaomicron*	1×10^9	3×10^8	12×10^8	5×10^9	3×10^{10}	5×10^9
	distasonis	1×10^{10}	6×10^7	3×10^9	4×10^8	1×10^9	
	vulgatus	13×10^8	4×10^9	1×10^{10}	3×10^8	2×10^{10}	6×10^9
	ovatus				1×10^7	1×10^7	
	fragilis		3×10^7			7×10^9	
B. putredinis		1×10^9					1×10^7
Bacteroides sp.			1×10^8				
F. necrophorum			4×10^7				1×10^7
F. varium		9×10^6	2×10^4		6×10^8		5×10^6
Total Anaerobes		1×10^{10}	4×10^9	5×10^{10}	5×10^9	11×10^{10}	11×10^9
Total *Bacteroides* sp.		1×10^{10}	4×10^9	1×10^{10}	5×10^9	6×10^{10}	11×10^9
Total *Fusobacterium* sp.		9×10^6	4×10^7	$<10^2$	6×10^8	$<10^2$	1×10^7
National Origin		Filipino	Japanese	Japanese 2nd generation (Japanese-American)	Japanese 1st generation (Japanese-American)	Japanese 3rd generation (Japanese-American)	Japanese
Length of residence in U.S.		6.5 years	7 days	43 years	55 years	15 years	2 months

TABLE XIII-VI

DISTRIBUTION OF BACTEROIDACEAE IN FECES OF CAUCASIANS AND NEGROES IN THE U.S.

Organism \ Specimen		1	2	3	4	5	6	7	8	9	10	11
B. fragilis, ss.	*thetaiotaomicron*	4×10^8	7×10^9	1×10^8	6×10^8	1×10^9	1×10^9	2×10^9	2×10^9	2×10^9		2×10^{10}
	distasonis	3×10^8			3×10^9	7×10^8	10×10^7	1×10^9	10×10^7	2×10^8	2×10^8	
	vulgatus	5×10^9	2×10^{10}	1×10^8	4×10^9	2×10^9	1×10^8		3×10^7	1×10^8	5×10^9	1×10^{10}
	ovatus	1×10^7		20×10^7		4×10^6						
	fragilis							1×10^8	3×10^4			3×10^7
B. putredinis							2×10^8	36×10^7	1×10^6			
B. capillosus			6×10^8					8×10^7	1×10^7			
B. furcosus								1×10^8				
Bacteroides sp.		8×10^7*	41×10^7	1×10^8*		1×10^7*				1×10^8*	3×10^8	
F. varium					4×10^4			2×10^5		5×10^6		
F. nucleatum							30×10^4				5×10^6	9×10^3
Total Anaerobes		6×10^9	3×10^{10}	7×10^{10}	8×10^9	4×10^9	2×10^9	14×10^9	5×10^9	3×10^9	5×10^9	3×10^{10}
Total *Bacteroides* sp.		6×10^9	3×10^{10}	4×10^8	8×10^9	4×10^9	1×10^9	4×10^9	2×10^9	2×10^9	5×10^9	3×10^{10}
Total *Fusobacterium* sp.		$<10^2$	$<10^2$	$<10^2$	4×10^4	$<10^2$	3×10^5	2×10^5	$<10^2$	5×10^6	5×10^6	9×10^3

*Unidentified large pleomorphic rod

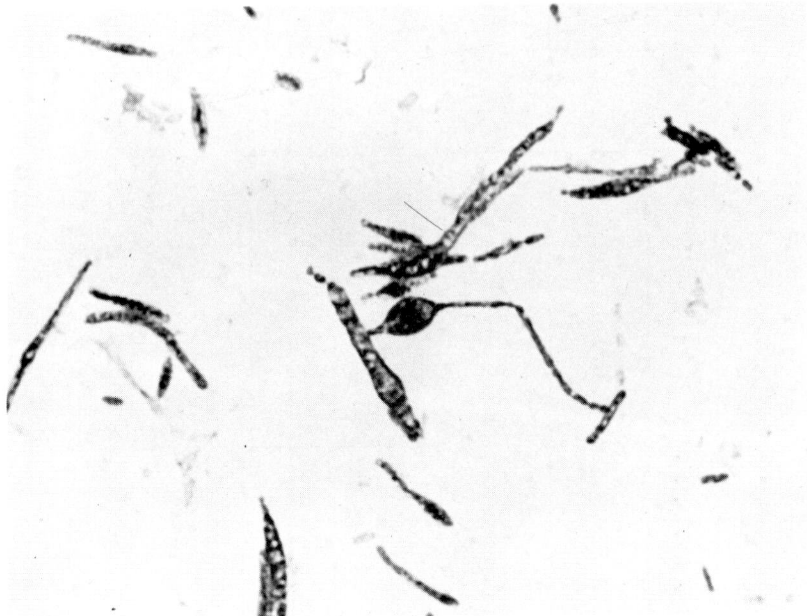

Figure XIII-1. Unidentified large pleomorphic bacteroides strain isolated from feces of Caucasians and Negroes residing in the U.S.

of simple presumptive identification by means of the antibiotic disc method (Sutter *et al.*, 1972), they would be categorized as *B. fragilis*, but the biochemical characteristics were different from those of *B. fragilis*. Details concerning this strain will be reported separately.

Two of the subjects who showed *F. varium* were Caucasians who had stayed in Viet Nam and Japan for about 2 and 7 years, respectively. One has rice in his regular diet. The other, who has a Japanese wife, eats Japanese food regularly more than 3 times a week. (His wife and 9-year-old son also showed *F. varium* in their stools.) The third subject with *F. varium* in the feces had no history of foreign travel.

Fusobacterium nucleatum, which was not found in the feces of Japanese in Japan or Japanese-Americans, was isolated from 3 of the 11 subjects in the Caucasian and Negro group. Werner and Seeliger (1963) reported finding *F. nucleatum* in counts of 10^7 to 10^8 per gram.

Change in Distribution of Bacteroidaceae in Relation to Diet

One male subject, who is Japanese, came from Japan to the U.S. in December 1971. He ate western foods almost exclusively and Japanese foods very infrequently for about 6 months, until his family came to the U.S. in the middle of June 1972. He then ate Japanese foods once a day (supper). The change in distribution of the family Bacteroidaceae in his feces was observed during 10 months of his stay in the U.S. Fusobacteria were found at a level of 10^7 per gram at the time he came to the U.S., but decreased to less than 10^2 per gram (the lowest detectable level) after 3 months. When he began eating Japanese foods regularly, the count of fusobacteria increased again.

As shown in Figure XIII-2, among *Bacteroides fragilis* sub-

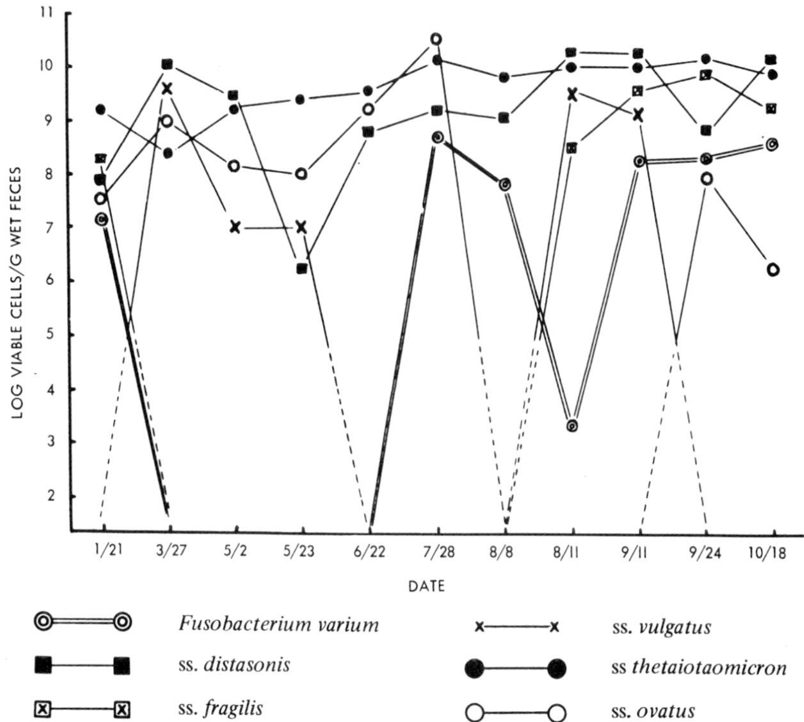

Figure XIII-2. Bacterial count of *Fusobacterium varium* and subspecies of *Bacteroides fragilis* in the feces of a Japanese subject residing in the U.S. during 10-month period.

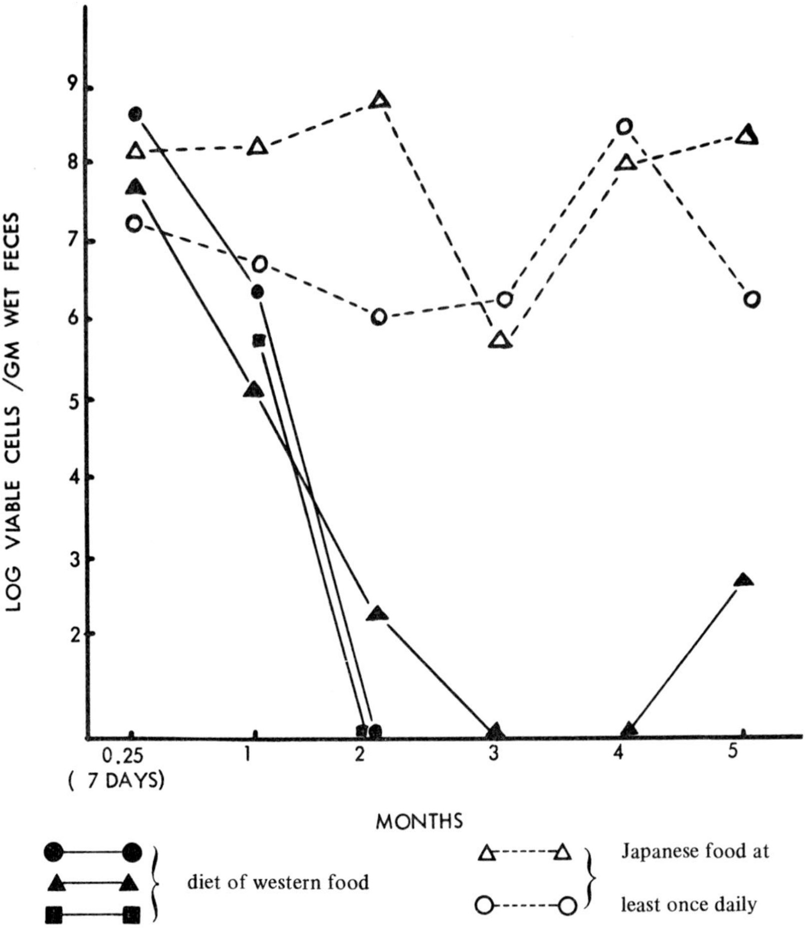

Figure XIII-3. Effect of diet on bacterial counts of fusobacteria (sphaerophorus) in feces of Japanese residing in the U.S. during a 5-month period.

species, ss. *thetaiotaomicron* and ss. *distasonis* were always predominant, with almost no change in count despite the change in diet. In contrast, the counts of ss. *ovatus*, ss. *vulgatus*, and ss. *fragilis* changed markedly. Of particular interest was the parallel in change of counts of *B. fragilis* ss. *fragilis* with change in counts of fusobacteria. We think that both may be influenced by diet. There was a reciprocal relationship between counts of *F. varium* and *B. fragilis* ss. *vulgatus*, and it is not known whether diet affects both or affects one, which then in turn affects the other organism.

Among Japanese subjects who had just come to the U.S., we selected 3 who usually ate only western foods after arriving in this country, eating Japanese foods infrequently, and 2 others who continued to eat Japanese foods regularly, *i.e.* at least once a day. We examined the change in count of fusobacteria in their feces, making counts one week after they arrived in the U.S. and then once a month for 5 months (Fig. XIII-3). The 3 who ate chiefly western foods showed a marked decrease in fusobacteria counts to less than 10^2 per gram; one subject later had a count of 5×10^2 of *F. varium* at the fifth month. In contrast, the 2 subjects who ate Japanese food showed variations in the count over the period, but no marked decrease.

REFERENCES

Barnes, E.M., and Goldberg, H.S.: The relationship of bacteria within the family *Bacteroidaceae* as shown by numerical taxonomy. *J Gen Microbiol, 51:*313, 1968.

Breed, R.S., Murray, E.G.D., and Smith, N.R.: *Bergey's Manual of Determinative Bacteriology.* Baltimore, Williams and Wilkins, 1957.

Buttiaux, R., Beerens, H., and Tacquet, A.: *Manuel de techniques bactériologiques.* Paris, Flammarion, 1966.

Finegold, S.M.: Interaction of antimicrobial therapy and intestinal flora. *Am J Clin Nutr, 23:*1466, 1970.

Finegold, S.M.: Studies on antibiotics and the normal intestinal flora. *Tex Rep Bio Med, 9:*432, 1951.

Finegold, S.M., Harada, N.E., and Miller, L.G.: Lincomycin: Activity against anaerobes and effect on normal human fecal flora. In: *Antimicrobial Agents and Chemotherapy—1965.* Ann Arbor, American Society for Microbiology, 1966, p. 659.

Finegold, S.M., and Miller, L.G.: Normal fecal flora of adult humans. In: *Bacteriological Proc.—1968.* Abstr. #M164, p. 93. Ann Arbor, American Society for Microbiology, 1968.

Finegold, S.M., Miller, A.B., and Posnick, D.J.: Further studies on selective media for *Bacteroides* and other anaerobes. *Ernährungsforschung, 10:*517, 1965.

Finegold, S.M., Sugihara, P.T., and Sutter, V.L.: Use of selective media for isolation of anaerobes from humans. In: *Isolation of Anaerobes.* (The Society for Applied Bacteriology Technical Series, No. 5.) London, Acad Pr, 1971.

Haenel, H.: Zur mikrobiologischen Ökologie des Menschen. *Zentralbl Bakteriol* [*Orig*], *172:*73, 1958.

Haenel, H.: Some results in the ecology of intestinal microflora of man. *J Appl Bacteriol*, *24:*242, 1961.

Holdeman, L.V., and Moore, W.E.C., Eds.: *Anaerobe Laboratory Manual.* Blacksburg, Virginia Polytechnic Institute and State Univ., 1972.

Kozakai, N., and Suzuki, S.: *Anaerobes in Clinical Medicine.* Tokyo, Igaku Shoin, 1968.

Mitsuoka, T.: Eine verbesserte Methodik der qualitativen und quantitativen Analyse der Darmflora von Menschen und Tieren. *Zentralbl Bakteriol* [*Orig*], *195:*455, 1965.

Moore, W.E.C., Cato, E.P., and Holdeman, L.V.: Quantitation and speciation of the normal human fecal flora. Abstr. # M3, p. 80. Paper presented at 72nd Ann. Mtg. American Society for Microbiology, Phila., April 23–28, 1972.

Ninomiya, K., Ohtani, F., Koosaka, S., Ueno, K., Suzuki, S., and Inoue, T.: Simple and expedient methods of differentiation among *Bacteroides*, *Sphaerophorus* and *Fusobacterium. Jap J Med Sci Biol*, *25:*63, 1972.

Ohtani, F.: Selective media for the isolation of gram-negative anaerobic rods. II. Distribution of gram-negative anaerobic rods in feces of normal human beings. *Jap J Bacteriol*, *25:*292, 1970.

Prévot, A.R.: *Manuel de classification et de détermination des bactéries anaérobiés.* Paris, Masson, 1957.

Prévot, A.R.: *Les bactéries anaérobiés.* Paris, Dunod, 1967.

Rosebury, T.: *Microorganisms Indigenous to Man.* New York, McGraw, 1962.

Sabbaj, J., Sutter, V.L., and Finegold, S.M.: Urease and deaminase activities of fecal bacteria in hepatic coma. In: *Antimicrobial Agents and Chemotherapy—1970.* Bethesda, American Society for Microbiology, 1971, p. 181.

Sugihara, P.T., Attebery, H.R., Sutter, V.L., and Finegold, S.M.: *Fusobacterium* spp. in feces of healthy humans. Abstr. M5, p. 80. Paper presented 72nd Ann. Mtg. American Society for Microbiology, Phila., April 23–28, 1972.

Sutter, V.L., Attebery, H.R., Rosenblatt, J.E., Bricknell, K., and Finegold, S.M.: *Anaerobic Bacteriology Manual.* Los Angeles, Univ. of Calif., Los Angeles Ext. Div., 1972.

Sutter, V.L., Sugihara, P.T., and Finegold, S.M.: Rifampin-blood-agar as a selective medium for the isolation of certain anaerobic bacteria. *Appl Microbiol*, *22:*777, 1971.

Werner, H.: *Die gram negativen anaeroben sporelosen Stäbchen des Menschen.* Munchen, Veb Gustav Fischer Verlag, Jena, 1968.

Werner, H., and Seeliger, H.P.R.: Kulturelle Untersuchungen über den Keimgehalt der Appendix unter besonderer Berücksichtigung der Anaerobier. *Zentralbl Bakteriol*, *188:*345, 1963.

Zubrzycki, L. and Spaulding, E.H.: Studies on the stability of the normal human fecal flora. *J Bacteriol*, *83:*968, 1962.

Chapter XIV

ENTEROTOXIN OF *CLOSTRIDIUM PERFRINGENS*

A.H.W. HAUSCHILD

ABSTRACT: *Clostridium perfringens type A food poisoning results from ingestion of food contaminated with about* 10^6 *to* 10^7 *vegetative cells of the microorganism per g, subsequent multiplication and sporulation of* C. perfringens *in the lumen of the small intestine, enterotoxin production which is associated with the early stages of sporulation, release of the toxin by cell lysis, and excess fluid movement from the blood stream into the intestinal lumen caused by the enterotoxin.*

The enterotoxin is lethal, antigenic, emetic, causes fluid accumulation in ligated intestinal loops, increases the permeability of blood capillaries, causes erythema in the skin of guinea pigs and rabbits, and is inactivated at 60°C or by treatment with proteolytic enzymes. The purified toxin is free of nucleic acids, lipids and reducing sugars and has all the characteristics of a protein. Its molecular weight is 33,000 to 36,000, its Stokes radius 2.6 nm, and its isoelectric point pH 4.3. Its specific toxicity to mice is about 2,100 MLD/mg N.

FOOD POISONING BY *CLOSTRIDIUM PERFRINGENS* TYPE A

Clostridium perfringens type A is one of the most common causes of bacterial food poisonings in humans (Bryan, 1969; C.D.C., 1971, 1972; Vernon, 1970). The illness is relatively mild and is characterized by profuse diarrhea and light abdominal cramps, and the absence or near absence of vomiting. The symptoms appear about 6 to 12 hours after ingestion of food contaminated with large numbers of vegetative *C. perfringens* cells, and generally subside during the subsequent 12 hours (Hobbs, 1965). The foods most commonly implicated are beef, poultry, stews and meat pies (C.D.C., 1972; Vernon, 1970) containing about 10^6 to 10^7 vegetative *C. perfringens* cells per g. After passage through the gastric juice, the viable cells multiply in the lumen of the small intestine, produce spores, and induce excess fluid movement from the blood capillaries into the intestinal lumen (Duncan and Strong, 1969b; Duncan *et al.*, 1968; Hauschild *et al.*, 1967, 1971b).

Experimental *C. perfringens* type A enteritis has been produced in human volunteers (Dische and Elek, 1957; Hauschild and Thatcher, 1967) and in animals (Duncan and Strong, 1969a, 1971; Hauschild *et al.*, 1967), but a simpler method for studying the etiology of the disease is the ligated intestinal loop technique in rabbits and lambs (Duncan *et al.*, 1968; Hauschild *et al.*, 1969), which was originally developed for the study of cholera (De and Chatterje, 1953). The suitability of the method for the study of *C. perfringens* type A enteritis has been demonstrated in a number of publications (Duncan and Strong, 1969a, 1969b, 1971; Hauschild *et al.*, 1968, 1970b; Strong *et al.*, 1971).

PRODUCTION OF ENTEROTOXIN *IN VITRO* AND *IN VIVO*

Sporulating cultures of *C. perfringens* type A produce a factor capable of causing diarrhea and fluid accumulation in ligated intestinal loops of rabbits and lambs (Duncan and Strong, 1969b; Hauschild *et al.*, 1970a, 1970b) and of producing erythema in the skin of rabbits and guinea pigs (Hauschild, 1970). In young cultures, the factor is located exclusively in the sporulated cell, outside the spore (Hauschild *et al.*, 1970a, 1970b), and can be extracted from the cell by sonic lysis (Duncan and Strong, 1969b; Hauschild *et al.*, 1970a). An erythemal factor was also found to be synthesized by *C. perfringens* in ligated intestinal loops of lambs (Hauschild *et al.*, 1971b). Comparisons of this *in vivo*-produced factor with the *in vitro*-produced factor leave little doubt that the two are identical.

The role of the erythemal factor in the enteric disease could also be demonstrated by the ligated intestinal loop technique in lambs (Hauschild *et al.*, 1971b). The loops were first injected with vegetative cells of *C. perfringens.* When they had swollen by accumulation of fluid, their contents were harvested and centrifuged. The supernatant fluids and the sonic extracts of the corresponding sediments were injected into new recipient loops and found to again cause accumulation of fluid. However, their activity was completely neutralized when the preparations were mixed with an antiserum that was specific to the erythemal factor. The enterotoxic action of the contents of the intestinal lumen was therefore due entirely to this erythemal factor. Because of its role in the enteric disease, but also due to its formation *in vivo*, its apparently direct action on the intestine, and its similarities to the enterotoxins of *Vibrio cholerae* and *Escherichia coli*, the factor has been termed "*C. perfringens*" enterotoxin (Hauschild *et al.*, 1971b).

The enterotoxic action of this toxin has been confirmed in primates. Cells and supernatant fluids from sporulated cultures of *C. perfringens*, administered into the stomachs of rhesus monkeys, or ingested by human volunteers, were shown to induce emesis and diarrhea (Duncan and Strong, 1971; Strong *et al.*, 1971). The same symptoms were induced in cynomolgus monkeys with purified enterotoxin, while no response was obtained when the toxin was first neutralized with enterotoxin-specific antiserum (Hauschild *et al.*, 1971c).

Emesis is a rare symptom in *C. perfringens* type A food poisoning (Despaul, 1966). The emetic response of the monkeys may be explained by the fact that, in these experiments, a preformed toxin was administered into the stomach, while in natural food poisonings the enterotoxin is formed in the small intestine and is unlikely to come in contact with the stomach.

Attempts to demonstrate the enterotoxin in vegetative *C. perfringens* cultures have all failed (Duncan and Strong, 1969b; Hauschild, 1970), which would suggest a possible relationship between the processes of sporulation and toxinogenesis. In the author's experience, occasional low spore yields in *C. perfringens* cultures were always accompanied by low enterotoxin yields. A correlation was also demonstrated between the degree of sporulation and enterotoxin production by individual strains of *C. perfringens* (Hauschild *et al.*, 1970b).

The best evidence for such a correlation was recently published by Duncan *et al.* (1972) who showed that spore-negative mutants, derived from sporulating and enterotoxin-producing parent strains, became enterotoxin-negative. Revertants from a spore-negative mutant which regained the ability to sporulate also regained the ability to produce enterotoxin. These results correlate to human food poisoning, as *C. perfringens* cells readily sporulate in the intestine.

GENERAL CHARACTERISTICS OF THE ENTEROTOXIN

The toxin has been purified by combinations of gel filtration and ion exchange chromatography (Hauschild and Hilsheimer, 1971) and RNase treatment (Stark and Duncan, 1972). Purified preparations were free of nucleic acids, fatty acids, phosphatides and reducing sugars. The UV absorption spectrum had a maximum at 278 to 280 nm and a minimum at 250 nm (Hauschild and Hilsheimer, 1971). Enterotoxin is destroyed at 60°C (10 min) and by treatment with pronase or *B. subtilis* protease, but not by trypsin, chymotrypsin or papain (Duncan and Strong, 1969b; Hauschild and Hilsheimer, 1971). Hydrolysis of the toxin recently yielded 19 common amino acids; among these, aspartic acid, serine, glutamic acid and leucine were the predominant components (Hauschild and Hilsheimer, to be published). All these data suggest that we are dealing with a simple protein. Molecular weight determinations were 36,000 (Hauschild and Hilsheimer, 1971) and 34,000 (Stark and Duncan, 1972) by the gel filtration method and 35,000 by sedimentation equilibrium (Stark and Duncan, 1972). A Stokes radius of 2.6 nm was calculated from Kav values on Sephadex G-100. A relatively low isoelectric point of pH 4.3 was determined by electrofocusing in a pH gradient of 3 to 6 (Hauschild and Hilsheimer, 1971).

ACTIONS OF THE ENTEROTOXIN *IN VIVO*

All known *in vivo* actions of the toxin are relatively fast. Within 90 min after injection of enterotoxin or saline into the lumen of ligated intestinal loops in rabbits, the fluid volumes of the toxin-treated loops were signifcantly larger than the fluid volumes of the control loops (Hauschild *et al.*, 1971a). Cynomolgus monkeys started to vomit within 30 min after gastric administration of the enterotoxin (Hauschild *et al.*, 1971c). Erythema in guinea pigs and rabbits developed within about 60 min after intradermal injection of enterotoxin. This reaction is now being used for rapid and accurate assay of the toxin (Hauschild, 1970). Since the erythema reaction is free of necrosis, it can be easily distinguished from the reactions of all other known toxins of *C. perfringens* (Sterne and Warrack, 1964). The enterotoxin differs also from these toxins by its cellular nature; all the others are true exotoxins (Hauschild, 1971a). The erythema reaction in the guinea pig skin is preceded by increased permeability of the blood capillaries within 15 to 20 min after intradermal injection of enterotoxin (Niilo *et al.*, 1971). Mice (Hauschild and Hilsheimer, 1971) and guinea pigs (Niilo *et al.*, 1971) were killed within the same period after intravenous injection of toxins. The specific toxicities of purified enterotoxin determined by intravenous injection of mice were 2,100 to 2,200 MLD/mg N (Hauschild and Hilsheimer, 1971; Stark and Duncan, 1972).

The enterotoxin is antigenic in rabbits (Hauschild *et al.*, 1970) and sheep (Niilo *et al.*, 1971). Crude antisera have been purified to enterotoxin specificity by adsorption with cell extracts of a nonsporulating strain of *C. perfringens* or with extracts of vegetative cells of a homologous strain. In agar gel immunodiffusion against crude extracts of sporulated cells, the adsorbed antisera produced a single precipitin band (Hauschild *et al.*, 1971b; Niilo *et al.*, 1971).

Rabbits could be protected against the lethal action of enterotoxin by immunization with extracts of sporulated *C. perfringens* cells (Hauschild, 1971b), but immunization of rabbits and lambs had no effect on the accumulation of fluid induced by enterotoxin in ligated intestinal loops of these animals (Hauschild, 1971b; Niilo *et al.*, 1971). Circulating antibody, therefore, seems to have little or no neutralizing effect on the enterotoxin in the intestine.

The mode of action of the enterotoxin *in vivo* is not as yet understood. Possible mechanisms have been discussed in a recent, more detailed review of *C. perfringens* enterotoxin and its role in food poisoning (Hauschild, 1973).

REFERENCES

Bryan, F.L.: What the sanitarian should know about *Clostridium perfringens* foodborne illness. *J Milk Food Technol, 32:*381, 1969.

C.D.C.: *Foodborne Outbreaks. Annual Summary 1970.* Atlanta, Center for Disease Control, 1971.

C.D.C.: *Foodborne Outbreaks. Annual Summary 1971.* Atlanta, Center for Disease Control. 1972.

De, S.N., and Chatterje, D.N.: An experimental study of the mechanism of action of *Vibrio cholerae* on the intestinal mucous membrane. *J Pathol Bacteriol, 66:*559, 1953.

Despaul, J.E.: The gangrene organism—a food poisoning agent. *J Am Diet Assoc, 49:*185, 1966.

Dische, F.E., and Elek, S.D.: Experimental food poisoning by *Clostridium welchii. Lancet, II:*71, 1957.

Duncan, C.L., and Strong, D.H.: Experimental production of diarrhea in rabbits with *Clostridium perfringens. Can J Microbiol, 15:*765, 1969a.

Duncan, C.L., and Strong, D.H.: Ileal loop fluid accumulation and production of diarrhea in rabbits by cell-free products of *Clostridium perfringens. J Bacteriol, 100:*86, 1969b.

Duncan, C.L., and Strong, D.H.: *Clostridium perfringens* type A food poisoning. I. Response of the rabbit ileum as an indication of enteropathogenicity of strains of *Clostridium perfringens* in monkeys. *Infect Immunol, 3:*167, 1971.

Duncan, C.L., Strong, D.H., and Sebald, M.: Sporulation and enterotoxin production by mutants of *Clostridium perfringens. J Bacteriol, 110:*378, 1972.

Duncan, C.L., Sugiyama, H., and Strong, D.H.: Rabbit ileal loop response to strains of *Clostridium perfringens. J Bacteriol, 95:*1560, 1968.

Hauschild, A.H.W.: Erythemal activity of the cellular enteropathogenic factor of *Clostridium perfringens* type A. *Can J Microbiol, 16:*651, 1970.

Hauschild, A.H.W.: *Clostridium perfringens* toxins types B, C, D and E. In: Kadis, S., Montie, T.C., and Ajl, S.J., (eds.): *Microbial Toxins,* New York, Acad Pr. 1971a, Vol. IIA, p. 159.

Hauschild, A.H.W.: *Clostridium perfringens* enterotoxin. *J Milk Food Technol, 34:*596, 1971b.

Hauschild, A.H.W.: Food poisoning by *Clostridium perfringens. Can Inst Food Sci Technol J, 6:*106, 1973.

Hauschild, A.H.W., and Hilsheimer, R.: Purification and characteristics of the enterotoxin of *Clostridium perfringens* type A. *Can J Microbiol, 17:*1425, 1971.

Hauschild, A.H.W., Hilsheimer, R., and Rogers, C.G.: Rapid detection of *Clostridium perfringens* enterotoxin by a modified ligated intestinal loop technique in rabbits. *Can J Microbiol, 17:*1475, 1971a.

Hauschild, A.H.W., Niilo, L., and Dorward, W.J.: Experimental enteritis with food poisoning and classical strains of *Clostridium perfringens* type A in lambs. *J Infect Dis, 117:*379, 1967.

Hauschild, A.H.W., Niilo, L., and Dorward, W.J.: *Clostridium perfringens* type A infection of ligated intestinal loops in lambs. *Appl Microbiol, 16:*1235, 1968.

Hauschild, A.H.W., Niilo, L., and Dorward, W.J.: Enteropathogenic factors of food poisoning *Clostridium perfringens* type A. *Can J Microbiol, 16:*331, 1970a.

Hauschild, A.H.W., Niilo, L., and Dorward, W.J.: Response of ligated intestinal loops in lambs to an enteropathogenic factor of *Clostridium perfringens* type A. *Can J Microbiol, 16:*339, 1970b.

Hauschild, A.H.W., Niilo, L., and Dorward, W.J.: The role of enterotoxin in *Clostridium perfringens* type A enteritis. *Can J Microbiol, 17:*987, 1971b.

Hauschild, A.H.W., Walcroft, M.J., and Campbell, W.: Emesis and diarrhea induced by enterotoxin of *Clostridium perfringens* type A in monkeys. *Can J Microbiol, 17:*1141, 1971c.

Hauschild, A.H.W., and Thatcher, F.S.: Experimental food poisoning with heat-susceptible *Clostridium perfringens* type A. *J Food Sci, 32:*467, 1967.

Hobbs, B.C.: *Clostridium welchii* as a food poisoning organism. *J Appl Bacteriol, 28:*74, 1965.

Niilo, L.: Mechanism of action of the enteropathogenic factor of *Clostridium perfringens* type A. *Infect Immunol, 3:*100, 1971.

Niilo, L., Hauschild, A.H.W., and Dorward, W.J.: Immunization of sheep against experimental *Clostridium perfringens* type A enteritis. *Can J Microbiol, 17:*391, 1971.

Stark, R.L., and Duncan, C.L.: Purification and biochemical properties of *Clostridium perfringens* type enterotoxin. *Infect Immunol, 6:*662, 1972.

Sterne, M., and Warrack, G.H.: The types of *Clostridium perfringens. J Pathol Bacteriol, 88:*279, 1964.

Strong, D.H., Duncan, C.L., and Perna, G.: *Clostridium perfringens* type A food poisoning. II. Response of the rabbit ileum as an indication of enteropathogenicity of strains of *Clostridium perfringens* in human beings. *Infect Immunol 3:*171, 1971.

Vernon, E.: Food poisoning and *Salmonella* infections in England and Wales, 1968. *Publ Health (London), 84:*239, 1970.

CHAPTER XV

INCIDENCE OF INTESTINAL ANAEROBES IN BLOOD CULTURES

ALEX C. SONNENWIRTH

ABSTRACT: *The reported incidence of intestinal anaerobes in blood cultures and bacteremia has increased over the last decade to the extent that they now occur in approximately 8 to 15 percent of all positive blood cultures, constituting 8 to 17.5 percent of all bacteremias. Much of the increase is accounted for by the rise in the rate of isolation of* Bacteroides *species, especially* B. fragilis. *The reported incidence of clostridia and anaerobic cocci in blood cultures is much lower (ca. 1%) and their frequency does not seem to have increased appreciably. Bifidobacteria and eubacteria are rarely isolated from blood cultures. The increase in isolation rates of anaerobes from blood cultures reflects, in part, improvements in anaerobic methodology and the use of improved media, as well as the general increase in the rate of all bacteremias. Among the major predisposing factors likely responsible for the increased incidence of anaerobes in blood cultures are the ever-spreading use of complex surgical and instrumental procedures, especially in individuals already debilitated by underlying conditions (malignancies, hepatic disease, etc), and the lowered resistance of patients due to aging or to various modalities of treatment, especially radiation, steroid and antimetabolic therapy.*

The association of anaerobes with infections in man was recognized by Veillon and Zuber as early as 1897 but anaerobic infections, with the exception of clostridial diseases, continued to be overlooked and their importance was rarely credited during the next six decades. While there was some continuing activity in this area among clinicians and bacteriologists in Europe in the years before World War II (Teissier *et al.*, 1931; Pham, 1935; Lemierre, 1936), Dack's excellent review of the medical importance of nonsporeforming anaerobic bacteria, published in 1940, was for all practical purposes ignored in the United States. In 1958, Stokes demonstrated a striking incidence of anaerobes in all types of clinical material and the value of routine anaerobic culture of various clinical specimens. Such observations, coupled with the development of improved culture techniques (Sonnenwirth, 1972) and progress in the taxonomy, nomenclature and identification of anaerobes, resulted in the last few years in a spectacular increase of our knowledge concerning anaerobic infections in man (Finegold, 1969; Moore *et al.*, 1969).

A steady, worldwide rise in the frequency of bacteremias, particularly those due to intestinal gram-negative facultative bacilli, was noted especially within the last decade and has been extensively documented (Finland, 1970; Sonnenwirth, 1973). While reports of anaerobes, especially of bacteroides, in bacteremia were published during the 1930's and sporadically thereafter (Thompson and Beaver, 1932; Dixon and Deuterman, 1937; Gunn, 1956), it was only recently that the increased incidence and importance of anaerobic bacteremia was recognized and documented (Felner and Dowell, 1971; Bodner and Goodman, 1970; Gelb and Seligman, 1970; Marcoux *et al.*, 1970; Wilson *et al.*, 1972; Alpern and Dowell, 1969, 1971; Bornstein *et al.*, 1964; Tynes and Frommeyer, 1962).

It should be pointed out here that until a few years ago, blood specimens were not routinelv cultured anaerobically in many otherwise proficient and reputable clinical laboratories, e.g. the laboratory of Massachusetts General Hospital (as noted by Bornstein *et al.* in their 1964 report on anaerobic infections encountered during the period of 1960–1963 at that institution). In contrast, Bartlett's 1971 survey of blood culture practices in 21 proficient U.S. clinical laboratories directed by well-known microbiologists and clinical pathologists showed that only one laboratory did not use routinely some anaerobic method for culture of blood (Bartlett, 1973).

The purpose of this report is to review the incidence of intestinal anaerobes in blood cultures in the literature and in my own laboratory as a contribution to the better understanding of the association of the intestinal flora with disease. The difficulties anticipated in such retrospective studies were, in fact, encountered. A meaningful comparison of published reports was hampered by: widely divergent criteria employed by various authors concerning the criteria for and clinical significance of bacteremia; lack of data in numerous reports on the total patient population from which blood cultures with anaerobes were obtained ("incidence"); frequent lack of speciation of isolates; the confusing plethora of genus and species names applied until recently; and the variety of isolation methods employed. Nevertheless, the data presented here indicate a steadily increasing incidence of certain intestinal anaerobes in bacteremia.

Intestinal Anaerobes

In man, the upper small intestine contains few bacteria, with anaerobes isolated occasionally in small numbers (Finegold, 1969; Williams *et al.*, 1971; Gorbach, 1971). In the lower small intestine the number and variety of anaerobes increase slowly until the terminal ileum is reached, where anaerobes are qualitatively similar to those of feces (Finegold, 1969; Drasar and McLeod, 1969). Obligate anaerobes constitute 90 to 95 percent of the bacterial flora of feces. Table XV-I lists the types and counts per gram of wet feces of identified anaerobic bacteria found in the normal human feces in five studies. These data require little further elaboration here except for the following comments: (a) The predominant and most prevalent organisms in the fecal flora are *B. fragilis* and bifidobacteria; (b) clostridia and propionibacteria represent very minor components of the fecal flora; (c) there is

TABLE XV-I.

ANAEROBES IN NORMAL FECAL FLORA*

Type of Organism	*Count/gram Wet Feces*
Bacteroides fragilis†	10^{10}–10^{11}
Bifidobacterium† sp.	10^{9}–10^{11}
Other Bacteroidaceae spp.	10^{4}–10^{9}
B. melaninogenicus	
Sphaerophorus	
Fusobacterium	
Others	
Other gram-positive nonsporing rods	10^{4}–10^{8} (?)
Eubacterium sp.	
Peptostreptococcus sp.	10^{4}–10^{8}
Veillonella sp.	10^{4}–10^{6}
Clostridium sp.	10^{3}–10^{8}

* From Finegold (1969); Moore *et al.* (1969); Williams *et al.* (1971); Drasar and McLeod (1969); Werner (1966).

† Invariably present.

a paucity of published counts for gram-positive nonsporeformers other than bifidobacteria in the fecal flora.

Incidence of Intestinal Anaerobes in Blood Cultures and in Bacteremic Patients

As mentioned earlier, many of the reported series of anaerobic infections and of bacteremias fail to provide the numbers of all blood cultures, of positive blood cultures, or of the patient population, data that would allow authentication of the true incidence of intestinal anaerobes in blood cultures or in clinically significant bacteremias. In calculating and comparing frequencies of intestinal anaerobes culled from suitable reports in the literature and from the records of my laboratory (shown in Tables XV-II–XV-VI), blood cultures that yielded only *Corynebacterium*, *Propionibacterium*, or *Bacillus* species or *Staphylococcus epidermidis* were omitted. This arbitrary decision was made because it is difficult to distinguish between contaminating and significant isolates of these organisms (Washington, 1971). However, it should be kept in mind that such organisms have been repeatedly cited in the literature as etiologic agents of bacteremia.

As shown in Table XV-II, intestinal anaerobes are found in 1.9 to 14.2 percent of *all positive blood cultures* (not includ-

TABLE XV-II

INCIDENCE OF INTESTINAL ANAEROBES IN BLOOD CULTURES

Period	*Positive Blood Cultures**	*Cultures Yielding Anaerobes**	*Percent Yielding Anaerobes*	*References*
1951–56	85	9	10.5	Stokes (1958)
1958–65	308	16	5.2	Watt and Okubadejo (1967)
1951–65	1,429	31†	2.1	Dalton and Allison (1967)
1960–67	403	8	1.9	Crowley (1970)
1960	120	6	5.0	Sonnenwirth (unpublished)
1970	303	28	9.2	"
1968–69	3,103	341	11.0	Washington (1971)
1971	847	113	13.3	Washington (1972)
1970–71	1,368	195	14.2	Wilson *et al.* (1972)

* Cultures positive only for *Staphylococcus epidermidis* or *Corynebacterium, Propionibarterium* or *Bacillus* species are omitted.

† *Bacteroides* and *Clostridium* species only.

ing the organisms listed above). The lowest incidence was reported from an English hospital (Crowley, 1970), with the next lowest from a large southern U.S. hospital (Dalton and Allison, 1967) in a period covering 15 years (1951–1965). In my laboratory, intestinal anaerobes occurred in 5 percent of all positive blood cultures in 1960, but 9.2 percent in 1970. A similar increase occurred at the Mayo Clinic (Wilson *et al.*, 1972; Washington, 1971, 1972).

The incidence of intestinal anaerobes as etiologic agents in *bacteremic patients* is shown in Table XV-III. Between 1948 and 1959, at the Mayo Clinic only 2.2 percent of all bacteremias were due to intestinal anaerobes (McHenry *et al.*, 1961). By 1971, these organisms were responsible for 17.5 percent of all bacteremias at the same institution (Washington, 1972).

The experience with bacteremia in the years 1960 to 1970 at The Jewish Hospital of St. Louis, a general, university-affiliated, 530-bed hospital, is shown in Table XV-IV. The number of blood cultures rose from 88.3/1000 admissions in 1960 to 199.8/1000 in 1970, and the number of bacteremias per 1000 admissions increased from 5.63 to 9.75. Intestinal anaerobes were involved in 3.8 percent of all bacteremias in 1960. By 1970, they represented 9.9 percent of all bacteremias, an increase of 167 percent.

For a study of conditions predisposing to anaerobic infections, including bacteremia, see the recent exhaustive review

TABLE XV-III

INCIDENCE OF INTESTINAL ANAEROBES IN BACTEREMIC PATIENTS

Period	*Total No. Bacteremias**	*No. Anaerobic Bacteremias (%)**	*References*
1948–59	1,026	23 (2.2)	McHenry *et al.* (1961)
1960–67	248	8 (3.2)	Crowley (1970)
1960	79	3 (3.8)	Sonnenwirth (1973)
1970	151	15 (9.9)	"
1971	337	59 (17.5)	Washington (1972)

* Those yielding only *Corynebacterium, Propionibacterium,* and *Bacillus* species or *Staphylococcus epidermidis* are omitted.

TABLE XV-IV

INTESTINAL ANAEROBES IN BACTEREMIA AT THE JEWISH HOSPITAL OF ST. LOUIS

Year	*Blood Cultures* Total No.	*Blood Cultures* No. per 1000 Admissions	Bacteremia Per 1000 Admissions*	No. Bacteremic Patients*	Bacteroidaceae No. (%)†	Clostridia No. (%)†	Peptostreptococci No. (%)†	Intestinal Anaerobes %†
1960	1,238	88.3	5.63	79	1 (1.3)	2 (2.5)	0 (0)	3.8
1961	1,229	85.2	5.49	78	2 (2.6)	2 (2.5)	2 (2.5)	7.6
1962	1,281	87.7	4.99	73	2 (2.7)	2 (2.7)	0 (0)	5.4
1963	1,147	75.0	4.31	66	1 (1.5)	1 (1.5)	0 (0)	3.0
1964	1,268	82.7	4.69	72	1 (1.4)	2 (2.7)	0 (0)	4.1
1965	1,457	93.9	5.15	81	5 (6.2)	2 (2.5)	1 (1.2)	9.9
1966	1,616	102.4	5.48	86	5 (5.8)	2 (2.3)	1 (1.2)	9.2
1967	1,668	110.8	5.71	89	6 (6.7)	1 (1.1)	2 (2.2)	10.1
1968	2,055	130.3	6.59	104	8 (7.7)	0 (0)	0 (0)	7.7
1969	2,647	167.16	8.27	131	7 (5.3)	3 (2.3)	0 (0)	7.6
1970	3,094	199.8	9.75	151	10 (6.6)	4 (2.6)	1 (0.7)	9.9

* *Corynebacterium* spp., *Propionibacterium* spp., *S. epidermidis*, *Bacillus* spp. not included.
† % of bacteremic patients.

by Finegold *et al.* (1970) on anaerobic infections in the compromised host.

Bacteroidaceae in Blood Cultures and in Bacteremic Patients

Bacteroides bacteremia is usually associated with prior surgery (most often intra-abdominal), uro-gynecologic manipulations, septic abortion, malignant neoplasms, pyogenic liver abscess, and antibiotic, steroid, immunosuppressive or cytotoxic therapy (Felner and Dowell, 1971; Bodner and Goodman, 1970; Gelb and Seligman, 1970; Marcoux *et al.*, 1970; Wilson *et al.*, 1972; Sabbaj *et al.*, 1972). The review by Gunn (1956) of the world literature covering the period of 1899 to 1954 included 173 cases of bacteroides bacteremia. Recently, large series of Bacteroidaceae bacteremia were reported in considerable detail by Felner and Dowell (1971), Bodner and Goodman (1970), Gelb and Seligman (1970) and Marcoux *et al.* (1970). These reports, except for the last-named, do not include data suitable for establishing the actual incidence and comparative frequency of Bacteroidaceae in blood cultures and in bacteremias.

Table XV-V lists data on the incidence of Bacteroidaceae in blood cultures obtained from reports on the evaluation of blood culture media and on bacteremia in various hospitals. Reports from two British hospitals (Watt and Okubadejo, 1967;

TABLE XV-V

INCIDENCE OF BACTEROIDACEAE, CLOSTRIDIA AND PEPTOSTREPTOCOCCI IN BLOOD CULTURES IN SEVERAL HOSPITALS

Period	*No. Positive Blood Cultures**	*Bacteroidaceae No. (%)*	*Clostridia No. (%)*	*Peptostreptococci No. (%)*	*References*
1958–65	308	8 (2.6)	4 (1.3)	4 (1.3)	Watt and Okubadejo (1967)
1951–67	1,429	12 (0.8)	19 (1.3)	0 (0)	Dalton and Allison (1967)
1960–67	403	4 (0.9)	3 (0.7)	0 (0)	Crowley (1970)
1960	120	4 (3.3)	2 (1.7)	0 (0)	Sonnenwirth (1973)
1970	303	23 (7.6)	4 (1.3)	2 (0.7)	"
1968–69	3,103	265 (8.5)	44 (1.4)	27 (0.8)	Washington (1971)
1971	847	92 (10.8)	7 (0.8)	9 (1.0)	Washington (1972)

* Those yielding only *S. epidermidis* or *Corynebacterium, Propionibacterium* or *Bacillus* spp. are omitted.

Crowley, 1970) and from a large southern U.S. hospital (Dalton and Allison, 1967) show a relatively small incidence (2.6%, 0.9% and 0.8%) of these intestinal anaerobes in blood cultures in the period 1958 to 1967. By 1969 to 1971, a higher incidence was reported in my laboratory and at the Mayo Clinic. Table XV-VI shows a corresponding increase in the incidence of Bacteroidaceae as etiologic agents of bacteremia in patients. By 1970 at The Jewish Hospital of St. Louis, Bacteroidaceace were involved in 6.6 percent of all bacteremias, representing 10 percent of all gram-negative bacteremias and 63 percent of all anaerobic bacteremias for the year.

In several large series of Bacteroidaceae bacteremias (Bodner and Goodman, 1970; Gelb and Seligman, 1970; Marcoux *et al.*, 1970), the organisms isolated were not classified according to genus or species. Wherever speciation was performed, organisms of the genus *Bacteroides* occurred much more frequently than those of the genus *Fusobacterium*. Of the Bacteroidaceae bacteremias observed at The Jewish Hospital (Sonnenwirth, 1973), 90 percent involved *Bacteroides* and only 10 percent were due to *Fusobacterium* species. Of the 71 patients with Bacteroidaceae bacteremia reported from the Mayo Clinic in 1972 by Wilson *et al.*, 65 involved classifiable bacteroides and only 3 (4%) had fusobacteria (1 *F. fusiforme* and 2 *F. necrophorum*).

TABLE XV-VI

INCIDENCE OF BACTEROIDACEAE BACTEREMIA IN PUBLISHED SERIES

Period	*Bacteroidaceae Bacteremia*	*% of All Bacteremias**	*References*
1934–36	6	4.1	Gunn (1956)
1937–38	8	3	Gunn (1956)
1948–59	14	1.4	McHenry *et al.* (1961)
1958–66	35	4.0	DuPont and Spink (1969)
1960	1	1.2	Sonnenwirth (1973)
1970	10	6.6	"
1968–70	145	11.6	Washington (1971)
1970–71	71	5.4	Wilson *et al.* (1972)

* Those yielding only *S. epidermidis* or *Corynebacterium, Propionibacterium* or *Bacillus* spp. are omitted.

B. fragilis is the most common species of intestinal anaerobe isolated from bacteremia (Felner and Dowell, 1971; Wilson *et al.*, 1972; Finegold *et al.*, 1970) as well as other anaerobic infections (Finegold *et al.*, 1970). Data concerning the comparative incidence of the five subspecies of *B. fragilis*, i.e. ss. *fragilis* ss. *distasonis*, ss. *ovatus*, ss. *thetaiotaomicron*, ss. *vulgatus*, in blood cultures, bacteremias or other anaerobic infections have not yet been reported. Werner (personal communication, 1971) found that 80 to 90 percent of strains isolated from various clinical specimens in his laboratory are *B. fragilus* and 10 to 15 percent are ss. *thetaiotaomicron*. In Dowell's experience (personal communication, 1972), the predominant organism among isolates from clinical material is *B. fragilis* ss. *fragilis*. *B. fragilis* ss. *thetaiotaomicron*, ss. *vulgatus* and ss. *distasonis* occur much less frequently and *B. fragilis* ss. *ovatus* does not seem to occur in systemic infections. Among 250 isolates of anaerobic, gram-negative, nonsporulating bacilli from blood cultures, submitted to the Anaerobe Bacteriology Laboratory of the Center for Disease Control between 1963 and 1969, Felner and Dowell (1971) identified 140 *B. fragilis*, compared with only 13 *B. oralis* (5.2%), 11 *B. variabilis* (4.4%), 8 *B. incommunis* (3.2%), 4 *B. melaninogenicus* (1.6%) and 4 *B. terebrans* strains. In addition, there were 21 *F. fusiforme* (8.4%), 20 *F. necrophorum* (8%), 8 *F. girans* (3.2%) and 5 unspeciated (2%) fusobacterial strains. In Wilson *et al.*'s series of Bacteroidaceae bacteremias (1972), the 65 *Bacteroides* isolates included 57 *B. fragilis* (88%), 2 *B. oralis*, 2 *B. terebrans*, and 4 others listed as *Bacteroides* sp.

Clostridia in Blood Cultures and in Bacteremic Patients

Clostridial blood stream invasion, most commonly due to *Clostridium perfringens* and *C. septicum*, may occur in anaerobic myonecrosis, but is much more common in postabortal and puerperal infections of the uterus (Willis, 1969). Clostridial bacteremia, especially with *C. perfringens*, has also been reported in patients with acute cholecystitis, after biliary and intestinal tract surgery, perforation of a gastric or duodenal ulcer, appendiceal abscess, ulcerations of the small intestine and necrotizing enterocolitis (Willis, 1969). There is a definite association between clostridial sepsis and malignancy (Finegold *et al.*, 1971), especially with *C. septicum* (Alpern and Dowell, 1969). A large series of nonhistotoxic clostridial bacteremias, in most cases consistent with an endogenous origin of the clostridia, was recently reported by Alpern and Dowell (1971). As noted earlier, a few reports in the literature contain data concerning the actual incidence of clostridia in blood cultures or bacteremias.

Table XV-V lists some data on the incidence of clostridia in blood cultures. Clostridia constituted 0.7 to 1.6 percent of all positive blood cultures reported by two British and three U.S. institutions. Another report (Wilson *et al.*, 1972) shows that at the Mayo Clinic in a recent 15-month period, gram-positive sporeforming rods were isolated from blood cultures of 12 patients, representing 0.9 percent of all patients with positive blood cultures and 4.5 percent of all patients with anaerobes in their blood cultures. The 12 isolates included three *C. septicum*, three *C. perfringens*, two *C. paraputrificum* and four unclassified strains. Only six of the 12 patients (four with gastrointestinal malignancy, one with a wound infection and one with omphalitis) were considered to have clinically significant bacteremia. At The Jewish Hospital of St. Louis clostridial bacteremia constituted 2.5 percent of all bacteremia in 1960 (Table XV-IV); a decade later its incidence was virtually unchanged.

The data shown in Tables XV-IV and XV-V are noteworthy because of the remarkably small year-by-year variation and the lack of any significant rise in the incidence of blood cultures yielding clostridia.

Anaerobic Cocci in Blood Cultures and in Bacteremic Patients

The difficulties encountered with the foregoing two groups of anaerobes in establishing their incidence in blood cultures and bacteremic patients as reported in the literature were even more pronounced with the anaerobic cocci. Until recent years, little or no speciation of these organisms was attempted and most reports in the literature refer simply to anaerobic cocci, anaerobic streptococci, anaerobic micrococci, or, sporadically, veillonella. Therefore, the column headed "Peptostreptococci" in Table XV-V includes data on organisms that were listed in the literature either as peptostreptococci, anaerobic streptococci or anaerobic cocci. Since only a few reports specifically mention isolation of *Peptococcus* species from blood cultures and bacteremias (Wilson *et al.*, 1972; Washington, 1971, 1972; Finegold *et al.*, 1968), while others likely include them in the general category of anaerobic cocci or streptococci, the true incidence of peptococci in blood cultures could not be determined.

Invasion of the blood stream with anaerobic streptococci has been reported in peritonitis, endometritis, septic abortion, acute and subacute bacterial endocarditis, puerperal fever, and various abscesses, including liver abscess (Wilson *et al.*, 1972; Bornstein *et al.*, 1964; Hare, 1967; Finegold *et al.*, 1968; McDonald *et al.*, 1937; Pien *et al.*, 1972; Sabbaj *et al.*, 1972).

Table XV-V lists some data on the incidence of peptostreptococci in blood cultures. As shown in Table XV-IV, in a period of 11 years at The Jewish Hospital of St. Louis, anaerobic streptococci were involved in 7 cases of bacteremia, representing 0.69 percent of all cases of bacteremia and 9.2 percent of all cases of anaerobic bacteremia. In another series of 113 patients with cultures positive for anaerobes (propionibacteria not included), reported by Wilson *et al.* (1972) from the Mayo Clinic, cultures from nine patients yielded peptostreptococci, five had peptococci, and one had both peptococci and peptostreptococci. These 15 patients constituted 13.2 percent of all patients whose blood cultures yielded anaerobes (excepting propionibacteria).

Bifidobacterium and Eubacterium in Blood Cultures

Despite their prevalence in the fecal flora (Table XV-I), bifidobacteria are rarely isolated from blood cultures or other clinical material (Moore *et al.*, 1969). In one series of 3,103 positive blood cultures reported by Washington (1971), only five (0.1%) yielded bifidobacteria. Similarly, in another of his series (1972), only one of 847 positive cultures yielded bifidobacteria (0.1%). In a group of 113 patients with blood cultures positive for anaerobes, one patient had a clinically significant bacteremia with a bifidobacterium and *Eubacterium lentum* (Wilson *et al.*, 1972).

Likewise, eubacteria are rarely reported from blood cultures (Moore *et al.*, 1969). In the two series reported by Washington (1971, 1972) comprising 3,950 positive blood cultures, seven cultures (0.17%) yielded eubacteria.

Polymicrobial Isolations From Blood Cultures

The factors which predispose patients to bacteremia with anaerobes also predispose to polymicrobial bacteremia (Wilson *et al.*, 1972), defined as a bacteremic episode due to at least two different organisms isolated from the same blood sample (Hermans and Washington, 1970). Polymicrobial bacteremia with anaerobes or facultative anaerobes was present in 31 percent of patients with Bacteroidaceae bacteremia in the series of Wilson *et al.* (1972), a much higher rate than the 6 percent reported by Hermans and Washington (1970) for all patients with bacteremia at the Mayo Clinic and the 7.3 percent rate reported from The Jewish Hospital of St. Louis (Sonnenwirth, 1973). The true incidence of anaerobes in polymicrobial isolations from blood cultures cannot be determined with certainty from the few studies presently available. The experience of von Graevenitz and Sabella (1971) indicates that mixtures of gram-negative facultatively anaerobic and obligately anaerobic rods in blood cultures may be missed unless subcultures are performed on differential anaerobic media containing antibiotics. Such media inhibit the facultatively anaerobic component of the mixture but allow growth and differentiation of the anaerobes.

REFERENCES

Alpern, R.J., and Dowell, V.R.: *Clostridium septicum* infections and malignancy. *JAMA, 209*:385, 1969.

Alpern, R.J., and Dowell, V.R.: Nonhistotoxic clostridial bacteremia. *Am J Clin Pathol, 55*:717, 1971.

Bartlett, R.C.: Contemporary blood culture practices. In: Sonnenwirth, A.C., (Ed.): *Bacteremia—Laboratory and Clinical Aspects*. Springfield, Thomas, 1973.

Bodner, S.J., and Goodman, J.S.: Bacteremic bacteroides infections. *Ann Intern Med, 73:*537, 1970.

Bornstein, D.L., Weinberg, A.N., Swartz, M.N., and Kunz, L.J.: Anaerobic infections—review of current experience. *Medicine (Baltimore), 43:*207, 1964.

Crowley, N.: Some bacteraemias encountered in hospital practice. *J Clin Pathol, 23:*166, 1970.

Dack, G.M.: Non-sporeforming anaerobic bacteria of medical importance. *Bacteriol Rev, 4:*227, 1940.

Dalton, H.P., and Allison, M.J.: Etiology of bacteremia. *Appl Microbiol, 15:*808, 1967.

Dixon, C.F., and Deuterman, J.L.: Postoperative *Bacteroides* infection: Report of six cases. *JAMA, 108:*181, 1937.

Drasar, B.S., and McLeod, G.M.: Studies on the intestinal flora. *Gastroenterology, 56:*71, 1969.

Dupont, H.L., and Spink, W.W.: Infections due to gram-negative organisms: An analysis of 860 patients with bacteremia at the Univ. of Minnesota Medical Center, 1958–1966. *Medicine (Baltimore), 48:*307, 1969.

Ellner, P.D.: System for inoculation of blood in the laboratory. *Appl Microbiol, 16:*1892, 1968.

Felner, J.M., and Dowell, V.R.: "Bacteroides" bacteremia. *Am J Med, 50:*787, 1971.

Finland, M.: Changing ecology of bacterial infections as related to antibacterial therapy. *J Infect Dis, 122:*419, 1970.

Finegold, S.M.: Intestinal bacteria—the role they play in normal physiology, pathologic physiology, and infection. *Calif Med, 110:*455, 1969.

Finegold, S.M., Marsh, V.H., and Bartlett, J.G.: Anaerobic infections in the compromised host. In: *Proceedings of the International Conference on Nosocomial Infections* (Center for Disease Control, Aug. 3–6, 1970). Chicago, Amer. Hosp. Assoc., 1971.

Finegold, S.M., Miller, A.B., and Sutter, V.L.: Anaerobic cocci in human infection. In: *Bact. Proc. 1968*. Abstr. # M166 p. 94. Ann Arbor, Am. Soc. for Microbiology, 1969.

Gelb, A.F., and Seligman, S.J.: Bacteroidaceae bacteremia—Effect of age and focus of infection upon clinical course. *JAMA, 212:*1038, 1970.

Goodman, J.S.: Bacteroides sepsis: Diagnosis and therapy. *Hosp Prac, 6:*121, 1971.

Gorbach, S.L.: Intestinal microflora. *Gastroenterology, 60:*1110, 1971.

Gunn, A.A.: Bacteroides septicaemia. *J R Coll Surg Edinb, 2:*41, 1956.

Hare, R.: The anaerobic cocci. In: Waterson, A.P., (Ed.): *Recent Advances in Medical Microbiology*. Boston, Little, 1967.

Hermans, P.E., and Washington, J.A.: Polymicrobial bacteremia. *Ann Intern Med, 73:*387, 1970.

Holdeman, L.V., and Moore, W.E.C., (Eds.): *Anaerobe Laboratory Manual.* Blacksburg, Va., Anaerobe Laboratory, Virginia Polytechnic Institute and State University, 1972.

Lemierre, A.: On certain septicaemias due to anaerobic organisms. *Lancet, I:*701, 1936.

Marcoux, J.A., Zabransky, R.J., Washington, J.A., Wellman, W.E., and Martin, W.J.: Bacteroides bacteremia. *Minn Med, 53:*1169, 1970.

McDonald, J.R., Henthorne, J.C., and Thompson, L.: Role of anaerobic streptococci in human infections. *Arch Pathol, 23:*230, 1937.

McHenry, M.C., Wellman, W.E., and Martin, W.J.: Bacteremia due to bacteroides. *Arch Intern Med, 107:*572, 1961.

Moore, W.E.C., Cato, E.P., and Holdeman, L.V.: Anaerobic bacteria of the gastrointestinal flora and their occurrence in clinical infections. *J Infect Dis, 119:*641, 1969.

Pham, H.C.: *Les septicémies dues au Bacillus funduliformis.* Thèse pour le Doctorat en Médicine, Paris, 1935.

Pien, F.D., Thompson, R.L., and Martin, W.J.: Clinical and bacteriologic studies of anaerobic gram-positive cocci. *Mayo Clin Proc, 47:*251, 1972.

Sabbaj, J., Sutter, V.L., and Finegold, S.M.: Anaerobic pyogenic liver abscess. *Ann Intern Med, 77:*629, 1972.

Sonnenwirth, A.C.: Bacteremia—Extent of the problem. In: Sonnenwirth, A.C. (Ed.): *Bacteremia—Laboratory and Clinical Aspects.* Springfield, Thomas, 1973.

Sonnenwirth, A.C.: Evolution of anaerobic methodology. *Am J Clin Nutr, 25:*1295, 1972.

Sonnenwirth, A.C.: Unpublished data.

Stokes, E.J.: Anaerobes in routine diagnostic cultures. *Lancet, I:*668, 1958.

Sutter, V.L., Attebery, H.R., Rosenblatt, J.E., Bricknell, K.S., and Finegold, S.M.: *Anaerobic Bacteriology Manual.* Los Angeles, Anaerobic Bacteriology Lab, Wadsworth Hospital Center (VA), 1972.

Teissier, P., Reilly, J., Rivalier, E., and Stefanesco, V.: Les septicémies primitives dues au *Bacillus funduliformis.* Étude clinique, bactériologique et expérimentale. *Ann Méd, 30:*97, 1931.

Thompson, L., and Beaver, D.C.: Bacteremia due to anaerobic gram-negative organisms of the genus *Bacteroides. Med Clin North Am, 15:*1611, 1932.

Tynes, B.S., and Frommeyer, W.B.: Bacteroides septicemia—Cultural, clinical, and therapeutic features in a series of twenty-five patients. *Ann Intern Med, 56:*12, 1962.

Veillon, A., and Zuber, A.: Sur quelques microbes strictement anaerobies et leur role dans la pathologie humaine. *Soc de Biol, 49:*253, 1897.

von Graevenitz, A., and Sabella, W.: Unmasking additional bacilli in gram-negative rod bacteremia. *J Med (Basel), 2:*185, 1971.

Washington, J.A.: Comparison of two commercially available media for detection of bacteremia. *Appl Microbiol, 22:*604, 1971.

Washington, J.A.: Evaluation of two commercially available media for detection of bacteremia. *Appl Microbiol, 23:*956, 1972.

Watt, P.J., and Okubadejo, O.A.: Changes in incidence and aetiology of bacteraemia arising in hospital practice. *Br Med J, 1:*210, 1967.

Werner, H.: The gram-positive nonsporing anaerobic bacteria of the human intestine with particular reference to the corynebacteria and bifidobacteria. *J Appl Bacteriol, 29:*138, 1966.

Williams, R.E.O., Hill, M.J., and Drasar, B.S.: The influence of intestinal bacteria on the absorption and metabolism of foreign compounds. *J Clin Pathol, 24, Suppl (Roy Coll Path), 5:*125, 1971.

Willis, A.T.: *Clostridia of Wound Infection.* London, Butterworths, 1969.

Wilson, W.R., Martin, W.J., Wilkowski, C.J., and Washington, J.A.: Anaerobic bacteremia. *Mayo Clin Proc, 47:*639, 1972.

CHAPTER XVI

IMMUNE RESPONSE TO ANAEROBIC INFECTIONS

DAN DANIELSSON,
DWIGHT W. LAMBE, JR. AND SVEN PERSSON

ABSTRACT: *The pathogenic significance of anaerobic bacteria isolated from the blood of patients with septicemia has been well documented, but it has been more difficult to assess the clinical importance of specific anaerobic isolates from other sources. In the present study, antigens were prepared from anaerobic isolates from seven patients with septicemia and five with abscesses, wound infections or intestinal fistulas. Serologic assays, including tube agglutination, indirect immunofluorescence, agar gel diffusion and passive hemagglutination tests, were performed with the patients' sera against homologous and heterologous strains. Sera from ten blood donors were used as controls. The results demonstrated an active immune response against anaerobic bacteria in patients with septicemia and also in patients with other types of infections, which lends support to the concept that these organisms do have pathogenic significance in such infections. The significance of various reactions with heterologous strains was discussed.*

During the last ten years there has been an increasing interest in anaerobic bacteria. Recently improved techniques for the cultivation, isolation and identification of anaerobes have shown that these bacteria are by far the most numerous of the indigenous flora of man. Anaerobes are present on the mucous membranes of the upper respiratory tract, the oral cavity and genito-urinary tract, and comprise the majority of the bacteria in the intestinal tract. The relative frequency of anaerobic bacteria in clinical infections varies in different bacteriologic studies from less than 1 percent to more than 10 percent (Dack, 1940; DuPont and Spink, 1969; Mattman *et al.*, 1958; McHenry *et al.*, 1961; Stokes, 1958). In a recent study by Zabransky (1970) anaerobes were isolated in approximately one-fourth of 1,223 specimens investigated. In this and several other studies (DuPont and Spink, 1969; Felner and Dowell, 1971; Finegold, 1968; McHenry *et al.*, 1961; Pearson and Andersson, 1967), members of the genera *Bacteroides* and *Fusobacterium* accounted for more than half of the anaerobes isolated from the various types of clinical specimens investigated. Our knowledge of the host-parasite interactions involved in infections with anaerobic bacteria is rather limited. It is known that pure cultures of anaerobic bacteria are found in only one-fourth of infections with anaerobes but in mixtures with aerobic bacteria in approximately two-thirds and with other anaerobes, with or without aerobes, in the rest of the cases.

Very little has been reported in the literature about the immune response of humans to anaerobic bacteria. We demonstrated an antibody response to *Bacteroides fragilis* ss. *fragilis* and *Clostridium difficile* in a patient with a perirectal abscess from which these bacteria were isolated (Danielsson *et al.*, 1972). The present report is from an extended serologic investigation of the immune response to *Fusobacterium* species and various species and subspecies of *Bacteroides* in patients with infections from which these bacteria were isolated.

MATERIALS AND METHODS

Patients

Twelve patients were included in this study. Of these, seven were treated at the Central County Hospital in Orebro, Sweden and five at the Emory University Hospital in Atlanta. Sex, age, type of infection, underlying disease, source of isolates, and bacteria isolated are summarized in Table XVI-I.

Anaerobic Culture and Identification

Materials from infected wounds, abscesses and fistulas were cultured under anaerobic conditions onto freshly prepared blood agar (BA) plates and laked blood agar. The BA plates were made from brain heart infusion (BHI) agar base with the addition of 0.5 percent yeast extract and 5 percent sheep blood. At the Orebro Hospital the blood agar was also enriched with menadione and hemin. Laked blood agar was made from BHI agar base with the addition of 4 percent agar, 5 percent laked sheep blood, menadione (0.1 mg/liter) and hemin (5 mg/liter). The media were used within 2 hours after preparation, incubated under strict anaerobic conditions

TABLE XVI-I

COLLATION OF PATIENTS INCLUDED IN THE STUDY

Patient (Sex & Age)	*Diagnosis and/or Underlying Disease*	*Source of Isolates*	*Anaerobic Bacteria Isolated*	*Other Bacterial Isolates*
T.K. (M, 18)	Septicemia (Peritonsillitis)	Blood	*Fusobacterium necrophorum*	None
S.M. (F, 86)	Septicemia	Blood	*Bacteroides fragilis* ss. *fragilis*	None
E.C. (F, 65)	Septicemia	Blood	*B. fragilis* ss. *fragilis* *Peptococcus asaccharolyticus*	None
J.G. (M, 34)	Septicemia	Blood	*B. fragilis* ss. *fragilis*	None
B.P. (F, 34)	Septicemia	Blood	*B. fragilis* ss. *fragilis*	None
K.V. (M, 61)	Septicemia (Rectal cancer)	Blood	*B. fragilis* ss. *fragilis*	*Escherichia coli* from wound infection
F.H. (M, 40)	Septicemia	Blood	*B. fragilis* ss. *thetaiotaomicron*	None
B.L. (F, 15)	Crohn's disease	Perirectal abscess	*B. fragilis* ss. *fragilis* *Clostridium difficile*	*E. coli* *Staphylococcus epidermidis* *Propionibacterium acnes*
R.A. (F, 17)	Appendicitis	Abdominal abscess	*B. fragilis* ss. *fragilis*	None
E.I. (M, 65)	Inguinal hernia	Abdominal abscess	*B. fragilis* ss. *fragilis*	None
G. K-A. (M, 59)	Rectal cancer	Wound infection	*B. fragilis* ss. *thetaiotaomicron*	*E. coli*
J.K. (F, 62)	Chronic sigmoiditis	Intestinal fistula	*Fusobacterium varium* *Bacteroides fragilis* ss. *fragilis* Anaerobic streptococci	*Proteus mirabilis* *Escherichia dispar*

immediately after inoculation, and examined for growth after 2 and 7 days.

For blood cultures, the medium used at Emory University Hospital was prereduced anaerobically sterilized (PRAS) BHI broth (Scott Laboratories). At the Orebro Hospital the procedure utilized a bottle containing a chocolate agar slant and enrichment broth with added penicillinase with p-aminobenzoic acid; air was replaced by nitrogen gas. Blood cultures were inspected for growth every second day during a 10-day period.

The methods outlined by Cato *et al.* (1970) were followed for the final identification of anaerobic isolates. These examinations were carried out at the Emory University Hospital. Fermentation tests were performed in PRAS media (Scott Laboratories). Gas chromatography was performed with the Beckman GC-2A gas chromatograph from an ether extract of a 48-hour glucose broth culture. On certain occasions gas chromatography was performed on extracts of cultures enriched with threonine or lactate. The biochemical reactions of the anaerobic strains studied are shown in Table XVI-II.

Aerobic Culture

Blood agar, MacConkey agar and phenylethyl alcohol agar were used for aerobic culture, and blood agar for culture in CO_2 atmosphere. Standard laboratory procedures were followed for the identification of aerobes.

Serologic Assays

Blood specimens were drawn from the patients as early as possible after anaerobic bacteria were detected. A second blood specimen was collected two or more weeks after the first. Additional blood specimens were collected from 2 patients at various intervals during 21 and 22 week periods, respectively. Serum specimens were frozen and stored at −20° to −40°C until used.

Twelve strains of *Bacteroides* species, three strains of *Fusobacterium* species and one strain of *Clostridium difficile* were used for the production of antigens. For preparation of antigens the strains were cultured for 48 to 72 hours on

TABLE XVI-II
BACTERIOLOGIC CHARACTERISTICS OF ANAEROBIC MICROORGANISMS USED AS ANTIGENS

Patient	Anaerobic Microorganism	Fructose	Glucose	Lactose	Maltose	Mannitol	Sucrose	Rhamnose	Trehalose	Xylose	Gelatin	Milk	Nitrate Reduction	Indole	Esculin hydrolysis	Starch acid	Threonine → Propionate	Lactate → Propionate	Bile Stimulation	End Products from glucose
T.K.	*Fusobacterium necrophorum*	–	–	–	–	–	–	–	–	–	–	–	–	+	–	–	+	+	–	A,P,B
S.M.,E.C., J.G.,B.P., K.V.,R.A., E.I.	*Bacteroides fragilis* ss. *fragilis*	A*	A	A	A	–	A	–	–	A	–	C	–	–	+	+	ND	ND	S	A,P,(IB),IV
F.H., G.K.-A.	*Bacteroides fragilis* ss. *thetaiota-omicron*	A	A	A	A	–	A	A	A	A	–	C	–	+	+	A	ND	ND	S	A,P,IV
B.L.	*B. fragilis* ss. *fragilis*	A	A	A	A	–	A	–	–	A	–	C	–	–	+	+	ND	ND	S	A,P,IV
	Clostridium difficile†	ND	A	–	–	A	–	ND	–	ND	+	–	–	–	ND	ND	ND	ND	ND	A,IB,B,IV,V,IC
J.K.	*B. fragilis* ss. *fragilis*	A	A	A	A	–	A	–	–	A	–	C	–	–	+	+	ND	ND	S	A,P,IB,IV
	Fusobacterium varium	A	A	–	–	–	–	–	–	–	ND	ND	–	–	–	–	+	–	–	A,P,B

*Code:
A = acid
ND = not done
C = curd
S = stimulated

†Mannose A; lipase and lecithinase negative

freshly prepared BA and then harvested in saline. Whole bacterial cells treated with 3 percent formalin or boiled for 1 hour in a water bath were used as antigens in agglutination tests. Untreated whole cells were used as antigens in indirect immunofluorescence (IFL) tests. Bacterial cells, used as antigens in agar gel diffusion (AGD) tests, were disrupted with an MSE 100 Watt ultrasonic apparatus operated at maximum efficiency. The procedures for all these preparations are described in detail elsewhere (Danielsson *et al.*, 1972). The procedures described by Neter (1956) were followed for the preparation of antigens for passive hemagglutination (HA) tests. Similar results were obtained with the technique of Edwards and Driscoll (1967).

Agglutination, indirect IFL and AGD tests were performed as described elsewhere (Danielsson *et al.*, 1972). Sensitization of sheep red blood cells and performance of passive HA tests were done as described by Edwards and Driscoll (1967) using a standard micro technique with volumes of 0.025 ml diluted serum and 0.025 ml sensitized sheep blood cells in V-shaped micro-titer plates. Appropriate controls were included in each test.

RESULTS

The Immune Response to Anaerobic Bacteria in Patients with Septicemia

The serological findings in seven patients with *Fusobacterium* or *Bacteroides* septicemia are summarized in Table XVI-III and XVI-IV. The results in Table XVI-III refer to an 18-year-old male patient (T.K.) from whom *Fusobacterium necrophorum* was isolated from blood. We had the opportunity of examining five serum samples from this patient over a 21-week period, and this case can serve as an illustrative example of the immune response to a clinically significant infection with anaerobic bacteria. Briefly, the patient presented the following history:

> On May 1, 1971, he had a headache, a sore throat and difficulty in swallowing. The following day he had a pronounced feeling of fatigue and on May 3, 1971, he had a high fever and increasing headache. Four days later he appeared at the emergency clinic

TABLE XVI-III

SEROLOGIC FINDINGS WITH AGGLUTINATION (AGGL.), PASSIVE HEMAGGLUTINATION (HA), INDIRECT IMMUNOFLUORESCENCE (IFL) AND AGAR GEL DIFFUSION (AGD) IN AN 18-YEAR-OLD MAN (T.K.) WITH *FUSOBACTERIUM NECROPHORUM* SEPTICEMIA

Serum Specimens (Time After Symptoms Started)	F. necrophorum *1* *Patient Strain*				F. necrophorum *2* *Stock Lab. Strain*				Fusobacterium varium *Stock Lab. Strain*			
	Aggl.	*HA*	*IFL*	*AGD*	*Aggl.*	*HA*	*IFL*	*AGD*	*Aggl.*	*HA*	*IFL*	*AGD*
Patient T.K.:												
7 Days	Spont aggl	<10*	10*	0†	Spont aggl	<10	10	0	<10	<10	<10	0
10 Days	″	<10	40	0	″	<10	40	0	<10	<10	<10	0
3 Weeks	″	320	320	2	″	80	160	1	<10	<10	<10	0
4 Weeks	″	320	320	1	″	80	160	1	<10	<10	<10	0
21 Weeks	″	10	40	1	″	10	20	0	<10	<10	<10	0
Ten blood donors	″	≤10	≤10 (20–40)	0	″	≤10 (Occ. 20)	≤10 (20–40)	0	<10	<10	<10	0

* Reciprocal titers.
† Number of precipitin lines.

of the Orebro Hospital with high fever (41.5°C) and chills. On examination he had peritonsillitis, leucocytosis, hematuria and proteinuria. He also complained of marked tenderness of his right hip, but an X-ray was normal. On this occasion a blood culture was done and treatment with penicillin was started. Three days later (May 10) his conjunctivae were slightly icteric and the total bilirubin 6.8 mg percent. The next day strict anaerobic gram-negative rods were demonstrated in his blood culture. On May 13, he complained of pain in the epigastrium and breathing was painful. X-ray revealed a few hazel-nut-sized areas of parenchymatous densities in his left lung which were regarded as septic emboli. X-ray of his abdomen showed that his spleen and liver were slightly enlarged. He developed anemia (7.7 grams hemoglobin per 100 ml), a systolic heart murmur and a pathologic electrocardiogram. His symptoms did not respond to penicillin, which was replaced by tetracyclines. Within 2 weeks his symptoms disappeared, and on June 5, 1971, he felt quite well and could leave the hospital. The anaerobic gram-negative bacillus isolated from his blood was later identified as *Fusobacterium necrophorum*.

Five serum specimens, the first two obtained 7 and 10 days after his symptoms appeared and the last three collected after 3, 4 and 21 weeks, were examined by tube agglutination, HA, indirect IFL and AGD tests against antigens prepared from the *F. necrophorum* strain isolated from the patient and also from stock laboratory strains of *F. necrophorum* and *F. varium* (Table XVI-III). The patient's first two serum specimens were negative in both HA and AGD tests, while the indirect IFL titer increased from 1:10 to 1:40. In the specimens obtained 3 and 4 weeks after symptoms appeared, both the HA and indirect IFL titers showed a marked increase (1:320), and one or two precipitin bands were noted in AGD tests. In the last specimen (21 weeks) the HA and indirect IFL titers had dropped to 1:10 and 1:40, respectively; a weak precipitin line could still be detected in the AGD test. Positive results were also obtained with the antigens prepared from a stock laboratory strain of *F. necrophorum*, but the reactions were somewhat weaker. No reactions at all were obtained with antigens prepared from *F. varium*.

In contrast to the results in Patient T.K., serum specimens from 10 blood donors gave negative results or low titers in HA and indirect IFL tests (Table XVI-III).

The results presented in Table XVI-IV are from 6 patients with *Bacteroides* septicemia: five with *B. fragilis* ss. *fragilis* and one with *B. fragilis* ss. *thetaiotaomicron* infection. Five of the patients (S.M., E.C., J.G., B.P. and F.H.) were treated at the Emory University Hospital and one (K.V.) at the Orebro Hospital. Paired serum specimens were obtained from four of the patients but only one sample from two. These sera were examined by tube agglutination and AGD tests with antigens prepared from the strains isolated from the patients. Significant, i.e. at least fourfold, increases in the titers were detected in the second serum sample from two of the patients (B.P. and K.V.) with the agglutination test using homologous strains. The highest titers of these sera were 1:40 and 1:160. The sera of the other patients (S.M., E.C., J.G. and F.H.) gave agglutination titers of 1:80 or 1:160.

Agglutination tests of the patients' sera with heterologous strains, i.e. the *Bacteroides fragilis* strains isolated from other patients, usually gave titers of 1:10 or 1:20. A few patients' sera, however, gave titers of 1:40 with heterologous strains, *e.g.* the serum from patient S.M. with the *B. fragilis* strain from patient B.P., and the serum specimens from patients E.C. and B.P. using the *B. fragilis* strain isolated from patient J.G. However, no increases or decreases of the titers were noted in these sera. Serum specimens from ten blood donors gave agglutination titers of less than 1:10.

In AGD tests the patients' sera gave 1, 2 or 4 precipitin lines with antigens from their homologous strains but only on one occasion with antigens from a heterologous strain, i.e. when sera from patient B.P. were tested with the *B. fragilis* strain isolated from patient E.C. However, the serum from patient E.C. did not produce any precipitin lines with the *B. fragilis* strain from patient B.P. No precipitin lines were obtained with the blood donor sera.

The Immune Response to Anaerobic Bacteria in Patients with Abscesses, Wound Infections and Intestinal Fistula

The antibody responses of patients with abscesses, wound infection and intestinal fistula are summarized in Tables XVI-

TABLE XVI-IV

SEROLOGIC FINDINGS WITH AGGLUTINATION (AGGL.) AND AGAR GEL DIFFUSION (AGD) IN SIX PATIENTS WITH *BACTEROIDES FRAGILIS* SEPTICEMIA

		Patient Strains of Subspecies of Bacteroides fragilis											
		ss. fragilis *S.M.*		*ss.* fragilis *E.C.*		*ss.* fragilis *J.G.*		*ss.* fragilis *B.P.*		*ss.* fragilis *K.V.*		*ss.* thetaiotaomicron *F.H.*	
Patient	*Serum Number*	*Aggl.*	*AGD*	*Aggl.*	*AGD*	*Aggl.*	*AGD*	*Aggl.*	*AGD*	*Aggl.*	*AGD*	*Aggl.*	*AGD*
S.M.	1	80*	2†	<10	0	<10	0	40	0	N.D.	N.D.	10	0
	2	160	1	10	0	10	0	40	0	″	″	10	0
E.C.	1	20	0	80	1	40	0	20	0	″	″	20	0
J.G.	1	20	0	20	0	80	4	20	0	″	″	20	0
B.P.	1	20	0	20	1	40	0	40	1	″	″	<10	0
	2	20	0	10	1	40	0	160	1	″	″	<10	0
K.V.	1	N.D.§	N.D.	N.D.	N.D.	N.D.	N.D.	N.D.	N.D.	10	0	N.D.	N.D.
	2	″	″	″	″	″	″	″	″	40	1	″	″
F.H.	1	20	0	20	0	20	0	10	0	N.D.	N.D.	80	1
	2	20	0	10	0	20	0	10	0	″	″	40	2
Ten Blood Donors		≤10	0	≤10–20	0	≤10	0	≤10	0	≤10	0	≤10	0

* Reciprocal titers.
† Number of precipitin lines.
§ Not done.

V, XVI-VI, and XVI-VII. The results in Table XVI-V refer to a 15-year-old girl (B.L.) with chronic Crohn's disease (confirmed by X-ray) in the terminal ileum and left colon. She was hospitalized because of a perirectal abscess. The abscess was drained surgically and from the purulent material *B. fragilis* ss. *fragilis* and *C. difficile* were isolated and used in the serologic study. Growth of *Staphylococcus epidermidis, Escherichia coli* and *Propionibacterium acnes* was also obtained, although these latter three organisms were not included in the serologic study. The infection was regarded as a mixed aerobic and anaerobic infection. Serum specimens were obtained from the patient 2, 4, 12 and 22 weeks after diagnosis of the perirectal abscess. These were tested with tube agglutination, indirect IFL and AGD tests against antigens from the *B. fragilis* and *C. difficile* isolates and from a stock laboratory strain of *B. fragilis.* In addition, HA tests were performed with antigens from the *B. fragilis* strains.

The data from Table XVI-V show that the first three serum specimens gave titers of 1:80, 1:160 and 1:320 in the agglutination, HA and IFL tests respectively with the homologous strain of *B. fragilis;* one or two percipitin lines were obtained in AGD tests. The fourth serum specimen taken 22 weeks after diagnosis showed a fourfold decrease in these titers, and no precipitin line was noted in the AGD test. No reactions were noted with the stock laboratory strain of *B. fragilis.* The patient's four serum specimens gave no reactions in the tube agglutination tests and AGD tests with antigens of *C. difficile.* However, in indirect IFL tests the first three serum specimens gave titers of 1:80 to 1:160, while the fourth specimen showed a fourfold decrease to 1:20. All of the blood donor sera were negative in tube agglutination tests and AGD test with all three anaerobic bacteria. Titers of less than 1:10 or occasionally 1:20 were noted in HA and indirect IFL tests.

The serologic findings in Table XVI-VI are from two patients (R.A. and E.I.) with abscesses due to single anaerobes and from a patient (G.K-A.) with a mixed aerobic and anaerobic wound infection. Acute and convalescent phase serum specimens (Se 1 and Se 2 respectively) were tested with tube agglutination, indirect IFL and AGD tests against homologous

TABLE XVI-V

RESULTS OF AGGLUTINATION (AGGL.), HEMAGGLUTINATION (HA), INDIRECT IMMUNOFLUORESCENCE (IFL) AND AGAR GEL DIFFUSION (AGD) IN A PATIENT (B.L.) WITH A PERIRECTAL ABSCESS DUE TO A MIXED AEROBIC AND ANAEROBIC INFECTION

Serum	*Patient Strains*							*Stock Laboratory Strain*	
	Bacteroides fragilis *ss* fragilis				Clostridium difficile			B. fragilis *ss.* fragilis	
Patient B. L.: (*Weeks after Diagnosis*)	*Aggl.*	*HA*	*IFL*	*AGD*	*Aggl.*	*IFL*	*AGD*	*Aggl. HA IFL*	*AGD*
2 Weeks	80*	160	320	2†	<10	80	0	≤10	0
4 Weeks	80	160	320	2	<10	160	0	≤10	0
12 Weeks	80	160	320	1	<10	80	0	≤10	0
22 Weeks	20	40	80	0	<10	20	0	≤10	0
Ten blood donors	<10	≤10 (occ. 20)	≤10 (occ. 20)	0	<10	≤10 (occ. 20)	0	≤10 (occ. 20)	0

* Reciprocal titers.
† Number of precipitin lines.

TABLE XVI-VI

RESULTS OF AGGLUTINATION (AGGL.), INDIRECT IMMUNOFLUORESCENCE (IFL) AND AGAR GEL DIFFUSION (AGD) IN TWO PATIENTS (R.A. AND E.I.) WITH STRICT ANAEROBIC INFECTIONS (ABSCESSES), AND ONE PATIENT (G. K-A.) WITH A MIXED AEROBIC AND ANAEROBIC WOUND INFECTION

Patients and Serum Specimens		Bacteroides fragilis *Strains Isolated from:* *Patient R.A.* (*ss.* fragilis)			*Patient E.I.* (*ss.* fragilis)			*Patient G. K-A.* (*ss.* thetaiotaomicron)		
		Aggl.	*IFL*	*AGD*	*Aggl.*	*IFL*	*AGD*	*Aggl.*	*IFL*	*AGD*
R.A. (Abscess after appendectomy)	Se 1*	10†	40	N.D.§	N.D.	N.D.	N.D.	N.D.	N.D.	N.D.
	Se 2	40	80	"	"	"	"	"	"	"
E.I. (Abscess after hernia operation)	Se 1	N.D.	N.D.	N.D.	10	40	1#	<10	10	0
	Se 2	"	"	"	40	80	1	<10	10	0
G. K-A. (Wound infection after intestinal surgery)	Se 1	N.D.	N.D.	N.D.	<10	10	0	<10	40	0
	Se 2	"	"	"	<10	10	0	20	40	1
Ten blood donors		<10	≤10 (occ. 20)	N.D.	<10	≤10 (occ. 20)	0	<10	≤10 (occ. 20)	0

* Se 1 = acute phase serum; Se 2 = convalescent phase serum.
† Reciprocal titers.
§ N.D. = not done.
# Number of precipitin lines.

strains of *B. fragilis*. Sera from patients E.I. and G.K-A. were also tested against heterologous strains. A fourfold increase of the agglutination titer was obtained with the serum from each patient with homologous strains, but there was no reaction with antigens from heterologous strains. The titers in indirect IFL were 1:40 or 1:80, but no significant (fourfold) increase was noted in the convalescent sera. Serum specimens from patients E.I. and G.K-A. gave precipitin lines in AGD tests.

All of the blood donor sera were negative in agglutination and AGD tests. Positive reactions with titers of 1:10 to 1:20 were occasionally obtained in indirect IFL tests. On these occasions, however, less than 10 percent of the organisms were stained while in the positive tests with serum from patients all organisms on the slide were stained.

Table XVI-VII shows results of serologic testing in a patient (J.K.) with chronic sigmoiditis with intestinal fistulas. Antigens were prepared from strains of *F. varium*, *B. fragilis* ss. *fragilis*, *Escherichia dispar* and *Proteus mirabilis* isolated from the patient's fistula, and serum specimens from the patient and from ten blood donors were tested against them. In agglutination tests blood donor sera gave negative reactions with the anaerobic bacteria, while the patient's serum gave titers of 1:1280 and 1:80 with *Fusobacterium* and *Bacteroides* strains, respectively. With *F. varium*, the patient's HA titer was only 1:10 despite the high tube agglutination titer, when the latter was obtained with formalin treated organisms. No agglutination was observed with heat treated cells. The patient's serum gave a 1:640 IFL titer with *F. varium*. With the *Bacteroides* strain, the patient's serum gave a 1:640 HA titer and a 1:160 IFL titer. Blood donor sera gave negative HA and IFL reactions to both *Fusobacterium* and *Bacteroides*.

In tests with the aerobic isolates, the patient's serum gave an agglutination titer of 1:160 with the *P. mirabilis* strain while the blood donor titers were less than 1:10. Despite the agglutination titer, the patient's HA titer was less than 1:10 with this antigen. With *E. dispar*, the patient's agglutination and HA titers were 1:160 and 1:640, respectively, but

TABLE XVI-VII

RESULTS OF AGGLUTINATION (AGGL.) PASSIVE HEMAGGLUTINATION (HA) AND INDIRECT IMMUNOFLUORESCENCE (IFL) IN A PATIENT (J.K.) WITH CHRONIC SIGMOIDITIS COMPLICATED WITH FISTULAS

Antigens Prepared from Bacterial Strains Isolated Patient J.K. (Fistula)	*Serum from Patient J.K.*			*Serum from Ten Blood Donors*		
	Aggl.	*HA*	*IFL*	*Aggl.*	*HA*	*IFL*
Fusobacterium varium	1280*	10	640	≤10	<10	≤10 (occ. 20)
Bacteroides fragilis ss. *fragilis*	80	640	160	<10	≤10 (occ. 40)	≤10 (occ. 20)
Escherichia dispar	160	640	N.D.†	10–160	40–640	N.D.
Proteus mirabilis	160	<10	N.D.	<10	≤10 (occ. 20)	N.D.

* Reciprocal titer.
† Not done.

corresponding titers were also obtained with some of the blood donor sera.

DISCUSSION

The clinical and pathogenic significance of anaerobic bacteria isolated from the blood of patients with septicemia has been well documented by several workers (DuPont and Spink, 1969; Felner and Dowell, 1971; Gelb and Seligman, 1970; McHenry *et al.*, 1956). However, it has been more difficult to assess the importance of specific anaerobic isolates from other sources, i.e. abscesses, wound infections and fistulas. Recent improved techniques have shown that anaerobes occur more frequently than is usually recognized, and in many instances multiple anaerobes occur in one infection with or without aerobic bacteria. The findings presented in this report show that an active immune response against anaerobic bacteria occurs in patients with septicemia and also in patients with other types of infections, which gives support to the pathogenic significance of anaerobes in infections other than septicemia.

Tube agglutination tests were suitable for screening antibodies to the *Bacteroides* species tested, since these cell suspensions had no tendency to spontaneous agglutination as occurred with the cells of one isolate of *F. necrophorum* from a patient and a stock culture of *F. necrophorum* (Table XVI-III). Agglutination titers were usually one or two dilution steps higher in patients with septicemia than in patients with other types of infections. A tube agglutination titer of 1:40 may indicate a past or present infection with that particular strain, since reactions with the serum of blood donors usually were negative or occasionally gave titers of only 1:10 or 1:20. On the other hand, titers of 1:40 were obtained on a few occasions in cross agglutination tests of sera from patients with septicemia with the heterologous *Bacteroides fragilis* strains tested. These agglutination reactions could indicate an antigenic relationship between the various *Bacteroides* isolates used in the tests. However, no increase in titers between the first and the second serum specimens (see patients S.M. and B.P., Table XVI-IV) were recorded with antigens of the heterologous strains in the same way as with antigens of homologous strains; nor (with one exception) were any precipitin lines obtained with heterologous antigens in the AGD tests. Tests with rabbit hyper-immune sera also indicate that these strains probably do not share antigens responsible for agglutination (Lambe *et al*, Data to be published).

Agglutination reactions with the heterologous *Bacteroides* isolates could be interpreted as the result of a past infection with these strains. This is supported by several isolations of *Bacteroides* in some of the patients over a long period of time. The agglutination reactions could also be due to normally occurring antibodies. These usually belong to the IgM class of antibodies, and tests should therefore be carried out to elucidate if antibodies of this Ig class are responsible for the reactions. Preliminary results with anti-human IgG in indirect IFL tests indicate that this may be true in at least some of the patients (Lambe *et al.*, Data to be published).

It is of interest that sensitive tests such as passive HA and indirect IFL can be used to demonstrate antibodies to *Bacteroides* and *Fusobacterium* species. These two tests and tube agglutination are complementary to each other, for the passive HA technique used here measures antibodies to thermostable lipopolysaccharides, while agglutination and indirect IFL tests measure antibodies to one or more surface antigens, both thermolabile and thermostable antigens. The Ouchterlony technique is also a valuable tool in these types of analyses, since it allows a qualitative analysis of antigens taking part in the reaction.

A close antigenic relationship was demonstrated with HA, indirect IFL and AGD tests between the *F. necrophorum* strain isolated from patient T.K. and the stock laboratory strain of the same species. From a bacteriological point of view these two strains differed from each other, as the laboratory strain was hemolytic on blood agar but the strain from the patient was not. The findings merit a closer serologic investigation of various fusobacteria.

It was noted that the sera of blood donors did not react

in tube agglutination tests or AGD tests with the various *B. fragilis* antigens used in the present investigation. However, some of the blood donor sera gave reactions at dilutions of 1:10 to 1:40 with indirect IFL and passive HA tests; these latter two serologic tests are more sensitive than tube agglutination and AGD tests. At the present time we do not know if the antibodies responsible for these reactions reflect an infection in the past, or if they are due to cross reacting antibodies. Investigations are under way to elucidate this.

Chronic inflammatory intestinal diseases, i.e. Crohn's disease, chronic sigmoiditis and diverticulitis, are a great problem in gastroenterology. The serologic findings with both aerobic and anaerobic bacteria in a patient with chronic sigmoiditis and fistulas are therefore noteworthy. This patient was operated on and the pathological anatomical examination showed granulomatous inflammatory changes similar to Crohn's disease. A more complete serologic investigation will be presented elsewhere (Danielsson *et al.*, Data to be published).

During recent years several attempts have been made to classify serologically various species and subspecies of *Bacteroides* (Beerens *et al.*, 1971; De La Cruz and Cuadra, 1971; Reinhold, 1971; Sonnenwirth, 1960; Werner, 1969). The findings in the present paper point out the urgent need for such a classification to permit a serological comparison of clinical isolates of these organisms from different hospitals and laboratories.

REFERENCES

Beerens, H., Wattre, P., Shinjo, T., and Romond, C.: Premiers Results d'un essai de classification serologique de 131 souches de *Bacteroides* de group *fragilis* (*Eggerthella*). *Ann Inst Pasteur, 121:*187, 1971.

Cato, E.P., Cummins, C.S., Holdeman, L.V., Johnson, J.L., Moore, W.E.C., Simbert, R.M., and Smith, L.DS.: *Outline of Clinical Methods in Anaerobic Bacteriology.* Blacksburg, Virginia Polytechnic Institute and State University, 1970.

Dack, G.M.: Non-sporeforming anaerobic bacteria of medical importance. *Bacteriol Rev, 4:*227, 1940.

Danielsson, D., Lambe, D.W., and Persson, S.: The immune response in a patient to an infection with *Bacteroides fragilis* ss. *fragilis* and *Clostridium difficile. Acta Pathol Microbiol Scand* [*B*], *80:*709, 1972.

Danielsson, D., Kjellander, J., Persson, S., and Wallensten, S.: The immune response to anaerobic and aerobic microorganisms in a patient with chronic sigmoiditis sive Crohn's disease. (To be published in *Acta Chir Scand.*)

De La Cruz, E., and Cuadra, C.: Antigenic characteristics of five species of human *Bacteroides. J Bacteriol, 100:*1116, 1971.

DuPont, H.L., and Spink, W.W.: Infections due to gram-negative organisms: An analysis of 860 patients with bacteremia at the University of Minnesota Medical Center, 1958–1966. *Medicine, 48:*307, 1969.

Edwards, E.A., and Driscoll, W.S.: Group-specific hemagglutination test for *Neisseria meningitidis* antibodies. *Proc Soc Exp Biol Med, 126:*876, 1967.

Felner, J.M., and Dowell, V.R.: *Bacteroides* bacteremia. *Am J Med, 50:*787, 1971.

Finegold, S.M.: Infections due to anaerobes. *Med Times, 96:*174, 1968.

Gelb, A.F., Seligman, S.J.: *Bacteroidaceae* bacteremia. Effect of age and focus of infection upon clinical course. *JAMA, 212:*1038, 1970.

Lambe, D.W., Danielsson, D., Arauz, C., and Carver, K.: Serologic studies of the antibody response in humans with *Bacteroides fragilis* septicemia and in immunized rabbits. I. Agglutination tests and immunofluorescence tests. (To be published.)

Mattman, L.H., Senos, G., and Barrett, E.D.: The anaerobic micrococci: Incidence, habitat, growth requirements. *Am J Med Tech, 24:*167, 1958.

McHenry, M.C., Wellman, W.E., and Martin, W.J.: Bacteremia due to *Bacteroides. Arch Intern Med, 107:*572, 1961.

Neter, E.: Bacterial hemagglutination and hemolysis. *Bacteriol Rev, 20:*166, 1956.

Pearson, H.E., and Andersson, G.V.: Perinatal deaths associated with *Bacteroides* infections. *Obstet Gynecol, 30:*486, 1967.

Reinhold, L.: Serologische Untersuchungen an Stäummen von *Bacteroides thetaiotaomicron* und *Bacteroides fragilis* im Agargelprazipitationstest. *Zentralbl Bakteriol, I Abt Orig, 216:*219, 1971.

Sonnenwirth, A.C.: *A Study of Certain Gram-negative, Non-sporulating Anaerobic Bacteria Indigenous to Man, with Special Reference to their Classification by Serological Means.* Thesis. St. Louis, Mo., Washington University, 1960.

Stokes, E.J.: Anaerobes in routine diagnostic cultures. *Lancet, 1:*668, 1958.

Werner, H.: Das serologische Verhalten von stammen der Species *Bacteroides convexus, B. thetaiotaomicron, B. vulgatus* and *B. distasonis. Zentralbl Bakteriol, I Abt Orig, 210:*192, 1969.

Zabransky, R.J.: Isolation of anaerobic bacteria from clinical specimens. *Mayo Clin Proc, 45:*256, 1970.

PART III

ANAEROBIC INFECTIONS— GENERAL CONSIDERATIONS

CHAPTER XVII

SOME INFECTIONS DUE TO ANAEROBIC SPOREFORMING BACILLI

A.T. Willis

ABSTRACT: *Most clinicians are familiar with the classical conditions of anaerobic myonecrosis (gas gangrene), which may be caused by any of a number of pathogenic clostridia but most commonly* Clostridium perfringens, *and tetanus, caused by* C. tetani. *The clinical features of these diseases are reviewed briefly. Clostridial myonecrosis may follow clean elective surgery as well as accidental trauma, especially surgery of the lower limb.* C. perfringens *septicemia following cholecystectomy is a rare but devastating event; a case is described in which the patient's collapse was sudden and dramatic, and death occurred within eight hours after the onset of symptoms. Postabortal clostridial myonecrosis is reviewed briefly. Clinical, bacteriological and postmortem findings in a case of clostridial meningitis following craniotomy illustrate the characteristics of this rare disease.*

INTRODUCTION

Clostridial infections of man are usually of traumatic origin and have been encountered most commonly in times of war. Their comparative rarity in peace time, however, in no way reduces their importance, nor does it provide an excuse for their neglect in the clinical microbiology laboratory.

Most clinicians are familiar with the classical conditions of anaerobic myonecrosis (gas gangrene) and tetanus. Both of these syndromes most commonly occur following accidental trauma and result from contamination of the wound by the appropriate organism. Tetanus is specifically caused by *Clostridium tetani;* anaerobic myonecrosis may be caused by any of a number of pathogenic clostridia of which the commonest is *C. perfringens (C. welchii).* These anaerobic sporeforming bacilli are widely distributed in nature, being

---

I am greatly indebted to Mr J. Nayman and Dr J. Forbes for allowing me to reproduce Figures 2 and 3, and 5 and 7 respectively.

I thank the Editors of the *Lancet*, the *Journal of Medical Microbiology*, and *Medicine* for permission to publish copyright material, and Messers Butterworth and Co. for allowing me to quote from *Clostridia of Wound Infection* (Willis, 1969). I am also indebted to Mr J. Harrison for his help in preparation of the figures, and to my wife for her secretarial assistance.

present in soil, dust, clothing and so on, and in the intestinal tract of animals and man. Their very ubiquity ensures that a large proportion of accidental wounds are exposed to the risk of contamination at the time of injury. However, mere contamination of a wound is not inevitably followed by the development of tetanus or gas gangrene, because the very special conditions of anaerobiosis must prevail in the lesion before the organisms can multiply. In the absence of an anaerobic environment, the spores of toxigenic clostridia are dormant, but potentially dangerous contaminants.

The most important single factor which enables clostridia to flourish in a wound is a low oxygen tension. This may be obviously present in severe and extensive wounds, such as open fractures, and those caused by high velocity missiles. Here, anaerobiosis is initiated not only by the presence of necrotic tissue, blood clot and foreign bodies at the site of wounding, but also by more distant vascular damage which greatly reduces the blood supply to the part. Although it is this type of injury which is the classical precursor of gas gangrene, early surgical intervention after wounding ensures that serious anaerobic infections rarely develop.

ANAEROBIC CELLULITIS

Unlike clostridial myonecrosis, anaerobic cellulitis is not a serious disease, and it carries a good prognosis. Typically, poorly toxigenic and relatively nonpathogenic clostridia of limited invasive power are restricted to the depths and crevices of the infected wound and to the interconnecting tissue spaces; muscle tissue is not involved. The infection may vary in extent from a limited "gas abscess" to the extensive involvement of a limb (Fig. XVII-1). The onset of the disease is more gradual than that of gas gangrene, and there is usually no associated systemic toxemia or shock. Locally, the dirty wound exudes a brownish seropurulent discharge and gives off an offensive odor. Gas is a constant and prominent feature. The skin is rarely discolored and there is little or no edema.

CLOSTRIDIAL MYONECROSIS

There is always a recent history and clinical evidence of wounding, and although the preceding trauma is usually

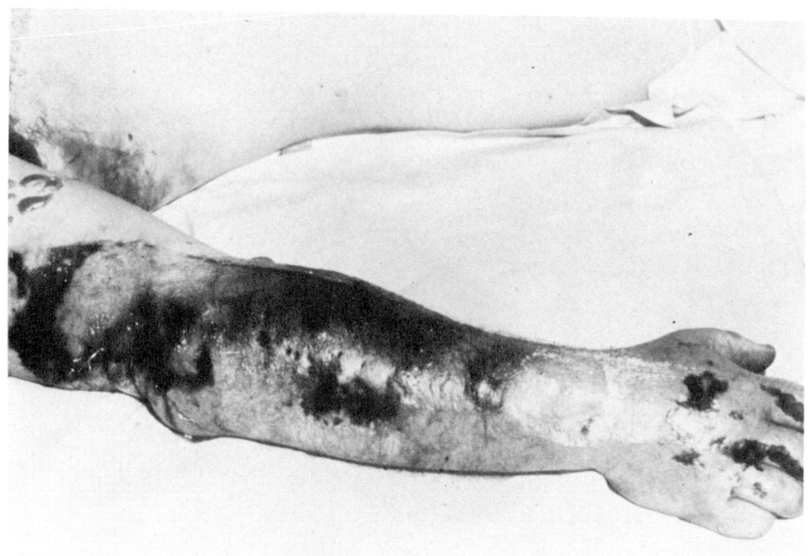

Figure XVII-1. *C. perfringens* cellulitis following injury to the elbow.

extensive and severe, anaerobic infections can follow minor injuries (e.g. hypodermic injections).

The incubation period of gas gangrene after injury may be as short as 7 hours or as long as 6 weeks, but in the great majority of cases the disease develops within 7 days of wounding. We may note here that the average incubation periods of the three main types of gas gangrene are: with *C. perfringens*, 10 to 48 hours; with *C. septicum*, 2 to 3 days; and with *C. novyi (C. oedematiens)*, 5 to 6 days.

One of the earliest prodromal signs is the development of pain in the region of the wound, which may be quite sudden in onset; but more commonly it develops gradually and steadily increases in intensity. Accompanying the pain there is a progressive swelling and edema of the affected part. Concomitant with increasing pain, there is a steep rise in the pulse rate, and sometimes a rise in temperature, although pyrexia is not great as a rule and is often absent.

Locally, there is obvious edema in the neighborhood of the wound, and the area is extremely tender. Early in the infection there is a thin watery discharge, but as the disease progresses there is a profuse serous or serosanguinous discharge which may be so copious as to cause hemoconcentration. The area of edema rapidly increases, bubbles of gas may be seen escaping from the wound, and the tissues may become crepitant. With increasing swelling and tension the skin becomes white and marbled. In untreated cases the disease process extends rapidly and inexorably, not even ceasing with the patient's death.

Despite the term "gas gangrene," gas may not be an obvious feature of the disease, either because it is not produced in any considerable quantity, as in infections due to *C. novyi* and *C. histolyticum*, or because its presence in the tissues is masked by the intense swelling and edema. In any event, overt gas formation in cases of clostridial myonecrosis is usually a late manifestation of the disease. Moreover, the presence of gas in a wound, even an infected wound, does not necessarily indicate clostridial infection. Thus, infections due to other organisms, especially *Escherichia coli*, may be associated with copious gas formation; air may be forced into the tissues at the time of wounding by a high-velocity missile due to the cavitational effects, and in industrial accidents with compressed air (Potts, 1944; Kemp and Vollum, 1946; McDonald, 1947; Desmond, 1947; Altemeier and Culbertson, 1948; Culbertson, 1958; Filler *et al.*, 1968; Chopra and Mukherjee, 1970; Brightmore, 1971; Anderson *et al.*, 1972).

The characteristic muscle changes are seen to best advantage at operation. In the early stages of infection, there may be little apart from edema and pallor. Later, however, involved muscle is slate-blue, brick-red or purplish in color. It is noncontractile and poorly or nonbleeding, and gas may be demonstrated. Still later, with the development of diffuse myonecrosis, muscle masses become friable and dark purple or black in color. The appearance of the tissues does show considerable variation from one case to another, depending on the nature of the infecting organisms.

The most serious and characteristic effects of the disease are a profound toxemia and prostration, about which Macfarlane and MacLennan (1945) published the following classical account:

> The patient lies collapsed and obviously desperately ill. He has a livid pallor, the extremities are cold, and sometimes the veins

> cannot be filled sufficiently to make venepuncture possible. The pulse is often impalpable; it is feeble and irregular, and we have noticed it to be markedly dicrotic in some cases. The blood-pressure, particularly the diastolic pressure, is low. In some cases there is a very large pulse-pressure, with systolic readings of about 100 mm of mercury, while the diastolic pressure is too low to be recorded. Mentally, the patient is unusually alert and clear, anxious, even terrified, and apparently fully aware of his danger. Sometimes he lapses into coma or delirium before death, but more often he dies suddenly, particularly during some disturbance such as being moved or anaesthetised. Death appears to be due to circulatory failure.

It is most unusual for clostridial wound infections to be complicated by intravascular hemolysis. This event is much more commonly associated with gas-gangrenous infections of the uterus, and of the biliary tract following surgery.

POSTOPERATIVE CLOSTRIDIAL INFECTIONS

While gas-gangrenous infections of wounds are most commonly associated with accidental trauma, they sometimes follow clean elective surgical procedures (Gye *et al.*, 1961; Heineman and Braude, 1961; Bornstein *et al.*, 1964; Parker, 1969) and have followed the intramuscular injection of adrenaline and other substances (Cooper, 1946; Bowie, 1956; Tonge, 1957; Koons and Boyden, 1961; Maguire and Langley, 1967; Harvey and Purnell, 1967). In most of these cases, the offending organism has been *C. perfringens.* Studies of the anaerobic flora of hospital and operating theater air (Lowbury and Lilly, 1958; Gye *et al.*, 1961) have shown the ubiquity of the organism in these situations, and normal contamination of the skin by *C. perfringens* derived from the large bowel, especially at such sites as the buttocks and thighs, is well known.

In the case of anaerobic infections following clean surgical procedures, the infecting organisms may be derived from exogenous sources as a result of some breakdown in theater sterility or in the theater ventilating system; or, perhaps more commonly, they are of endogenous origin, occurring on the patient's skin as contaminants that are implanted into the underlying tissues during the course of surgery.

Elective Limb Surgery

The majority of cases of gas gangrene that have developed as a complication of "clean" surgery have followed operations on the lower limb, the skin of which is likely to be more constantly and more heavily contaminated with clostridia than the skin in other areas. Midthigh amputations for vascular disease, especially in diabetes, and orthopedic operations involving the insertion of a foreign body, e.g. pinning of neck of femur, are procedures that seem to carry a special risk of postoperative gas gangrene (Eliason *et al.*, 1937; Gye *et al.*, 1961; Parker, 1969; Ayliffe and Lowbury, 1969; Drewett *et al.*, 1972). Consequently, careful attention to preoperative skin preparation is of the utmost importance in patients who are to be submitted to lower limb surgery, and for this purpose Lowbury *et al.* (1964) recommended the application of compresses soaked with one of the iodophores, substances that are effective in destroying bacterial spores. Recognizing the hazard of gas gangrene after amputations through the thigh in patients with arterial insufficiency, Taylor (1960) recommended the routine use of prophylactic penicillin, and Parker (1969) has suggested extension of its use to lower limb orthopedic surgery in general, and especially for operations involving the hip. (Fig. XVII-2).

Cholecystectomy

Invasion of the bloodstream by clostridia may occur as a late manifestation of anaerobic myonecrosis, and is most commonly due to *C. perfringens*. However, *C. perfringens* septicemia is more commonly associated with gas-gangrenous postabortal and puerperal infections of the uterus, and with *C. perfringens* infections that occur as a complication of biliary tract surgery.

C. perfringens septicemia following elective cholecystectomy is a rare but devastating event. The condition differs in no essential way from clostridial invasion of the bloodstream that can complicate anaerobic infections at other sites, being characterized by its dramatic onset and overwhelmingly rapid course. As long ago as 1948, Brown and Milch considered that "*C. welchii* infection is a considerable factor in

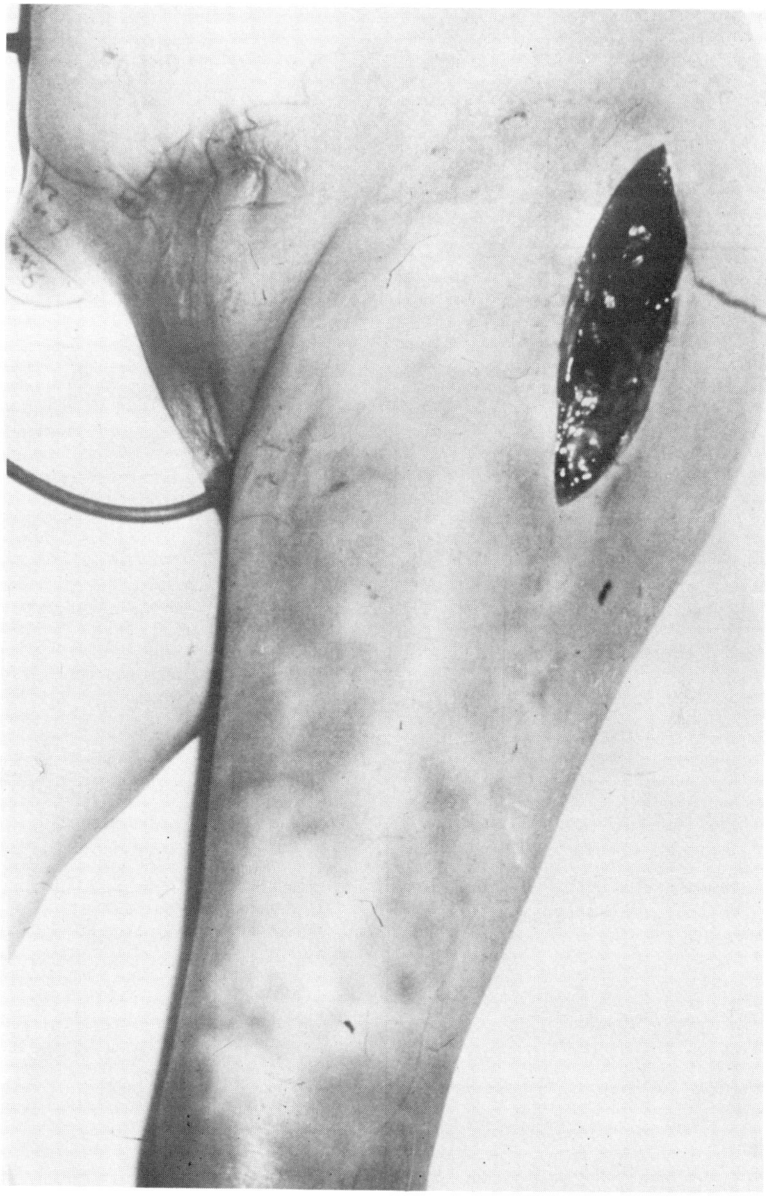

Figure XVII-2. Postoperative *C. perfringens* gas gangrene following elective pinning of neck of femur.

mortality in biliary tract surgery, occurring more frequently than has been generally recognized." Case reports are to be found in the publications of Pyrtek and Bartus (1962), Plimpton (1964), Turner (1964) and Yudis and Zucker (1967), and it has been wisely suggested (Annotation, 1962) that "*biliary surgery* is not '*clean surgery*', but should be classified as *potentially contaminated.*" With the development of septicemia, there is usually extensive intravascular hemolysis with hemoglobinuria, hemoglobinemia and rapidly deepening jaundice. The organism is not only recoverable from the bloodstream by culture, but can sometimes also be seen in direct films of the blood stained by Gram's method.

Case Report

An example of this syndrome (Guest, personal communication) was seen in a 60-year-old woman following elective cholecystectomy for cholelithiasis. A sample of blood was collected from the patient for a routine blood examination at 12 noon on the first postoperative day. At that time the patient was well and was preparing to have her lunch. The hemoglobin was 15.6 g/100 ml. Because this was a routine examination, without indication for urgency, the blood films were not examined until 2:15 P.M., when the hematologist noted that the erythrocytes showed marked spherocytosis and that there was a neutrophil leukocytosis. These unexpected findings prompted the hematologist to request an immediate repeat blood examination. In the meantime, the patient had suddenly collapsed; she was pyrexial with a rapid pulse, a falling blood pressure, and cyanosis and jaundice were evident. Repeat hematological examinations at 4 P.M. and 5 P.M. showed marked hemolysis of erythrocytes, a pronounced leukocytosis and the presence of numerous microorganisms resembling *C. perfringens.* The hemoglobin was 9.7 g/100 ml, while the free plasma hemoglobin was 8 g/100 ml. Despite intensive treatment, which included blood transfusions and intravenous penicillin and gas-gangrene antitoxin, the patient's condition rapidly deteriorated, and death occurred at 8:20 P.M.

Abortion

Almost all cases of clostridial infection of the uterus are due to *C. perfringens,* although Hill (1936) reported two cases in which the offending organism was *C. septicum.* The organism gains access to the gravid or postgravid uterus from either endogenous or exogenous sources. *C. perfringens* may possibly occur as a normal inhabitant of the lower vagina, although there is considerable variation in the incidence reported by different workers.

The manner of entry of clostridia into the uterine cavity in cases of criminal abortion is not difficult to understand, exogenous and endogenous contamination occurring as the result of unskillful manipulations and the use of unsterile and unclean "instruments" and abortifacients. Apart from a conventional dilatation and curettage that might be used by a well-informed and practiced abortionist, favorite methods have included the insertion of rubber catheters and tubes, and irrigations with solutions of quinine and soap. There is, however, almost no limit to the variety of household and other effects that have from time to time been introduced through the cervical os in an attempt to terminate pregnancy. Stem pessaries, uterine packs, pieces of slippery elm and garden hose, tail combs, coat hangers, hair clips and meat skewers have all been pressed into service. Once the interior of the gravid uterus has been contaminated by a minimum infective number of clostridia, fragments of blood clot and necrotic tissue (fetal, placental or maternal) and damaged maternal tissue provide conditions that are favorable for the multiplication of the contaminating organisms (Fig. XVII-3).

Descriptions of *C. perfringens* uterine infections have been published by Hill (1964); Strum *et al.* (1968); Smith *et al.* (1971); and Pritchard and Whalley (1971). The patient with severe post-abortal infection due to *C. perfringens* often presents a characteristic picture. Following manipulation, there is the onset of fever, chills and lower abdominal pain, often associated with vomiting and occasionally with diarrhea. The patient is usually alert and well orientated, and the pulse is disproportionately high in relation to the degree of fever. When septicemia and hemolysis occur, the patient develops jaundice which in some cases rapidly deepens to a curious mahogany color due to concurrent vascular collapse and cyanosis. A profound anemia may develop very rapidly due to hemolysis, and excretion of hemoglobin gives the urine a burgundy-wine color. The blood plasma is also discolored, and there is a marked leukocytosis. It is noteworthy that infec-

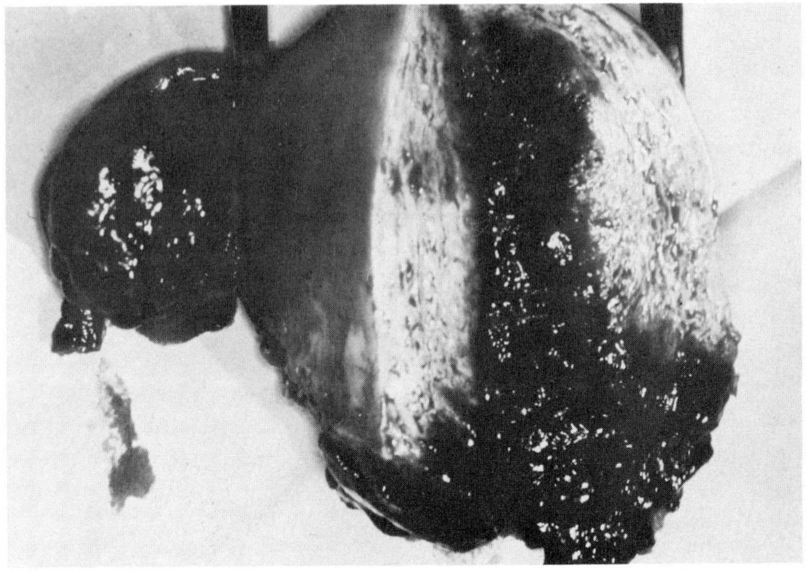

Figure XVII-3. *C. perfringens* infections of the uterus following criminal abortion.

tion with jaundice may proceed to circulatory failure without signs of hemolysis in the blood or urine, and that patients with hemolysis whose infection is controlled face the inevitable danger later of renal failure.

Craniotomy

As might be expected, intrathecal infections due to clostridia are encountered more frequently in war surgery than in civilian practice. In two systematic studies by Russian workers during World War II (Report by a Committee of Soviet Scientists, 1943; Grashchenkov, 1945–46), there are reports of 53 anaerobic infections of the brain following penetrating cranial wounds, and there are a further 18 cases of brain abscess reported upon in detail by Cairns *et al.* (1947). In the latter series, the clostridia isolated were *C. sporogenes* from 12 cases, *C. perfringens* from 11 cases, *C. bifermentans* from two, and *C. capitovale* and *C. hastiforme* from one case each. In most of these patients the clostridia were mixed with aerobic organisms such as *Streptococcus viridans*, *S.*

faecalis and coliform bacilli, but in four of them *C. perfringens* was present as the sole infecting organism. In all these cases of intracranial anaerobic infection occurring among battle casualties, and in four civilian cases (Lovering and Craig, 1941; Otenasek and Chambers, 1945; Gilbert *et al.*, 1961; Russell and Taylor, 1962–63), infection followed cranial wounding and usually developed as a brain abscess. Acute pyogenic meningitis due to clostridia was not observed as a primary condition, although it could occur as a rare complication of intracerebral abscess.

There are only five reports in the literature of primary purulent meningitis due to clostridia occurring as an independent disease, all of them due to *C. perfringens* (Henderson *et al.*, 1954; Møller, 1955; Colwell *et al.*, 1960; Willis and Jacobs, 1964; Mackay *et al.*, 1971). In three of these cases infection was preceded by small penetrating puncture wounds of the head, one was unassociated with head injury or brain abscess, and one (see below) followed elective craniotomy.

Case Report

The patient, a male aged 73 yr, was admitted to hospital for elective surgical treatment of trigeminal neuralgia. On admission his general state of health was good. Neurological examination showed right-sided hyperesthesia over the cutaneous distribution of the first and second branches of the fifth cranial nerve and, as the result of previous left trigeminal section, sensation was absent over that half of the face.

Operation was performed on the day of admission. A right posterior fossa approach was made to expose the fifth nerve which was painted with 5% phenol in glycerol. The wound was closed and the patient was returned to the ward in a satisfactory condition at 11:30 A.M.

Examination of the patient in the late afternoon of the same day showed that cerebrospinal fluid was leaking from the wound. A pad of cottonwool had been placed over the dressing by the ward staff, and this was soaked with the fluid. The wound was immediately resutured and sprayed with Polybactrin.

The patient was free from trigeminal pain, and progressed well until 4 P.M. the following day, i.e. about 30 hr after operation. He then became mentally confused, and his temperature rose to 38.9°C. His condition deteriorated during the next 4 hr, and although there were no clinical signs of meningitis, a lumbar puncture was per-

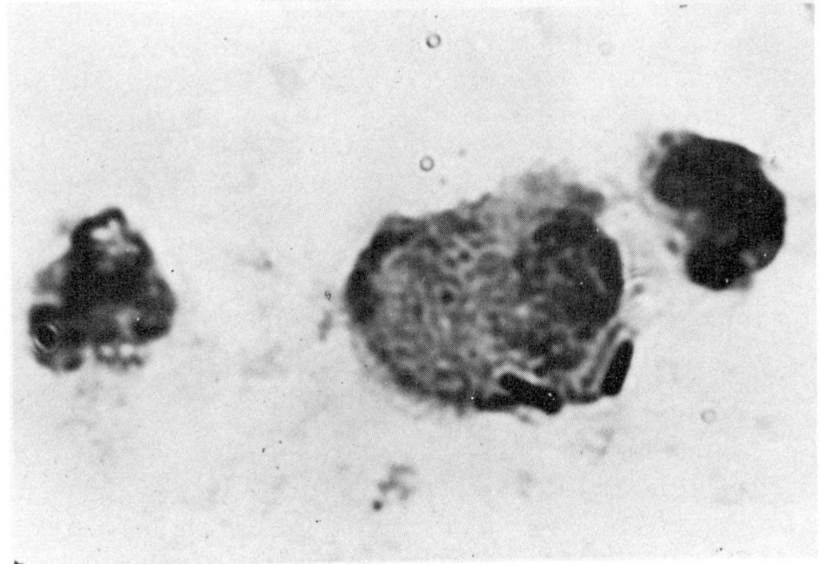

Figure XVII-4. Gram-stained smear of centrifuged deposit of cerebrospinal fluid showing pus cells and large gram-positive bacilli X3000.

formed at 8 P.M., and 20,000 units of penicillin were given intrathecally. The cerebrospinal fluid was cloudy and xanthochromic, and microscopic examination of a centrifuged deposit showed the presence of numerous pus cells and large gram-positive bacilli. The bacilli were morphologically indistinguishable from *C. perfringens* (Fig. XVII-4) and this organism was subsequently isolated in pure and heavy culture. On the basis of the microscopic findings the patient was immediately given 2 g sulphadiazine orally and 1,000,000 units of soluble penicillin intramuscularly. Subsequently he was given 0.5 g sulphadiazine orally and 100,000 units of penicillin intramuscularly every 6 hr, and an intrathecal injection of 20,000 units of penicillin 12-hourly.

In spite of this treatment his condition deteriorated and he became unconscious at 2 A.M. on the second postoperative day, and showed marked stiffness of the neck. At this time the cerebrospinal fluid showed protein and sugar levels of 2600 mg and 5 mg per 100 ml, respectively. Because of lack of clinical improvement, 100 mg of streptomycin, 5,000 units of bacitracin and 50,000 units of polymyxin were added, on an empirical basis, to the intrathecal treatment regimen.

The patient's condition now began to improve, and by 10 A.M. on the third postoperative day his pyrexia had resolved and he was quite rational and well orientated. However, the cerebrospinal fluid protein and sugar levels were unchanged, and the cell count was 6,800 polymorphonuclear leukocytes per mm³.

At 4 P.M. on the same day he suddenly collapsed and became comatose. His blood pressure was unrecordable, his respiration rate rose to 100 per min and he became hyperpyrexial. As there was no physical evidence of pulmonary disease, these manifestations were attributed to involvement of the central nervous system; the patient died within 3 hr. Shortly before death, blood cultures were taken and the wound was aspirated to obtain material for bacteriological examination. The blood cultures were sterile, but the wound aspirate grew a pure culture of *C. perfringens.*

At no time during the postoperative period was there any clinical evidence of wound infection.

Necropsy findings. The necrospy was performed by Dr. D.G.F. Harriman, who kindly made available a summary of the relevant findings.

Respiratory system. The trachea and bronchi contained much mucopurulent material, and the entire lower lobe of the right lung was consolidated. The left lung merely showed hypostatic friability.

Skull and central nervous system. An operation aperture was present in the right squamous occipital bone, and a little watery pus could be expressed from the tissues immediately beneath the surgical incision. There was, however, no evidence of wound infection. The subdural space over the dorsal aspect of the cerebellar hemispheres contained a film of thick, green purulent material, and this extended ventrally to the anterior aspect of the medulla and inferiorly over the dorsum of the cervical cord. A small amount of pus was present in the subarachnoid space. The dura showed hemorrhagic patches and the meninges over the cerebrum were grossly congested. A cerebellar pressure cone was present and the convolutions of the cerebral hemispheres were flattened. Microscopic examination of sections of the occipital meninges and subdural pus showed an exudate containing large numbers of pus cells and large gram-positive bacilli morphologically indistinguishable from *C. perfringens.*

Bacteriology. The organism causing the meningitis, *C. perfringens* type A, was isolated in pure culture on three occasions: first, from the initial specimen of cerebrospinal fluid upon which the diagnosis of *C. perfringens* meningitis was made, then from the material aspirated from the surgical wound shortly before death, and finally from a sample of subdural pus collected at necropsy. All the cultures were identical and showed the morphological, cultural and biochemical characteristics of *C. perfringens.* However, no hemolysis was produced by cultures growing on horse blood agar, and this absence of theta-toxin production was confirmed in the detailed antigenic analysis of a representative filtrate, kindly examined by Professor C.L. Oakley. The organism proved to be a poorly toxigenic variety of *C. perfringens* type A.

TETANUS

Although tetanus is more likely to develop in patients with severe, badly soiled wounds than in those with clean, superficial injuries, nowadays tetanus more commonly follows mild injuries. This is due to the protective measures that are routinely instituted in cases of severe trauma, and are so frequently omitted in cases of mild wounding. Small lesions that have been incriminated as the foci of infection in cases of clinical tetanus may have been so minor at the time of wounding that the patient did not then seek medical advice. Indeed, in cases of so-called idiopathic tetanus, no focus of infection can be found to account for the disease. It is presumed that the wound was so trivial as to have completely healed before evidence of intoxication developed.

A slight penetrating wound produced by a rusty nail, a splinter of wood or a thorn, or even a dirty abrasion are the types of lesions which most commonly precede the development of clinical tetanus (Moynihan, 1956; Ashley and Bell, 1969; Laforce *et al.*, 1969; National Communicable Disease Center, 1969). In a series of 33 cases of tetanus reported from Sheffield by Cox *et al.* (1963), 10 followed abrasions, 7 were due to lacerations, 2 to varicose ulcer of the leg, 2 to penetrating wounds of the foot, 1 to a thorn in the finger, 1 to cracks on the hands, and in 6 cases no injury was found. Other minor lesions from which tetanus may develop are plaster sores, boils and epistaxis (Cole, 1951), ear-piercing (Rey *et al.*, 1967), and paronychia (Cormie, 1962). Otogenic tetanus complicating acute and chronic infections of the external auditory meatus has been reported by Shah (1955), and Sherman (1970) has discussed tetanus in the burn patient.

Injection and Addict Tetanus

It is uncommon these days for infection to follow the injection of substances such as vaccines and therapeutic agents. However, in Bombay 20 cases were reported in which the portal of entry was a smallpox vaccination site, and a further 32 cases in which the tetanus bacillus was apparently introduced during intramuscular inoculation of a drug (Patel *et al.*, 1960a,b). Although drug addiction has been regarded as a noteworthy but minor cause of tetanus, more recently emphasis has been placed on it as a cause of the disease in urban populations (Levinson *et al.*, 1955; Cherubin, 1967, 1971). It was reported that between 1955 and 1965, nearly 75 percent of all known cases of tetanus in New York City occurred in narcotic addicts. There were no fewer than 102 cases of addict tetanus with a mortality rate of almost 90 percent.

Neonatal and Uterine Tetanus

In modern communities, postabortal and puerperal tetanus are uncommon and tetanus neonatorum is rare. Uterine tetanus follows infection of the genital tract, and is most frequently due to criminal interference. Neonatal tetanus results from contamination of the cut surface of the umbilical cord and is a common cause of neonatal death in underdeveloped communities (Stahlie, 1960; Schofield *et al.*, 1961; Daramola, 1968).

Recurrent Tetanus

Recurrent tetanus is uncommon, but by no means unknown (Beare, 1953; Aguileiro Moreira *et al.*, 1960; Patel *et al.*, 1961; Pace and Busuttil, 1971).

Postoperative Tetanus

Postoperative tetanus is fortunately rare, but devastating when it does occur. It is usually the result of some breakdown in theater sterility (Sevitt, 1949; Ministry of Health, 1958).

Clinical Features of Tetanus

Although infection of extensive accidental wounds of the uterus, of the cut umbilical cord and of elective surgical wounds by *C. tetani* is extremely dangerous and of an overwhelming nature, tetanus more often follows very minor injuries. In the presence of minimal tissue necrosis, the

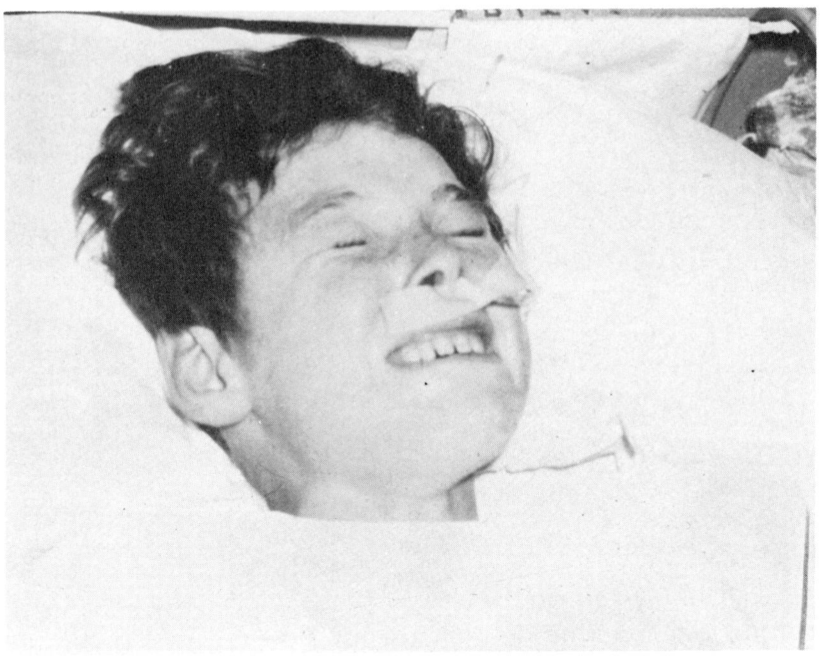

Figure XVII-5. Risus sardonicus in a boy with clinical tetanus.

tetanus bacillus can proliferate and produce its neurotoxin in lethal amount from a comparatively small site.

Since the organism produces no tissue-destroying enzymes and is noninvasive, lesions infected by it are quite unremarkable, and symptoms and signs are absent. The incubation period of the disease proceeds uneventfully but relentlessly. On about the tenth day, too late for prophylaxis to be effective, evidence of intoxication develops.

The clinical features of tetanus have been fully discussed by a number of workers. The publications of Knott and Cole (1952) Adams *et al.* (1969) are of particular value.

The commonest early symptom is trsmus, often combined with pain and stiffness in the neck, back and abdomen. Occasionally dysphagia appears first. These signs increase slowly or rapidly according to the severity of the attack. Twenty-four hours after the onset a patient with a moderately severe attack has a characteristically anxious expression, the "risus sardonicus", in which the eyebrows and the corners of the mouth are drawn up (Figs. XVII-5, XVII-6). There is a varying degree

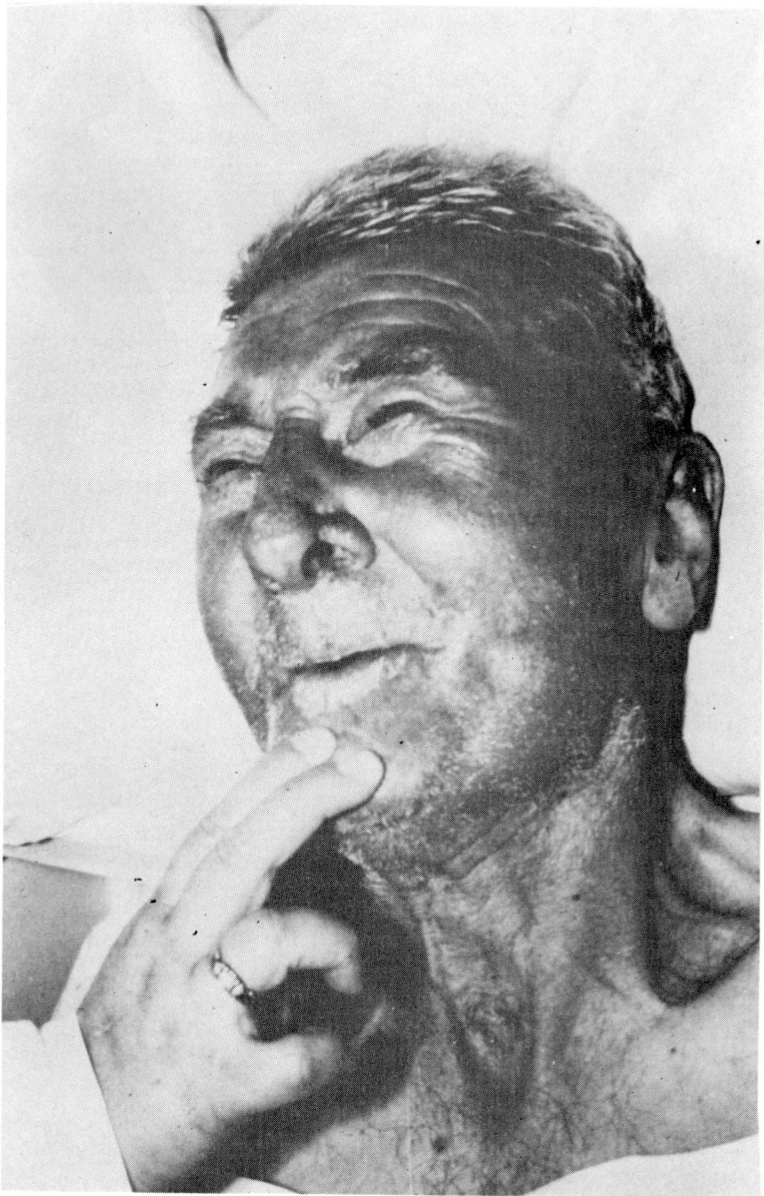

Figure XVII-6. Risus sardonicus in a man with clinical tetanus.

of rigidity of the muscles of the neck and trunk, and the back is usually slightly arched. Board-like abdominal rigidity appears early, but there is little tenderness. The limbs are

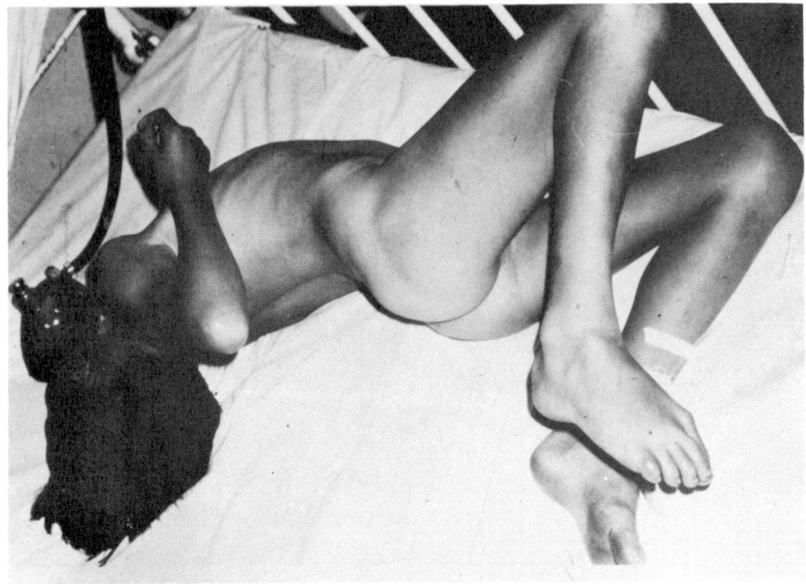

Figure XVII-7. Boy with clinical tetanus showing opisthotonos.

comparatively relaxed and the reflexes normal or increased. The patient is comfortable except for occasional pain in the neck or back, which tends to be made worse by movement. Manipulation of a limb or palpation of any part of the body tends to increase the rigidity and may bring on cramp-like pain. This tendency becomes more pronounced as the stage of reflex spasms is approached.

In very severe cases reflex spasms may begin 12 hr after the onset of clinical tetanus. In moderately severe cases they begin after 2 to 3 days, and in milder ones after 5 or more days. At first they are brought on by some external stimulus, such as moving the patient or knocking the bed. Later they occur spontaneously at regular and increasingly short intervals, until the height of the disease is reached.

The spasms often begin with a sudden jerk, every muscle in the body being thrown into intense tonic contraction. The jaws are tightly clenched, and the head is retracted, the back arched, the chest and abdomen fixed, and the limbs are usually extended (Fig. XVII-7). A severe attack stops respiration, and intense cyanosis may develop. Spasms may last a few seconds or several minutes. When they occur frequently they lead rapidly to exhaustion and sometimes to death from asphyxiation.

Workers at the United Oxford Hospitals have drawn attention to a characteristic syndrome of sympathetic nervous overactivity which may develop in patients with severe tetanus (Kerr *et al.*, 1968; Corbett *et al.*, 1969; Prys-Roberts *et al.*, 1969). These observations are of considerable importance for it seems clear that this syndrome is a factor in the present mortality of the treated disease.

In milder cases the temperature is normal and the pulse and respiration are not much raised except during the spasm. However, in severe attacks they all tend to increase, and there is profuse perspiration. In patients who recover, the reflex spasms gradually diminish but they may last for 3 weeks or more. The remaining rigidity then slowly passes off until recovery is complete.

In fatal cases, death may result from a number of causes: respiratory failure from prolonged spasm of respiratory muscles, cardiac failure secondary to exhaustion, hypotension with pulmonary edema and disordered temperature control due to brain-stem involvement. Aspiration pneumonia is a common contributory cause of death.

Less common manifestations of the disease include local contracture of muscles in the neighborhood of the wound, which may precede the more generalized forms of involvement. This is called local tetanus and may be a result of immunization. In *cephalic tetanus*, irritation or paralysis of cranial nerves appears early and dominates the picture. The facial nerve is most frequently affected, but trismus and dysphagia may also be present. This condition, which is a type of local tetanus, follows wounds of the head and face, and the symptoms often appear first on the injured side.

Tetanus neonatorum is always severe and tends to run a fulminating course. There is often a history of continuous crying for a number of hours, followed by cessation of sucking and crying. Other principal presenting symptoms are convulsions and fever. Spasms in the newborn always tend to be violent, so that dysphagia is marked. Severe spasm of the respiratory muscles is a common cause of death.

An interesting account of an attack of tetanus has been

published in which the course of the illness was described both by the clinicians and the patient (Cole *et al.*, 1968).

REFERENCES

Adams, E.B., Laurence, D.R., and Smith, J.W.G.: *Tetanus*. Oxford, Blackwell, 1969.

Aguileiro Moreira, J.M., Braneiro, R.L., and Ghigliazza, H.M.: A case of a second attack of clinical tetanus. (Spanish) *Semana Med, 117:*969, 1960.

Altemeier, W.A., and Culbertson, W.R.: Acute nonclostridial crepitant cellulitis. *Surg Gynecol Obstet, 87:*206, 1948.

Anderson, C.B., Marr, J., and Jaffe, B.M.: Anaerobic streptococcal infections simulating gas gangrene. *Arch Surg, 104:*186, 1972.

Ashley, M.J., and Bell, J.S.: Tetanus in Ontario: A review of the epidemiological and clinical features of 102 cases occurring in the 10-year period 1958-1967. *Can Med Assoc J, 100:*798, 1969.

Ayliffe, G.A.J., and Lowbury, E.J.L.: Sources of gas gangrene in hospital. *Br Med J, 2:*333, 1969.

Beare, F.: Some observations on tetanus. *Med J Aust, 2:*949, 1953.

Bornstein, D.L., Weinberg, A.N., Swartz, M.N., and Kunz, L.J.: Anaerobic infections—review of current experience. *Medicine, 43:*207, 1964.

Bowie, J.H.: Gas gangrene after injection. *Lancet, 2:*997, 1956.

Brightmore, T.: Nonclostridial gas infection. *Proc Roy Soc Med, 64:*1084, 1971.

Brown, R.K., and Milch, E.: Gallbladder gas gangrene. *Gastroenterol, 10:*626, 1948.

Cairns, H., Calvert, C.A., Daniel, P., and Northcroft, G.B.: Complications of head wounds, with especial reference to infection. *Br J Surg, War Surg Suppl, 1:*198, 1947.

Cherubin, C.E.: Infectious disease problems of narcotic addicts. *Arch Intern Med, 128:*309, 1971.

Cherubin, C.E.: Urban tetanus. The epidemiologic aspects of tetanus in narcotic addicts in New York City. *Arch Environ Health, 14:*802, 1967.

Chopra, I.B., and Mukherjee, P.: Nonclostridial crepitant cellulitis. *J Indian Med Assoc, 54:*177, 1970.

Cole, L.B.: Tetanus immunization. *Practitioner, 167:*247, 1951.

Cole, L.B., Youngman, H.R., and Gandy, A.P.: An attack of tetanus. *Lancet, 2:*567, 1968.

Colwell, F.G., Sullivan, J., Shuman, H.H., and Cohen, J.R.: Acute purulent meningitis due to *Clostridium perfringens*. *N Engl J Med, 262:*618, 1960.

Cooper, E.V.: Gas gangrene following injection of adrenaline. *Lancet, 1:*459, 1946.

Corbett, J.L., Kerr, J.H., Prys-Roberts, C., Smith, A.C., and Spalding, J.M.K.:

Cardiovascular disturbances in severe tetanus due to overactivity of the sympathetic nervous system. *Anaesthesia, 24:*198, 1969.

Cormie, J.: Unusual presentation of tetanus. *Br Med J, 1:*31, 1962.

Cox, C.A., Knowelden, J., and Sharrard, W.J.W.: Tetanus prophylaxis. *Br Med J, 2:*1360, 1963.

Culbertson, W.R.: Acute nonclostridial crepitant cellulitis. *Arch Surg, 77:*462, 1958.

Daramola, T.: Tetanus in Lagos. *W Afr Med J, 17:*136, 1968.

Desmond, A.M.: Surgical emphysema due to compressed air. *Br Med J, 1:*842, 1947.

Drewett, S.E., Payne, D.J.H., Tuke, W., and Verdon, P.E.: Skin distribution of *Clostridium welchii:* use of iodophor as sporicidal agent. *Lancet, 1:*1172, 1972.

Eliason, E.L., Erb, W.H., and Gilbert, P.D.: The *Clostridium welchii* and associated organisms. A review and report of 43 new cases. *Surg Gynecol Obstet, 64:*1005, 1937.

Filler, R.M., Griscom, N.T., and Pappas, A.: Post-traumatic crepitation falsely suggesting gas gangrene. *N Engl J Med, 278:*758, 1968.

Gilbert, A.I., Tolmach, R.S., and Farrell, J.J.: Gas gangrene of the brain. *Am J Surg, 101:*366, 1961.

Gaschenkov, N.I.: Anaerobic infection of the brain. *Ann Rev Sov Med, 3:*5, 1945–46.

Gye, R., Rountree, P.M., and Lowenthal, J.: Infection of surgical wounds with *Clostridium welchii. Med J Aust, 1:*761, 1961.

Harvey, P.W., and Purnell, G.V.: Fatal case of gas gangrene associated with intramuscular injections. *Br Med J, 1:*744, 1967.

Heineman, H.S., and Braude, A.I.: Shock in infectious diseases. *DM,* p. 24, October, 1961.

Henderson, J.K., Kennedy, W.F.C., and Potter, J.M.: Recovery from acute *Clostridium welchii* meningitis. *Br Med J, 2:*1400, 1954.

Hill, A.M.: Why be morbid? Paths of progress in the control of obstetric infection. *Med J Aust, 1:*101, 1964.

Hill, A.M.: Post-abortal and puerperal gas gangrene. *J Obstet Gynaecol Br Emp, 43:*201, 1936.

Kemp, F.H., and Vollum, R.L.: Anaerobic cellulitis due to actinomyces, associated with gas production. *Br J Radiol, 19:*248, 1946.

Kerr, J., Corbett, J.L., Prys-Roberts, C., Smith, A.C., and Spalding, J.M.K.: Involvement of the sympathetic nervous system in tetanus. Studies on 82 cases. *Lancet, 2:*236, 1968.

Knott, F.A., and Cole, L.B.: In: Horder, Lord, (Ed.): *British Encyclopaedia of Medical Practice,* 2nd Ed., Vol. 12. London, Butterworths, 1952, p. 40.

Koons, T.A., and Boyden, G.M.: Gas gangrene from parenteral injection. *JAMA, 175:*46, 1961.

Laforce, F.M., Young, L.S., and Bennett, J.V.: Tetanus in the United States (1965–66). Epidemiologic and clinical features. *N Engl J Med, 280:*569, 1969.

Levinson, A., Marske, R.L., and Shein, M.K.: Tetanus in heroin addicts. *JAMA, 157:*658, 1955.

Lovering, J., and Craig, W.M.: Compound comminuted fracture of the skull complicated by gas bacillus infection and brain abscess. *Proc Staff Meet Mayo Clin, 16:*660, 1941.

Lowbury, E.J.L., and Lilly, H.A.: The sources of hospital infection of wounds with *Clostridium welchii. J Hyg (Camb), 56:*169, 1958.

Lowbury, E.J.L., Lilly, H.A., and Bull, J.P.: Methods for disinfection of hands and operation sites. *Br Med J, 2:*531, 1964.

Macfarlane, R.G., and MacLennan, J.D.: The toxaemia of gas gangrene. *Lancet, 2:*328, 1945.

Mackay, N.N.S., Grüneberg, R.N., Harries, B.J., and Thomas, P.K.: Primary *Clostridium welchii* meningitis. *Br Med J, 1:*591, 1971.

Maguire, W.B., and Langley, N.F.: Gas gangrene following an adrenaline-in-oil injection into the left thigh with survival. *Med J Aust, 1:*973, 1967.

McDonald, E.J.: The clinical significance of gas shadows in x-ray examinations of compound wounds. *Med Ann DC, 16:*595, 1947.

Ministry of Health: An outbreak of tetanus. In: *1957 Report of the Ministry of Health.* Part 2. London, 1958, p. 73.

Møller, B.: Purulent *Clostridium welchii* meningitis originating from a penetrating cranial wound. *Acta Chir Scand, 109:*395, 1955.

Moynihan, N.H.: Tetanus prophylaxis and serum sensitivity tests. *Br Med J, 1:*260, 1956.

National Communicable Disease Center: *Tetanus Surveillance Report No.* 2. Atlanta, US Dept. of Health, 1969.

New England Journal of Medicine (Editorial): Biliary-tract surgery: clean or contaminated? *N Engl J Med, 266:*732, 1962.

Otenasek, F.J., and Chambers, J.W.: Fulminating gas gangrene of the brain: report of a case in a civilian. *J Neurosurg, 2:*539, 1945.

Pace, J.B., and Busuttil, A.: Recurrent tetanus. *St. Luke's Hosp Gaz, 6:*48, 1971.

Parker, M.T.: Postoperative clostridial infections in Britain. *Br Med J, 3:*671, 1969.

Patel, J.C., Aiyar, A.A., Mehta, B.C., and Nanavati, B.H.: Dosage of antitetanus serum in the treatment of tetanus. *Indian J Med Sci, 14:*855, 1960a.

Patel, J.C., Dhirawani, M.K., Mehta, B.C. and Verdhachari, N.S.: Tetanus following vaccination against small-pox. *Indian J Paediat, 27:*251, 1960b.

Patel, J.C., Mehta, B.C., Dhirawani, M.K., and Mehta, V.R.: Relapse and recurrence of tetanus. *J Assoc Physicians India, 1:*1, 1961.

Plimpton, N.C.: *Clostridium perfringens* infection. A complication of gallbladder surgery. *Arch Surg, 89:*499, 1964.

Potts, W.J.: Battle casualties in a South Pacific evacuation hospital. *Ann Surg, 120:*886, 1944.

Pritchard, J.A., and Whalley, P.J.: Abortion complicated by *Clostridium perfringens* infection. *Am J Obstet Gynecol, 111:*484, 1971.

Prys-Roberts, C., Corbett, J.L., Kerr, J.H., Smith, A.C., and Spalding, J.M.K.: Treatment of sympathetic overactivity in tetanus. *Lancet, 1:*542, 1969.

Pyrtek, L.J., and Bartus, S.H.: *Clostridium welchii* infection complicating biliary-tract surgery. *N Engl J Med, 266:*689, 1962.

Report by a Committee of Soviet Scientists (162): Gas infection of the brain as one form of the serious complications of cerebrocranial injuries. *Br Med J, 1:*785, 1943.

Rey, M., Armengaud, M., and Mar, I.D.: In: Eckmann, L., (Ed.): *Proceedings of The International Conference on Tetanus, Bern 1966—Principles on Tetanus.* Bern, Huber, 1967, p. 49.

Russell, J.A., and Taylor, J.C.: Circumscribed gas-gangrene abscess of the brain. Case report together with an account of the literature. *Br J Surg. 50:*434, 1962–63.

Schofield, F.D., Tucker, V.M., and Westbrook, G.R.: Neonatal tetanus in New Guinea. Effect of active immunization in pregancy. *Br Med J, 2:*785, 1961.

Sevitt, S.: Source of two hospital-infected cases of tetanus. *Lancet, 2:*1075, 1949.

Shah, N.J.: Study of otogenic tetanus. *Indian J Med Sci, 9:*52, 1955.

Sherman, R.T.: The prevention and treatment of tetanus in the burn patient. *Surg Clin North Am, 50:*1277, 1970.

Smith, L.P., McLean, A.P.H., and Maughan, G.B.: *Clostridium welchii* septicotoxemia. *Am J Obstet Gynecol, 110:*135, 1971.

Stahlie, T.D.: The role of tetanus neonatorum in infant mortality in Thailand. *J Trop Pediatr, 6:*15, 1960.

Strum, W.B., Cade, J.R., Shires, D.L., and de Quesada, A.: Postabortal septicemia due to *Clostridium welchii. Arch Intern Med, 122:*73, 1968.

Taylor, G.W.: Preventive use of antibiotics in surgery. *Br Med Bull, 16:*51, 1960.

Tonge, J.I.: Gas gangrene following the injection of adrenaline in oil. *Med J Aust, 2:*936, 1957.

Turner, F.P.: *Fatal Clostridium welchii* septicemia following cholecystectomy. *Am J Surg, 108:*3, 1964.

Willis, A.T.: *Clostridia of Wound Infection.* London, Butterworths, 1969.

Willis, A.T., and Jacobs, S.I.: A case of meningitis due to *Clostridium welchii. J Pathol Bacteriol, 88:*312, 1964.

Yudis, M., and Zucker, S.: *Clostridium welchii* bacteraemia: a case report with survival and review of the literature. *Postgrad Med J, 43:*487, 1967.

CHAPTER XVIII

INFECTIONS DUE TO ANAEROBIC NONSPOREFORMING BACILLI

JAY S. GOODMAN

ABSTRACT: *Infections caused by nonsporulating anaerobic bacilli have a clinical spectrum which varies from minor superficial infection to deep abscesses or septicemia with a high mortality. The most common serious anaerobic infections observed in hospitals today are caused by gram-negative bacilli of the family Bacteroidaceae. Cultivation of many important species is not difficult and their recognized role in disease will undoubtedly expand as anaerobic laboratory techniques improve and become routine. Awareness of the abundance of these bacteria in the normal microflora and the settings in which they are likely to become pathogenic is extremely important.* B. fragilis *attracts the most concern because it is the most frequent species isolated and manifests predictable resistance to many commonly used antimicrobial drugs. A combination of medical and surgical therapy is often necessary for proper management. Improved antimicrobial agents presently under investigation may favorably alter the outlook for patients with these infections.*

Dramatic illnesses caused by anaerobic *sporeforming* bacilli commanded attention for hundreds of years before their microbial etiology was entertained. Though sporeformers have lost none of their dreaded potential, recent interest in anaerobic culturing techniques has spotlighted the surprising prevalance of *nonsporeforming* anaerobic bacilli in clinical specimens. The taxonomy of these microorganisms has gone through several changes and is still expanding. More than 70 different nonsporeforming anaerobic bacilli have been identified in exudates or blood (Holdeman and Moore, 1972).

GRAM-POSITIVE BACILLI

Of the *gram-positive* nonsporeforming bacilli inhabiting man, those familiar to clinical laboratories in most major hospitals are *Propionibacterium*, *Actinomyces*, and *Lactobacillus* species. With newer anaerobic methodology, species of these genera, as well as of *Eubacterium* and *Bifidobacterium*, can also be identified. Propionibacteria, especially *P. acnes* and *P. granulosum*, are common on the skin and their occurrence as contaminants in blood cultures is not surprising. However, these "anaerobic diphtheroids" are a rare but well-recognized cause of endocarditis, especially on prosthetic valves and occasionally cause other types of infection in altered hosts (Johnson and Kaye, 1970). Actinomyces, sometimes referred to as fungi because of their morphologic appearance, are normally present in the mouth. These microorganisms (particularly *A. israelii*), can produce well-known suppurative infections of the cervicofacial region, thorax, or abdominal cavity.

Most of the time, gram-positive nonsporeforming bacilli are found in exudates mixed with other anaerobic or facultative bacteria, and the infections are closely related to the skin or mucous membranes of the oropharynx, bowel, or vagina, where anaerobic nonsporeforming bacilli are normal inhabitants. With the exceptions noted above, their role as pathogens is often difficult to assess.

GRAM-NEGATIVE BACILLI

Anaerobic *gram-negative* nonsporeforming bacilli comprise a larger and potentially more dangerous microbial population. These microorganisms account for the vast bulk of normal human microflora. Extensive clinical experience has consistently implicated a few species of the family Bacteroidaceae as the most frequent clinical gram-negative anaerobic isolates (Table XVIII-I). The pathogenic potential of these microorganisms is not disputed. Although they not infrequently coexist in infections with other bacteria, they are often found in exudates as the predominant organism or in blood in pure culture. *Bacteroides* species, and *B. fragilis* in particular, appear to cause infection more often than the other bacteria of this group and will be the main focus of this overview. One contributing factor may be their numerical superiority within the body. For instance, normal feces contains 10^{11} *Bacteroides* species per gram, most of which are

TABLE XVIII-I

CHARACTERISTICS OF CLINICALLY IMPORTANT BACTEROIDACEAE

	Pleomorphism	*Sensitivity to:* Penicillin (2μ)	*Sensitivity to:* Colistin (10 μg)	*Esculin Hydrolysis*
Bacteroides fragilis	−	R	R	+
B. oralis	±	S/R	S/R	+
B. melaninogenicus	+	S	S/R	±
Fusobacterium nucleatum	+	S	S	−
F. fusiforme	+	S	S	−

B. fragilis (Broido *et al.*, 1972). This microorganism also forms part of the normal anaerobic flora of the oral and vaginal mucosa, though not in the ratio found in the bowel. Neither *B. fragilis* nor *B. oralis* appear particularly fastidious and can usually be grown using minimăl anaerobic culturing techniques (*eg* thioglycollate broth). This characteristic might also contribute to the frequency of isolation of *B. fragilis*, but its proclivity toward causing serious infection in many different tissues cannot be ignored. It may well be a more effective invader than other commensals of this group. The importance of this microorganism is further underscored by a unique antimicrobial susceptibility pattern which sets it apart from other anaerobes. This feature, which can be useful in identification (Sutter and Finegold, 1971), has crucial clinical implications.

PATHOGENESIS AND PREDISPOSING FACTORS

Infections caused by nonsporeforming anaerobic bacilli are nearly always endogenous and exotoxins do not appear important in their pathogenicity. In general, invasion by these microorganisms originates in areas where they exist as abundant commensals. It stands to reason that disease due to normal flora would have significant prerequisites. Local predisposing factors are usually obvious in infections emanating from the abdominal cavity or female pelvic structures. Trauma (including surgery, perforating injuries, delivery, or abortion) and vascular insufficiency interrupt the integrity of mucosal surfaces harboring large numbers of anaerobes. Tissue anaerobiosis (low Eh) may result, promoting further multiplication and invasion of these microorganisms with development of intraabdominal or pelvic abscesses, peritonitis, and/or bacteremia. Luminal obstruction which may occur from tumors, fecaliths, or gallstones also appears important. Malignancy may additionally predispose to anaerobic bacterial invasion by offering a necrotic milieu or causing perforation of a hollow viscus. *B. fragilis* is the most frequent isolate in these situations, but other nonsporeforming bacilli, as well as anaerobic cocci and facultative bacteria, may be copathogens.

Dental and gingival disease associated with poor oral hygiene and tissue necrosis favor overgrowth of the normal anaerobic mouth flora including the Bacteroidaceae. These microorganisms may invade locally, producing all types of dental infections including necrotizing ulcerative gingivitis. This latter process, which appears to involve fusobacterial species along with other indigenous mouth organisms, may progress to a severe necrotizing pharyngitis (Vincent's angina). Anaerobic nonsporeforming bacilli can find their way into the paranasal sinuses, middle ear, or mastoid, causing chronic suppuration in these areas. Brain abscess is a well-known consequence. Aspiration of anaerobic mouth flora accompanied by bronchial obstruction from mouth debris or tumor may lead to necrotizing pneumonia, putrid lung abscess, or empyema. *B. melaninogenicus* and fusobacteria are the Bacteroidaceae isolated most frequently from anaerobic respiratory infections and tend to be sensitive to penicillin. However, brain abscess and especially lung abscess often harbor a variety of microorganisms, making it difficult to pinpoint a primary pathogen. Bacteremia is uncommonly detected in either.

Systemic conditions, such as diabetes mellitus, leukemia, lymphoma, and corticosteroid and immunosuppressive therapy, which alter resistance to many microbial invaders, seem to predispose to anaerobic infections as well. Such underlying conditions have been noted frequently in patients with serious anaerobic nonsporeforming bacillary infections reported by several investigators (Bodner *et al.*, 1970; Finegold *et al.*, 1971). Prior therapy with antimicrobials is particularly common in patients with these infections. However, this factor is difficult to evaluate because *Bacteroides* species, the most frequent isolates, are often resistant to those antibiotics likely to be given when anaerobic infection is unsuspected, as is frequently the case. Orally administered aminoglycosides which commonly precede bowel surgery, and to which *B. fragilis* is indifferent, may alter susceptibility to this microorganism by their effect on other intestinal bacteria. This supposition is attractive but has not been tested.

BACTEREMIA

Specific disease syndromes caused by anaerobic nonsporeforming bacilli are covered in greater detail elsewhere in this symposium. A notable feature touching many of these syndromes is bacteremia. *Gram-negative* nonsporeforming anaerobic bacilli are by far the most frequent cause of clinically significant anaerobic bacteremia, reflecting the importance of these microorganisms in serious anaerobic infections (Finegold *et al.*, 1971; Wilson *et al.*, 1972). *Bacteroides* species were the fourth most common cause of hospital-acquired bacteremia at the University of Maryland Hospital in 1971. Anaerobic bacteremia often has diagnostic as well as prognostic importance. Although blood stream invasion may occur without an apparent portal of entry, a positive blood culture is often the first clue that a serious anaerobic infection exists and should spur the search for its primary site. Bacteremia may also signal additional complications.

Table XVIII-II summarizes the primary focus of infection or precipitating event in 58 cases of anaerobic gram-negative rod bacteremia. A brief analysis of this series illustrates the types of organisms invading the blood stream and precipitating events which can be expected in any general hospital. All of the microorganisms were identified as *Bacteroides* species; the vast majority were *B. fragilis*. A major proportion of these bacteremic patients had gastrointestinal or gynecological disease; this association has been emphasized in several

TABLE XVIII-II

PRIMARY LOCUS OF INFECTION OR PRECIPITATING EVENT IN 58 PATIENTS WITH BACTEROIDES BACTEREMIA*

GASTROINTESTINAL	
Colectomy/colostomy (11 CA of colon)	14
Small or large bowel perforation (3 GSW of abd.)	5
Appendectomy (2 appendiceal abscesses)	4
Cholecystectomy/choledochoduodenostomy	3
Mesenteric artery thrombosis	2
Diverticulitis	1
Rectal abscess	1
Subphrenic abscess	1
Transrectal prostate biopsy	1
Gastroenteritis of unknown cause	1
GYNECOLOGICAL	
Puerperal/postabortal sepsis	4
Tuboovarian abscess	3
Hysterectomy (CA cervix or uterus, fibroids)	4
Choriocarcinoma	1
Bartholin abscess	1
MISCELLANEOUS	
Chronic mastoiditis	1
Chronic sinusitis	1
Tricuspid valve endocarditis	1
Postlaminectomy meningitis	1
Abd. aortic prosthesis abscess	1
Necrotizing fasciitis (perineum to axilla associated with Gr A strep)	1
Unknown site of origin	6

* Findings in 39 patients were reported previously (Bodner *et al.*, 1970).

recent reviews (Bodner *et al.*, 1970; Wilson *et al.*, 1972; Marcoux *et al.*, 1970; Felner and Dowell, 1971). Colonic surgery was the single most frequent precipitating event; 11 of 14 patients in that category had bowel carcinoma. Malignancy was also a common occurrence in the gynecological group. One patient with a *B. fragilis* subphrenic abscess may have acquired this microorganism from portal blood following traumatic rupture of the dome of the liver. Though not seen in this series, pyogenic liver abscess has been recognized as frequently containing anaerobic bacteria and should be considered when bacteroides (or other anaerobic) bacteremia occurs in the absence of an obvious portal of entry (Sabbaj *et al.*, 1972).

Twelve patients in this series had miscellaneous primary infections not ostensibly related to the gastrointestinal or pelvic organs. One patient with mastoiditis and one with sinusitis were the only cases with a proven primary cephalic locus.

The portal of entry of the *B. fragilis* causing tricuspid valve endocarditis could not be discovered even at autopsy. Bacteroides endocarditis is associated with a high mortality because of its infrequent recognition and the lack of bactericidal agents available for treatment (Felner and Dowell, 1970). The bacteroides meningitis following lumbar laminectomy may have reflected contamination from the patient's skin. The patient who developed a bacteroides infection of an abdominal aortic prosthesis had undergone colectomy for diverticulitis ten months previously and may have had a persisting nidus of infection at the time of his aortic surgery. Details of these latter two cases are reported elsewhere (Bodner *et al.*, 1970). Six additional patients had no apparent primary site of infection and a portal of entry could not be determined. It is notable that three of these had disseminated malignancy and one presented with diabetic ketoacidosis. Eighteen patients had an identical bacteroides isolated from wound or abscess drainage. Although other anaerobic and/or aerobic microorganisms were frequently grown from exudates, only six patients had polymicrobial bacteremia.

The mortality in this series of patients was 35 percent, underscoring the serious nature of these infections. As expected, the presence of shock, severity of underlying disease, use of inappropriate antimicrobials, failure to drain abscesses or localize the primary site of infection, and the occurrence of superinfection adversely influenced outcome. These factors have been emphasized in other studies and are known to alter the prognosis in sepsis due to many different microorganisms (Bodner *et al.*, 1970; Wilson *et al.*, 1972). Certainly, early awareness of the possibility of anaerobic infection will favor successful therapy.

FEATURES SUGGESTING GRAM-NEGATIVE ANAEROBIC BACILLARY INFECTION

Hectic fever, chills, and other symptoms which varied with the primary locus of infection were not sufficiently helpful in differentiating these patients from those with infections due to aerobic enteric pathogens. Shock, a well-known complication of any gram-negative sepsis, occurred in 20 patients in the above series. Nevertheless, there are important clinical features that should strongly suggest the presence of anaerobic gram-negative bacillary infection. Foremost is the occurrence of sepsis in the setting of the intraabdominal or pelvic conditions exemplified in Table XVIII-II. It can be additionally helpful to learn that the patient is already receiving antibiotics, such as cephalothin and gentamicin, to which he is not responding. *Bacteroides* species are known by many infectious disease clinicians as the "surgeon's bug." Unfortunately, this sobriquet is not appreciated by enough surgeons. Brain abscess is another condition which, from the standpoint of management, must be assumed to contain *Bacteroides* species (Heineman and Braude, 1963; Heineman *et al.*, 1971).

Foul-smelling exudate from abscesses or wounds provides a potent, though not infallible clue to the presence of anaerobic infection (Bodner *et al.*, 1970). Gram staining, a simple but grossly underused procedure, can be effectively applied to such exudate and may even be helpful for presumptive identification of the organism seen. It would be significant, for instance, to observe gram-negative rods which may not have been present in routine aerobic cultures. Pleomorphism and pale staining are additional characteristics of these microorganisms on smear, but it must be stressed that bacteroides are often morphologically indistinguishable from aerobic enteric bacilli (Bodner *et al.*, 1970; Goodman, 1971). If organisms with tapered ends are observed, they are likely to be fusobacteria, a finding which can be of help in selecting antibiotics.

One recognized feature of bacteroides infections which can be a helpful diagnostic clue as well as a roadblock to therapeutic success is the tendency for these microorganisms to invade regional veins and produce septic thrombophlebitis. Bacteroides have been shown to elaborate heparinase (Gesner and Jenkins, 1961); whether this enzyme is causally important is unknown but it provides an attractive explanation for the phenomenon. Thrombophlebitis and/or emboli were detected in only eight of the cases included in Table XVIII-II,

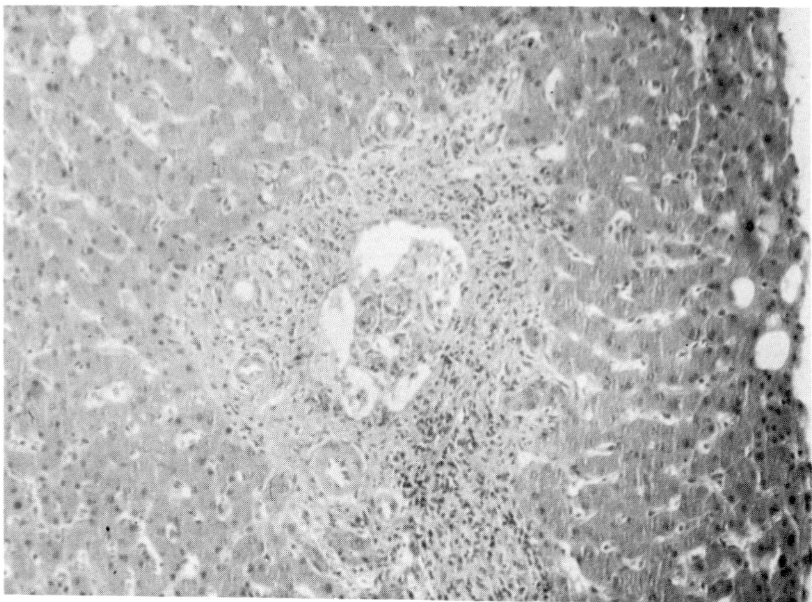

Figure XVIII-1. Liver biopsy showing an embolus in a portal vein tributary from a patient with bacteroides bacteremia complicating diverticulitis and septic thrombophlebitis of the mesenteric veins.

but have been reported in as many as 30 percent of other series (Felner and Dowell, 1971). Septic venous involvement may reveal itself by shedding pulmonary emboli; these may cause severe respiratory embarrassment in an already gravely ill patient. Metastatic pneumonia, lung abscess, or empyema can also result (Bodner *et al.*, 1970; Tillotson and Lerner, 1968). The occurrence of such complications following gastrointestinal or gynecological surgery should strengthen the suspicion of bacteroides sepsis. Emboli from the portal venous system may lodge in the liver. Figure XVIII-1 shows a liver biopsy with an embolus in a portal tributary from a patient with bacteroides sepsis who had undergone colectomy for diverticulitis. At operation, thrombophlebitis of some of the mesenteric veins had been noted.

The following two case histories illustrate the problems which may arise in managing patients with bacteroides sepsis complicated by septic thrombophlebitis.

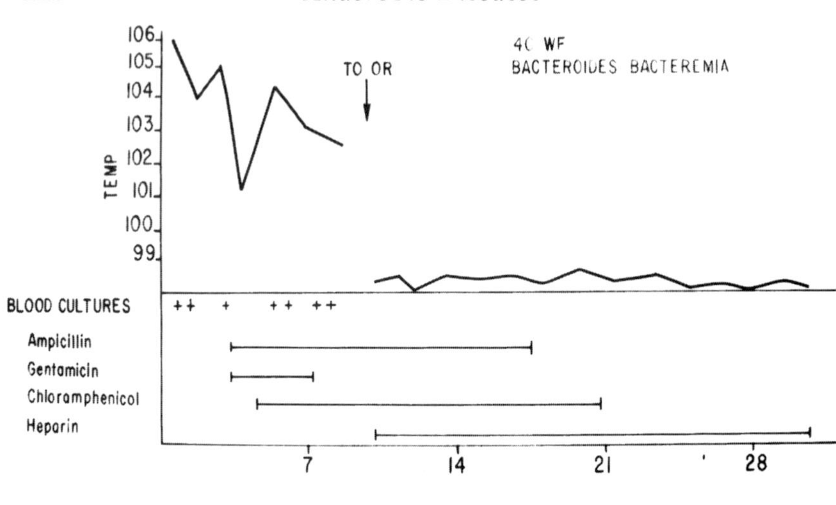

Figure XVIII-2. Course summary of Case 1

Case 1: A 40-year-old diabetic female had the onset of fever and chills beginning two months after an uneventful hysterectomy and left oophorectomy. Examination revealed a temperature of 105.8°F and right abdominal tenderness. Ampicillin and gentamicin were begun and chloramphenicol, 3 g/day, was added when several blood cultures were found to contain *B. fragilis* (Fig. XVIII-2). Bacteroides bacteremia persisted despite therapy, and swelling of the entire right leg and signs of pulmonary embolism appeared. Right ileofemoral venous thrombectomy, salpingo-oophorectomy, and vena caval clipping were performed. Culture of the thrombus, a microscopic section of which is shown in Figure XVIII-3, and periaortic lymph nodes revealed *B. fragilis.* Heparin and antibiotics were continued and the patient was afebrile for the remainder of her hospital course and recovered completely.

Case 2: A 35-year-old waitress presented with a six week history of malaise, chills, and fever. Diarrhea had been present early in her illness. Physical examination was unrevealing but leukocytosis was present and a blood culture was positive for an *Enterobacter* species. There was no response to ampicillin or cephalothin therapy, and her fever became hectic with spikes to 105°F (Fig. XVIII-4). Two blood cultures from the 16th hospital day plus one from the second hospital day were found to contain *B. fragilis.* Chloramphenicol, 4 g/day, was begun, but without clinical response. Blood cultures taken three days later revealed *B. fragilis*, an enterobacter, and *Proteus mirabilis*, all sensitive to antibiotics the patient was receiving. An intensive search for the origin of

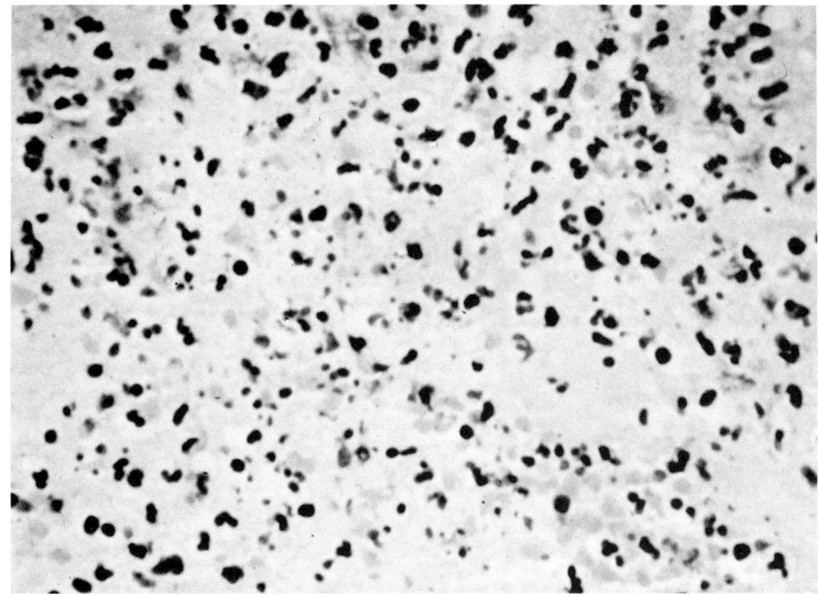

Figure XVIII-3. Section of the ileofemoral thrombus from Case 1 showing microorganisms (H & E).

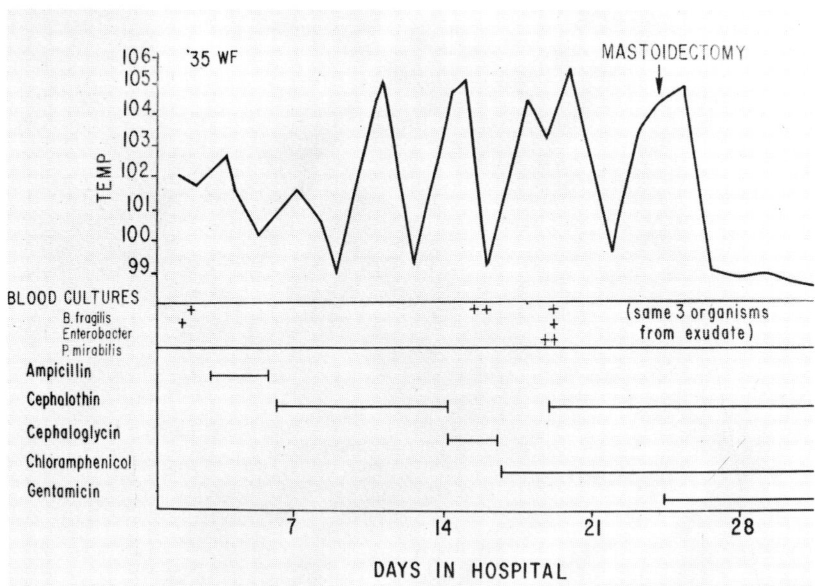

Figure XVIII-4. Course summary of Case 2.

the polymicrobial sepsis, including upper gastrointestinal, large bowel, and gallbladder radiography, liver and brain scan, EEG, and lumbar puncture uncovered no abnormalities. Exploratory laparotomy was strongly considered but a repeat physical examination revealed a perforation in the right tympanic membrane. X-ray studies showed a lytic area in the right mastoid. Mastoidectomy was performed and cholesteatomas were found as well as an abscessed thrombus in the right lateral sinus. Exudate from this lesion yielded the three microorganisms recovered in the patient's blood cultures. Postoperative recovery was uneventful.

COMMENT. In both of these patients, bacteremia persisted in the face of seemingly appropriate antimicrobial therapy. This phenomenon, which can also be seen in patients with bacteroides endocarditis, is not unusual in cases of septic thrombophlebitis due to other microorganisms involving intravenous infusion sites (Stein and Pruitt, 1970). One could argue that larger doses of antibiotics might have been more effective but high titers of microorganisms present in intravascular sites must be at least equally important. The necessity for surgical intervention in such cases seems clear. The second patient demonstrated that bacteremic bacteroides infections are not restricted to below the diaphragm. That *B. fragilis*, presumably of oral origin, can cause serious intracranial infection points out that portal of entry cannot be relied upon for selection of initial antimicrobials. (The prevalence of other *Bacteroides* species and fusobacteria in orally derived anaerobic infections may explain the frequent success of penicillin therapy in such settings). In this case, the persistent bacteremia with *B. fragilis* and other enteric bacilli in the absence of cephalic signs diverted attention elsewhere in the search for a primary site of infection. Of course, a thorough initial physical examination would have provided an earlier diagnosis.

BACTERIOLOGY

Suitable techniques for isolation and identification of anaerobic bacteria are discussed elsewhere in this symposium. The best of these techniques are time-consuming and require specialized equipment. Most hospital laboratories performing routine anaerobic cultures presently use thioglycollate broth and a GasPak or other type of anaerobe jar for incubation of blood agar plates. Prereduced enriched media in sealed tubes is available commercially and may be superior (McMinn and Crawford, 1970). However, many clinically important anaerobic microorganisms, *eg* the Bacteroidaceae, can be grown in an anaerobe jar or in thioglycollate broth. Simple modifications of these materials can increase their efficiency (Finegold, 1970; Martin, 1971). *B. melaninogenicus* often requires enriched media containing hemin and vitamin K. Whatever the media used, delay must be avoided in the delivery of specimens to the laboratory. A degassed transport tube has been recommended for this purpose (Finegold, 1970). Once inoculated, anaerobic cultures should be retained for at least 14 days before considering then negative.

When collecting specimens for anaerobic culture, extreme care must be taken to exclude normal flora. For example, lochia or expectorated sputum would be expected to contain the normal anaerobes of the vagina and mouth, respectively. Misinterpretation of results of such cultures can lead to unnecessarily aggressive therapy.

Definitive speciation of nonsporeforming anaerobic bacilli involves extensive biochemical testing and analysis of metabolic end products by gas chromatography. While these procedures are beyond the capacity of most hospital laboratories, traditional techniques are still valuable and will usually suffice for presumptive identification. Colonial and gram morphology, as well as the presence of pigmentation or hemolysis on blood agar, are inexpensive clues to identification. Some useful characteristics relevant to classification of the Bacteroidaceae are shown in Table XVIII-I. This abbreviated schema makes use of antibiotic disc sensitivity testing as well as the Gram stain and esculin hydrolysis. For instance, an isolate with rounded ends which is esculin positive and resistant to penicillin is very likely *B. fragilis*.

Antibiotic susceptibility testing of anaerobic bacteria for therapeutic purposes is extremely important, but no standardized technique is yet in general use. The clinical (and laboratory) axiom that gram-positive anaerobic bacilli can be assumed to be penicillin sensitive is still mainly true. Most fusobacteria and some *Bacteroides* species are also susceptible to penicillin. However, the assumption that anaerobic gram-negative bacilli are tetracycline sensitive has become a dangerous anachronism (Bodner *et al.*, 1970; 1972). Agar and tube dilution, as well as disc sensitivity studies, have shown that 30 to 60 percent of *B. fragilis* strains are tetracycline resistant (Bodner *et al.*, 1972; Thornton and Cramer, 1970). This microorganism is also consistently resistant to penicillins, cephalosporins, aminoglycosides, and polymixins. Many strains are, in addition, resistant to lincomycin and erythromycin. So far, nearly all anaerobic nonsporeforming bacilli appear susceptible to chloramphenicol, clindamycin, and rifampin. Modifications of the single disc method used for aerobic bacteria have been found applicable to anaerobic microorganisms and can be helpful in determining sensitivity to many of these drugs (Bodner *et al.*, 1972; Wilkins *et al.*, 1972).

TREATMENT

The mainstay of treatment for anaerobic nonsporeforming bacillary infections (as well as other anaerobic infections) is drainage of abscesses and debridement of necrotic tissue. These time-honored procedures may alone lead to cure in some cases. On the other hand, the value of antibiotics should not be understated. Many patients with post-surgical bacteroides sepsis have been treated effectively with antibiotics alone, an outcome which vindicates the surgeon's natural reluctance to reopen a recently operated patient. Furthermore, success with medical therapy has recently been reported in six patients with presumed anaerobic intracranial suppuration, thus raising doubt about the necessity for traditional craniotomy in all such cases (Heineman *et al.*, 1971). Between these poles are the many patients who have loculated abscesses, septic phlebitis, or necrotic tissue requiring surgical intervention as well as antimicrobial therapy for cure.

Selection of initial antimicrobial drugs requires a high index of suspicion of anaerobic infection. It is even less expedient to await culture results before initiating therapy in presumed anaerobic infections, in which culture results may be greatly delayed, than in presumed aerobic bacterial sepsis. Examination of a Gram-stained exudate is a rapid and potentially rewarding maneuver. Many times, however, exudate is not available and treatment must be initiated empirically. In these cases, *B. fragilis* must be assumed to be present because of its known frequency in these infections and its broad antimicrobial resistance. Chloramphenicol is presently the drug of choice when sensitivities are unknown and should be given in doses of 3 to 4 g/day or more. Clindamycin, which appears more active *in vitro* against *Bacteroides* species, has shown promise in preliminary clinical trials against various anaerobic infections (Bodner *et al.*, 1972; Bartlett *et al.*, 1972). A parenteral form of clindamycin has recently become available and may provide an alternative to chloramphenicol in this setting. Other potentially effective agents such as rifampin and metronidazole are under investigation (Nastro and Finegold, 1972). When choosing antimicrobials for anaerobic nonsporulating bacilli, the probability of mixed infection should be considered, unless there is evidence to the contrary. Anaerobic streptococci and aerobic enteric bacilli are frequent copathogens. Thus, a regimen containing an antibacteroides agent plus penicillin or an aminoglycoside may be appropriate. A possible exception to these therapeutic recommendations is putrid lung abscess secondary to aspiration. Here, penicillin, even in low doses, and postural drainage are usually adequate for cure. This may reflect the greater susceptibility of oropharyngeal anaerobes (including *Bacteroides* species) to penicillin or the existence of synergistic infection where elimination of one susceptible co-pathogen suffices therapeutically. Except in unusual circumstances, anaerobic (transtracheal) sputum cultures are probably not necessary.

REFERENCES

Bartlett, J.G., Sutter, V.L., and Finegold, S.M.: Treatment of anaerobic infections with lincomycin and clindamycin. *N Engl J Med*, *287*:1006, 1972.

Bodner, S.J., Koenig, M.G., and Goodman, J.S.: Bacteremic bacteroides infections. *Ann Intern Med, 73:*537, 1970.

Bodner, S.J., Koenig, M.G., Treanor, L.L., and Goodman, J.S.: Antibiotic susceptibility testing of *Bacteroides. Antimicrob Agents Chemother, 2:*57, 1972.

Broido, P.W., Gorbach, S.L., and Nyhus, L.M.: Microflora of the gastrointestinal tract and the surgical malabsorption syndromes. *Surg Gynecol Obstet, 135:*449, 1972.

Felner, J.M., and Dowell, V.R.: Anaerobic bacterial endocarditis. *N Engl J Med, 283:*1188, 1970.

Felner, J.M., and Dowell, V.R.: Bacteroides bacteremia. *Am J Med, 50:*787, 1971.

Finegold, S.M.: Isolation of anaerobic bacteria. In: Blair, J.E., Lennett, E.H., and Truant, J.P. (Eds.): *Manual of Clinical Microbiology.* Bethesda, Williams Wilkins, 1970.

Finegold, S.M., Marsh, V.H., and Bartlett, J.G.: Anaerobic infections in the compromised host. In: *Proceedings International Conference on Nosocomial Infections.* Chicago, American Hosp. Assoc., 1971.

Gesner, B.M., and Jenkins, S.R.: Production of heparinase by bacteroides. *J Bacteriol, 81:*595, 1961.

Goodman, J.S.: Bacteroides sepsis: diagnosis and therapy. *Hosp Pract, 6:*121, Jan. 1971.

Heineman, H.S., and Braude, A.I.: Anaerobic infection of the brain. *Am J Med, 35:*682, 1963.

Heineman, H.S., Braude, A.I., and Osterholm, J.L.: Intracranial suppurative disease. *JAMA, 218:*1542, 1971.

Holdeman, L.V., and Moore, W.E.C., Eds., *Anaerobe Laboratory Manual.* Blacksburg, Virginia Polytechnic Institute and State Univ., 1972.

Johnson, W.D., and Kaye, D.: Serious infections caused by diphtheroids. *Ann NY Acad Sci, 174:*568, 1970.

Marcoux, J.A., Zabransky, R.J., Washington, J.A., Wellman, W.E., and Martin, W.J.: Bacteroides bacteremia. *Minn Med, 53:*1169, 1970.

Martin, W.J.: Practical method for isolation of anaerobic bacteria in the clinical laboratory. *Appl Microbiol, 22:*1168, 1971.

McMinn, M.T., and Crawford, J.J.: Recovery of anaerobic microorganisms from clinical specimens in prereduced media versus recovery by routine clinical laboratory methods. *Appl Microbiol, 19:*207, 1970.

Nastro, L.J., and Finegold, S.M.: Bactericidal activity of five antimicrobial agents against *Bacteroides fragilis. J Infect Dis, 126:*104, 1972.

Sabbaj, J., Sutter, V.L., and Finegold, S.M.: Anaerobic pyogenic liver abscess. *Ann Intern Med, 77:*629, 1972.

Stein, J.M., and Pruitt, B.A.: Suppurative thrombophlebitis. *N Engl J Med, 282:*1452, 1970.

Sutter, V.L., and Finegold, S.M.: Antibiotic disc susceptibility tests for rapid presumptive identification of gram-negative anaerobic bacilli. *Appl Microbiol, 21:*13, 1971.

Thornton, G.F., and Cramer, J.A.: Antibiotic susceptibility of *Bacteroides* species. In: *Antimicrob. Agents and Chemother.—1970.* Ann Arbor, Am Soc. for Microbiology, 1971, p. 509.

Tillotson, J.R., and Lerner, A.M.: Bacteroides pneumonias. *Ann Intern Med, 68:*308, 1968.

Wilkins, T.D., Holdeman, L.V., Abramson, I.J., and Moore, W.E.C.: Standardized single-disc method for antibiotic susceptibility testing of anaerobic bacteria. *Antimicrob Agents Chemother, 1:*451, 1972.

Wilson, W.R., Martin, W.J., Wilkowske, C.J., and Washington, J.A.: Anaerobic bacteremia. *Mayo Clin Proc, 47:*639, 1972.

CHAPTER XIX

THE AGENTS OF HUMAN ACTINOMYCOSIS

Lucille K. Georg

Abstract: *Four* Actinomyces *species,* A. israelii, A. naeslundii, A. viscosus, *and* A. odontolyticus, *as well as a morphologically similar organism,* Arachnia propionica, *exist as commensals in the human oral cavity. All of these species have the ability to induce suppurating lesions. In some instances, they may cause disease that has all the clinico-pathological stigmata of classical actinomycosis.*

The important characteristics and the tests useful for the laboratory identification of these organisms are described. The evidence for their etiologic relationships to disease in man is discussed.

Bollinger, in a talk in 1876, was the first to describe granules in the suppurative exudate from lesions of cattle having the disease commonly known to veterinarians as "lumpy jaw." In a paper published in 1877, he stated that material from such bovine lesions had been sent to the botanist Harz for study, and that Harz (1879) had given the name *Actinomyces bovis* (literally, "ray-fungus of the cow") to the filamentous organisms that had been observed microscopically in the granules. The disease was called actinomycosis. The organism had not been isolated in culture from bovine clinical material.

A similar disease of the cervicofacial areas of human beings was recognized at about this same time. Because of the similarity to the disease of cattle, described by Bollinger and Harz, it was called actinomycosis also. A filamentous organism was isolated in pure culture and characterized from human clinical material by Wolff and Israel in 1891. This organism was named *Streptothrix israeli* by Kruse in 1896, but in 1898 it was placed in the genus *Actinomyces* as *Actinomyces israelii* by Lachner-Sandoval.

For many years it was believed that *Actinomyces bovis* and *A. israelii* actually represented the same organism, and the name *A. bovis* was used by many workers for the etiologic agent of both animal and human actinomycosis, since this name had priority. However, in 1940, Erikson in Scotland carried out a careful comparative study of bovine and human isolates and presented evidence that these were not the same species. Although the two organisms appeared the same in the granules, i.e. as highly filamentous branched forms, in culture the cattle isolates were largely diphtheroidal in form and produced smooth colonies, whereas human isolates remained highly filamentous and produced rough colonies. Besides these morphological differences, she pointed out differences in biochemical reactions as well as serological differences.

Subsequent studies in the United States confirmed the work of Erikson (Thompson, 1950; Pine, Howell and Watson, 1960; Lambert, Brown and Georg, 1967), and it is now accepted that these are indeed two distinct species. The name *A. bovis* is reserved for the species usually isolated from bovine infections, and *A. israelii* for the species usually isolated from human infections. These two species are not, however, host specific, since *A. israelii* has been isolated at least twice* from bovine infections. To our knowledge, *A. bovis* has not been demonstrated in, or isolated from, human material.

We now recognize that human actinomycosis is not caused solely by *A. israelii,* but that the disease has a multiple etiology. The agents of human actinomycosis include at least four *Actinomyces* species, as well as an organism of a second genus, *Arachnia.* This latter organism, *Arachnia propionica,* is morphologically similar to *Actinomyces israelii* and produces clinically similar disease. A new and separate genus was created for it because of basic metabolic and cell wall differences from members of the genus *Actinomyces. Arachnia propionica,* as the name suggests, produces large amounts of propionic acid in its metabolism of glucose. Also different from members of the genus *Actinomyces,* it contains diaminopimelic acid in its cell walls. It was the general recommendation of the Subcommittee on Taxonomy of Actinomycetes that the genus *Arachnia* be included in the order Actinomycetales and the family *Actinomycetaceae.*

* Two bovine isolates of *A. israelii* have been deposited in the American Type Culture collection as #13031 and #19038.

Members of the genera *Actinomyces* and *Arachnia* are facultative or anaerobic gram-positive organisms. With the exception of *Actinomyces viscosus*, which is facultative, they grow best under anaerobic conditions at 37°C. The addition of CO_2 stimulates the growth of all isolates. They are highly variable in morphology, but are usually either diptheroidal or filamentous in form. Bacillary and coccoid forms occur also. Branching is characteristic, but often difficult to demonstrate. Cell walls, similar to those of other Schizomycetes, consist of sugars, glucosamines, and amino acids. Bacteriologic techniques now well standardized for the isolation and identification of the anaerobic bacteria are directly applicable to the study of this group of organisms. The diagnostic procedures for the isolation and identification of the etiologic agents of actinomycosis have been recently reviewed by Georg (1970).

The agents of human actinomycosis, in the general order of their importance in human disease, are as follows: *Actinomyces israelii, Arachnia propionica, Actinomyces naeslundii, Actinomyces viscosus,* and *Actinomyces odontolyticus.* I shall describe each of these briefly, citing some of the important characteristics useful in identifying them, and discuss the evidence we have for their etiologic relationship to disease. Tables XIX-I and XIX-II list the differential biochemical reactions; morphologically similar organisms are included in these tables for comparison.

Actinomyces israelii (Kruse) Lachner-Sandoval, 1898

Actinomyces israelii, as well as the other agents of human actinomycosis, exists as a commensal in the human mouth and throat. A number of workers have isolated *A. israelii* from sputum and tooth surfaces of apparently normal individuals. In a survey to determine the prevalence of *A. israelii* in selected areas of the oral cavity and saliva, Howell *et al.* (1962) recovered *A. israelii* from 48 percent of 368 nonsalivary samples from 90 individuals, and from 30 percent of 90 salivary samples from 55 individuals. This species is also commonly found in tonsils removed during routine tonsillectomies. In one series *A. israelii* was detected in approx-

TABLE XIX-I

Biochemical Characteristics of *Actinomyces*, *Arachnia*, *Corynebacterium*, and *Propionibacterium* species

Biochemical Reactions [a]	Actinomyces bovis	Actinomyces israelii	Actinomyces naeslundii	Actinomyces odontolyticus	Actinomyces viscosus	Arachnia propionica	Corynebacterium pyogenes	Propionibacterium acnes
Catalase	O	O	O	O	+	O	O	+
Esculin hyd.	+	$+^{O}$	$+^{O}$	V	+	O^{+}	O	O
Gelatin hyd.	O	O	O	O	O	O^{+}	+	+
Indol	O	O	O	O	O	O	O	$+^{O}$
Milk	A^{OC}	A^{C}	A^{OC}	A^{OC}	A^{OC}	A^{OC}	P	A^{C}
Nitrate red.	O^{+}	V	$+^{O}$	+	+	+	O	$+^{O}$
Starch hyd. [b]	+	O^{+}	O^{+}	O	O^{+}	O^{+}		O
Urease	O	O	O	O	O	O	O	O
Reaction on blood agar HIBA [c]	NH	NH	NH	Red colonies after 3-5 days	NH	NH	Beta hemolysis	NH
Prod. of propionic acid [d]	O	O	O	O	O	+	O	+
Diamino-pimelic acid (DAP) in cell walls [e]	O	O	O	O	O	+	O	+

a. Reactions read at 7 days. Symbols: A=acid, a=weak acid, C=coagulated, D=digested, H=hemolytic, NH=non-hemolytic, P=peptonized, V=variable, Superscripts=occasional reactions.

b. Starch hydrolysis: + = wide zone of clearing, O^{+} = negative, or a narrow zone of clearing (less than 10 mm D).

c. HIBA = Heart Infusion Agar + 5% rabbit blood.

TABLE XIX-II

Fermentation of Carbohydrates by Actinomyces, Arachnia, Corynebacterium, and Propionibacterium species

Reactions[a]	Actinomyces bovis	Actinomyces israelii	Actinomyces naeslundii	Actinomyces odontolyticus	Actinomyces viscosus	Arachnia propionica	Corynebacterium pyogenes	Propionibacterium acnes
Adonitol	O	O	O	O^A	O	A		V
Arabinose	O^A	V	O^A	V	O	O	O^a	O
Dulcitol	O	A^O	O	O	O	O		O
Glucose	A	A	A	A	A	A	A	A
Glycerol	O^A	O	V	A^O	A^O	O^A	a^O	A^O
Inositol	A^O	V	A^O	V	A^O	O^A	a^O	O
Inulin	O	O^A	V	O	A^O	O		O
Lactose	A	A^O	A^O	A^O	A^O	A	A	O
Maltose	A^O	A	A^O	A	A	A	A	O
Mannitol	O^A	V	O	O	O	A	O	V
Raffinose	O	V	A^O	V	A^O	A		O
Rhamnose	O^A	O^A	O^A	V	O	O		O
Salicin	O^A	V	V	A^O	A^O	O^A	O	O
Sorbitol	O	O^A	V	O	O	A		V
Starch	A	O^A	a^O	A	A	A^O	A	O
Sucrose	A^O	A	A^O	A^O	A	A	A	O
Trehalose	V	V	A^O	A^O	V	A		O
Xylose	V	A^O	O	V	O	O	A	O

[a] Reactions read at 7 days by removing aliquots and adding indicator. Late reactions read at 10 to 14 days in culture medium.

Symbols: A=Acid (yellow with bromthymol blue), a=weak or late acid. O=negative, V=variable. Superscripts=occasional reactions.

imately 20 percent of the specimens studied (Blank and Georg, 1968).

The ability of *A. israelii* to produce suppurative lesions spontaneously in man or after inoculations of pure cultures into experimental animals is well documented (Georg and Coleman, 1970; Brown and von Lichtenberg, 1970). There is no doubt that it is the common and most important cause of actinomycosis in humans. Clinical lesions range from small abscesses on the mucous membranes of the mouth known as "gumboils" that frequently heal spontaneously when they rupture to deeply imbedded abscesses of the glands or subcutaneous tissues, or widespread disseminated lesions involving the major organs of the body. In early or acute infections the organisms may appear free in the exudate as gram-positive filamentous forms. In well developed or chronic lesions, the characteristic granules are usually found.

In culture, *A. israelii* usually is filamentous. Colonies are rough or "R" type. In broth it has a fine or discrete coarsely granular growth. On the surface of agar medium incubated anaerobically for 24 to 48 hours, a filamentous microcolony develops—the characteristic "spider colony." Macrocolonies observed at 5 to 10 days are usually heaped and lobulated, and often present a characteristic "molar tooth-like" form. Smooth variants, however, do occur. In a study of 64 isolates of *A. israelii* by Slack *et al.* (1969), about a third of the cultures produced compact, entire, smooth or "S" type microcolonies and macrocolonies. Nine of the 64 isolates gave a diffuse to viscous type growth in broth.

Although the morphology of *A. israelii* in culture is frequently diagnostic, definitive identification depends either on determination of oxygen requirements and biochemical reactions or on the use of specific serological reagents. The identification of the *Actinomyces* species by serological methods such as agglutination or complement fixation, as applied by early workers, presented difficulties. However, by such methods, Holm in 1930 suggested the existence of two serological groups of *A. israelii*, and Erikson in 1940 clearly separated *A. israelii* from *A. bovis* serologically. The fluorescent antibody (FA) technique was first applied to the *Actinomyces* species by Slack *et al.* in 1961. The specificity of the FA technique and the existence of two serotypes of *A. israelii* were confirmed by Lambert *et al.* in 1967.

Brock and Georg (1969a) reviewed over a hundred clinical isolates of *A. israelii* submitted to the CDC for identification and found that approximately 95 percent were serotype 1. *A. israelii* serotype 2 constituted the remaining 5 percent. Brock and Georg (1969b) also compared serotype 1 and serotype 2 strains. They found that "S" type morphology was most commonly encountered in serotype 2 isolates and that all of 14 serotype 2 cultures failed to ferment arabinose, a sugar that is usually fermented by serotype 1 isolates. Holm in 1948 had previously described similar characteristics of *A. israelii* serotype 2, which he called "Group II." Slack *et al.* (1969) have reported similar findings with regard to *A. israelii* serotypes 1 and 2.

Cummins (1970) has compared the cell wall constituents of *A. israelii* serotype 1 and 2 cultures (Holm's Group I and Group II cultures) and reported that isolates of both serotypes had alanine, glutamic acid, lysine, and ornithine as their major amino acid components. Diaminopimelic acid was not detected in any isolate. In a determination of cell wall sugars, Cummins found galactose to be the major sugar in all isolates. In addition, all but three of 17 of Holm's Group II isolates had an appreciable amount of rhamnose in their cell walls.

Arachnia propionica (Buchanan and Pine) Pine and Georg, 1969

Arachnia propionica, originally described as *Actinomyces propionicus* by Buchanan and Pine in 1962, was placed in the new genus *Arachnia* in 1969 because of important differences in cell wall composition and general metabolism. When first isolated and described from a case of lacrimal canaliculitis (Pine and Hardin, 1959) it was believed to be a rare bacterium. It is now recognized as a common inhabitant of the human mouth and throat where it exists around the teeth and gums (Slack *et al.*, 1971). Like *A. israelii* it is also an important agent of human infections. In 1967 Gerencser and Slack reported three isolations from human infections, and over the past 8 years the Mycology Branch of the Center for Disease Control (CDC) has identified 16 additional isolates from human disease. These were from both localized and systemic infections. Clubbed granules composed of branched filaments were observed in the suppurative exudates from a number of these cases. The clinical picture produced by *Ar. propionica** can be identical to that seen in actinomycosis caused by *A. israelii*. Brock *et al.* (1973) have recently reviewed the disease actinomycosis as caused by *Ar. propionica*.

---

* In order to distinguish *Arachnia* from *Actinomyces* in this paper, the abbreviation *Ar.* will be used for *Arachnia*.

Table XIX-III lists the clinical sources of *Ar. propionica* isolates. This species has been isolated from a wide range of clinical entities ranging from lacrimal canaliculitis to severe disseminated suppurative disease.

Similar to the *Actinomyces* species, *Ar. propionica* appears to exist as a commensal in the human mouth. It produces disease when it has an opportunity to invade injured or otherwise diseased tissues. The fact that *Ar. propionica* was isolated from a finger wound caused by human bite provides evidence for this source of infection. Similar lesions, called "punch actinomycosis," have been previously reported to be caused by *A. israelii* (Burrows, 1945).

Ar. propionica generally is morphologically indistinguishable from *A. israelii* in clinical materials. It appears as gram-positive branched filaments that occur either free in the exudate or confined in distinct granules. The filaments in granules may or may not be clubbed (Brock *et al.*, 1973). Figure XIX-1 illustrates a granule from a mycetoma of the knee. In this particular granule, the filaments (stained by the Brown and Brenn method) are clearly seen in the periphery of the granule.

Ar. propionica is also similar to *A. israelii* in culture and

TABLE XIX-III

CLINICAL SOURCES OF *ARACHNIA PROPIONICA* ISOLATES STUDIED AT CDC, 1965–1972

Clinical Sources	*Isolates Identified Elsewhere*	*Isolates Identified at CDC*	*Total*
Lacrimal canaliculitis	2*	3	5
Cervico-facial abscess	1	3	4
Pulmonary abscess	0	4	4
Sputum (case of pulmonary abscess)	0	1	1
Plug from bronchus	0	1	1
Kidney abscess	0	1	1
Brain abscess	0	1	1
Wound of finger (following human bite)	0	1	1
Mycetoma of knee	0	1	1
Totals	3	16	19

* This includes the type strain (ATCC #14157, isolated by Pine and Hardin, 1959).

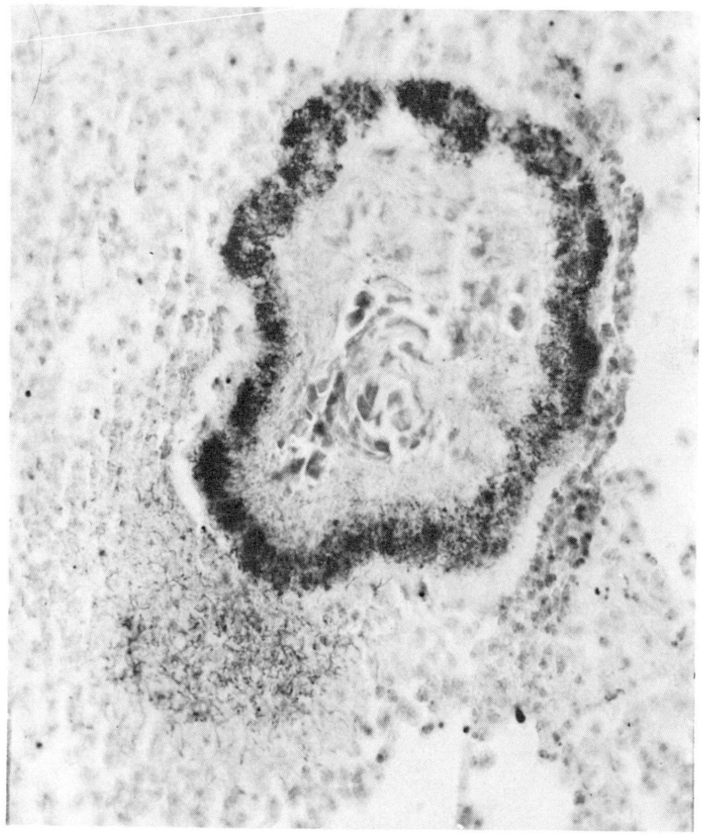

Figure XIX-1. Granule in tissue produced by *Arachnia propionica*. Brown and Brenn stain. Mg 400X.

in its routine biochemical reactions. Both species are facultatively anaerobic and grow best under anaerobic conditions. Both produce a granular or "bread crumb" type of growth in broth, and filamentous, "spider" microcolonies on solid agar media (Figs. XIX-2 and XIX-3). Macrocolonies are more variable. A few isolates of *Ar. propionica* produce the characteristic rough "molar tooth" colonies, as usually seen with *A. israelii* (Fig. XIX-4); however, most produce somewhat smoother macrocolonies.

Routine biochemical reactions also are similar for the two organisms. Both are catalase, indol and urease negative;

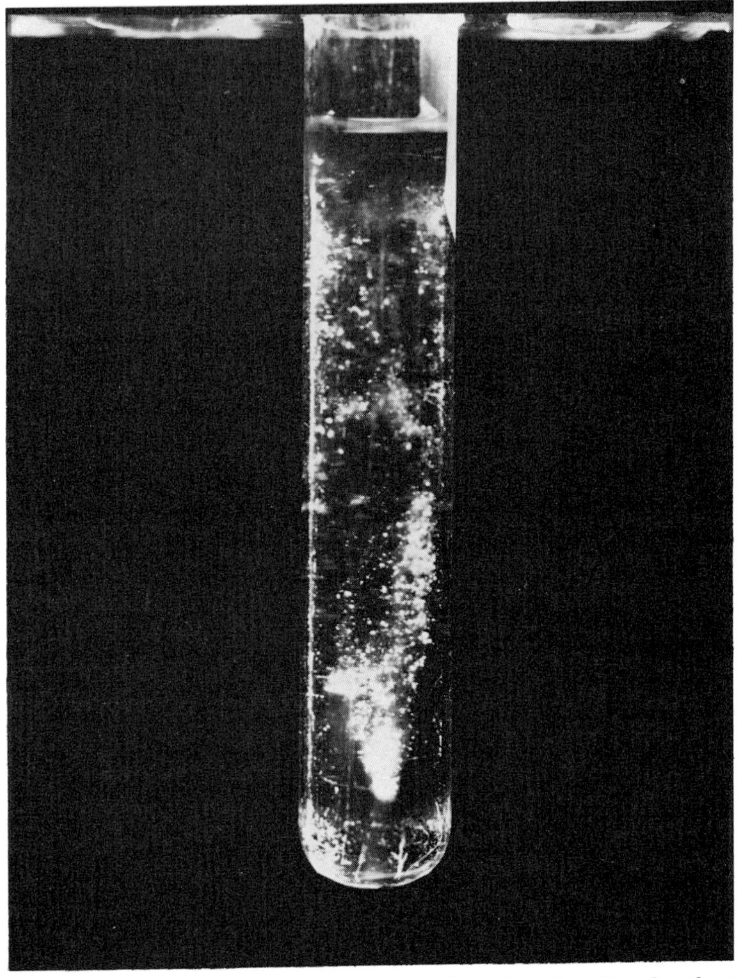

Figure XIX-2. *Arachnia propionica.* Growth in thioglycollate broth at 37° C.

neither is proteolytic. Sugar fermentations are similar, with the exception that *Ar. propionica* isolates do not ferment arabinose and xylose. *A. israelii* ferments either or both of these sugars.

Because of morphological similarities and irregular reactions in routine biochemical tests, *Ar. propionica* is indistinguishable from *A. israelii* on the basis of conventional laboratory methods. Special diagnostic procedures are

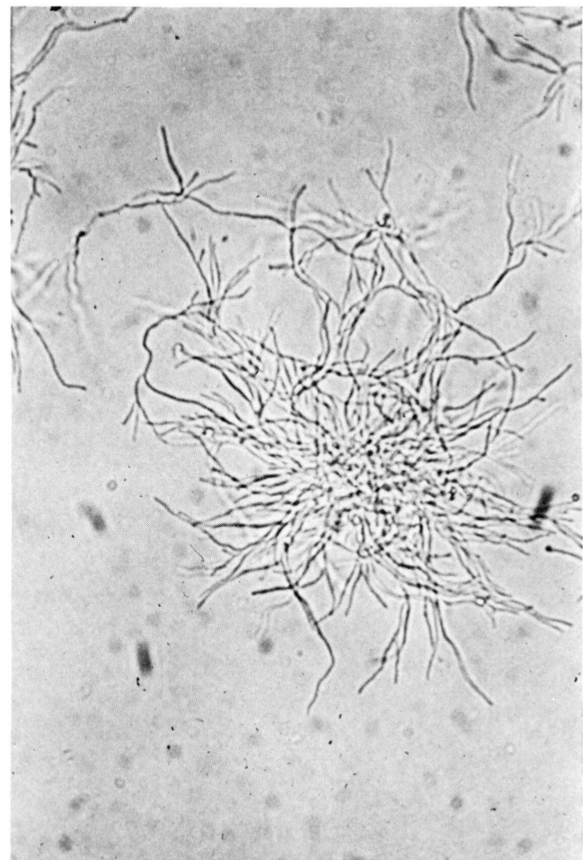

Figure XIX-3. *Arachnia propionica.* Characteristic filamentous microcolony, "spider colony", on Brain Heart Infusion (BHI) agar after 48 hrs anaerobic incubation at 37° C. Mg 380X.

required for identification. The demonstration of diaminopimelic acid (DAP) and the determination of metabolic end products are such procedures. DAP may be demonstrated in the cell walls of *Ar. propionica* by a method in which whole cell hydrolysates and simple paper chromatography are used (Boone and Pine, 1968). Propionic acid may be demonstrated in the volatile acid products by gas chromatography methods as recommended in the Virginia Polytechnic Institute *Outline of Clinical Methods in Anaerobic Bacteriology* (Cato *et al.*, 1970).

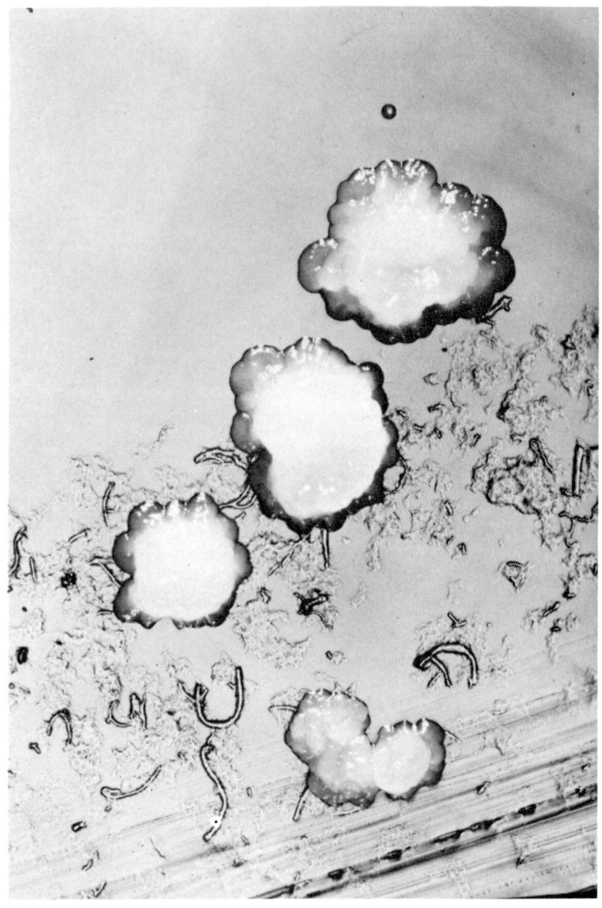

Figure XIX-4. *Arachnia propionica.* Rough "molar tooth" macrocolonies on BHI agar after 7 days anaerobic incubation at 37° C. Mg7X.

The use of specific fluorescent antibody (FA) reagents for these species is a particularly useful tool for their separate identification. Figure XIX-5 illustrates the results of direct FA staining with an *Ar. propionica* conjugate on an exudate from a draining sinus of a pulmonary lesion. Our studies as well as those of Slack *et al.* (1971) have shown that isolates fall into two antigenic groups and probably represent two serotypes of this species. To identify these two types, it is recommended that a bivalent diagnostic FA reagent be used that contains antibody to both serotypes.

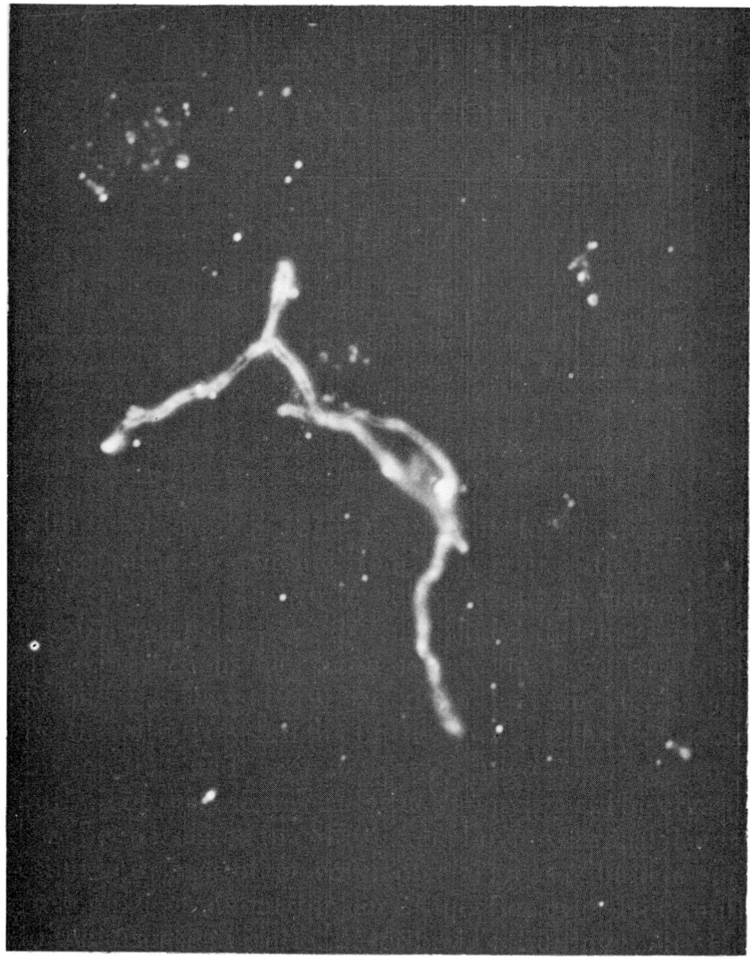

Figure XIX-5. *Arachnia propionica*. Direct FA staining of the organism in exudate from a pulmonary lesion. Mg 1125X.

Actinomyces naeslundii—Thompson and Lovestedt, 1951

A. naeslundii was probably first observed in 1925 by Naeslund, who described it as a "facultative Actinomyces-like organism" from human dental tartar. Thompson and Lovestedt described similar isolates from the oral cavity in 1951 and proposed the name *A. naeslundii*. Further characterization and a comparison with *A. israelii* were made by Howell *et al.* (1959). All of these workers considered *A. naeslundii* a harmless saprophyte commonly occurring in the oral cavity of man. Recently, however, Coleman *et al.* (1969) described eight isolates of *A. naeslundii* from pathological clinical materials. In two instances, where *A. naeslundii* was isolated in pure culture, this actinomycete was identified in the exudates or tissues by specific FA staining.

On the basis of the human clinical materials studied, it appears that *A. naeslundii* does not form granules as readily as does *A. israelii*, but develops freely in the tissues. The pathogenic potential of *A. naeslundii* has been further supported by animal inoculation studies. In a study of the comparative pathogenicity of *A. naeslundii* and *A. israelii* by Coleman and Georg (1969), *A. naeslundii* produced significant lesions in 89.7 percent of the mice inoculated. In comparison, *A. israelii* produced infections in 95.8 percent. In many instances, lesions produced by *A. naeslundii* were as severe as those produced by *A. israelii*.

Socransky *et al.* (1970) reported inducing periodontal destruction in gnotobiotic rats with an isolate of *A. naeslundii* from the mouth of a human. Further studies are currently underway in many laboratories to determine the possibility of an etiologic relationship between *A. naeslundii* and human periodontal disease.

A. naeslundii, in contrast to *A. israelii* and *Ar. propionica*, will grow well under either aerobic or anaerobic conditions. However most isolates do require CO_2 for good growth, and should be considered as facultative only in the presence of CO_2. Broth cultures of this species develop rapidly and usually show soft, diffuse growth throughout the lower seven-eighths of the broth medium. Microcolonies produced on solid media at 24 to 48 hr are well developed, heavily fringed "spider colonies," usually with dense centers (Fig. XIX-6). Because of the rapid growth of this species, agar plates should be examined at 18 to 24 hr in order to observe the characteristic microcolonies. Macrocolonies are usually smooth.

The biochemical reactions of *A. naeslundii* are quite similar to those of *A. israelii*, and are not particularly useful in the separation of these two closely related species (Coleman *et al.*, 1969). The two species also share common antigens, and

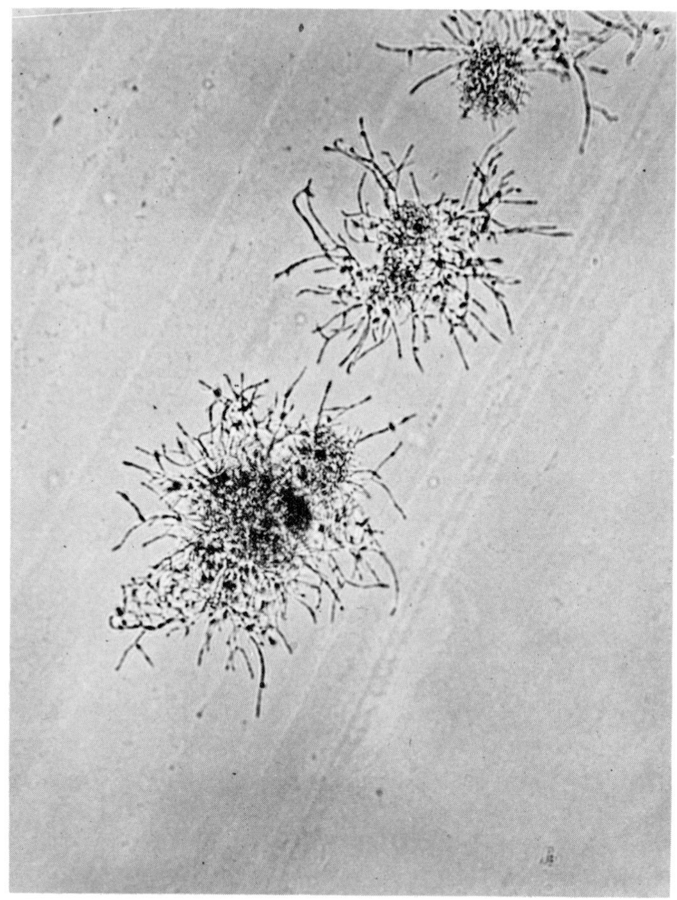

Figure XIX-6. *Actinomyces naeslundii.* Characteristic microcolonies, "spider colonies" with dense centers, on BHI agar after 48 hrs anaerobic incubation at 37° C. Mg 380X.

therefore cross-reactions occur in immunologic tests. However, *A. naeslundii* may be specifically stained by using properly diluted or adsorbed FA conjugates (Lambert *et al.*, 1967). Recent studies suggest the existence of two serotypes of *A. naeslundii* (Bragg *et al.*, 1972).

Actinomyces viscosus (Howell, Jordan, Georg and Pine) Georg, Pine and Gerencser, 1969

A. viscosus was first isolated from the gingival plaques of hamsters with periodontal disease by Howell in 1963. The organism was shown to induce experimental periodontal disease in hamsters by Jordan and Keyes (1964). This organism, first known simply as the "hamster organism" and later as *Odontomyces viscosus* (Howell *et al.*, 1965), is morphologically similar to the *Actinomyces* species. It differs from the other *Actinomyces* species chiefly by being catalase positive, and by its ability to grow well aerobically with or without added CO_2. Recently, with a redefinition of the genus *Actinomyces* to include catalase positive organisms, it has been placed in this genus.

The actinomycete now known as *A. viscosus*, has been found to exist in the oral cavities of animals other than the hamster as well as in man. Gerencser and Slack (1969) have reported numerous isolations from human dental calculus and we at CDC occasionally identify the organism in cultures from extra-oral human clinicopathological materials.

Recently *A. viscosus* was described as an agent of actinomycosis in the dog (Georg *et al.*, 1972). In one case it was found in pure culture in a liver abscess. In a second case, it was found in an exudate from a disseminated, mixed infection of the abdomen. In both cases clubbed granules characteristic of classical actinomycosis were formed.

The canine isolates, as well as hamster and human isolates, have been shown to have the ability to produce suppurative lesions when inoculated intraperitoneally into mice (Georg and Coleman, 1970). In 1969 Gerencser and Slack reviewed the distinguishing characteristics of *A. viscosus*. It resembles *A. naeslundii*, but differs by being catalase positive and more aerobic. Like *A. naeslundii*, its growth is stimulated by CO_2. It is related antigenically to *A. naeslundii*, but can be distinguished serologically. Two serotypes were described.

A. viscosus grows readily and diffusely in brain heart infusion or trypticase soy broth and produces a moderate to heavy sediment that is usually somewhat viscous. In thioglycollate broth the growth may be finely granular and is generally heavier in the upper portion of the tube. Like other *Actinomyces* species, it will not grow on a simple medium such as Sabouraud dextrose agar, and does not produce aerial mycelium. These latter characteristics and the fact that it ferments a wide variety of carbohydrates differentiate it from the *Nocardia* species, with which it has occasionally been mistaken because of its preference for aerobic growth.

Both smooth and rough isolates occur. According to Gerencser and Slack (1969), freshly isolated cultures are usually highly filamentous and produce "spider" microcolonies with dense centers, similar to those of *A. naeslundii*. Macrocolonies are usually smooth.

Actinomyces odontolyticus—Batty, 1958

A. odontolyticus was described by Batty in 1958 as a new *Actinomyces* species that occurred commonly in the human oral cavity. All of Batty's 200 isolates were derived either from scrapings of carious teeth or from saliva. Although *A. odontolyticus* seemed to be associated with dental caries, there was no evidence of an etiologic relationship. It does appear, however, to have the ability to produce suppurative lesions in man. During the past several years the CDC Mycology Branch has identified at least a dozen isolates from various extra-oral pathological lesions. These included an isolation in pure culture from a gangrenous appendix and one from a brain abscess. The organism may be present in distinct granules or free in the exudate. In attempts to produce infections in mice experimentally, however, only minimal lesions have been seen (Georg and Coleman, 1970). The true pathogenic potential of this organism remains to be established.

Actinomyces odontolyticus is morphologically and physiologically similar to the other *Actinomyces* species. Outstanding characteristics of this species are the very delicate beaded filaments observed in cultures, and the ability of the organism to produce colonies that develop a dark red color after blood agar plates have been removed from anaerobe or CO_2 jars and allowed to stand at room temperature for 7 to 14 days.

REFERENCES

Batty, I.: *Actinomyces odontolyticus*, a new species of actinomycete regularly isolated from deep carious dentine. *J Pathol Bacteriol*, *75*:544, 1958.

Blank, C.H., and Georg, L.K.: The use of fluorescent antibody for the detection and identification of *Actinomyces* species in clinical material. *J Lab Clin Med, 71:*283, 1968.

Bollinger, O.: Ueber eine neue pilzkrankheit beim rinde. *Zentralbl Med Wissen, 15:*481, 1877.

Boone, C.J., and Pine, L.: Rapid method for characterization of actinomycetes by cell wall composition. *Appl Microbiol, 16:*279, 1968.

Bragg, S., Georg, L., Ibrahim, A.: Determination of a new serotype of *Actinomyces naeslundii.* Abstr. #G48, p. 38. Paper presented at 72nd Ann. Mtg. American Society for Microbiology, Philadelphia, April 23–28, 1972.

Brock, D.W., and Georg, L.K.: Determination and analysis of *Actinomyces israelii* serotypes by fluorescent-antibody procedures. *J Bacteriol, 97:*581, 1969a.

Brock, D.W., and Georg, L.K.: Characterization of *Actinomyces israelii* serotypes 1 and 2. *J Bacteriol, 97:*589, 1969b.

Brock, D.W., Georg, L.K., Brown, J.M., and Hicklin, M.W.: *Arachnia propionica* as an agent of actinomycosis. *Am J Clin Pathol, 59:*66, 1973.

Brown, J.R., and von Lichtenberg, F.: Experimental actinomycosis in mice. *Arch Pathol, 90:*391, 1970.

Buchanan, B.B., and Pine, L.: Characterization of a propionic acid producing actinomycete, *Actinomyces propionicus,* sp. nov. *J Gen Microbiol, 28:*305, 1962.

Burrows, H.J.: Actinomycosis from punch injuries. With a report of a case affecting a metacarpal bone. *Br J Surg, 32:*506, 1945.

Cato, E.P., Cummins, C.S., Holdeman, L.V., Johnson, J.L., Moore, W.E.C., Smibert, R.M., and Smith, L.D.S.: *Outline of Clinical Methods in Anaerobic Bacteriology.* Blacksburg, Virginia Polytechnic Institute and State Univ., 1970.

Coleman, R.M., and Georg, L.K.: Comparative pathogenicity of *Actinomyces naeslundii* and *Actinomyces israelii. Appl Microbiol, 18:*427, 1969.

Coleman, R.M., Georg, L.K., and Rozzell, A.R.: *Actinomyces naeslundii* as an agent of human actinomycosis. *Appl Microbiol, 18:*420, 1969.

Cummins, C.S.: *Actinomyces israelii* type 2. In Prauser, H. (Ed.): *Proceedings of the International Symposium on Taxonomy of Actinomycetales, 1968.* Jena, German Democratic Republic, Gustav Fisher, 1970, pp. 29–34.

Erikson, D.: *Pathogenic Anaerobic Organisms of the Actinomyces Group.* Br. Med. Res. Council, Special Report Series, No. 240, pp. 1–63, 1940.

Georg, L.K.: Diagnostic procedures for the isolation and identification of the etiologic agents of actinomycosis. In: *Proceedings of the International Symposium on Mycoses.* Washington, D.C., Sci. Publ. Pan Amer. Health Org. No. 205, pp. 71–81, 1970.

Georg, L.K., Brown, J.M., Baker, H.J., and Cassell, G.H.: *Actinomyces viscosus* as an agent of actinomycosis in the dog. *Am J Vet Res, 33:*1457, 1972.

Georg, L.K., and Coleman, R.M.: Comparative pathogenicity of various *Actinomyces* species. In Prauser, H. (Ed.), *op. cit.*, 1970, pp. 35–45.

Georg, L.K., Pine, L., and Gerencser, M.A.: *Actinomyces viscosus,* comb. nov., a catalase positive, facultative member of the genus *Actinomyces. Int J Syst Bacteriol, 19:*291, 1969.

Gerencser, M.A., and Slack, J.M.: Identification of human strains of *Actinomyces viscosus. Appl Microbiol, 18:*80, 1969.

Gerencser, M.A., and Slack, J.M.: Isolation and characterization of *Actinomyces propionicus. J Bacteriol, 94:*109, 1967.

Harz, C.O.: *Actinomyces bovis,* ein neuer Schimmel in den Geweben des Rindes. *Jahresbericht der K. Central-Thierarztliche hochschule in München für 1877–1878, 5:*125, 1879.

Holm, P.: Comparative studies on some pathogenic anaerobic *Actinomyces. Acta Pathol Microbiol Scand, 20:*151, 1930.

Holm, P.: Some investigations into the penicillin sensitivity of human-pathogenic actinomycetes. *Acta Pathol Microbiol Scand, 25:*376, 1948.

Howell, A., Jr.: A filamentous microorganism isolated from periodontal plaque in hamsters. I. Isolation, morphology and general cultural characteristics. *Sabouraudia, 3:*81, 1963.

Howell, A., Jr., Jordan, H.V., Georg, L.K., and Pine, L.: *Odontomyces viscosus,* gen. nov. sp. nov., a filamentous microorganism isolated from periodontal plaque in hamsters. *Sabouraudia, 4:*65, 1965.

Howell, A., Jr., Murphy, W.C., III, Paul, F., and Stephan, R.M.: Oral strains of *Actinomyces. J Bacteriol, 78:*82, 1959.

Howell, A., Jr., Stephan, R.M., and Paul, F.: Prevalence of *Actinomyces israelii, A. naeslundii, Bacterionema matruchotti,* and *Candida albicans* in selected areas of the oral cavity and saliva. *J Dent Res, 41:*1050, 1962.

Jordan, H.V., and Keyes, P.H.: Aerobic, gram-positive, filamentous bacteria as etiologic agents of experimental periodontal disease in hamsters. *Arch Oral Biol, 9:*401, 1964.

Kruse, W.: Systematik der Streptothriceen. In: Flugge, C. (Ed.), *Die Mikroorganismen.* Leipzig, F.C.W. Vogel, 1896, Ch. 2, pp. 48–66.

Lachner-Sandoval, V.: *Uber Strahlenpilze.* Inaugural dissertation. Universität, Strassburg, 1898, pp. 1–5.

Lambert, F.W., Jr., Brown, J.M., and Georg, L.K.: Identification of *Actinomyces israelii* and *Actinomyces naeslundii* by fluorescent antibody and agar-gel diffusion techniques. *J Bacteriol, 94:*1287, 1967.

Naeslund, C.: Studies of *Actinomyces* from the oral cavity. *Acta Pathol Microbiol Scand, 2:*110, 1925.

Pine, L., and Georg, L.K.: Reclassification of *Actinomyces propionicus. Int J System Bacteriol, 19:*267, 1969.

Pine, L., and Hardin, H.: *Actinomyces israelii,* a cause of lacrimal canaliculitis in man. *J Bacteriol, 78:*164, 1959.

Pine, L., Howell, A., Jr., and Watson, S.J.: Studies of the morphological, physiological, and biochemical characters of *Actinomyces bovis. J Gen Microbiol, 23:*403, 1960.

Slack, J.M., Landfried, S., and Gerencser, M.A.: Morphological, biochemical, and serological studies on 64 strains of *Actinomyces israelii*. *J Bacteriol*, *97:*873, 1969.

Slack, J.M., Landfried, S., and Gerencser, M.A.: Identification of *Actinomyces* and related bacteria in dental calculus by the fluorescent antibody technique. *J Dent Res*, *50:*78, 1971.

Slack, J.M., Winger, A., and Moore, D.W., Jr.: Serological grouping of *Actinomyces* by means of fluorescent antibodies. *J Bacteriol*, *82:*54, 1961.

Socransky, S.S., Hubersak, C., and Propas, D.: Induction of periodontal destruction in gnotobiotic rats by human oral strain of *Actinomyces naeslundii*. *Arch Oral Biol*, *15:*993, 1970.

Thompson, L.: Isolation and comparison of *Actinomyces* from human and bovine infections. *Proc Mayo Clin*, *25:*81, 1950.

Thompson, L., and Lovestedt, S.A.: An actinomyces-like organism obtained from the human mouth. *Proc Mayo Clin*, *26:*169, 1951.

Wolff, M., and Israel, I.: Ueber Reinculture des *Actinomyces* und seine Uebertragbarkeit auf Thiere. *Arch Pathol Anat Physiol*, *126:*11, 1891.

CHAPTER XX

INFECTIONS DUE TO ANAEROBIC COCCI

ARNOLD N. WEINBERG

ABSTRACT: *Principles of pathogenicity related to the anaerobic cocci are discussed briefly in this paper. The anaerobic cocci which cause infections in man are part of the normal flora of the skin and upper respiratory, oral, large intestinal and female genital areas in association with other obligate anaerobes and facultative anaerobes. When trauma or other factors upset the normal balance between bacterial growth factors and host defenses, "opportunistic" infection may result. Typically the anaerobic cocci are isolated in association with other anaerobes or facultative organisms, and it is difficult to define their role in many types of mixed infections. Increasing awareness of the importance of anaerobes in infection will lead to better definition as appropriate specimen collection and culture techniques are used more widely.*

The remarks of many of the participants in this conference have focused on the endogenous nature of most of the anaerobic bacteria, and other papers describe specific diseases caused by these members of man's microflora. In the role of a generalist, I will discuss primarily principles of pathogenicity as they relate to the anaerobic cocci.

CLASSIFICATION OF ANAEROBIC COCCI

The anaerobic cocci that produce diseases in people reside on the skin and mucous membranes and as part of the fecal flora (Rosebury, 1962). In simplistic terms, most are gram-positive and arranged in pairs and chains (streptococci) or in pairs, tetrads and clusters (staphylococci). The gram-negative cocci are arranged in pairs, resembling neisseria, and irregular clusters like staphylococci. With such morphological similarities, the nomenclature has been confused. Some spe-

The author has benefited greatly in formulating many of the ideas expressed in this paper from discussions with Dr. Lawrence J. Kunz, Chief of the Bacteriology Laboratories at Massachusetts General Hospital. His wisdom and interest and friendship are gratefully acknowledged.

cies are less intolerant of oxygen than others and are referred to as microaerophilic streptococci or staphylococci. Many of these organisms are in fact more dependent on carbon dioxide as well as less intolerant of oxygen and probably should be referred to as capnophilic. The strict anaerobes have also been classified as *Micrococcus* sp., independent of Gram staining characteristics, and gram-negative anaerobes have been called *Neisseria* sp. This terminology is unacceptable now that appropriate methods of identification are available (Holdeman and Moore, 1972).

Table XX-I lists a simple classification of the anaerobic cocci that produce disease in man. Actual speciation into the seven species of peptococci, five species of peptostreptococci and two species of veillonella requires considerable sophistication not essential to good patient care (Sutter *et al.*, 1972). However the efforts of the research laboratory are important to establish the ecological relationships of the various species of anaerobes and define the conditions and mechanisms of pathogenicity for individual members of the anaerobic cocci.

ECOLOGY OF ANAEROBIC COCCI

Humans are essentially the only natural hosts for the vast number of nonsporeforming obligate anaerobes that cause disease in man. Like other segments of the microbiota, the anaerobic cocci live in association with other obligate anaerobes and facultative anaerobes. The major locations of the anaerobic cocci as part of the normal flora correspond closely to the locations of most of the other anaerobes. Figure XX-1 illustrates the principal areas of distribution. Both gram-

TABLE XX-I

CLASSIFICATION OF ANAEROBIC COCCI PRODUCING DISEASE IN MAN

Order: EUBACTERIALES
 Family: *Peptococcaceae**—gram-positive cocci
  Genera: *Peptococcus*—cluster
   Peptostreptococcus—chain
 Family: *Neisseriaceae*—gram-negative cocci
  Genus: *Veillonella*—cluster

* Rogosa (1971).

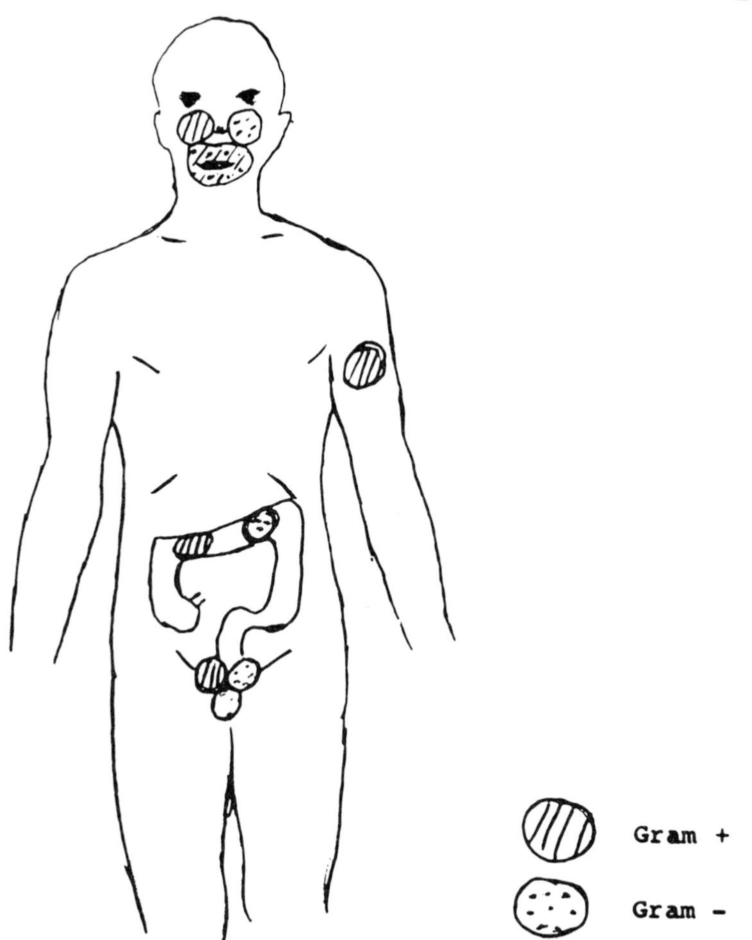

Figure XX-1. Normal distribution of anaerobic gram-positive and gram-negative cocci.

positive and gram-negative cocci are located in the upper respiratory, oral, large intestinal and female genital regions, while only gram-positive representatives are normally found on the skin and external genital areas. Thus, the peptococci and peptostreptococci are more widely distributed and only in the oral cavity do the veillonella predominate (Finegold *et al.*, 1972).

The actual biological circumstances that allow strict anaerobes to persist and thrive in healthy tissues remain speculative. However, the close association with facultative bacteria in epidermal pores, sulci of teeth, tonsillar crypts, large bowel and vagina suggests that the utilization of oxygen by these facultative organisms is one important factor. Supply of essential growth supplements, such as vitamin K, was mentioned by others. Whatever the conditions which promote local growth, they are balanced by host defense factors which essentially maintain a healthy relationship we describe as the "normal flora."

Factors which disturb that balance may lead to local and disseminated infection by one of many members of this heterogeneous group of bacteria, and these infections are "opportunistic." Circumstances promoting improved growth primarily determine whether or not regional infections and/or disseminated disease will occur. The multiplicity of isolates often found locally or in the blood stream suggests that bacterial virulence and invasiveness may be subordinate to local host conditions, at least at the time of initiation of infection.

CHARACTERISTICS OF ANAEROBIC COCCAL DISEASE

The vast majority of known infections involving anaerobic cocci are associated with the gram-positive peptococci and peptostreptococci (Pien *et al.*, 1972). When the organism is isolated in pure culture, pathogenic significance is certain. However, in as many as 50 to 75 percent of cases, other anaerobic and/or facultative organisms are found concurrently, and there is uncertainty in causation. When eight different bacterial species, including peptococci and/or peptostreptococci, are isolated from a subdiaphragmatic abscess, who can ascribe pathogenic significance to any one of the species? Undoubtedly many of these infections are synergistic. The animal and human studies of Meleney (1931) on synergistic gangrene and burrowing ulcer lend support to the concept that more than a single bacterial species may be essential in the genesis of certain infections.

In synergistic gangrene, a facultative organism, *Staphylococcus aureus*, may utilize oxygen locally to the point where associated peptostreptococci or microaerophilic (?capnophilic) streptococci can begin to grow actively. The anaerobes cause local necrosis and vascular thrombosis (by mechanisms which are not completely known) and this diminishes further the oxygen supply to the local area. Thus, an autocatalytic process may ensue and both *S. aureus* and the peptostreptococcus can thrive, the result being a necrotic ulcer. It should be emphasized that the organisms involved in synergistic gangrene are part of the normal skin microflora, and that most often the initial lesion is located at a surgical stay-suture or some other locally traumatized area. Host mechanisms of defense have been compromised locally and the *S. aureus* and peptostreptococcus act as opportunists rather than primary pathogens.

The association of bacteroides with peptostreptococci in local and disseminated infections provides another example of the difficulty of assessing the importance of each isolate to the pathogenicity of the disease. Some data suggest that these organisms act synergistically in-experimental animal infections (Hite *et al.*, 1949). Certainly this combination is often found in clinical specimens from surgical wounds (Pien *et al.*, 1972), septic abortions (Rotheram and Schick, 1969), and brain abscesses (Heineman and Braude, 1963); in our own general experience (Bornstein *et al.*, 1964) this combination has also been found. Among 77 cases of anaerobic streptococcal infections at Massachusetts General Hospital, the anaerobic streptococci were found in pure culture in 38 and mixed culture in 39; and in 21 of the 39, bacteroides were found in the mixed culture.

In identifying an infection due to anaerobes, odor and gas production may be helpful hints, but the site of infection and the presence of multiple morphologic and staining varieties of bacteria are even more suggestive. For example, in a patient with brain abscess, pus removed from the abscess at surgery revealed at least two morphologic and staining types of bacteria. *Bacteroides* sp. and a peptostreptococcus were recovered on anaerobic culture. In an elderly diabetic lady with creptitant and bullous cellulitis and subcutaneous abscess of the leg, fluid aspirated from the subcutaneous abscess contained three types of bacteria on stained smears. Cultures grew peptostreptococci, *Bacteroides* sp. and *Escherichia coli*.

Based on the normal distribution of the anaerobic cocci and past clinical experience, we can predict with some certainty where infections with these organisms are most likely to arise. Anticipation of suspected isolates can lead to rapid handling of specimens to ensure the best possible chances of isolating anaerobic bacteria, even in a crowd of facultatives. The typical patterns of spread of anaerobic coccal infections from their characteristic locations are illustrated by Figure XX-2. Differences in factors influencing the flora may produce widely varying types of infections. These include physical factors, e.g. trauma, surgery or radiation; chemical factors, e.g. diabetic or uremic acidosis, local toxins, steroids and antimetabolites; and biological factors, such as malignancy, facultative bacterial growth, microaspiration and childbirth. The most common infectious problems arising from the anaerobic cocci in the cutaneous, oral, respiratory, intestinal and female genital areas are as follows: local cellulitis or abscess; local septic thrombophlebitis; diffuse contiguous infection; distant abscess; and endocarditis.

Table XX-II illustrates the variety of underlying problems that are common in anaerobic coccal infections. The data are based on our experience at Massachusetts General Hospital over a 45-month period (Bornstein *et al.*, 1964). The hospital bacteriology laboratory was not specifically programmed for isolation of anaerobes, and also many specimens undoubtedly lingered in the operating room and elsewhere, exposed to ambient oxygen. The lack of anaerobic coccal isolates from brain abscesses reflects failure of our neurosurgical colleagues to handle such specimens appropriately; sterile abscesses were frequently encountered. The absence of isolates of anaerobic cocci from lower respiratory tract infections (pneumonia, abscess) similarly indicates a lack of aggressiveness in obtaining material from which anaerobes surely would be isolated.

COMMENTS

Awareness of the importance of anaerobic organisms in human infection is at the top of the list of factors leading

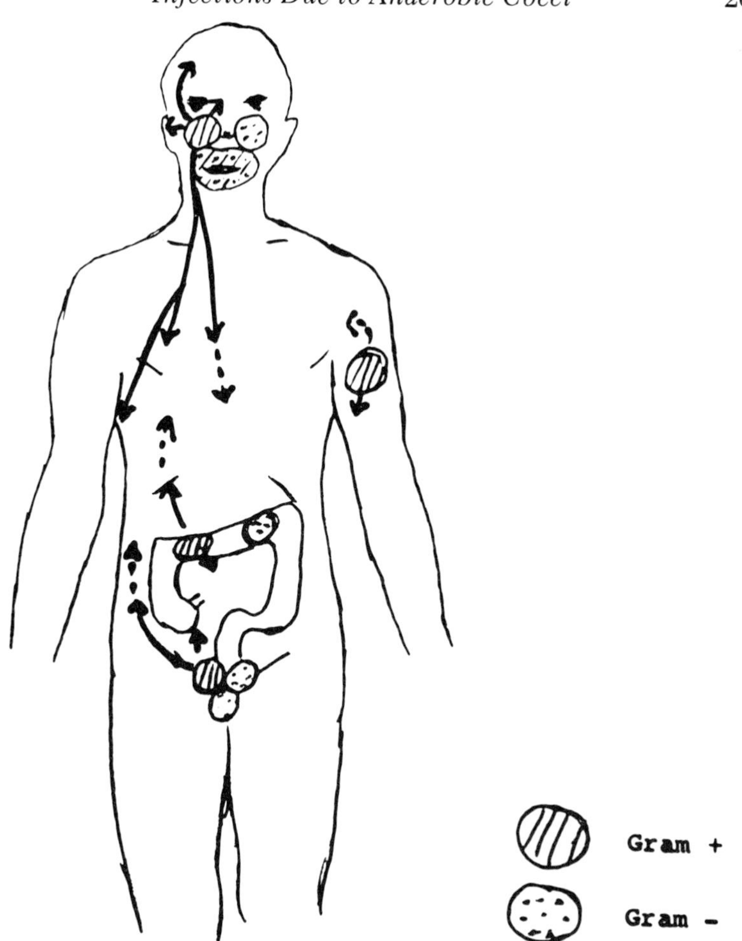

Figure XX-2. Local (solid lines) and distant (broken lines) spread of anaerobic cocci from normal sites of distribution to skin and subcutaneous tissues, head, thorax and abdominal areas.

to widespread identification of anaerobes in human infections. For example, many surgeons now specifically request primary anaerobic plating and Gram staining of pus evacuated from the variety of lesions characteristically involving anaerobes. Publicity given to such infectious problems by clinicians and bacteriologists will broaden the data base we use to study in detail the biology, pathogenesis and treatment of such infections, and we hope it will lead to methods of

TABLE XX-II

ANAEROBIC STREPTOCOCCAL INFECTIONS (77 CASES)*

Site of Culture	*Site of Underlying Disease*	
Blood (2)	Cholangitis	1
	Septic abortion	1
Abscess (67)	Perirectal	13
	Abdominal wall	3
	Intraabdominal	8
	Superficial—pilonidal or sebaceous cyst, boil	31
	Finger	1
	Breast	3
	Female pelvic	3
	Empyema	3
	Neck and dentoalveolar	2
Wound (8)	Skull and head	2
	Extremity	2
	Abdominal wall—synergistic gangrene	3
	Perineal	1

* Reprinted from *Medicine, 43:* 207, 1964.

avoiding the breakdown of host defenses that results in these opportunistic infections.

Lest we become too proud of the advances in our knowledge of anaerobic coccal infections, it might be well to remind ourselves that the lessons from the early descriptions of infections due to anaerobic cocci (Veillon, 1893; Schottmuller, 1910; Schwarz and Dieckmann, 1927) were largely forgotten until brought back into focus by recent investigators (Heineman and Braude, 1963; Rotheram and Schick, 1969; Bartlett and Finegold, 1972). I am sure that the lessons we are learning here will not be lost to history. As current efforts to learn more about the anaerobic cocci, their exact speciation, distribution, pathogenicity and factors that activate them in infection, continue in the special anaerobic laboratory, we should also continually modify and simplify procedures done as routine diagnostic studies in hospital settings. In this way the clinician and hospital bacteriologist will clearly see their roles as being achievable and practical.

REFERENCES

Bartlett, J.G., and Finegold, S.M.: Anaerobic pleuropulmonary infections. *Medicine, 51*:6, 1972.

Bornstein, D.L., Weinberg, A.N., Swartz, M.N., and Kunz, L.J.: Anaerobic infections—review of current experience. *Medicine, 43:*207, 1964.

Finegold, S.M., Rosenblatt, J.E., Sutter, V.L., and Attebery, H.R.: *Anaerobic Infections.* Scope Monograph. Kalamazoo, Upjohn, 1972.

Heineman, H.S., and Braude, A.I.: Anaerobic infection of the brain. Observations on eighteen consecutive cases of brain abscess. *Am J Med, 35:*682, 1963.

Hite, K.E., Locke, M., and Hesseltine, H.C.: Synergism in experimental infections with nonsporulating anaerobic bacteria. *J Infect Dis, 84:*1, 1949.

Holdeman, L.V., and Moore, W.E.C., Eds.: *Anaerobe Laboratory Manual.* Blacksburg, Virginia Polytechnic Institute and State Univ., 1972.

Meleney, F.L.: Bacterial synergism in disease processes with a confirmation of the synergistic bacterial etiology of a certain type of progressive gangrene of the abdominal wall. *Ann Surg, 94:*961, 1931.

Pien, F.D., Thompson, R.L., and Martin, W.J.: Clinical and bacteriologic studies of anaerobic gram-positive cocci. *Mayo Clin Proc, 47:*251, 1972.

Rogosa, M.: *Peptococcaceae*, a new family to include the gram-positive anaerobic cocci of the genera *Peptococcus, Peptostreptococcus,* and *Ruminococcus. Int J Syst Bacteriol, 21:*234, 1971.

Rosebury, T.: *Microorganisms Indigenous to Man.* New York, McGraw-Hill, 1962.

Rotheram, E.B., Jr., and Schick, S.E.: Nonclostridial anaerobic bacteria in septic abortion. *Am J Med, 46:*80, 1969.

Schottmuller, H.: Zur bedentung einiger anaeroben in der pathologie, insbesondere bei peuperalen erkrankungen. *Mitt Grenzget Med Chir, 21:*450, 1910.

Schwarz, O.H., and Dieckmann, W.J.: Puerperal infection due to anaerobic streptococci. *Am J Obstet Gynecol, 13:*467, 1927.

Sutter, V.L., Attebery, H.R., Rosenblatt, J.E., Bricknell, K.S., and Finegold, S.M.: *Anaerobic Bacteriology Manual.* Los Angeles, Univ. of California, LA Ext. Div., 1972.

Veillon, M.A.: Sur un microcoque anaerobie trouve dams des suppurations feltides. *C R Soc Biol (Paris), 45:*807, 1893.

CHAPTER XXI

ASPECTS OF THE PATHOGENICITY AND ECOLOGY OF THE INDIGENOUS ORAL FLORA OF MAN

RONALD J. GIBBONS

ABSTRACT: *One aspect of pathogenicity of anaerobic bacteria reviewed in this paper is the key role of* Bacteroides melaninogenicus *in certain synergistic experimental infections initiated by the indigenous oral flora. Addition of* B. melaninogenicus *transformed avirulent mixtures of indigenous bacteria into infectious complexes. Studies on the localization of* B. melaninogenicus *and other bacterial species in the mouth have led to recognition of an ecological determinant which has implications for all mucosal pathogens: in environments which contain surfaces exposed to a fluid flow, bacteria must either adhere to the surface, or grow at a rate which exceeds the dilution rate resulting from the fluid flow, if they are to colonize successfully. Since indigenous bacteria on mucosal surfaces are likely to be growing slowly, their ability to adhere to a freely exposed surface or become entrapped in a protected niche is essential for their colonization. There are numerous examples of positive correlations between virulence and adherence properties. Recognition of the specificity of bacterial adherence to mucosal surfaces and its function as a prerequisite for colonization has provided a novel hypothesis to explain how secretory immunoglobulins present in secretions bathing mucous membranes can interfere with microbial colonization and provide protective immunity which serves as a primary defense mechanism. Secretory immunoglobulin antibodies can specifically bind to the surfaces of bacteria to which they are directed, causing their agglutination, thus interfering with adherence of bacteria to mucosal surfaces and limiting colonization.*

INFECTIOUS POTENTIAL OF THE FLORA INDIGENOUS TO MAN

The indigenous bacterial flora of the oral cavity, the gastrointestinal canal and other mucous surfaces of man is complex. *Bacteroides* species, anaerobic vibrios, anaerobic cocci, fusobacteria, anaerobic diphtheroids and spirochetes together with facultative streptococci and diphtheroids are usually prominent (Socransky and Manganiello, 1971; Donaldson, 1964). Because these organisms are normal residents of mucous surfaces, they too often are considered to be innocuous. However, it is clear that man's indigenous flora has infectious potential. This can be demonstrated by inoculating experimental animals subcutaneously with bacterial scrapings from the oral cavity, or with human fecal suspensions (Shpuntoff and Rosebury, 1949; Rosebury *et al.*, 1950; Macdonald *et al.*, 1963; Socransky and Gibbons, 1965; Courant *et al.*, 1965). Necrotic infections, characterized by a mixed bacterial flora in which nonsporulating anaerobes are prominent, invariably develop. These infections may be of two types (Macdonald *et al.*, 1963; Socransky and Gibbons, 1965). One type remains localized as a purulent abscess, while the second is characterized by gangrenous spreading usually accompanied by bacteremia. Exudate from such experimentally induced lesions contains a complex mixture of bacteria analogous to that present in the original inoculum. Such exudate may be used to serially transmit the infection to additional animals. Throughout serial transmission, the anaerobic segments of the flora persist.

Necrotic infections characterized by a mixed bacterial flora in which nonsporulating anaerobes are prominent may occur spontaneously throughout the body. Such mixed anaerobic infections include peritonitis, appendicitis, lung abscesses, necrotic lesions of the skin, and infected surgical and bite wounds (Oliver and Wherry, 1921; Cohen, 1932; Altemeier, 1938; Weiss, 1943; Heinrich and Pulverer, 1960). These endogenous infections attest to the pathogenic potential of the bacterial flora indigenous to man's mucous surfaces.

Most investigators have found that pure cultures of indigenous bacteria lack infectious potential when inoculated subcutaneously in experimental animals. Similarly, pure cultures of organisms isolated from naturally occurring mixed anaerobic infections are almost always noninfective (Hite *et al.*, 1949; Macdonald *et al.*, 1954; Meleney, 1931; Altemeier, 1942), although their inoculation may produce caseous nodules. However, the entire bacterial complex consistently displays pathogenic potential, and a number of investigators have successfully reproduced transmissible infections in experimental animals using defined mixtures of bacteria (Macdonald *et al.*, 1963; Socransky and Gibbons, 1965; Hite *et al.*, 1949; Meleney, 1931; Macdonald *et al.*, 1956; Kestenbaum *et al.*, 1964; Altemeier, 1942). Thus, these mixed anaerobic infections provide a clear-cut example of bacterial synergism in the production of disease.

Several investigators have arrived at different defined combinations of bacteria which initiate experimental anaerobic infections. In addition, a variety of bacterial species have been isolated from naturally occurring necrotic lesions. Consequently, these infections give the impression that they are bacteriologically nonspecific. However, there are experimental data which suggest that they may exhibit more bacterial specificity than is currently recognized, and that *Bacteroides melaninogenicus* may be a key pathogen in many infections initiated by the indigenous flora of man.

Key Role of *B. melaninogenicus* in Experimental Synergistic Anaerobic Infections

Organisms of the *B. melaninogenicus* group are commonly found in the intestinal canal, and on male and female genitalia (Burdon, 1928; van Houte and Gibbons, 1966). This species is ubiquitously present in the oral cavity of adults where it comprises approximately 5 percent of the total bacteria cultivable from the gingival crevice area (Gibbons *et al.*, 1963). The organism is regularly recovered from experimental mixed infections initiated in animals inoculated with bacterial scrapings from the gingival crevice area (Macdonald *et al.*, 1963; Socransky and Gibbons, 1965; Macdonald *et al.*, 1956) or with suspensions of fecal material (Socransky and Gibbons, 1965). It has been isolated from a wide variety of naturally occurring mixed anaerobic infections of the head and neck, the chest, the abdomen and extremities (Oliver and Wherry, 1921; Weiss, 1943; Heinrich and Pulverer, 1960; Burdon, 1928). The organism is rarely recovered in pure culture; rather, it is present with other anaerobic and facultative forms. Most strains of *B. melaninogenicus* require hemin for growth and many isolates also require an analogue of vitamin K (Lev, 1959; Gibbons and Macdonald, 1960; Rizza *et al.*, 1968). Although compounds of the vitamin K group are produced by several types of bacteria, they are not present in sufficient quantities in most bacteriological culture media to support growth of the organism. Consequently, on primary isolation on blood agar plates, *B. melaninogenicus* is frequently found exhibiting satellite growth around colonies of bacteria which elaborate vitamin K (Burdon, 1928; Gibbons and Macdonald, 1960). However the organism is readily lost upon subculture in media which are not supplemented with hemin and a vitamin K analogue. Thus the organism may easily be missed in routine bacteriological examinations, even those which utilize adequate anaerobic techniques.

The key role of *B. melaninogenicus* in certain mixed anaerobic infections was suggested by the fact that defined infectious mixtures of indigenous bacteria delineated by several investigators contained it as a common component (Hite *et al.*, 1949; Macdonald *et al.*, 1956; Kestenbaum *et al.*, 1964). Subsequently, experiments were designed to compare the infectivity of complex mixtures of indigenous bacteria which contained *B. melaninogenicus* with mixtures which did not. Several approaches have been used to delete this organism from naturally occurring mixtures. In one experiment, all of the bacterial colonies present on an anaerobically incubated blood agar plate inoculated with a high dilution of gingival crevice debris were isolated (Socransky and Gibbons, 1965). Each colony was subcultured, and the mixture recombined after colonies of *B. melaninogenicus* were specifically deleted. A mixture of 56 strains of indigenous bacteria, which included peptostreptococci, *Veillonella* species, fusobacteria, anaerobic vibrios, anaerobic diphtheroids and facultative streptococci and diphtheroids, was found to be noninfective. However, when a strain of *B. melaninogenicus* was added to this mixture, the resulting complex proved uniformly infective (Socransky and Gibbons, 1965). Animals inoculated with the *B. melaninogenicus* strain alone failed to develop lesions, thereby demonstrating the synergistic nature of this infectious mixture.

Other methods used to delete *B. melaninogenicus* from naturally occurring mixtures of bacteria were based upon the organism's requirement for hemin (Macdonald *et al.*, 1963;

Socransky and Gibbons, 1965). Mixed suspensions of bacteria from the gingival crevice area, or from human feces, were inoculated as drops on plates of heart infusion agar with and without the addition of blood. The resulting mixed bacterial growth was serially transferred on the blood-free medium to achieve the elimination of *B. melaninogenicus* and other hemin requiring organisms. However these organisms persisted in the bacterial mixtures growing on blood agar. A comparison of the infectivity of these mixed bacterial growths indicated that mixtures containing naturally occurring *B. melaninogenicus* were capable of producing transmissible necrotic abscesses in experimental animals. However, complex mixtures of indigenous oral bacteria lacking this organism were avirulent (Macdonald *et al.*, 1963; Socransky and Gibbons, 1965). It is, of course, likely that the mixtures selected by growth on these two media contained microbial differences other than simply the presence or absence of *B. melaninogenicus*. However, since the addition of a pure culture of *B. melaninogenicus* to the avirulent mixture transformed it into an infectious complex, these studies again suggest that complex mixtures of indigenous bacteria are usually noninfective in the absence of this organism. The hypothesis emerging from these studies implies that *B. melaninogenicus* plays a dominant and likely an essential role in certain mixed anaerobic infections initiated by the indigenous flora of man.

While most strains of *B. melaninogenicus* prove noninfective when tested in pure culture (Macdonald *et al.*, 1956, 1963; Hite *et al.*, 1949; Socransky and Gibbons, 1965; Burdon, 1928), a few strains have been found which are clearly pathogenic by themselves (Macdonald *et al.*, 1963; Takazoe *et al.*, 1971; Pulverer and Heinrich, 1960). One of these strains has been reported to elaborate an antigenic capsule which appeared to be related to the organism's infectivity (Takazoe *et al.*, 1971). This structure was not apparent on typical strains of *B. melaninogenicus*.

Organisms considered to be *B. melaninogenicus* represent a heterogeneous group (Oliver and Wherry, 1921; Sawyer *et al.*, 1962; Courant and Gibbons, 1967; Schwabacher *et al.*, 1947; Pulverer and Heinrich, 1960; Weiss, 1937; Reddy and

Bryant, 1967). Some strains fail to ferment carbohydrates, while others exhibit diversity in the number and type of sugars fermented. Strains also exhibit serological heterogeneity (Courant and Gibbons, 1967) and various strains possess different cytochromes (Reddy and Bryant, 1967; Rizza *et al.*, 1968). Other examples of the heterogeneous nature of the group are the differences in nutritional requirements mentioned above, and differences in the reported proteolytic activity of strains. Regarding pathogenicity, fermentative and nonfermentative strains, representing different serotypes, can substitute for each other in defined mixtures of indigenous bacteria which initiate synergistic infections (Macdonald *et al.*, 1963). However, the number of strains tested have been too few to ascribe pathogenic potential to all isolates designated as *B. melaninogenicus*.

The organism has been frequently identified primarily on the basis of forming distinctive black-pigmented colonies when grown on blood agar plates. Oliver and Wherry (1921) believed the pigment to be melanin, and hence the organism's descriptive epithet. Later, Schwabacher *et al.* (1947) presented spectral data suggesting that the pigment was a hemin derivative, and recommended the organism be renamed *B. nigrescens*. Recently, Tracy (1969) has reported that strains of *B. melaninogenicus* and some other *Bacteroides* species may, under certain conditions, produce a black chromogen consisting of ferrous sulfide. Strains of *B. fragilis* and *B. necrophorus* frequently produce this chromogen in mixed culture, but not in pure culture, whereas pure colonies of *B. melaninogenicus* consistently form pigment on blood agar. Tracy (1969) suggested that the pigments observed by Schwabacher *et al.* (1947) were likely to be cytochrome components, and that other *Bacteroides* species may have been erroneously considered *B. melaninogenicus* upon primary isolation by investigators in the past.

It is not clear how *B. melaninogenicus* and associated bacteria cause infection. Strains of *B. melaninogenicus* have been shown to contain potent endotoxins (Mergenhagen *et al.*, 1961; Hofstad, 1968, 1970) and to produce deoxyribonuclease, hydrogen sulfide, indol, and copious amounts of ammonia

(Schwabacher *et al.*, 1947; Macdonald *et al.*, 1963; Sawyer *et al.*, 1962). The organisms grow well in peptone-containing media in the absence of carbohydrates, and they vigorously ferment nitrogenous constituents (Wahren and Gibbons, 1970). Most strains are highly proteolytic, exhibiting activity on a variety of proteins including casein, fibrin and native collagen (Weiss, 1943; Sawyer *et al.*, 1962; Pulverer and Heinrich, 1960; Burdon, 1932). The collagenolytic activity of the organism is of special interest, since it has been suspected of playing a role in the tissue destruction resulting from anaerobic infection. Strains of *B. melaninogenicus* hydrolyze native collagen from rat tails, guinea pig skin, human gingiva and human dentine (Gibbons and Macdonald, 1961; Waldvogel and Swartz, 1969; Hausmann *et al.*, 1967; Kerebel and Sedallian, 1972). Strains with high collagenolytic activity produce more acute infections in combination with anaerobic vibrios than strains with weak collagenolytic activity (Kestenbaum *et al.*, 1964). Recently, Kaufman *et al.* (1972) demonstrated that cell-free extracts of *B. melaninogenicus* possessing collagenolytic activity when given with a live *Fusobacterium* species produced nontransmissible lesions in rabbits which were more severe than those resulting from injection of either the extract or the organism separately. Additional studies of the biological effects of the collagenolytic enzyme system of *B. melaninogenicus* and of other metabolites governing the organism's pathogenic potential should prove profitable.

Studies of the Localization of Bacteria in Various Sites in the Mouth

It has been recognized for several years that certain bacterial species preferentially colonize different sites in the mouth (Table XXI-I). For example, *B. melaninogenicus* and spirochetes are found in highest proportions in the gingival crevice area, whereas they comprise a much lower proportion of the bacteria present on the tongue and cheek surfaces (Loesche and Gibbons, 1966). The gingival localization of *B. melaninogenicus*, together with its demonstrated path-

TABLE XXI-I

PROPORTIONAL DISTRIBUTION OF BACTERIA ON VARIOUS ORAL SURFACES

Organism	*Gingival Crevice*	*Coronal Tooth Surface*	*Tongue*	*Cheek*
B. melaninogenicus	4–8*	<1	<1	<1
Veillonella species	5–15	1–3	10–15	<1
S. mutans	?	0–40	<1	<1
S. sanguis	5–10	8–11	2–6	7–14
S. miteor	5–10	8–11	4–8	55–65
S. salivarius	<1	<1	15–25	7–14

* Data calculated as % of total flora cultivable on anaerobic blood agar. From van Houte *et al.*, 1970, 1971; Liljemark and Gibbons, 1971, 1972; Gibbons *et al.*, 1963, 1964; Loesche and Gibbons, 1966; Krasse, 1954.

ogenic potential, has suggested its possible involvement in suppurative periodontal disease (Macdonald *et al.*, 1963; Macdonald and Gibbons, 1962). The populations of oral streptococci (Gibbons *et al.*, 1964; Krasse, 1954; Carlsson, 1967; van Houte *et al.*, 1971; Lijemark and Gibbons, 1972), as well as of *Veillonella* species (Liljemark and Gibbons, 1971) also exhibit marked patterns of localization on various oral surfaces. The reasons for this selective bacterial colonization in various oral niches have never been clear. It was generally attributed to differences in nutrient availability which were postulated to exist between various sites in the mouth. However, recent studies of the nutritional requirements of oral streptococci have indicated an overall nutritional similarity among species (Carlsson, 1972), and the data do not substantiate nutritional parameters as being primarily responsible for the localization of these organisms within the mouth.

Our laboratories have recently found that bacteria differ widely in their ability to adhere to teeth and to various oral epithelial surfaces (van Houte *et al.*, 1971; Liljemark and Gibbons, 1971, 1972; van Houte *et al.*, 1970; Gibbons and van Houte, 1971). The relative adherence of all of several species studied for any given oral surface has been found to correlate with the proportions of the organisms found indigenously. Three experimental approaches have been used to study bacterial adherence to oral surfaces. In *in vitro* models, human buccal epithelial cells or hamster cheek pouches have been incubated with suspensions of bacteria under standardized conditions (Gibbons and van Houte, 1971; Gibbons *et al.*, 1972). The number of bacteria which attached to the epithelial cells has been determined by either direct microscopy, or by using isotopically-labeled bacteria. An *in vivo* approach has entailed the introduction of mixtures of streptomycin-resistant bacteria into the mouths of volunteers for short periods of time (Gibbons and van Houte, 1971; van Houte *et al.*, 1971; Liljemark and Gibbons, 1971, 1972). After the mouth is thoroughly rinsed, the ratios of labeled bacteria present on various oral surfaces have been determined using culture media supplemented with streptomycin, and compared to the ratios of the organisms contained in the original mixture. Finally, with some species which are naturally present in significant proportions in saliva, it has been possible to compare their adherence to teeth which have had their bacterial populations reduced to negligible proportions (van Houte *et al.*, 1970; Liljemark and Gibbons, 1971). With each method, and with all bacterial species studied, it has been found that the relative adherence of an organism to a given surface correlates directly with proportions of that organism found indigenously on the surface. For example, *Streptococcus sanguis* and *S. mitis* have been found to possess a marked ability to adsorb to teeth, and these species are found indigenously in high proportions on tooth surfaces (van Houte *et al.*, 1970, 1971; Lijemark and Gibbons, 1972). In contrast, *S. salivarius* and *Veillonella* species have been observed to adhere weakly to the tooth surface, correlating with their low proportions present naturally on teeth (van Houte *et al.*, 1970, 1971; Liljemark and Gibbons, 1971). However, these organisms adhere well to the tongue surface, reflecting their oral localization.

These studies have revealed a simple but hitherto unappreciated ecological determinant. In environments which contain surfaces exposed to a fluid flow, bacteria must either adhere to the surface or grow at a rate which exceeds the dilution rate resulting from the fluid flow, if they are to colonize successfully. Otherwise, the organisms are simply washed away. A flowing stream, the oral cavity, the nasopharyngeal area and portions of the intestinal canal represent examples of such environments. Since indigenous bacteria on mucosal surfaces are likely to be growing slowly (Gibbons, 1964; Gibbons and Kapsimalis, 1967), their ability to adhere to a freely exposed surface or to become mechanically entrapped in a protected niche is essential for their colonization. In the mouth, the periodontal pocket represents an example of a protected, stagnant area, and this appears to be the site where highest proportions of vibrios, spirochetes, and other motile forms are found (Socransky *et al.*, 1963; Loesche and Gibbons, 1966). The relatively slow overall rate of bacterial growth which is believed to occur in many natural environments, including the mouth and gastrointestinal canal (Gibbons, 1964; Gibbons and Kapsimalis, 1967; El-Shazly and Hungate, 1965), would tend to minimize microbial population shifts due to different growth rates of various species. This would be particularly true for epithelial surfaces which are continuously shed, and consequently bacteria may not proliferate for extended periods of time on any given cell surface. Rather, organisms must continuously reattach to these ever-renewing surfaces to colonize. Hence, the relative proportions of bacteria found on epithelial surfaces would seem almost entirely dependent upon the number of cells available for attachment and the organism's inherent capacity to attach.

Adherence in the Ecology of Mucosal Pathogens

The importance of bacterial adherence in the ecology of indigenous oral bacteria implies that it should also be important in the colonization of overt pathogens on the mucous surfaces of man. Several investigators have reported a correlation between the ability to adhere to the lumen of the intestine and the virulence of strains of *Shigella* and *E.coli* (Labrec *et al.*, 1964; Arbuckle, 1970; Drucker *et al.*, 1967; Bertschinger *et al.*, 1972). Pathogenic strains of *Mycoplasma* species have also been found to adhere to tracheal cells while nonpathogenic strains generally do not (Sobeslansky *et al.*, 1970). However, the involvement of adherence as a prerequisite for colonization of these organisms has not been appreciated. Similarly, almost all of the gram-negative bacteria isolated from bladder infections have been found to be piliated (Brinton, 1967). Although pili are well recognized to impart adhesive qualities to bacteria, the likely involvement of pili in facilitating attachment of these organisms to the epithelial lining of the bladder so as to foster colonization has not been considered.

We have recently found that virulent strains of *Streptococcus pyogenes* adhered well to human oral epithelial cells (Ellen and Gibbons, 1972). However, an avirulent strain lacked this ability. *S. pyogenes* possesses an antigenic surface component, designated M protein, which serves as the type specific antigen for this species (Lancefield, 1962). It has been known for years that the presence of M protein correlates with the virulence of *S. pyogenes* strains. We have found that tryptic removal of M protein from virulent strains of *S. pyogenes* markedly impaired their ability to adhere to epithelial surfaces (Ellen and Gibbons, 1972). An avirulent mutant of *S. pyogenes* lacking M protein also adhered feebly, and this avirulent strain was cleared rapidly from the mouths of mice in relation to its M positive virulent parent strain (Ellen and Gibbons, 1972). These observations, plus the finding that type specific anti-M serum is able to inhibit the adherence of *S. pyogenes* to epithelial surfaces, indicate that the M protein surface component of this organism functions in its adherence to epithelial surfaces, thereby fostering its colonization. It would appear that the virulence of a wide range of mucosal pathogens is influenced by their ability to adhere to epithelial surfaces. This influences the extent to which they may colonize and subsequently produce disease.

Inhibition of Bacterial Adherence by Secretory Immunoglobulins

Recognition of the specificity of bacterial adherence to mucosal surfaces and its function as a prerequisite for colonization has provided a novel hypothesis to explain how immunoglobulins present in bathing secretions may interfere with microbial colonization, and provide protective immunity. Secretory immunoglobulin A (S-IgA) is the predominant class of antibody in secretions which bathe mucous membranes (Thomasi, 1970). S-IgA has been demonstrated to serve a protective function against viral infections, but the mechanism by which S-IgA antibodies may function in the disposal of bacterial antigens is not understood. This is because S-IgA is not generally considered to be bactericidal, to mediate complement dependent bacterial lysis, or to enhance phagocytosis (Thomasi, 1970). Since mucinous glycoproteins in secretions generally possess anti-complement activity, and few active leukocytes survive in the hypotonic environment of secretions, it would seem a priore that complement fixing or opsonizing activities would not likely be important activities for immunoglobulins in secretions. However, S-IgA antibodies are capable of specifically binding to the surfaces of bacteria to which they are directed, and of causing their agglutination. Consequently they have the potential to interfere with the adherence of bacteria to mucosal surfaces, thereby limiting colonization. We have recently shown that S-IgA isolated from human parotid saliva does specifically inhibit the adherence of various streptococcal species to epithelial cells *in vitro* (Williams and Gibbons, 1972). Because continued colonization of a mucosal surface *in vivo* requires bacterial reattachment, it seems likely that if S-IgA antibodies inhibited the adherence of one serotype of an organism relative to another, it would lead to the rapid elimination of the affected organism. Because the effect of a continuous small degree of inhibition in adherence would be multiplied by the need for reattachment on the ever-renewing epithelial surface, this mechanism of bacterial elimination would be highly efficient, and require only small amounts of S-IgA. This may explain why only low titers of S-IgA and other immunoglobulins are usually detected in secretions.

The studies of Freter (1969, 1970) also substantiate the concept that secretory antibodies may exert a protective action by inhibiting bacterial adherence. He observed that induced coproantibodies (which are primarily S-IgA) were protective against infection by *Vibrio cholerae*. When immune animals were challenged with *V. cholerae*, the organisms tended to remain free in the lumen whereas in nonimmune animals they adsorbed to the mucosal lining. Watson and coworkers (1946) have obtained data concerning *Streptococcus pyogenes* which are also of interest in this regard. They observed that when monkeys were inoculated intranasally with virulent strains of *S. pyogenes*, the organisms colonized the animals for weeks before being eliminated. After the infection resolved, and the animals were again challenged with the same M type of *S. pyogenes*, the organism was cleared rapidly and was generally undetected 24 hrs after inoculation. However, if the immune animals were challenged with a different M type of *S. pyogenes*, the serologically different strain colonized for several weeks. These studies therefore indicate that a type-specific immunity affecting bacterial colonization occurs on the mucosal surfaces of primates, and that the basis of this immunity apparently entails the more rapid clearance of affected organisms, thereby preventing their colonization. Our recent data concerning the involvement of M protein in the adherence of *S. pyogenes* to epithelial cells, and the observation that type-specific antiserum inhibits adherence (Ellen and Gibbons, 1972), provide additional support for this view.

A logical question arises concerning the colonization of indigenous bacteria on mucosal surfaces. Should not their colonization also be influenced by S-IgA, since it is recognized that secretions contain low titers of immunoglobulins directed against indigenous bacteria (Sirisinha, 1970)? In fact, it appears that most indigenous bacteria are in a state of suppressed adherence (Williams and Gibbons, 1972). If one examines epithelial cells scraped directly from human cheeks, they are found to harbor only 10 to 15 bacteria per cell, even though they are continually exposed to concentrations of streptococci and other indigenous bacteria approaching 10^8 organisms per ml of saliva. If these epithelial cells are washed and incubated with suspensions of salivary streptococci grown *in vitro*, hundreds of bacteria attach to each epithelial cell within a few minutes (Gibbons and van Houte, 1971; Williams and Gibbons, 1972). Thus it appears that the adherence of many indigenous bacteria *in vivo* is suppressed, and it is also known that bacteria present in human saliva are coated with S-IgA (Brantzaeg *et al.*, 1968).

Most discussions of the ecology of infectious agents lead one to believe that small numbers of organisms are introduced as droplets on a mucous surface, and the organisms are presumed to overgrow indigenous bacteria present. However, there are few convincing data concerning most pathogens which would lead one to believe that either their nutritional requirements or their potential growth rates are sufficiently different from indigenous bacteria to produce this numerical predominance. If one considers the adherence of the indigenous flora to be partially suppressed by specific secretory antibodies, then exogenously introduced, serotypically distinct pathogens would be expected to enjoy a temporary immunologic advantage whereby they could adhere and colonize, unimpeded by secretory antibodies. This selective advantage could enable small numbers of exogenous pathogens to emerge to numerical predominance within a relatively short period of time. Once they attained a mass or induced tissue damage sufficient to strongly stimulate production of secretory antibodies, their adherence would be inhibited to a greater extent than that of the indigenous flora. Consequently, the pathogen would be selectively eliminated, and an immune state would exist. However, should colonization progress slowly and evoke only a weak antigenic challenge, the small amounts of S-IgA produced would affect the pathogen comparably to indigenous organisms, and a transient balanced or "carrier" state would result. This potential ability of secretory immunoglobulins to influence the ecology of both indigenous bacteria and overt pathogens also can explain the emergence and disappearance of specific serotypes or phagetypes of bacteria which has been observed to occur on mucous surfaces (Williams and Gibbons, 1972). This concept also sheds light as to why surface components of bacteria are frequently the most important antigens to which protective immunity is directed.

The hypothesis proposed can explain how secretory immunoglobulins may function in the disposal of bacterial antigens, and affect the ecology of indigenous and pathogenic bacteria, without possessing the bactericidal or opsonizing qualities of serum immune systems. The secretions bathing mucous surfaces have long been recognized to serve a cleansing function. The concepts advanced imply that immunoglobulins in these secretions facilitate the clearance of bacteria, and as such provide the basis of a primary defense mechanism operative on mucosal surfaces. Should organisms elude this defense and subsequently invade the host tissues, systemic defense mechanisms including serum antibodies, complement, phagocytosis, and other recognized systemic factors of immunity would then assume importance.

REFERENCES

Altemeier, W.A.: The bacterial flora of acute perforated appendicitis with peritonitis. *Ann Surg, 107:*517, 1938.

Altemeier, W.A.: The pathogenicity of the bacteria of appendicitis peritonitis. *Surgery, 11:*374, 1942.

Arbuckle, J.B.R.: The localization of *Escherichia coli* in the pig intestine. *Med Microbiol, 3:*333, 1970.

Bertschinger, H.V., Moon, H.W., and Whipp, S.C.: Association of *E. coli* with the small intestinal epithelium. II. Variations in association index and the relationship between association index and enterosorption in pigs. *Infect and Immunity, 5:*606, 1972.

Brantzaeg, P., Fjellander, J., and Gjeruldsen, S.T.: Adsorption of immunoglobulins onto oral bacteria *in vivo. J Bacteriol, 96:*242, 1968.

Brinton, C.C.: Contributions of pili to the specificity of the bacterial surface. In: Davis, B.D., and Warren, L. (Eds.): *The Specificity of Cell Surfaces.* Englewood Cliffs, Prentice Hall, 1967, p. 37.

Burdon, K.L.: *Bacterium melaninogenicum* from normal and pathologic tissues. *J Infect Dis, 42:*161, 1928.

Burdon, K.L.: Isolation and cultivation of *Bacterium melaninogenicum. Proc Soc Exp Biol Med, 29:*1144, 1932.

Carlsson, J.: Nutritional requirements of *Streptococcus sanguis. Arch Oral Biol, 17:*1327, 1972.

Carlsson, J.: Presence of various types of nonhaemolytic streptococci in dental plaque and in other sites of the oral cavity of man. *Odontol Revy, 18:*55, 1967.

Cohen, J.: The bacteriology of abscess of the lung and methods for its study. *Arch Surg, 24:*171, 1932.

Courant, P.R., and Gibbons, R.J.: Biochemical and immunological heterogeneity of *Bacteroides melaninogenicus. Arch Oral Biol, 12:*1605, 1967.

Courant, P.R., Paunio, I., and Gibbons, R.J.: Infectivity and hyaluronidase activity of gingival crevice debris. *Arch Oral Biol, 10:*119, 1965.

Donaldson, R.M.: Normal bacterial populations of the intestine and their relation to intestinal function. *N Engl J Med, 270:*938, 994, 1050, 1964.

Drucker, M.M., Yeivin, R., and Sacks, T.G.: Pathogenesis of *Escherichia coli* enteritis in the ligated rabbit gut. *Isr J Med Sci, 3:*445, 1967.

Ellen, R.P., and Gibbons, R.J.: M-protein associated adherence of *Streptococcus pyogenes* to epithelial surfaces: a prerequisite for virulence. *Infect and Immunity, 5:*826, 1972.

El-Shazly, K., and Hungate, R.E.: Fermentation capacity as a measure of net growth of rumen microorganisms. *Appl Microbiol, 13:*62, 1965.

Freter, R.: Mechanism of action of intestinal antibody in experimental cholera. II. Antibody-mediated antibacterial reaction at the mucosal surface. *Infect and Immunity, 2:*556, 1970.

Freter, R.: Studies on the mechanism of action of intestinal antibody in experimental cholera. *Tex Rep Biol Med, 27:*299, 1969.

Gibbons, R.J.: The bacteriology of dental caries. *J Dent Res, 43:*1021, 1964.

Gibbons, R.J., and Kapsimalis, B.: Estimates of the overall rate of growth of the intestinal microflora of hamsters, guinea pigs and mice. *J Bacteriol, 93:*510, 1967.

Gibbons, R.J., Kapsimalis, B., and Socransky, S.S.: The source of salivary bacteria. *Arch Oral Biol, 9:*101, 1964.

Gibbons, R.J., and Macdonald, J.B.: Degradation of collagenous substrates by *Bacteroides melaninogenicus. J Bacteriol, 81:*614, 1961.

Gibbons, R.J., and Macdonald, J.B.: Hemin and vitamin K compounds as required factors for cultivation of certain strains of *Bacteroides melaninogenicus. J Bacteriol, 80:*164, 1960.

Gibbons, R.J., Socransky, S.S., Sawyer, S., Kapsimalis, B., and Macdonald, J.B.: The microbiota of the gingival crevice area of man. II. The predominant cultivable organisms. *Arch Oral Biol, 8:*281, 1963.

Gibbons, R.J., van Houte, J., and Liljemark, W.F.: Some parameters effecting the adherence of S. *salivarius* to oral epithelial surfaces. *J Dent Res, 51:*424, 1972.

Gibbons, R.J., and van Houte, J.: Selective bacterial adherence to oral epithelial surfaces and its role as an ecological determinant. *Infect and Immunity, 3:*567, 1971.

Hausmann, E., Courant, P.R., and Arnold, D.S.: Conditions for the demonstration of collagenolytic activity of *Bacteroides melaninogenicus. Arch Oral Biol, 12:*317, 1967.

Heinrich, S., and Pulverer, G.: Uber den Nachweis der *Bacteroides melaninogenicus* in Krankheitsprozessen bei Mensch und Tier. *Z Hyg, 146:*331, 1960.

Hite, K.E., Locke, M., and Hesseltine, H.C.: Synergism in experimental infections with nonsporulating anaerobic bacteria. *J Infect Dis, 84:*1, 1949.

Hofstad, T.: Biological activities of endotoxins from *Bacteroides melaninogenicus. Arch Oral Biol, 15:*343, 1970.

Hofstad, T.: Chemical characteristics of *Bacteroides melaninogenicus* endotoxin. *Arch Oral Biol, 13:*1149, 1968.

Kaufman, E.J., Mashimo, P.A., Hausmann, E., Hanks, C.T., and Ellison, S.A.: Fusobacterial infection: enhancement by cell free extracts of *Bacteroides melaninogenicus* possessing collagenolytic activity. *Arch Oral Biol, 17:*577, 1972.

Kerebel, B., and Sedallian, A.: Action de *Bacteroides melaninogenicus* sur la dentine *in vitro. Rev mens suisse Odonto-stomatol, 82:*731, 1972.

Kestenbaum, R.C., Massing, J., and Weiss, S.: The role of collagenase in mixed infections containing *Bacteroides melaninogenicus.* Abstr. # 9. Int. Assoc. Dental Research 42nd Gen. Meeting, 1964.

Krasse, B.: The proportional distribution of *Streptococcus salivarius* and other streptococci in various parts of the mouth. *Odontol Revy, 5:*203, 1954.

Labrec, E.H., Schneider, H., Magnani, T.J., and Formal, S.B.: Epithelial cell penetration as an essential step in the pathogenesis of bacillary dysentery. *J Bacteriol, 88:*1503, 1964.

Lancefield, R.C.: Current knowledge of type-specific M antigens of Group A streptococci. *J Immunol, 89:*307, 1962.

Lev, M.: The growth promoting activity of compounds of the vitamin K group and analogues for a rumen strain of *Fusiformis nigrescens. J Gen Microbiol, 20:*697, 1959.

Liljemark, W.F., and Gibbons, R.J.: Ability of *Veillonella* and *Neisseria* species to attach to oral surfaces and their proportions present indigenously. *Infect and Immunity, 4:*264, 1971.

Liljemark, W.F., and Gibbons, R.J.: The proportional distribution and relative adherence of *Streptococcus miteor* (*mitis*) in the human oral cavity. *Infect and Immunity, 6:*852, 1972.

Loesche, W.J., and Gibbons, R.J.: Influence of nutrition on the ecology and cariogenicity of the oral microflora. In: Nizel, A.E. (Ed.): *The Science of Nutrition and its Application to Clinical Dentistry.* Philadelphia, Saunders, 1966, pp. 305–317.

Macdonald, J.B., and Gibbons, R.J.: The relationship of indigenous bacteria to periodontal disease. *J Dent Res, 41:*320, 1962.

Macdonald, J.B., Socransky, S.S., and Gibbons, R.J.: Aspects of the pathogenesis of mixed anaerobic infections of mucous membranes. *J Dent Res, 42:*529, 1963.

Macdonald, J.B., Sutton, R.M., and Knoll, M.L.: The production of fusospirochetal infections in guinea pigs with recombined pure cultures. *J Infect Dis, 95:*275, 1954.

Macdonald, J.B., Sutton, R.M., Knoll, M.L., Madlener, E.M., and Grainger, R.M.: The pathogenic components of an experimental fusospirochetal infection. *J Infect Dis, 98:*15, 1956.

Meleney, F.L.: Bacterial synergism in disease processes with a confirmation of the synergistic bacterial etiology of a certain type of progressive gangrene of the stomach wall. *Ann Surg, 94:*961, 1931.

Mergenhagen, S.E., Hampp, E.G., and Scherp, H.W.: Preparation and biological activities of endotoxins from oral bacteria. *J Infect Dis, 108:*304, 1961.

Oliver, W.W., and Wherry, W.B.: Notes on some bacterial parasites of the human mucous membranes. *J Infect Dis, 28:*341, 1921.

Pulverer, G., and Heinrich, S.: Infektionsversache an Laboratorumstieren und *in vitro* Untersuchungen zur Fermentausstaltung des *Bacteroides melaninogenicus*. *Z Hyg, 146:*341, 1960.

Reddy, C.A., and Bryant, M.P.: DNA base composition and cytochromes in certain species of the genus *Bacteroides*. In: *Bacteriol Proceedings*. Ann Arbor, Am Soc for Microbiology, 1967, p. 40 (G109).

Rizza, V., Sinclair, P.R., White, D.C., and Courant, P.R.: Electron transport system of the protoheme-requiring anaerobe *Bacteroides melaninogenicus*. *J Bacteriol, 96:*665, 1968.

Rosebury, T., Macdonald, J.B., and Clark, A.R.: A bacteriologic survey of gingival scrapings from periodontal infections by direct examination, guinea pig inoculation, and anaerobic cultivation. *J Dent Res, 29:*718, 1950.

Sawyer, S.H., Macdonald, J.B., and Gibbons, R.J.: Biochemical characteristics of *Bacteroides melaninogenicus*. *Arch Oral Biol, 7:*685, 1962.

Schwabacher, H., Lucas, D.R., and Rimington, C.: *Bacterium melaninogenicum*—a misnomer. *J Gen Microbiol, 1:*109, 1947.

Shpuntoff, H., and Rosebury, T.: Infectivity of fuso-spirochetal exudates for guinea pigs, hamsters, mice and chick embryos by several routes of inoculation. *J Dent Res, 28:*7, 1949.

Sirisinha, S.: Reactions of human salivary immunoglobulins with indigenous bacteria. *Arch Oral Biol, 15*, 551, 1970.

Sobeslansky, O., Prescott, B., and Chanock, R.M.: Adsorption of *Mycoplasma pneumoniae* to neuraminic acid receptors of various cells and possible role in virulence. *J Bacteriol, 96:*695, 1970.

Socransky, S.S., and Gibbons, R.J.: Required role of *Bacteroides melaninogenicus* in mixed anaerobic infections. *J Infect Dis, 115:*247, 1965.

Socransky, S.S., Gibbons, R.J., Dale, A.C., Bortnick, L., Rosenthal, E., and Macdonald, J.B.: The microbiota of the gingival crevice area of man. I. Total microscopic and viable counts, and counts of specific organisms. *Arch Oral Biol, 8:*275, 1963.

Socransky, S.S., and Manganiello, A.D.: The oral microbiota of man from birth to senility. *J Periodontol, 42:*485, 1971.

Takazoe, I., Tanaka, M., and Homma, T.: A pathogenic strain of *B. melaninogenicus*. *Arch Oral Biol, 16:*817, 1971.

Tomasi, T.B.: Structure and function of mucosal antibodies. *Ann Rev Med, 21:*281, 1970.

Tracy, O.: Pigment production in *Bacteroides*. *J Med Microbiol, 2:*309, 1969.

van Houte, J., and Gibbons, R.J.: Studies on the cultivable flora of normal human feces. *Antonie van Leeuwenhoek, 32:*212, 1966.

van Houte, J., Gibbons, R.J., and Banghart, S.: Adherence as a determinant of the presence of *Streptococcus salivarius* and *Streptococcus sanguis* on the tooth surface. *Arch Oral Biol, 15:*1025, 1970.

van Houte, J., Gibbons, R.J., and Pulkkinen, A.J.: Adherence as an ecological determinant for streptococci in the human mouth. *Arch Oral Biol, 16:*1131, 1971.

Waldvogel, F.A., and Swartz, M.N.: Collagenolytic activity of bacteria. *J Bacteriol, 98:*662, 1969.

Wahren, A., and Gibbons, R.J.: Amino acid fermentation by *Bacteroides melaninogenicus*. *Antonie van Leeuwenhoek, 36:*149, 1970.

Watson, R.F., Rothbard, S., and Swift, H.F.: Type specific protection and immunity following intranasal inoculation of monkeys with group A hemolytic streptococci. *J Exp Med, 84:*127, 1946.

Weiss, C.: Observations on *Bacterium melaninogenicum*. Demonstration of fibrinolysin, pathogenicity, and serological types. *Proc Soc Exp Biol Med, 37:*473, 1937.

Weiss, C.: The pathogenicity of *Bacteroides melaninogenicus* and its importance in surgical infections. *Surgery, 13:*683, 1943.

Williams, R.C., and Gibbons, R.J.: Inhibition of bacterial adherence by secretory immunoglobulin A: a mechanism of antigen disposal. *Science, 177:*697, 1972.

CHAPTER XXII

TREATMENT OF BOTULISM AND WOUND BOTULISM

MICHAEL CHERINGTON

ABSTRACT: *In addition to the usual therapeutic approaches to botulism with symptomatic treatment and antitoxin, two drugs have recently shown promise: guanidine hydrochloride and germine monoacetate. Thirteen patients have shown at least some response to guanidine therapy, while five others have been reported to fail to respond. Germine, which has been used with benefit in patients with myasthenia gravis, reverses the neuromuscular block of botulism in animals. In animals poisoned with botulinal toxin, a combination of guanidine and germine produced a greater effect than either drug used alone; the combination might be useful in treating human cases of botulism.*

Findings in cases of wound botulism are summarized briefly. Despite the ubiquitous nature of Clostridium botulinum, *this is a rare complication of traumatic wounds.*

Botulism is a paralyzing disease caused by one of the most potent of poisons (Lamanna, 1959), the estimated mouse lethal dose of type A toxin given intraperitoneally being as little as $10^{-5}\mu g$. The toxin is produced by an anaerobic organism, *Clostridium botulinum*. The discovery that botulism is caused by a toxin was made by Professor Emile Van Ermengem just before the turn of the century (Van Ermengem, 1896). An outbreak of botulism occurred in the small Belgian town of Ellezelles among persons who had eaten raw ham. Professor Ermengem isolated an anaerobic organism from the ham and then produced the disease in laboratory animals by injecting the toxin produced by the organism.

Most cases of botulism have resulted from ingesting contaminated protein-containing foods. Toxin in association with *C. botulinum* bacilli can be recovered from foods. Contaminated home-canned foods are the usual source of the toxin, but the danger of botulism from commercially canned

I am indebted to Dr. Michael H. Merson of the CDC for providing recent data on cases of wound botulism. I am also indebted to the Martin J. and Mary Anne O'Fallon Trust for partial support of this work.

foods remains a serious health problem (Lancet, 1971). During processing cans are ordinarily heated to 121.1°C for 30 minutes. Inadequate heating may kill vegetative forms of *C. botulinum*, but surviving spores can still germinate and produce toxin. There is greater risk with canned vegetables than with canned fruit, which has a high acid content, probably because toxin is not produced at a pH lower than 4.6. Heating or boiling canned food prior to eating it should protect the consumer, for the toxin is heat labile.

Symptoms and signs of botulism usually begin within 12 to 36 hours after ingestion of toxin. They consist of diplopia, dysphagia, nausea, dizziness and descending paralysis. The cause of death is frequently respiratory paralysis. During the years 1899 to 1969 there were 1,696 clinical cases of botulism reported in the United States, with 959 fatalities (Gangarosa *et al.*, 1971).

APPROACHES TO TREATMENT

The major form of treatment of botulism is good medical and nursing care, including respiratory monitoring in patients who require ventilatory assistance. Antitoxin has failed to alter the course of many patients with type A or type B botulism, but seems to be more effective in type E botulism (Koenig *et al.*, 1964). For optimal benefit antitoxin should be given early in the course of the disease. However, the diagnosis of botulism should be nearly certain before the antitoxin is administered, for the antitoxin, being of equine origin, is not free of risk. Early diagnosis can be accomplished by demonstration of toxin in the patient's serum. The suspected food should also be tested for mouse killing power. Specificity may be ascertained by protecting mice with type-specific botulinal antitoxin.

A treatment which seems promising on the basis of recent animal and human experience is administration of guanidine hydrochloride. Botulinal toxin causes a defect of neuromuscular transmission by interfering with release of acetylcholine, and guanidine, $HN = C(NH_2)_2$, enhances the release of acetylcholine from nerve terminals. As demonstrated by the use of electrophysiologic techniques, the electrical abnormality found in botulism resembles that seen in the myasthenic syndrome of Lambert and Eaton. Because of this similarity, guanidine, known to be of benefit in the myasthenic syndrome, was tried in botulism (Cherington and Ryan, 1967, 1970; Scaer *et al.*, 1969). The results in human cases of botulism suggest that guanidine is a useful adjunct in the treatment of botulism when combined with good medical care. To date the drug has been of benefit of varying degrees in 13 patients (Table XXII-I). Cherington and Ginsberg (1971) reported the first case of guanidine failure in a patient who died with type B botulism. Faich *et al.* (1971) described four patients with type A botulism who recovered, but who showed no noticeable effect from guanidine therapy.

The usual oral dose of guanidine is 15 to 40 mg/kg/day in divided doses. Because of a considerable number of side effects and potential toxicity, caution is required. Gastrointes-

TABLE XXII-I

SUMMARY OF EXPERIENCE WITH GUANIDINE IN THE TREATMENT OF BOTULISM

Source	*Age*	*Sex*	*Type of Botulism*	*Response to Guanidine*	*Outcome*
Cherington and Ryan (1968)	57	F	–	+	Expired
Cherington and Ryan (1970)	48	F	A	+	Recovered
Ricker and Doll (1970)	38	M	–	+	Recovered
	44	F	–	+	Recovered
	41	M	–	+	Recovered
	47	F	B	+	Recovered
	29	F	–	+	Recovered
Cherington and Ginsberg (1971)	41	M	B	0	Expired
Ryan and Cherington (1971)	39	M	A	+	Recovered
	40	M	A	+	Recovered
	39	M	A	+	Recovered
	42	F	A	+	Recovered
	52	M	A	+	Recovered
Robineau *et al.* (1971)	64	F	–	±	Recovered
Faich *et al.* (1971)	49	M	A	0	Recovered
	22	F	A	0	Recovered
	10	M	A	0	Recovered
	43	F	A	0	Recovered
Lahiri and Marcuse (1972)	39	M	–	+	Recovered

tinal side effects are usually the first to appear; guanidine, which denatures protein, may have a deleterious effect on gastrointestinal mucosa (New Engl J Med, 1968). Side effects arising from the nervous system include hyperirritability, tremors and muscle twitching. A few patients have developed the serious side effect of pancytopenia (Howard, 1972). Finally, cardiac arrhythmia and hypotension have occurred in conjunction with guanidine therapy (Nakano and Tyler, 1972).

Another pharmacologic agent which reverses the neuromuscular block of botulism in animals is germine monoacetate (Cherington and Greenberg, 1971). Germine-3-acetate is a veratrum derivative with the chemical structure depicted in Figure XXII-1. Veratrum alkaloids have an effect on skeletal muscles known as a veratrine response, and also can lower blood pressure. Esterification of germine with acetic acid produces an alkaloid which acts on skeletal muscles without causing hypotensive effects. Germine monoacetate

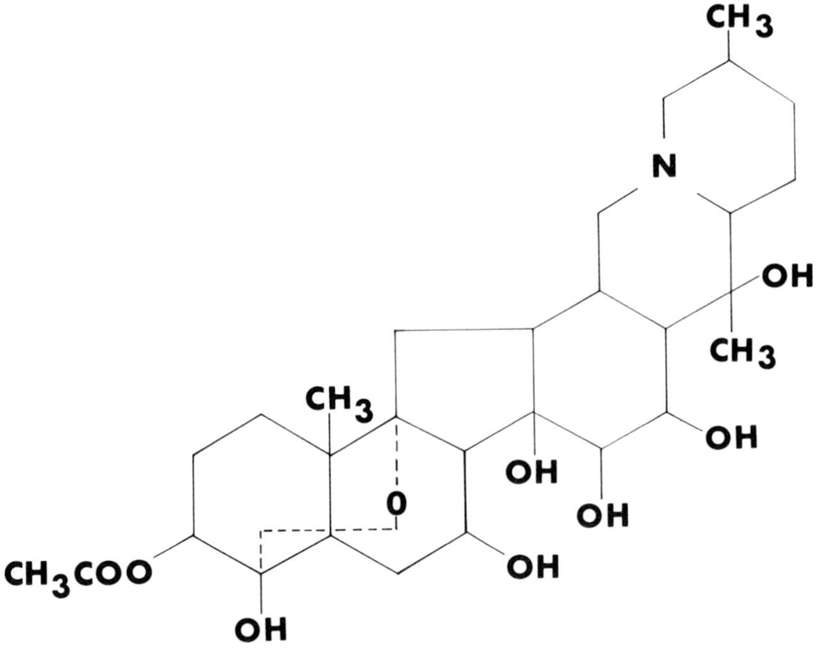

Figure XXII-1. Germine monoacetate

acts at least in part at the postsynaptic site of the myoneural junction, increasing muscle tension by converting the single muscle action potential evoked by a single nerve stimulus into a brief burst of repetitive potentials (Flacke, 1969). The mechanism by which it produces this repetitive firing is not known. Germine does not alter the processes involved in transmitter release, and quantum size, quantum content, and the time course of end-plate potentials are unchanged at germine dose levels which cause repetitive activity (Detwiler, 1972). Germine has been used with benefit in patients with myasthenia gravis (Flacke, 1963). Side effects from the drug include paresthesias and nausea.

The combination of guanidine and germine administered to animals poisoned with botulinal toxin produces a greater reversal of neuromuscular block than either drug used alone (Cherington *et al.*, 1972). Since the major action of guanidine is at a presynaptic site, and the action of germine is probably at a postsynaptic site, it is not surprising that when the drugs are used together, they produce a greater effect than either used alone. This combination might be useful in treating human cases of botulism.

WOUND BOTULISM

Human cases of botulism are almost always caused by ingestion of food in which *Clostridium botulinum* has been growing and producing toxin. The toxin can be absorbed from all mucous membrane surfaces, including the entire alimentary tract. It can also be absorbed from wounds and broken skin areas (Geiger, 1924), as well as from the respiratory tract. Three cases of botulism in laboratory workers were ascribed to inhalation of the toxin (Holzer, 1962).

The natural habitat of *C. botulinum* is the soil; it is a common soil anaerobe in the western states, and is more prevalent in virgin soils than in soil collected from animal corrals (Thomas *et al.*, 1951). Considering the ubiquitous nature of the organism and the consequent potential for contamination of traumatic wounds, it is surprising that wound botulism is rare. A review of the literature revealed only 12 cases (9

TABLE XXII-II

CASES OF WOUND BOTULISM REPORTED IN THE LITERATURE

Source	*Age*	*Sex*	*Wound*	*Type*	*Outcome*
Hall (1945)	42	M	Left leg	B	No clinical signs of botulism; recovered
	9	M	Left leg	A	No clinical signs of botulism; recovered
	38	M	Left axilla	A	No clinical signs of botulism; recovered
Davis *et al.* (1951)	15	F	Left leg	A	Expired 16th day
Thomas *et al.* (1951)	13	M	Compound fracture Left thigh	A	Expired 9th day
Hampson (1951)	—	M	Gunshot Right thigh	A	Expired 9th day
Condit and Defries (1968)	44	M	Electric saw Left wrist	—	Improved
Perlstein and Altrocchi (1971)	7	M	Compound fracture Left ulna	A	Recovered
Grizzle (1972)	12	F	Compound fracture Puncture wound of foot	A	Expired 8th day
Gutmann (1972)	38	M	Foot wound	—	Improved over period of weeks
Center for Disease Control (1972)	7	M	Motor bike accident Hematoma, left leg	—	Recovered
Center for Disease Control (1972)	45	M	Aspirated Compound fracture of hand	A	Recovered

with clinical signs of the illness) of wound botulism; the findings are summarized briefly in Table XXII-II.

The rarity of botulism resulting from wound infection is probably due to failure of *C. botulinum* spores to germinate readily in the tissues. In experimental animals botulism was produced by the injection of *C. botulinum* spores only if large numbers of spores were injected (Keppie, 1951). Toxin can be demonstrated in the cytoplasm of botulinum spores when they are destroyed by mechanical means.

Wound botulism is no less severe than the exogenous form. Of nine patients listed in Table XXII-II who had clinical signs of botulism, four expired. Although an uncommon complication, wound botulism should be considered in any wound case if the patient develops bulbar signs and muscle weakness.

REFERENCES

Cherington, M., and Ginsberg, S.: Type B botulism: Neurophysiologic studies. *Neurology, 21:*43, 1971.

Cherington, M., and Greenberg, H.: Botulism and germine. *Neurology, 21:*966, 1971.

Cherington, M., and Ryan, D.W.: Botulism and guanidine. *N Engl J Med, 278:*931, 1968.

Cherington, M., and Ryan, D.W.: Guanidine in botulism. *Lancet, 2:*1360, 1967.

Cherington, M., and Ryan, D.W.: Treatment of botulism with guanidine: Early neurophysiologic studies. *N Engl J Med, 282:*195, 1970.

Cherington, M., Soyer, A., and Greenberg, H.: Effects of guanidine and germine on the neuromuscular block of botulism. *Curr Ther Res, 14:*91, 1972.

Condit, P.K., and Defries, W.: Suspect wound botulism. *Morbidity and Mortality Weekly Report, 17:*199, 1968.

Davis, J.B., Mattman, L.H., and Wiley, M.: *Clostridium botulinum* in a fatal wound infection. *JAMA, 146:*646, 1951.

Detwiler, P.B.: The effects of germine-3-acetate on neuromuscular transmission. *J Pharmacol Exp Ther, 180:*244, 1972.

Faich, G.A., Graber, R.W., and Sato, S.: Failure of guanidine therapy in botulism A. *N Engl J Med, 285:*773, 1971.

Flacke, W.: Pharmacological activity of some esters of germine with acetic acid. *Arch Exp Path Pharmacol, 240:*369, 1969.

Flacke, W.: Studies on veratrum alkaloids. XXXVL. Action of germine monoacetate and germine diacetate on mammalian skeletal muscle. *J Pharmacol Exp Ther, 141:*230, 1963.

Gangarosa, E.J., Donadio, J.A., Armstrong, R.W., Meyer, K.F., Brachman, P.S., and Dowell, V.R.: Botulism in the United States. *Am J Epidemiol, 93:*93, 1971.

Geiger, J.C.: The possible danger of absorption of toxin of *B. botulinus* through flesh wounds and from mucous surfaces. *Am J Public Health, 14:*309, 1924.

Grizzle, C.O.: Botulism from a puncture wound. *Rocky Mt Med J, 69:*47, 1972.

Gutmann, L.: Wound botulism. Personal communication. 1972.

Hall, I.C.: The occurrence of *Bacillus botulinus*, Types A and B in accidental wounds. *J Bacteriol, 50:*213, 1945.

Hampson, C.R.: A case of botulism due to wound infection. *J Bacteriol, 61:*647, 1951.

Holzer, V.E.: Botulismus durch inhalation. *Med Klin, 57:*1735, 1962.

Howard, F.: Pancytopenia and guanidine. Personal communication. 1972.

Keppie, J.: The pathogenicity of the spores of *Clostridium botulinum. J Hyg, 49:*36, 1951.

Koenig, M.G., Spikard, A., Cardella, M.A., and Rogers, D.E.: Clinical and laboratory observations on Type E botulism in man. *Medicine, 43:*517, 1964.

Lahari, T.J., and Marcuse, E.K.: Guanidine in case of botulism. Personal communication. 1972.

Lamanna, C.: The most poisonous poison. *Science, 130:*763, 1959.

Lancet (Editorial): Botulism from canned soup. *Lancet, 2:*536, 1971.

Nakano, K.K., and Tyler, H.R.: Cardiovascular complications of guanidine hydrochloride. *Ann Intern Med, 77:*658, 1972.

New England Journal of Medicine (Editorial): On guanidine and the treatment of botulism. *N Engl J Med, 278:*963, 1968.

Perlstein, G., and Altrocchi, P.H.: Endogenous botulism. Personal communication. 1972.

Ricker, R., and Doll, W.: Guanidinbehandlung des botulismus. *Z Neurology. 198:*332, 1970.

Robineau, M., Modai, J., Laffay, J., and Domart, A.: Note preliminoire sur le traitement du botulisme par la guanidine. *La Presse Medical, 79:*1169, 1971.

Ryan, D.W., and Cherington, M.: Human Type A botulism. *JAMA, 216:*513, 1971.

Scaer, R.C., Tooker, J., and Cherington, M.: Effect of guanidine on the neuromuscular block of botulism. *Neurology, 19:*1107, 1969.

Thomas, C.G., Keleher, M.F., and McKee, A.P.: Botulism, a complication of Clostridium botulinum wound infection. *Arch Pathol, 51:*623, 1951.

Van Ermengem, E.: Recherches sur des cas d'accidents alimentaires. *Revue d'Hygiene, 18:*761, 1896.

Chapter XXIII

ENDOTOXINS OF ANAEROBIC GRAM-NEGATIVE MICROORGANISMS

TOR HOFSTAD

ABSTRACT: *Studies carried out during the last decade have established the presence of endotoxins in gram-negative anaerobic bacteria, which can be extracted and purified by conventional methods. In the electron microscope they appear as structured particles like endotoxins from aerobic bacteria. They are macromolecular complexes of carbohydrate, lipid and protein (or polypeptide), the exact chemical structure of which remains unknown. The carbohydrate moiety is rich in antigenic determinants. Tools are now available for classification of at least* Fusobacterium *and* Veillonella *species into serotypes. Such a classification may be of value in studies of immune factors in anaerobic infections.*

Although anaerobic bacteria have been isolated with increasing frequency from clinical infections, their pathogenic potential and role in disease are still far from clarified. Endotoxin may be one of several factors determining virulence. A thorough study of the host-reactive properties of endotoxins from anaerobic gram-negative bacteria is timely and much needed.

Endotoxins, or endotoxic lipopolysaccharides (LPS) constitute part of the outer cell membrane in gram-negative bacteria. As obtained by common extraction procedures, LPS are macromolecular structures which consist of polysaccharide, phospholipid and small quantities of protein. In aerobic gram-negative bacteria, like salmonella, the lipid moiety is covalently bound to a polysaccharide core region to which are linked oligosaccharide side chains (Lüderitz *et al.*, 1966). The high molecular weight complex incorporates the major somatic antigens (O-antigens) of the bacteria and a large number of toxic and other host reactive properties, collectively described as endotoxic.

Evidence for endotoxic activity in cultures of anaerobic gram-negative bacteria was described by Césari in 1912, and later supported by Kirchheiner (1940). Böe, in 1941, produced the local Shwartzman phenomenon in rabbits with cell-free culture filtrates of fusobacteria and leptotrichia. The recent interest in endotoxin from gram-negative anaerobic microorganisms started 20 years later with the pioneer work of Mergenhagen *et al.* (1961), who reported the isolation of endotoxins from selected strains of oral bacteria. Since that time LPS have been prepared from *Veillonella* (Mergenhagen *et al.*, 1962; Hofstad and Kristoffersen, 1970b; Hewitt *et al.*, 1971), *Fusobacterium* (de Araujo *et al.*, 1963; Jensen, 1967; Kristoffersen and Hofstad, 1970), *Leptotrichia* (de Araujo *et al.*, 1963; Gustafson *et al.*, 1966), *Bacteroides* (Hofstad, 1968; Hofstad and Kristoffersen, 1970a; 1971a) and *Sphaerophorus* species (Hofstad and Kristoffersen, 1971b). The data presented in this paper are principally derived from studies carried out in our laboratory.

ISOLATION AND PURIFICATION

In most cases LPS have been extracted with 45 percent aqueous phenol (Westphal *et al.*, 1952), and purified by sedimentation in the ultracentrifuge. LPS prepared in this way from acetone-dried whole cells of veillonella (Hofstad and Kristoffersen, 1970b) and from crushed, defatted and washed cells of fusobacteria (Kristoffersen and Hofstad, 1970) contained small amounts of protein and were free of nucleic acids. The same procedure had earlier been used for preparation of biologically active LPS from whole cells of *B. melaninogenicus* (Hofstad, 1968) and *B. fragilis* (Hofstad and Kristoffersen, 1970a). However, in the case of *B. melaninogenicus* and several other bacteroides strains (unpublished work) substantial amounts of glycogen and sometimes protein and nucleic acids were sedimented together with LPS in the ultracentrifuge. A new purification method was therefore adopted for *B. melaninogenicus* LPS (Hofstad and Kristoffersen, 1971a). The method included digestion of the supernatant fluid after ultracentrifugation of the water phase following phenol-water extraction with ribonuclease and deoxyribonuclease, gel filtration on Bio-Gel A-1.5 m and ion exchange chromatography on DEAE-cellulose. Biologically active LPS were purified in the same way from phenol-water extracts of *S. necrophorus* (Hofstad and Kristoffersen, 1971b).

Pilot experiments have suggested that extraction with cold trichloroacetic acid (Boivin and Mesrobeanu, 1935) may be used for preparation of LPS from *B. melaninogenicus*. Hewitt *et al.* (1971) obtained high yields of LPS from veillonella by extraction with phenol-chloroform-petroleum ether (Galanos *et al.*, 1969).

CHEMICAL COMPOSITION

The chemical composition of representative batches of LPS prepared in this laboratory is shown in Table XXIII-I. Table XXIII-II illustrates the difference in sugar patterns, as revealed by paper chromatography of acid hydrolysates, of LPS isolated from strains of the anaerobic bacteria studied. The results suggest that LPS isolated from strains within each of the genera *Veillonella*, *Fusobacterium*, and *Sphaerophorus* have a basic set of sugars and in addition individual sugar constituents which may be related to O-antigenic specificity. Glucosamine, 2-keto-3-deoxyoctonate (KDO) and an aldoheptose were present in all the strains examined from these three genera, although the KDO content in LPS from *Fusobacterium* and *Sphaerophorus* species was low. In contrast all LPS

TABLE XXIII-I

PERCENTAGE CHEMICAL COMPOSITION OF BATCHES OF LPS PREPARED FROM PHENOL-WATER EXTRACTS OF REPRESENTATIVE STRAINS OF VEILLONELLA, FUSOBACTERIA, *B. MELANINOGENICUS*, *B. FRAGILIS* AND *S. NECROPHORUS**

Source of LPS	*Neutral Sugar*	*Hexos-amine*	*KDO*†	*Fatty Acid Ester*	*Protein*	*Nitrogen*
Veillonella strain Ve5§	30.7	7.7	9.5	30.2	7.7	2.4
Fusobacterium strain F1	50.1	15.7	1.4	15.2	2.5	2.0
B. fragilis strain NCTC 9343	27.1	14.8	—	24.7	3.4	2.2
B. melaninogenicus strain B10	26.8	11.9	—	25.5	4.7	3.1
S. necrophorus strain N167	16.8	7.3	0.7	16.4	8.4	4.1

* From Hofstad and Kristoffersen (1970a; 1970b; 1971a; 1971b) and Kristoffersen and Hofstad (1970).

† KDO = 2-keto-3-deoxy-octonate; — = not detected.

§ Veillonella and fusobacterium purified by sedimentation in the ultracentrifuge; *B. fragilis*, *B. melaninogenicus* and *S. necrophorus* purified by gel filtration and ion exchange chromatography.

TABLE XXIII-II

SUGAR COMPONENTS IN LPS PREPARED FROM STRAINS OF VEILLONELLA, FUSOBACTERIA, *S. NECROPHORUS, B. MELANINOGENICUS* AND *B. FRAGILIS**

	Veillonella† (6)	*Fusobacteria* (7)	S. necrophorus (4)	B. melaninogenicus (2)	B. fragilis (5)
Glucosamine	++§	++	++	++	++
Galactosamine	++	–	–	++	++
Heptose	++	++	++	–	–
KDO	++	++	++	–	–
Glucose	++	+	++	++	++
Galactose	+	+	+	++	++
Mannose	–	+	+	++	++
Fucose	–	–	–	++	++
Ribose	+	–	–	–	–
Rhamnose	–	+	+	++	++

* From Hofstad and Kristoffersen (1970a; 1970b; 1971a; 1971b); Kristoffersen and Hofstad (1970) and unpublished work.

† Veillonella and fusobacteria purified by sedimentation in the ultracentrifuge; *S. necrophorus, B. melaninogenicus* and *B. fragilis* purified by gel filtration and ion exchange chromatography; numbers in ( ) = no. strains examined.

§ ++ = present in LPS of all strains examined; + = present in LPS of some strains (one or more); – = not present.

isolated from *B. fragilis* and *B. melaninogenicus* contained the same sugars, and no KDO or heptose was present. In view of the fact that KDO and heptose are essential sugar constituents in LPS of virtually all aerobic gram-negative bacterial species, their absence in LPS from *B. fragilis* and *B. melaninogenicus* may be worthy of attention.

Hewitt *et al.* (1971) using gravimetric analysis found that lipid comprised two thirds of LPS prepared from two strains of *V. parvula* and two strains of *V. alcalescens.* The major constituents of the lipid moiety were glucosamine, tridecanoic acid and 3-hydroxytridecanoic acid. Nothing is known about the lipid moiety of LPS isolated from *Fusobacterium, Bacteroides* or *Sphaerophorus* species.

ULTRASTRUCTURE

Bladen and Mergenhagen (1964) found that positively or negatively stained preparations of phenol-water extracts of veillonella contained particles in a variety of shapes, predominantly spherical or disc-like with diameters in the range 200 to 1400 Å. Similar disc-like particles, delimited by a dense line or by a triple-layered structure, were observed in positively stained preparations of purified *Fusobacterium* and *B. melaninogenicus* LPS as shown in Figure XXIII-1. (Selvig *et al.*, 1971). The preparations also contained rod-like particles 250 to 400 Å long with single or trilaminar borders. Rod-shaped structures were predominant in preparations of LPS from *B. fragilis* and *S. necrophorus* (Hofstad *et al.*, 1972). Similar particulate structures had earlier been observed in electron micrographs of LPS from aerobic bacteria (Shands, 1971).

IMMUNOLOGICAL CHARACTERISTICS

Purified endotoxic LPS from veillonella (Hofstad *et al.*, 1971), *B. melaninogenicus* (Hofstad, 1969) and *B. fragilis*

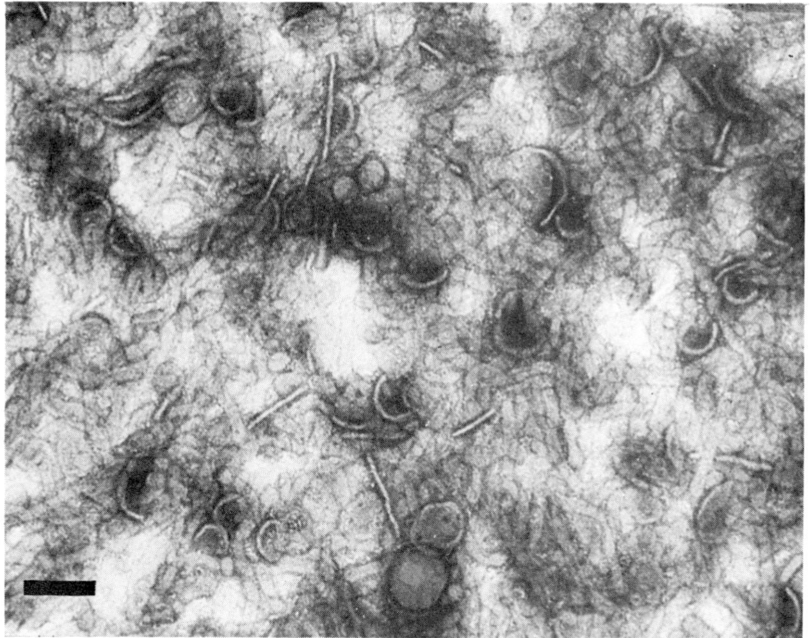

Figure XXIII-1. Lyophilized LPS extracted from a fusobacterium with phenol-water and purified by high speed centrifugation. Electron micrograph, magnification × 100,000. Horizontal bar represents 1000 Å.

(Hofstad, unpublished observations) are immunogenic in rabbits, giving rise to predominantly high-molecular-weight antibodies. The immunogenicity of fusobacteria LPS has not been examined.

The serological properties of LPS from anaerobic bacteria have mainly been studied by indirect hemagglutination and inhibition of hemagglutination. Mergenhagen and Varah (1963) and de Araujo *et al.* (1963) found that LPS from strains of veillonella and fusobacteria, respectively, exhibited a high degree of serological specificity. Kristoffersen *et al.* (1971), working with unabsorbed and absorbed antisera, showed the presence of at least three different antigenic specificities in LPS from three oral strains of fusobacteria. As shown in Table XXIII-III, one antigenic determinant was shared by all three strains and a second by two strains (F1 and F4), while the third was specific for one strain (Fev 1). Sheep cells sensitized with LPS from the *Fusobacterium* type strain ATCC 10953 were agglutinated by homologous antiserum only. Absorption with LPS from the type strain did not affect the titers of antibodies in antisera to strain F1, F4 and Fev 1. The serological activity remained unchanged following treatment of LPS

TABLE XXIII-III

EFFECTS OF ABSORPTION WITH VARIOUS LPS PREPARATIONS ON TITERS IN THE INDIRECT HEMAGGLUTINATION TEST OF RABBIT ANTISERA TO FUSOBACTERIUM STRAINS F1, F4, AND FEV 1*

Antiserum	Source of LPS Used for Absorption	Source of LPS Used for Sensitization		
		F1	F4	Fev 1
Anti-F1 unabsorbed		2048	2048	<16
Anti-F1	F1	<16	<16	<16
	F4	<16	<16	<16
	Fev 1	2048	2048	<16
Anti-F4 unabsorbed		1024	1024	<16
Anti-F4	F1	<16	<16	<16
	F4	<16	<16	<16
	Fev 1	1024	1024	<16
Anti-Fev 1 unabsorbed		128	128	2048
Anti-Fev 1	F1	<16	<16	2048
	F4	<16	<16	2048
	Fev 1	<16	<16	<16

* From Kristoffersen *et al.* (1971).

with pronase, but was completely destroyed by oxidation with periodate. Immunoprecipitates obtained by mixing LPS containing two different antigenic determinants with antibodies specific to one of the determinant groups contained both specificities. Analogous results were achieved when veillonella LPS were examined along the same lines (Hofstad *et al.*, 1971), but the immunological individuality seemed to be more pronounced. The findings clearly indicate that O-antigenic specificity is present in fusobacteria and veillonella, and that the antigenic determinants of any LPS are carried by the same molecular complex, assumed to be the carbohydrate moiety of the LPS. Results obtained in inhibition studies with monosaccharides and disaccharides support this concept (Kristoffersen *et al.*, 1971; Hofstad *et al.*, 1971).

The situation seems to be more complex in *B. melaninogenicus*. Hofstad (1969) found that sheep erythrocytes sensitized with LPS preparations from four strains purified by ultracentrifugation were agglutinated by both homologous and heterologous antisera. The presence of cross-reactivity was confirmed by hemagglutination inhibition. Only eight agglutinating doses of antiserum were used, so the results do not exclude the presence of low-titered type-specific antibodies. When the LPS preparations were examined by double diffusion in agar each preparation produced a principal line of precipitation with homologous antiserum not shared by any of the other LPS preparations. Treatment with periodate destroyed the serological activity. LPS purified by gel filtration and ion exchange chromatography exhibited the same serological specificity as LPS from the same strain sedimented by high speed centrifugation (Hofstad and Kristoffersen, 1971a). The results suggest that LPS from *B. melaninogenicus* possesses group- and type-specific antigenic determinants. However, the possibility can not be excluded that the cross-reactivity is due to a high-molecular-weight carbohydrate associated with the LPS preparations and shared by the *B. melaninogenicus* strains examined.

The serology of LPS isolated from *B. fragilis* or *S. necrophorus* has not been studied.

ENDOTOXIC ACTIVITY

A systematic study of all aspects of toxicity or host-reactive properties of LPS isolated from anaerobic gram-negative microorganisms has not yet been undertaken.

Mergenhagen *et al.* (1961) showed that crude preparations of LPS from oral strains of veillonella, fusobacteria, selenomonas and *B. melaninogenicus* were pyrogenic and produced the local Shwartzman phenomenon in rabbits. The rabbits responded with fever to submicrogram quantities of veillonella LPS and microgram amounts prepared for the Shwartzman reaction (Mergenhagen and Varah, 1963). Purified veillonella LPS is highly chemotactic for polymorphonuclear (Jensen *et al.*, 1966; Mergenhagen *et al.*, 1969) and mononuclear leukocytes (Hausman *et al.*, 1972), and has been used as a typical endotoxin in elucidating the effect of endotoxic LPS on the complement system (Gewurz *et al.*, 1971).

Purified LPS from fusobacteria proved to be a potent endotoxin when assayed for the capacity to alter the dermal reactivity in rabbits to epinephrine, but the toxicity for mice was relatively low, with LD_{50} ranging from 860 to 1000 μg (deAraujo *et al.*, 1963). Jensen (1967) studied the influence of fusobacteria LPS on host resistance in mice to systemic infection with oral streptococci. Intraperitoneal injection of LPS produced an initial period of susceptibility followed by increased resistance of somewhat longer duration. This biphasic effect on host resistance, which is characteristic of endotoxins, was paralleled by a decreased clearance of bacteria from the peritoneal cavity and increased clearance from blood.

Gustafson *et al.* (1966) compared endotoxin isolated from *L. buccalis* with *Escherichia coli* LPS by mouse lethality tests, the ability to cause febrile and leukocytic responses in rabbits, and production of the dermal Shwartzman reaction. *L. buccalis* endotoxin was highly active, and did not differ significantly from *E. coli* endotoxin in these tests with respect to biological activity.

The Shwartzman reaction was used for estimating the endotoxic activity of LPS from *B. fragilis* (Hofstad and Kristof-

TABLE XXIII-IV

DERMAL SHWARTZMAN REACTION ELICITED BY LPS FROM *B. MELANINOGENICUS*, STRAIN B10, *B. FRAGILIS*, STRAIN 9343 AND *S. NECROPHORUS*, STRAIN N167*

LPS	*Preparative Injections (μg)* 400	200	100	50	25	12.5	6.25	3.12	*Saline*
Exp. 1†									
LPS-B10	6/6§	6/6	3/6	1/6	1/6	0/6	0/6	0/6	0/6
LPS-*S. typhi* 0901				6/6	6/6	3/6	2/6	0/6	0/6
Exp. 2									
LPS-9343	6/6	6/6	3/6	1/6	0/6	0/6			0/6
Exp. 3									
LPS-N167			3/5	2/5	2/5	1/5	0/5	0/5	0/5

* From Hofstad (1970) and Hofstad and Kristoffersen (1970a; 1971b).
† Provocative injection 200 μg LPS-B10 + 50 μg LPS-*S. typhi* 0901 (Exp. 1), 400 μg LPS-9343 (Exp. 2) and 125 μg LPS-N167 (Exp. 3).
§ Positive Shwartzman reaction/total 18 hrs after provocative injection.

fersen, 1970a), *B. melaninogenicus* (Hofstad, 1970), and *S. necrophorus* (Hofstad and Kristoffersen, 1971b) (Table XXIII-IV). All LPS preparations were toxic but the endotoxic potency was low compared with endotoxin from *Salmonella typhi*. *B. melaninogenicus* LPS was lethal for mice, proved to be chemotactic for polymorphonuclear leukocytes, and had an infection-enhancing effect in mice when given together with infecting bacteria.

These scattered investigations have proved that LPS isolated from gram-negative anaerobic bacteria have endotoxic properties. Bacteroides LPS seem to be less toxic than LPS prepared from the other anaerobic bacteria. The difference in toxicity may be due to differences in chemical structure. Another possibility is that phenol-water extracted and purified LPS preparations from bacteroides are partly made up of high-molecular-weight compounds other than endotoxin.

REFERENCES

Araujo, W.C. de, Varah, E., and Mergenhagen, S.E.: Immunochemical analysis of human oral strains of *Fusobacterium* and *Leptotrichia*. *J Bacteriol*, *86*:837, 1963.

Bladen, H.A., and Mergenhagen, S.E.: Ultrastructure of *Veillonella* and morphological correlation of an outer membrane with particles associated with endotoxic activity. *J Bacteriol, 88:*1482, 1964.

Boivin, A, and Mesrobeanu, L.: Recherches sur les antigènes somatiques et sur les endotoxines des bactéries. I. Considérations générales et exposé des techniques utilisées. *Rev Immunol (Paris), 1:*553, 1935.

Böe, J.: *Fusobacterium:* Studies on its bacteriology, serology and pathogenicity. *Skr norske Videnskp—Akad, I Mat-nat Kl* No. 9, 1941, pp. 1–192.

Césari, E.: Études sur le bacille de Schmorl. *Ann Inst Pasteur, 26:*802, 1912.

Galanos, C., Lüderitz, O., and Westphal, O.: A new method for the extraction of R lipopolysaccharides. *Eur J Biochem, 9:*245, 1969.

Gewurz, H., Snyderman, R., Mergenhagen, S.E., and Shin, H.S.: Effects of endotoxic lipopolysaccharides on the complement system. In: Kadis, S., Weinbaum, G., and Ajl, S.J. (Eds.): *Microbial Toxins,* New York and London, Acad Pr, 1971, Vol. 5, pp. 127–149.

Gustafson, R.L., Kroeger, A.V., Gustafson, J.L., and Vaichulis, E.M.K.: The biological activity of *Leptotrichia buccalis* endotoxin. *Arch Oral Biol, 11:*1149, 1966.

Hausman, M.S., Snyderman, R., and Mergenhagen, S.E.: Humoral mediators of chemotaxis of mononuclear leukocytes. *J Infect Dis, 125:*595, 1972.

Hewett, M.J., Knox, K.W. and Bishop, D.G.: Biochemical studies of lipopolysaccharides of *Veillonella. Eur J Biochem, 19:*169, 1971.

Hofstad, T.: Chemical characteristics of *Bacteroides melaninogenicus. Arch Oral Biol, 13:*1149, 1968.

Hofstad, T.: Serological properties of lipopolysaccharide from oral strains of *Bacteroides melaninogenicus. J Bacteriol, 97:*1078, 1969.

Hofstad, T.: Biological activities of endotoxin from *Bacteroides melaninogenicus. Arch Oral Biol, 15:*343, 1970.

Hofstad, T., and Kristoffersen, T.: Chemical characteristics of endotoxin from *Bacteroides fragilis* NCTS 9343. *J Gen Microbiol, 61:*15, 1970a.

Hofstad, T., and Kristoffersen, T.: Chemical composition of endotoxin from oral *Veillonella. Acta Path Microbiol Scand* [*B*], *78:*760, 1970b.

Hofstad, T., and Kristoffersen, T.: Lipopolysaccharide from *Bacteroides melaninogenicus* isolated from the supernatant fluid after ultracentrifugation of the water phase following phenol-water extraction. *Acta Path Microbiol Scand* [*B*], *79:*12, 1971a.

Hofstad, T., and Kristoffersen, T.: Preparation and chemical characteristics of endotoxic lipopolysaccharide from three strains of *Sphaerophorus necrophorus. Acta Path Microbiol Scand* [B], *79:*385, 1971b.

Hofstad, T., Kristoffersen, T., and Maeland, J.A.: Serological properties of lipopolysaccharides from strains of oral *Veillonella. Acta Path Microbiol Scand* [*B*], *79:*615, 1971.

Hofstad, T., Kristoffersen, T., and Selvig, K.A.: Electron microscopy of endotoxic lipopolysaccharide from *Bacteroides, Fusobacterium* and *Sphaerophorus. Acta Path Microbiol Scand* [*B*], *80:*413, 1972.

Jensen, S.B.: *The influence of endotoxin from oral bacteria on host resistance.* Thesis, Arhus, 1967.

Jensen, S.B., Theilade, E., and Jensen, J.S.: Influence of oral bacterial endotoxin on cell migration and phagocytic activity. *J Periodont Res, 1:*129, 1966.

Kirchheiner, E.: Recherches biochimiques et immunologiques comparées sur *Sphaerophorus necrophorus* et *Sphaerophorus funduliformis. Ann Inst Pasteur, 64:*238, 1940.

Kristoffersen, T., and Hofstad, T.: Chemical composition of lipopolysaccharide endotoxin from human oral fusobacteria. *Arch Oral Biol, 15:*909, 1970.

Kristoffersen, T., Maeland, J.A., and Hofstad, T.: Serologic properties of lipopolysaccharide endotoxins from oral fusobacteria. *Scand J Dent Res, 79:*105, 1971.

Lüderitz, O., Staub, A.M., and Westphal, O.: Immunochemistry of O and R antigens of *Salmonella* and related *Enterobacteriaceae. Bacteriol Rev, 30:*192, 1966.

Mergenhagen, S.E., and Varah, E.: Serologically specific lipopolysaccharides from oral *Veillonella. Arch Oral Biol, 8:*31, 1963.

Mergenhagen, S.E., Hampp, S.E., and Scherp, H.W.: Preparation and biological activities of endotoxins from oral bacteria. *J Infect Dis, 108:*304, 1961.

Mergenhagen, S.E., Snyderman, R., Gewurz, H., and Shin, H.S.: Significance of complement to the mechanism of action of endotoxin. *Curr Top Microbiol Immunol, 50:*37, 1969.

Mergenhagen, S.E., Zipkin, I., and Varah, E.: Immunological and chemical studies on an oral *Veillonella* endotoxin. *J Immunol, 88:*482, 1962.

Shands, J.W., Jr.: The physical structure of bacterial lipopolysaccharides. In: Weinbaum, G., Kadis, S., and Ajl, S.J. (Eds.): *Microbial Toxins.* New York and London, Acad Pr, 1971, Vol. IV, pp 127–144.

Selvig, K.A., Hofstad, T., and Kristoffersen, T.: Electron microscopic demonstration of bacterial lipopolysaccharides in dental plaque matrix. *Scand J Dent Res, 79:*409, 1971.

Westphal, O., Lüderitz, O., and Bister, F.: Über die Extraktion von Bakterien mit Phenol/Wasser. Z *Naturforsch* [*B*], *7:*148, 1952.

PART IV

ANAEROBIC INFECTIONS—DISEASE SYNDROMES

CHAPTER XXIV

INFECTIONS OF THE CENTRAL NERVOUS SYSTEM

MORTON N. SWARTZ AND ADOLF W. KARCHMER

ABSTRACT: *An important role for a variety of anaerobic bacteria in the etiology of cerebral abscess has been suggested by bacteriologic data accumulated over the past several decades. Among 787 cases reported in 15 series between 1950 and 1967, anaerobic organisms were isolated in only 12 percent, but primary anaerobic cultures were not performed in some cases and when they were, the techniques were variable. "No growth" reported in 22 percent of the cases likely represented failure to isolate anaerobes in many cases. Among 30 cerebral abscesses studied at Massachusetts General Hospital over the past 9 years, 37 percent yielded anaerobic bacteria.*

Anaerobic bacteria, particularly streptococci, have also been involved in some cases of subdural empyema, for example, in 10 percent of 96 cases in 5 series. There is a need for further evaluation of specific anaerobic species in cerebral abscess, subdural empyema and also epidural abscess, employing the newer methods for isolation and identification of this group of microorganisms and for study of their susceptibility to antimicrobial agents.

Pyogenic infections of the central nervous system consist of five distinct clinical and pathological entities: 1. *bacterial meningitis* 2. *cerebral abscess* 3. *subdural empyema* 4. *epidural (cerebral or spinal) abscess* 5. *septic thrombophlebitis of the cortical veins and of the major cerebral venous sinuses.* A considerable body of information has accumulated concerning the bacteriology of the most common of these processes, pyogenic meningitis. The frequency with which anaerobic bacteria have been implicated appears to be quite low. In contrast, the role of anaerobic (or microaerophilic) species in cerebral abscess has been emphasized increasingly in recent years (Heineman and Braude, 1963). Subdural empyema occurs less frequently than cerebral abscess, and less data are available concerning its bacteriology. However, in view of the nature of the common initiating infections (chronic otitis media or infection of the nasal accessory sinuses), a role for anaerobic organisms would not be unexpected, and indeed this has proved to be the case. Cerebral epidural abscess (oftentimes a collection of infected granulations rather than a true abscess) usually represents extension of infection of the mastoid along the petrous ridge or extension of osteomyelitis involving the posterior wall of the frontal sinus. More bacteriologic data are needed before the role of anaerobes can be properly assessed in this infection. Septic phlebitis of the cortical veins or major dural sinuses usually occurs secondary to specific intracranial infections such as meningitis, subdural empyema or epidural abscess, or as the result of spread of infection about the face centrally along venous channels. Because of the limited accessibility of the involved tissues bacteriologic information has been meager. Since septic thrombophlebitis is most often secondary to infection in adjacent structures (meninges, subdural space, epidural area) the bacteria involved would likely be the same species as in the neighboring processes. Among the pyogenic central nervous system infections an etiologic role for anaerobes is clearest in cerebral abscess, and this will be the principal focus of this discussion.

CEREBRAL ABSCESS

Background Bacteriology

To provide a general overview of the microbial etiology of cerebral abscess, a summary of bacteriologic findings from 787 cases reported over a recent 18-year period has been compiled (Fig. XXIV-1) (Heineman and Braude, 1963; Ballantine and Shealy, 1959; Fog, 1958; Gates *et al.*, 1950; Gregory *et al.*, 1967; Kerr *et al.*, 1958; Kiser and Kendig, 1963; Newton, 1956; Levy, 1963; Liske and Weikers, 1964; Loeser and Scheinberg, 1957; Matson and Salam, 1961; McGreal, 1962; Newlands, 1965; Tutton, 1953). The bacteriologic techniques utilized, particularly in the search for anaerobic microorganisms, were far from uniform. More than one bacterial species was isolated from some of the abscesses. For comparison, the findings in these cases are contrasted with those reported in 207 patients with bacterial meningitis seen at the Massachusetts General Hospital (Swartz and Dodge,

1965). The bacterial species involved in the two processes are strikingly different. The various streptococci, members of the Enterobacteriaceae, and *Staphylococcus aureus* were most frequently isolated from the cerebral abscesses. These three groups of bacteria, relatively infrequent causes of bacterial meningitis, accounted for over 60 percent of the bacterial isolates. On the other hand, the three major species causing bacterial meningitis, *Hemophilus influenzae, Neisseria meningitidis,* and *Diplococcus pneumoniae* accounted for less than 8 percent of the isolates from the cerebral abscesses (Fig. XXIV-1). This is consonant with the view that cerebral abscess seldom if ever occurs as a complication of bacterial meningitis. The occurrence of the two processes simultaneously in a patient is usually the result of intraventricular leakage or rupture of a cerebral abscess. Occasionally, primary pyogenic meningitis is complicated by a ventricular empyema, a purulent collection that superficially mimics a cerebral abscess; but this in fact represents a different process with a different pathogenesis.

Two or more bacterial species were isolated from 14 percent of the 787 brain abscesses (Fig. XXIV-1). In contrast, mixed bacterial (involving two or more bacterial species simultaneously) meningitis is quite rare. The incidence of simultaneous mixed meningitis in children has been reported as high as 3.7 percent (Herweg *et al.*, 1963); in neonatal meningitis an incidence of 2.5 percent was noted in a summary of 478 cases from 17 series (Fosson and Fine, 1968). Mixed meningitis at the Massachusetts General Hospital accounts for less than 0.5 percent of the cases seen, encompassing all age groups.

Anaerobic organisms, principally anaerobic streptococci, members of the family Bacteroidaceae and *Actinomyces* species, accounted for 12 percent of the isolates in this composite series. This represents a minimum figure since primary anaerobic cultures were not performed in some cases, and, when such cultures were employed, the techniques were quite variable. The high percentage of cultures showing no growth (22%) in this compilation is likely due to the failure to isolate anaerobic organisms in many cases. Heineman and Braude (1963) were able to demonstrate the presence of anaerobic organisms in 16 of 18 consecutive cases of cerebral

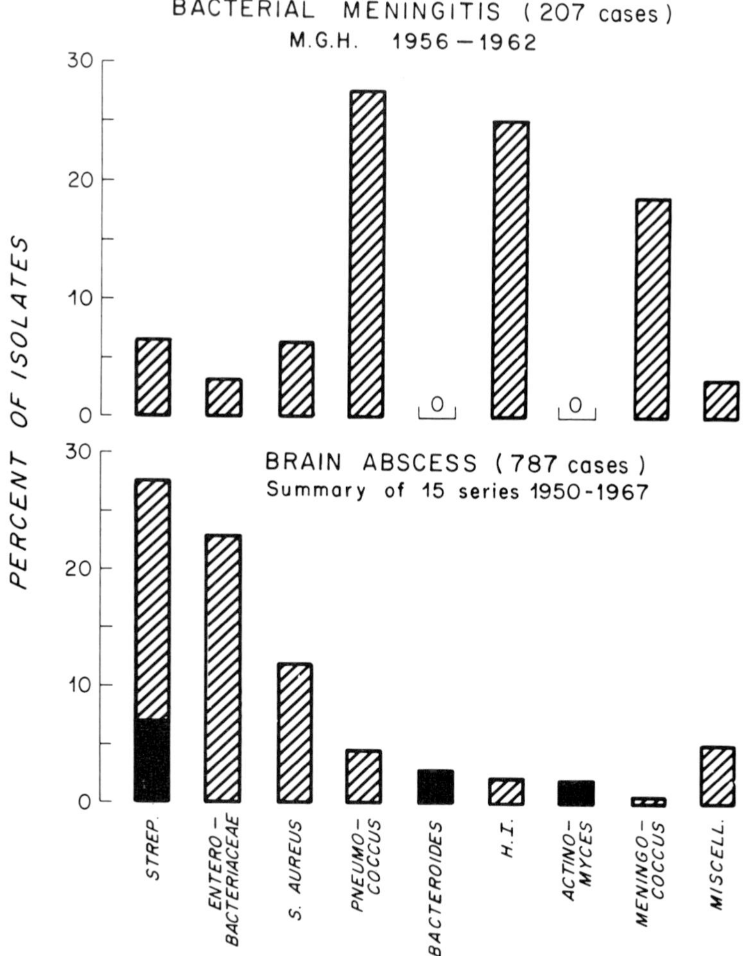

Figure XXIV-1. Comparison of the various bacteria involved in bacterial meningitis and cerebral abscess. Cases of bacterial meningitis were studied at the Massachusetts General Hospital from 1956 to 1962. Cases of brain abscess represent a compilation from 15 recently published series. The bars represent the percentage of the total isolates due to each group or species of bacteria. The solid black bars represent anaerobic bacterial species. A culture showing "no growth" was scored as a single isolate.

abscess by culturing abscess contents promptly and by utilizing primary anaerobic plating techniques. Anaerobic organisms alone were found in ten cases and mixed cultures of anaerobic and aerobic organisms were found in another six. Anaerobic streptococci were isolated most commonly; members of the family Bacteroidaceae were almost as frequent. The other anaerobic organisms isolated were *Actinomyces*, anaerobic corynebacteria, and *Veillonella* species. The isolation of anaerobic organisms from 88 percent of brain abscesses by these workers represents a considerably higher incidence than that reported in other series. It emphasizes the frequency with which these organisms can be implicated when appropriate care is given to their isolation and identification. The high incidence of anaerobic organisms in their series may have been related, to some extent, to the nature of the predisposing causes of the abscesses studied. None of these patients had penetrating head trauma, cyanotic congenital heart disease or pulmonary arteriovenous shunts, or demonstrated bacteremia. Half of the patients from whom anaerobic organisms were isolated had otogenic cerebral abscesses, and in another 20 percent of the patients the abscesses appeared to take origin in chronic infections of the nasal accessory sinuses. This is consistent with a prominent role for anaerobic organisms in chronic infections of the ears and sinuses.

Anaerobic organisms have been incriminated only rarely in cases of bacterial meningitis, even including neonatal meningitis where a wider range of bacteria may be involved. The role of anaerobes in meningitis may be underestimated since cerebrospinal fluid is not routinely cultured anaerobically, beyond the inoculation of thioglycollate broth media. However, in most cases of bacterial meningitis the initiating focus of infection and the route by which invasion occurs are quite different than in cases of cerebral abscess. Occasionally, meningitis due to a member of the family Bacteroidaceae occurs secondary to bacteremia or to spread of infection from chronic mastoiditis. Meningitis due to anaerobic (or microaerophilic) streptococci or to *Bacteroides* species may result from intraventricular leakage or rupture of a cryptic cerebral abscess. Indeed, the occurrence of meningitis due to such organisms should raise the question of a coexisting cerebral abscess. Rarely, meningitis due to *Clostridium perfringens* develops secondary to penetrating head trauma or as a post-craniotomy wound infection. Only one of the 207 cases of bacterial meningitis from the Massachusetts General Hospital involved an anaerobic organism, *C. perfringens.*

Current Series at the Massachusetts General Hospital

General Features

The diagnosis of cerebral abscess was established either at operation or at autopsy in 75 patients at the M.G.H. during the period 1957 to 1972. In 70 of these patients bacteriologic data were obtained. In only one quarter of the cases was the correct diagnosis made by the examining physician upon admission to the hospital (Table XXIV-I). Cerebral tumor or a cerebrovascular accident was suspected in some because the patient exhibited neither fever nor other manifestations of a pyogenic process. Other patients presented with the clinical features of bacterial meningitis and a cerebrospinal fluid pleocytosis. In retrospect, this was most often due to the intraventricular leakage of an abscess; but in a few instances the marked pleocytosis appeared to be associated with the presence of an abscess or bacterial cerebritis abutting on the

TABLE XXIV-I

INITIAL DIAGNOSIS IN 75 CASES OF BRAIN ABSCESS*

Diagnosis	*Percent*
Cerebral abscess	23
Meningitis	15
Brain tumor (1° or metastatic)	15
Cerebrovascular accident	10
Mastoiditis	10
Encephalitis	5
Subdural or epidural infection	4
Miscellaneous (hysteria, FUO, etc.)	18

* Massachusetts General Hospital, 1957–1972.

meninges. In a few patients with cerebral abscess the intracranial extension of the apparent mastoid infection was not appreciated initially.

The cerebral abscesses in this series were most frequently located in the temporal or frontal lobes (Fig. XXIV-2), undoubtedly related to the prominence of chronic ear (31%) and sinus (15%) infections in their genesis. Abscesses of bacteremic origin were sometimes more deeply situated and tended to involve the parietal lobes. However, otogenic abscesses sometimes extended up to involve the parietal lobe.

The mortality from brain abscess remains high despite advances in neurosurgical and anesthetic techniques and the availability of potent antimicrobial agents. This stems in part from the nature, location, and extent of the infection as well as from the not infrequent delay in establishing the correct diagnosis. In this series, if the cases with multiple, small abscesses associated with acute bacterial endocarditis and the bacteremic cases in patients dying of acute leukemia are excluded, the mortality was 40 percent. This is similar to the mortality rate observed in several recent large series (Ballantine and Shealy, 1959; Erasmus, 1966; Garfield, 1969).

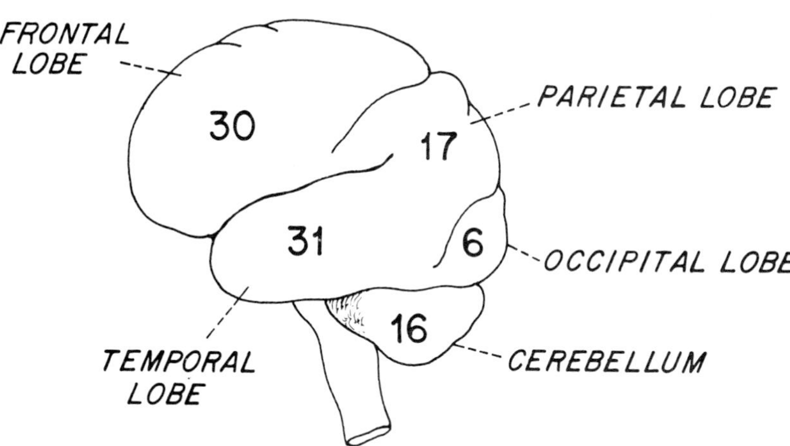

Figure XXIV-2. Sites of the 75 cerebral abscesses observed at the M.G.H. during the period 1957 to 1972. The numbers refer to the percent of the total located in each lobe and in the cerebellum.

Bacteriology

One or more bacterial species were isolated from 64 of the 70 patients with cerebral abscess who were studied bacteriologically. In 19 of the patients (27%) two or more species were identified on culture of the abscess (Table XXIV-II). In several patients as many as four different organisms were found.

TABLE XXIV-II

NUMBER OF BACTERIAL SPECIES ISOLATED FROM BRAIN ABSCESS*

Total number of cases 75
Cases with bacteriologic data 70

Number of Bacterial Species Isolated	*Number of Patients*	%
0	6	8
1	45	64
2	13	19
3	4	6
4	2	3

* Massachusetts General Hospital, 1957–1972.

Streptococci (including all aerobic and anaerobic species) were the leading cause of brain abscess and accounted for 37 percent of the isolates (Fig. XXIV-3). Streptococci belonging to groups H and D were the most common "groupable" streptococci found. *Staphylococcus aureus* strains comprised 21 percent of the isolates. *Bacteroides* species were third in frequency accounting for 10 percent of the total. Enterobacteriaceae, most commonly *Proteus* species and *Escherichia coli*, were usually found in association with abscesses of otogenic origin.

Etiologic Role of Anaerobic Bacteria

Anaerobic bacteria were isolated from 16 (23%) of the 70 cases of brain abscess (Table XXIV-III); however, among the 30 cerebral abscesses studied during the past 9 years, a period during which greater attention has been given to the isolation

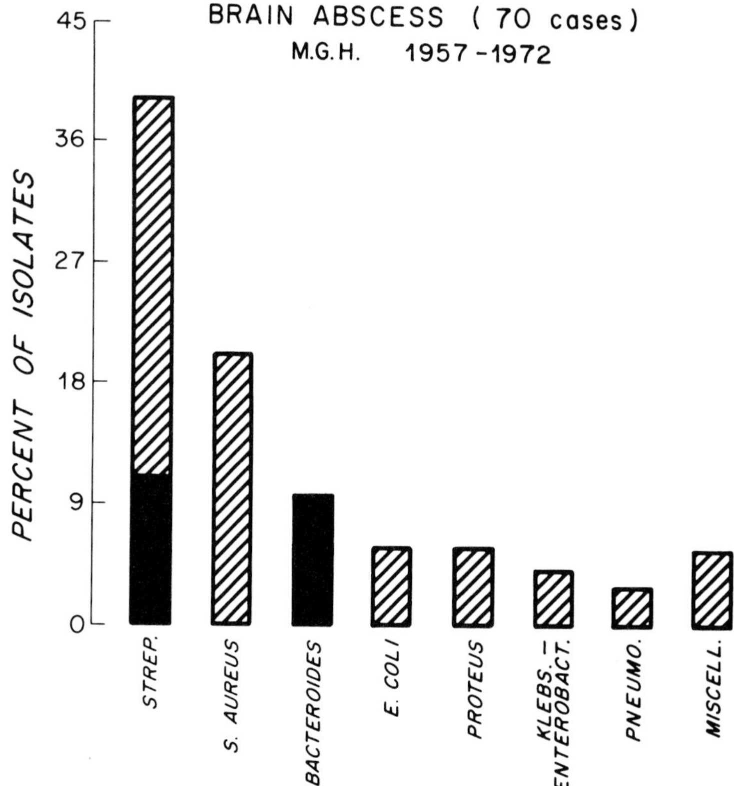

Figure XXIV-3. Various bacterial species isolated from 70 cases of cerebral abscess at the M.G.H. from 1957 to 1972. Each bar represents the percentage of the total isolates due to each species. The solid black bars represent anaerobic bacteria.

of anaerobic organisms, 37 percent yielded anaerobic bacteria. The incidence of anaerobic organisms in an earlier series from this hospital, covering the period 1946 to 1956, was 27 percent (Ballantine and Shealy, 1959). Anaerobic streptococci (including microaerophilic strains) were the anaerobic bacteria most frequently found in cerebral abscesses. Anaerobic streptococci were isolated alone or together with Bacteroidaceae. They were rarely isolated together with a facultatively aerobic organism. The association of anaerobic streptococci with Bacteroidaceae or other obligate anaerobic

TABLE XXIV-III

ANAEROBIC BACTERIA ISOLATED IN 70 CASES OF BRAIN ABSCESS*

Bacteria	*Number of Cases*
Anaerobic streptococci	6
Bacteroidaceae	4
Anaerobic streptococci and Bacteroidaceae	5
Anaerobic corynebacteria	1

* Massachusetts General Hospital, 1957–1972.

organisms was a prominent feature in the cases described by Heineman and Braude (1963).

Although the largest number of abscesses due to anaerobic bacteria were those of otogenic origin, anaerobes were responsible for abscesses of varied origins (Table XXIV-IV). Anaerobic bacteria were less often involved in abscesses secondary to disease of the nasal sinuses or to surgery and penetrating head trauma. The prominence of anaerobic bacteria, particularly anaerobic streptococci and Bacteroidaceae, in aspirational lung abscess and chronic pulmonary infections, probably accounts for their role in cerebral abscesses of pleuropulmonary origin. The occurrence of cerebral abscess as a complication of cyanotic congenital heart disease is well

TABLE XXIV-IV

NUMBER OF CASES OF BRAIN ABSCESS OF VARIOUS ORIGINS YIELDING ANAEROBIC ISOLATES IN SERIES OF 70 CASES*

Origin of Cerebral Infection	*Number of Cases§*		*Percent*
Otogenic	(22)†	6	27
Rhinogenic	(11)	1	9
Post-operative or associated with a foreign body	(8)	1	13
Pleuropulmonary	(6)	2	33
Congenital heart disease	(3)	2	66
Bacteremia	(11)	2	18
Cryptogenic	(9)	2	22

* Massachusetts General Hospital, 1957–1972.

† Numbers in parentheses represent the total number of abscesses in each category.

§ Numbers represent the number of abscesses in each category from which one or more species of anaerobic bacteria were isolated.

known (Newton, 1956; McGreal, 1962; Berthrong and Sabiston, 1951). A 4 to 6 percent incidence of hematogenous brain abscess in patients with congenital heart disease involving right-to-left shunts has been reported (Newton, 1956; Cohen, 1960). Such abscesses appear to have no special site of predilection, occurring with roughly equal incidence in frontal, parietal, and temporal lobes. These abscesses are most commonly solitary and are not associated with bacterial endocarditis. The high incidence of occlusive vascular disease affecting the nervous system, resulting from secondary polycythemia, and the right-to-left shunt, short-circuiting the normal pulmonary vasculature and the "filtering" role of the pulmonary capillary circulation, may be the major factors contributing to this predisposition. In this series, the three patients with brain abscess complicating congenital heart disease had the Tetralogy of Fallot. The etiologic agent in two of these was an anaerobic streptococcus. Newton (1956) has reviewed a total of 79 cases of cerebral abscess occurring in patients with cardiac malformations. Positive cultures of the contents of the abscesses were obtained in 22 of the cases. Nine of the 29 bacterial isolates appeared to be anaerobic organisms.

Among the 11 patients with bacteremic brain abscesses, four had acute endocarditis due to *S. aureus*. Another two patients had staphylococcal abscesses secondary to multiple furuncles. An additional two patients had bacteremic brain abscesses of odontogenic origin. From one of these both anaerobic streptococci and an anaerobic *Corynebacterium* species were isolated.

Only one of the brain abscesses occurring after craniotomy was due to an anaerobic organism (*Bacteroides* species). *S. aureus*, various streptococci, and *E. coli* were implicated in most of these cases. Two patients developed brain abscess many years following accidental penetration of the brain by a foreign body. One patient had sustained a penetrating injury from a shell fragment during World War II. A frontal lobe abscess related to this foreign body and due to *S. aureus* became manifest in 1968 and was removed surgically. The second patient was a seven-year-old child with a right frontal lobe abscess. At one year of age the child fell and injured his right eye with a stick. A foreign body was removed from the periorbital tissues but amblyopia resulted. At four years of age he began to have infrequent grand mal seizures. At age seven, during investigation of the seizure disorder, punctate calcifications were observed in the right frontal area on skull films, and cerebral angiography demonstrated a right frontal mass. At craniotomy a large thick-walled, chronic frontal lobe abscess was found, containing a piece of wood in its center. Four different bacteria (*S. aureus*, *Enterobacter*, *Citrobacter*, and *Hafnia* species) were isolated from the abscess. Three cases of cerebral abscess secondary to pencil point injuries with penetration of the cranial vault have been described, two due to *S. aureus* and one to *Diplococcus pneumoniae* (Horner *et al.*, 1964). Sir Hugh Cairns *et al.* (1947) have reported on their extensive experience in the British Army during World War II with cerebral abscess complicating head trauma. Twenty-three cases of acute intracranial abscess complicated 354 cases of penetrating brain injuries. Twelve of these contained clostridial species, principally, *C. perfringens*, *C. sporogenes* and *C. bifermentans*. Among the other nine patients from whom bacteriologic data were available, *S. aureus* was isolated from four and *S. albus* from another two. *E. coli*, streptococci, and pseudomonas were isolated from others, often in mixed cultures. The same authors studied 14 cases of chronic brain abscess occurring from three months to several years after a precipitating gunshot wound in the pre-penicillin era; straphylococci were the bacteria most commonly isolated (*S. aureus* in six and *S. albus* in five). One abscess contained anaerobic streptococci and *E. coli*. From their experience they concluded that organisms could survive for years not only in a chronic brain abscess but also in cerebral scars which later developed into acute abscesses.

Location of Abscesses Due to Anaerobic Bacteria

Anaerobic bacteria have been isolated from cerebral abscesses in each of the lobes of the brain. All anaerobic organisms isolated in this series, together with those whose localization was clearly defined in the report by Heineman and Braude (1963), have been summarized (Fig. XXIV-4). Since some of the abscesses contained more than one anaerobic species the total recorded is greater than the number of cases included. The number and variety of anaerobic species incriminated in temporal lobe abscesses is consistent with the prominence of anaerobic organisms in otogenic brain abscesses.

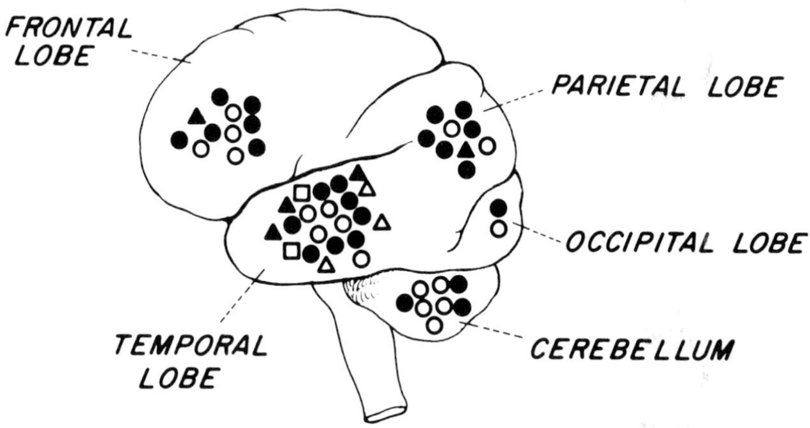

Figure XXIV-4. Sites of brain abscesses from which anaerobic organisms were isolated. The data are a composite from the cases at the M.G.H. (1957-1972) and the cases reported by Heineman and Braude (1963). Each symbol represents an abscess from which that organism was isolated. More symbols than abscesses are shown since some abscesses contained several anaerobic species. No attempt is made to accurately locate each symbol beyond placing it in the appropriate lobe.

SUBDURAL EMPYEMA

General Features

Subdural empyema is most commonly associated with infections of the ears and nasal sinuses, but it also develops as a complication of penetrating head trauma, intracranial surgery, hematogenous spread of infection to the subdural space (occasionally engrafted on a chronic subdural hematoma), and acute bacterial meningitis. Although in the early 20th century ear infection was the preponderant source of subdural empyema, antecedent sinus infection has achieved greater prominence in the past few decades (Kubik and Adams, 1943; Coonrod and Dans, 1972; Botterell and Drake, 1952). In our current series of 16 cases, antecedent ear and sinus disease were equally frequent. Subdural empyema, which accounts for about 20 percent of all cases of localized suppurative intracranial infection, is a life-threatening infection requiring both prompt surgical drainage and intensive antimicrobial therapy. It may occur together with cerebral abscess, meningitis, and epidural abscess. Cortical venous thrombophlebitis is a frequent complication, as these veins pass through the infected subdural space.

Bacteriology

To provide a sizeable sample the bacteriological data from four reported series of cases of subdural empyema (Botterell and Drake, 1952; Hitchcock and Andreadis, 1964; Bhandari and Sarkari, 1970; Wood, 1952) and an additional 16 cases seen at the Massachusetts General Hospital in the last 14 years have been summarized (Table XXIV-V). The total number of isolates exceeds the number of cases since more than one bacterial species were isolated from some of the empyemas. Anaerobic organisms, almost exclusively anaerobic streptococci, accounted for 10 percent of the total isolates. At this hospital the only anaerobic organisms isolated were anaerobic streptococci, accounting for 15 percent of the isolates. Half of the isolates in our cases were from empyemas associated with either ear or sinus disease. Two cases due

TABLE XXIV-V

BACTERIAL ISOLATES IN 96 CASES OF SUBDURAL EMPYEMA*

	Number of Isolates	*Percent*
Streptococci (other than Gp A)	27	25
Gp A streptococci	6	5
Anaerobic streptococci	10	9
S. aureus	18	16
Enterobacteriaceae	11	10
Corynebacterium species	4	4
H. influenzae	3	2
Pneumococci	3	2
Pseudomonas aeruginosa	2	2
C. perfringens	1	1
Miscellaneous	5	5
Negative cultures	20	20

* Compiled from 4 published series and 16 cases from Massachusetts General Hospital.

to *H. influenzae* occurred in infants and represented infected subdural effusions complicating *H. influenzae* meningitis.

EPIDURAL ABSCESS

We have only recently begun to look at the cerebral epidural infections at the Massachusetts General Hospital and have only very limited bacteriologic information. In the 13 cases studied, 8 had chronic otitis media with extension of infection through the mastoid. Various streptococci other than group A were isolated from six of the 13 cases. *S. aureus* was isolated from four. No anaerobic bacteria were found in this small group of cases. Obviously a more extensive search is needed before the role of anaerobic organisms in this process can be defined.

REFERENCES

Ballantine, H.T., Jr., and Shealy, C.N.: Role of radical surgery in treatment of abscess of the brain. *Surg Gynecol Obstet, 109:*370, 1959.

Berthrong, M., and Sabiston, D.C., Jr.: Cerebral lesions in congenital heart disease. *Bull Johns Hopkins Hosp, 89:*384, 1951.

Bhandari, Y.S., and Sarkari, N.B.S.: Subdural empyema: a review of 37 cases. *J Neurosurg, 32:*35, 1970.

Botterell, E.H., and Drake, C.G.: Localized encephalitis, brain abscess and subdural empyema. *J Neurosurg, 9:*348, 1952.

Cairns, H., Calvert, C.A., Daniel, P., and Northcroft, G.B.: Complications of head wounds, with especial reference to infection. *Br J Surg, War Surg Suppl, 1:*198, 1947.

Cohen, M.M.: The central nervous system in congenital heart disease. *Neurol, 10:*452, 1960.

Coonrod, J.D., and Dans, P.E.: Subdural empyema. *Am J Med, 53:*85, 1972.

Erasmus, J.F.P.: Cranial and spinal pyogenic disease—with notes on 415 personal cases. *S Afr Med J, 40:*(Part I) 416, (Part II) 433, 1966.

Fog, C.V.M.: Brain abscess. *Danish Med Bull, 5:*260, 1958.

Fosson, A.R., and Fine, R.N.: Neonatal meningitis. Presentation and discussion of 21 cases. *Clin Pediatr, 7:*404, 1968.

Garfield, J.: Management of supratentorial intracranial abscess. *Br Med J, 2:*7, 1969.

Gates, E.M., Kernohan, J.W., and Craig, W.M.: Metastatic brain abscess. *Medicine, 29:*71, 1950.

Gregory, D.H., Messner, R., and Zinneman, H.H.: Metastatic brain abscesses. *Arch Intern Med, 119:*25, 1967.

Heineman, H.S., and Braude, A.I.: Anaerobic infection of brain. *Am J Med, 35:*682, 1963.

Herweg, J.C., Middlekamp, J.N., and Hartmann, A.F., Sr.: Simultaneous mixed bacterial meningitis in children. *J Pediat, 63:*76, 1963.

Hitchcock, E., and Andreadis, A.: Subdural empyema: a review of 29 cases. *J Neurol Neurosurg Psychiat, 27:*422, 1964.

Horner, F.A., Berry, R.G., and Frantz, M.: Broken pencil points as a cause of brain abscess. *N Engl J Med, 271:*342, 1964.

Kerr, F.W.L., King, R.B., and Meagher, J.N.: Brain abscess—study of forty-seven consecutive cases. *JAMA, 168:*868, 1958.

Kiser, J.L., and Kendig, J.H.: Intracranial suppuration. Review of 139 consecutive cases with electron-microscopic observations on three. *J Neurosurg, 20:*494, 1963.

Kubik, C.S., and Adams, R.D.: Subdural empyema. *Brain, 66:18,* 1943.

Levy, L.F.: Intracranial abscess: the shape of things to come? *Br Med J, 1:*1455, 1963.

Liske, E., and Weikers, N.J.: Changing aspects of brain abscesses. *Neurol, 14:*294, 1964.

Loeser, E., Jr., and Scheinberg, L.: Brain abscesses: Review of ninety-nine cases. *Neurol, 7:*601, 1957.

Matson, D.E., and Salam, M.: Brain abscess in congenital heart disease. *Pediatrics, 27:*772, 1961.

McGreal, D.A.: Brain abscess in children. *Can Med Assoc J, 86:*261, 1962.

Newlands, W.J.: Otogenic brain abscess: Study of 80 cases. *J Laryngol Otol, 79:*120, 1965.

Newton, E.J.: Haematogenous brain abscess in cyanotic congenital heart disease. *Q J Med, XXV:*201, 1956.

Swartz, M.N., and Dodge, P.R.: Bacterial meningitis—review of selected aspects. *N Engl J Med, 272:*(Part I) 725, 779, 842, 898; (Part II) 954, 1003, 1965.

Tutton, G.K.: Cerebral abscess—the present position. *Ann Roy Coll Surg Engl, 13:*281, 1953.

Wood, P.H.: Diffuse subdural suppuration. *J Laryngol Otol, 66:*495, 1952.

CHAPTER XXV

ANAEROBIC PLEUROPULMONARY DISEASE: CLINICAL OBSERVATIONS AND BACTERIOLOGY IN 100 CASES

JOHN G. BARTLETT, VERA L. SUTTER
AND
SYDNEY M. FINEGOLD

ABSTRACT: *The clinical and bacteriological features of 100 cases of anaerobic pleuropulmonary disease are reviewed. The types of infections encountered included pneumonitis, pulmonary abscess, necrotizing pneumonitis and empyema. Suspected aspiration with infection in a dependent pulmonary segment was particularly common. Extrapulmonary anaerobic infection (periodontitis or abdominopelvic suppuration) and pulmonary lesions characterized by bronchial obstruction or pulmonary necrosis were additional, but less frequent, predisposing conditions. Seven patients had an underlying bronchogenic carcinoma; this emphasizes the need for proper investigation and follow-up evaluations.*

The clinical course was highly variable. One third of the patients had a relatively acute illness with symptoms prior to hospitalization for one week or less. Most of the remaining patients had prolonged symptoms which had sometimes persisted several months. Common presenting complaints included cough, putrid sputum, weight loss, fever and pleuritic chest pain.

The usual specimen sources utilized to establish the anaerobic etiology were pleural fluid when empyema was present, and percutaneous transtracheal aspiration when the infection was restricted to the pulmonary parenchyma. Blood cultures were positive in only two cases.

Organisms most commonly recovered were anaerobic gram-negative bacilli (Fusobacterium nucleatum, Bacteroides melaninogenicus, *and* Bacteroides fragilis) *and anaerobic gram-positive cocci. Aerobic potential pathogens were recovered concurrently in 33 cases.*

Therapy generally consisted of surgical drainage of empyemas and antimicrobial agents (usually penicillin or clindamycin) for parenchymal infections. Response to therapy was often delayed and prolonged antimicrobial therapy was usually required. Fourteen patients died from this infection and ten were left with residual symptomatic pulmonary disease.

Anaerobic bacteria are well established and relatively common pulmonary pathogens. In our experience these organisms are second only to *Diplococcus pneumoniae* in causing bacterial infections of the lung parenchyma and they are the most common isolates in empyema fluid. Unfortunately, precise incidence figures are not known and the vast majority of cases are never bacteriologically established. This results from the fact that clinicians are often not familiar with the clinical features of this infection. Even when this diagnosis is suspected it is seldom confirmed by proper culture of appropriate specimens. This led us to an analysis of our experience with anaerobic pleuropulmonary disease during the past 15 years. The clinical and bacteriological features of 100 well established cases are reviewed.

Actinomycosis is not included because this clinically distinctive anaerobic infection has been well described elsewhere (Warthin and Busheuff, 1958; McQuarrie and Hall, 1968). Also bronchiectasis was not considered unless accompanied by parenchymal infection or empyema.

TYPES OF INFECTIONS

The types of pulmonary infections encountered in this series are noted in Table XXV-I. Pulmonary abscesses were relatively large (>2 cm diameter), usually solitary cavitary lesions, while necrotizing pneumonitis was defined as multiple small cavities. These two overlapping pathologic processes were arbitrarily separated because of different prognostic implications.

No statement can be made regarding the true incidence of anaerobic pleuropulmonary disease. Although 100 cases

TABLE XXV-I

TYPES OF INFECTIONS

	CASES		
	1958–68	*1969–72*	*Total*
Pulmonary abscess without empyema	6	20	26
Pulmonary abscess with empyema	4	2	6
Necrotizing pneumonitis without empyema	7	10	17
Necrotizing pneumonitis with empyema	6	1	7
Empyema without abscess	20	8	28
Pneumonitis without empyema or abscess	0	15	15
Infected bronchogenic carcinoma cavity	1	0	1
Total	44	56	100
Total with empyema	30	11	41

were established over the 15-year period reviewed, more than one-half were diagnosed in the last 4 years. This reflects our greater awareness of this infection and improved diagnostic techniques during the latter part of the study. Nonetheless, even this more recent data probably represents only a fraction of the total cases at our hospital.

Anaerobes have been implicated in essentially all types of pulmonary infections. This diversity was also noted in the present series. One relatively consistent feature, however, was parenchymal necrosis resulting in cavitation (56 cases) or bronchopleural fistula underlying an empyema (documented in 14 cases). Fifteen patients had pneumonitis only. Ten of these were relatively acute illnesses in patients with symptoms less than one week. These may represent early stages of infections which would eventuate in pulmonary necrosis if allowed to continue untreated. Five patients, however, had chronic pneumonitis, a type of anaerobic infection which has been previously described (Smith, 1932).

As with other series (Tillotson and Lerner, 1967; Smith and Fekety, 1968; Bartlett and Finegold, 1972) the incidence of empyema reported here may be disproportionately high due to the relative ease of obtaining reliable specimens (pleural fluid) for anaerobic culture. This is particularly true of the first 11 years reviewed. The more frequent use of percutaneous transtracheal aspiration in the last four years has facilitated bacteriological documentation of anaerobic infections restricted to the pulmonary parenchyma. The types of infection reported during this latter period better represent the true nature of this disease.

CLINICAL FEATURES

The clinical features of anaerobic pleuropulmonary infection were best studied in the pre-antibiotic era (Smith, 1927, 1928, 1932; Guillemot *et al.*, 1904; Cohen, 1932; Lambert and Miller, 1927; Varney, 1920; Kline and Berger, 1935; Pilot and Davis, 1924; Allen and Blackman, 1936). At this time the natural history of the disease could be followed and autopsy or surgical specimens were commonly available for bacteriological confirmation. During this period more than 2,000 cases were reported, indicating the general awareness of these infections. Many of these reports referred to "fuso-spirochetal pulmonary disease." This appelation reflected the unique morphological characteristics of some of the isolates but tended to neglect the composite bacteriological findings. Most of the cases were thought to result from aspiration, because of the following evidence: organisms involved were those of the upper respiratory tract; many patients had preceding dental or pharyngeal anaerobic infections; typical pulmonary lesions could be produced in experimental animals with the intratracheal inoculation of gingivitis pus; many patients had conditions which predispose to aspiration, such as tonsillectomy, tooth extraction, general anesthesia, or other causes of compromised consciousness; and infections commonly occurred in dependent pulmonary segments. Nevertheless, some cases were clearly related to dissemination from other sites—particularly intraabdominal or pelvic infections with septic embolization or transdiaphragmatic spread.

With infections resulting from aspiration a typical sequence was described. Approximately one to two weeks after inoculation, there was an infiltrate on chest x-ray associated with the insidious onset of cough, nonfetid sputum and fatigue. Some cases were characterized by the early onset of fulminant symptoms with rapid death from bronchopneumonia. More commonly an indolent pneumonitis persisted for a week or more, with the eventual development of a solitary abscess, necrotizing pneumonitis, empyema, or combinations of these. Cavitation was often accompanied by copious amounts of fetid sputum and the appearance of an air-fluid level on chest x-ray. The subsequent course depended largely on the type of pulmonary lesion. Solitary abscesses went on to spontaneous cure in one-third of cases, were complicated by chronic residual symptomatic pulmonary disease in one-third, and were fatal in one-third. Necrotizing pneumonitis was usually (and sometimes rapidly) fatal. Empyemas not complicated by parenchymal cavitation generally responded when adequate surgical drainage was achieved, although the course was prolonged and often complicated. In all studies morbidity and mortality rates were high.

Although this sequence could be readily followed when the time of aspiration was known, *eg* after general anesthesia in patients or following inoculation of experimental animals, many patients lacked such a clearly defined preceding event. These latter patients usually presented in the mid-course of their disease with pernicious symptoms including cough, putrid sputum, weight loss, fever, and/or pleuritic chest pain. These symptoms and the x-ray changes often suggested tuberculosis, and many of these patients were mistakenly sent to sanitaria (Smith, 1967).

The clinical features of anaerobic pleuropulmonary disease in our experience are similar to these which have been described previously. Predisposing conditions in the present series are listed in Table XXV-II. Seventy-four percent of our patients had suspected aspiration. This was usually

TABLE XXV-II

PREDISPOSING CONDITIONS

Conditions	*Cases*
Aspiration	
Altered consciousness	63
Ethanolism	35
Cerebrovascular accident	9
Drug ingestion	6
Seizure disorder	5
General anesthesia	4
Miscellaneous causes	4
Dysphagia	5
Intestinal obstruction	5
Tooth extraction	2
Preceding anaerobic infection	
Periodontitis or gingivitis	30*
Intraabdominal	6
Local underlying conditions	
Bland pulmonary infarction	9
Bronchogenic carcinoma	7
Bronchiectasis	6
Foreign body	1
Systemic underlying conditions†	13
None apparent	15

* 62 patients had dental evaluations.

† Includes diabetes, extrapulmonary malignancy, corticosteroid therapy and immunosuppressive therapy.

associated with infection in a dependent pulmonary segment. Compromised consciousness from a variety of causes, especially ethanolism, was the usual precursor to aspiration. Although periodontitis was more common than in age-matched controls, this was an inconsistent finding. In fact, four patients with anaerobic infection clearly related to aspiration were edentulous. Five of the six patients with preceding intraabdominal suppuration had empyema formation above subphrenic collections, suggesting transdiaphragmatic spread. The previously described syndrome (Lemierre, 1936; Schottmueller, 1910; Thompson, 1931) of intraabdominal infection, suppurative thrombophlebitis and repeated septic emboli to various pulmonary segments was not observed. Pulmonary lesions which predisposed to anaerobic infection were those characterized by bronchial obstruction or necrosis. No apparent relationship was noted between anaerobic infection and other common types of pulmonary disease such as chronic bronchitis, emphysema or preceding pneumonitis caused by aerobic bacteria. Although 13 patients had systemic conditions which compromise host defenses, other previously noted predisposing conditions were also present in all but two. Finally, it should be noted that anaerobic pulmonary infection occurred in 15 patients with no apparent predisposing cause.

The clinical course of anaerobic pleuropulmonary disease was extremely variable. Seven patients had unsuspected infections detected on routine chest x-rays during hospitalization for other conditions, and five had fulminant infections which progressed to death in spite of optimal therapy. Most of the patients, however, presented with smoldering respiratory and constitutional symptoms similar to those previously noted. These were present for an average of four weeks prior to hospitalization. The average weight loss attributed to this infection was 15 lbs. Four patients were never febrile, three had peak temperatures greater than 105°F, and the average peak temperature was 102.6°F. The admission peripheral leukocyte count was less than 9,000/mm^3 in eight cases, greater than 25,000/mm^3 in 13 and averaged 18,000/mm^3.

Treatment usually consisted of prolonged antibiotic therapy for parenchymal infections and surgical drainage of empyemas. Duration of presenting symptoms, response to therapy, and outcome depended largely on the type of lesion (Table XXV-III). It is noted that patients with solitary abscesses had favorable courses; necrotizing pneumonitis was associated with excessive mortality; patients with empyema had prolonged recoveries with delayed healing of drainage sites; and patients with simple pneumonitis had relatively acute illnesses which responded rapidly to therapy.

DIAGNOSIS

The diagnosis of anaerobic pleuropulmonary infection is based on a high degree of clinical suspicion and collection of an appropriate specimen for careful anaerobic bacteriologi-

TABLE XXV-III

CLINICAL COURSE AND OUTCOME

	Abscess without Empyema	*Necrotizing Pneumonitis ± Empyema*	*Empyema*	*Pneumonitis only*	*Series*
Cases	*26*	*24*	*41*	*15*	*100*
Duration of symptoms prior to therapy	4*	2	4	0.5	3.5
Duration of fever after therapy	1.0	1.8	4	0.3	2.6
Duration of infections after therapy in patients followed to cure†	8	15	20	3	13
Outcome					
Cure	16§	13	22	8	56
Incompletely followed	5	3	10	2	16
Died of other cause	4	2	4	2	14
Died of this infection	1	6	5	3	14
Residual pulmonary disease	2	2	6	0	10

* Median duration in weeks.

† Patients were considered cured when chest x-rays were clear or stable, and empyema drainage sites were healed.

§ Total cases.

cal processing. Although clinical manifestations are highly variable, certain features should specifically suggest this diagnosis. Those which we have found most useful are listed in Table XXV-IV.

Most of our cases were characterized by tissue necrosis. This is in contrast to infections caused by *Diplococcus pneumoniae,* which seldom result in parenchymal necrosis. In fact, among 1,000 consecutive cases of untreated pneumococcal pneumonia followed by Cecil (1929) cavitation was never observed. Nevertheless, this must be considered nonspecific for anaerobic infection, since necrosis often characterizes pulmonary infections caused by *Staphylococcus aureus, Klebsiella pneumoniae,* fungi and mycobacteria, as well as certain noninfectious pulmonary diseases.

The insidious onset and indolent symptomatology which characterized the course in most of our cases would be unusual with pulmonary infections caused by most aerobic bacteria. This type of presentation, however, may lead to confusion with other types of pulmonary disease such as tuberculosis or bronchogenic carcinoma. This is illustrated by four of our cases in which thoracotomy was performed for suspected carcinoma, but only an anaerobic infection was discovered. Also, it should be noted that this type of prolonged smoldering

TABLE XXV-IV

CLUES TO ANAEROBIC PLEUROPULMONARY DISEASE (100 CASES)

	Cases
1. Pulmonary infection with tissue necrosis	
Abscess(es)	56
Bronchopleural fistula	14
2. Subacute or chronic presentation*	57
3. Underlying condition	
Suspected aspiration	74
Periodontitis or gingivitis	30
Intraabdominal infection	6
Preceding pulmonary lesion resulting in bronchial obstruction or pulmonary necrosis	19
4. Putrid discharge	53

* Symptoms present prior to hospitalization for more than two weeks or lesion discovered on routine chest x-ray.

course is not a consistent feature of anaerobic pulmonary infections. Thirty-three of our patients presented with symptoms which had been present for one week or less.

Putrid discharge is a valuable clue which is considered diagnostic of anaerobic infection. Occasionally, however, putrid sputum may result from oropharyngeal contamination of the expectorated sputum when the anaerobic infection is in the mouth rather than the lung. A greater problem is the tendency of some physicians to exclude the possibility of anaerobic infection if this finding is not present. In this series putrid discharge (sputum or empyema fluid) was not observed in nearly half of the cases. This is most likely to occur early in the course of the infection, with infections caused by anaerobic cocci, and with previously treated infections.

An additional clue to anaerobic infection may be the failure to recover a likely pathogen in aerobic cultures of expectorated sputum from patients with pulmonary infections. In our experience, however, expectorated sputum caused more confusion than clues. This reflects the established unreliability of these specimens in general (Barrett-Conner, 1971; Hoeprich, 1970). Sixty-seven patients in this series had expectorated sputum cultures, and aerobic potential pathogens were recovered in forty-four. Reliance on these specimens often resulted in delayed diagnosis and inappropriate therapy.

In summary, these clues may be helpful in determining when anaerobic pleuropulmonary disease is likely. None are specific and none are consistently present. If the clinician is presented with an alcoholic patient who has an abscess in a dependent pulmonary segment, typical chronic symptoms, putrid sputum, and no evidence of another explanation, the probability of anaerobic infection is quite obvious. Another patient may have anaerobic pneumonitis with the recent onset of symptoms and nonfoul sputum which yields *Escherichia coli* on culture. The diagnosis in this type of case will be made only if the clinician has a high degree of suspicion.

When anaerobic pleuropulmonary disease is suspected it is important to collect an appropriate specimen for careful anaerobic culture. In this series 14 percent of cases were fatal and 10 percent of patients were left with residual symptomatic pulmonary disease. This emphasizes the serious nature of this infection. Intensive and prolonged antimicrobial therapy is often required. This is optimally accomplished if the diagnosis is established and the pathogen is identified. Empirical use of antimicrobials may be unsatisfactory because susceptibility patterns of different anaerobes are variable (Martin *et al.*, 1972; Wilkins *et al.*, 1972; Sutter and Finegold, 1972). Important alternative diagnoses may be overlooked, and response may be difficult to evaluate because even properly treated cases often respond slowly. Once therapy has been initiated it may be difficult to recover the true pathogen, leaving the physician with a partially treated case and no definite diagnosis.

Specimen sources utilized to establish the diagnosis in this series are listed in Table XXV-V. Pleural fluid should be cultured anaerobically in all cases of empyema. With infections restricted to the pulmonary parenchyma a reliable culture source is less readily available. Any suitable specimen collection method in these cases must bypass the indigenous anaerobic flora of the upper respiratory passages (Rosebury, 1962) which "contaminate" anything which passes through. Thus, expectorated sputum, bronchoscopic aspiration and nasopharyngeal aspiration are unsatisfactory (Hoeprich, 1970). In cases not complicated by empyema we have found percutaneous transtracheal aspiration (Pecora, 1959) par-

TABLE XXV-V

CULTURE SOURCES UTILIZED TO ESTABLISH ANAEROBIC ETIOLOGY

	Cases
Pleural fluid	40 (7)*
Percutaneous transtracheal aspiration	57 (2)
Percutaneous transthoracic aspiration	1
Thoracotomy specimen	5 (2)
Blood culture	2 (2)
Distant metastatic site	1
Autopsy material	2 (1)

* Two or more of the sources listed yielded anaerobes.

ticularly useful. Our experience indicates that the tracheobronchial tree does not generally harbor anaerobes in the absence of anaerobic pulmonary infection. The procedure is safe when performed by experienced persons on patients with no contraindications (Kalinske *et al.*, 1967; Pecora, 1963), and we have experienced no serious complications with more than 300 aspirations. Other investigators have also found lung puncture aspiration useful in selected cases (Beerens and Tahon-Castel, 1965; Bandt *et al.*, 1972).

Following collection, specimens should be processed immediately or should be held in tubes containing oxygen-free gas (Attebery and Finegold, 1969). We attempt to limit air exposure of the specimen from the time of collection until incubation to 10 minutes. Gram stain should always be performed, since it will often permit a presumptive diagnosis and a guide to initial therapy.

Finally, bacteriological processing must be adequate to ensure optimal recovery (Bartlett and Finegold, 1972). Details of recommended anaerobic techniques have been described previously (Finegold, 1970; Sutter *et al.*, 1972; Cato *et al.*, 1970). We consider a minimal media requirement for anaerobes to be one all-purpose plate such as Brucella base blood agar, one selective plate such as kanamycin-vancomycin blood agar or neomycin blood agar (Finegold, 1970), and one all-purpose broth such as prereduced anaerobically sterilized chopped-meat glucose broth. Plates may be incubated in GasPak jars, which should not be opened for 48 hours. Although additional media and various anaerobic systems were often utilized during this study, the previous recommendations are generally adequate and realistic for most bacteriology laboratories.

BACTERIOLOGY

Bacteriological results of cases in the present series are listed in Table XXV-VI. In 25 cases a single anaerobe was recovered in pure culture, most commonly an anaerobic or microaerophilic gram-positive coccus. In 75 of the cases

TABLE XXV-VI

ANAEROBIC ISOLATES IN 100 CASES

	Number of Cases Isolated from	*Isolated in pure culture*
Anaerobic gram-negative bacilli		
Fusobacterium nucleatum	34	6
F. necrophorum	1	0
Bacteroides melaninogenicus	31	1
B. fragilis	21	2
B. oralis	16	0
Dialister pneumosintes (*Bacteroides pneumosintes*)	1	0
Unidentified	10	0
Anaerobic cocci		
Microaerophilic streptococci	26	12
Peptostreptococci	19	3
Peptococci	15	0
Veillonella species	7	0
Anaerobic gram-positive bacilli		
Propionibacterium species	9	0
Eubacterium species	5	0
Bifidobacterium species	3	0
Unidentified catalase-negative nonsporulating	10	1
Clostridium perfringens	5	2
Clostridium species	2	0
Concurrent aerobic potential pathogens	33	

there were two to nine anaerobic isolates. The predominant organisms were anaerobic gram-negative bacilli and anaerobic gram-positive cocci, which respectively accounted for 53 and 28 percent of the total anaerobes recovered. No anaerobic species was shown to be disproportionately associated with a specific type of pulmonary lesion or adverse course. Of particular interest is the recovery of *B. fragilis* in 21 cases, ten of which were associated with suspected aspiration. This may have important therapeutic implications, since this organism is often resistant to commonly used antibiotics such as penicillin, cephalosporins and tetracycline (Martin *et al.*, 1972; Wilkins *et al.*, 1972; Sutter and Finegold, 1972).

Aerobic potential pathogens were recovered in combination with anaerobes in 33 cases. These were *Staphylococcus aureus* (9 cases), *Pseudomonas aeruginosa* (7), *E. coli* (7), klebsiella-enterobacter (6), *Proteus* species (4), *D. pneumoniae* (4), *Hemophilus influenzae* (3), *Pseudomonas maltophilia* (1), *Eikonella corrodens* (1), group A β-hemolytic streptococcus (1) and *Providencia* species (1). When both aerobes and anaerobes are recovered it may be difficult to determine the true pathogen. Some cases may represent bacterial synergy with both types of organisms playing a contributing role. In other cases some of the isolates may be merely colonizing the infected site. In this series the following factors were helpful in establishing the etiologic role of the anaerobic isolates: putrid discharge; numerical dominance of anaerobes on direct Gram stain and in culture; type of pathologic process; recovery of anaerobes from additional reliable sources; failure of patients to respond to antibiotics directed exclusively against aerobes recovered; and good response to antimicrobial therapy directed exclusively against anaerobes recovered.

THERAPY

Recommended therapy for anaerobic pleuropulmonary disease generally consists of adequate drainage of empyemas and appropriate antimicrobial therapy for parenchymal infections.

Most of the patients in this series received penicillin G or clindamycin. Other antimicrobials which proved clinically useful include lincomycin, chloramphenicol, tetracycline, erythromycin and metronidazole. No conclusion can be made regarding the comparative therapeutic efficacy of these agents since many patients received antibiotic combinations, dosage and duration of therapy was highly variable, and outcome was more directly related to other factors such as drainage of empyemas, type of pulmonary infection and underlying condition. However, on the basis of our clinical experience with this disease and from *in vitro* susceptibility data a rational approach to antimicrobial therapy is summarized.

As previously noted, an appropriate specimen should be collected when this diagnosis is suspected. Direct Gram stain will often show the distinctive morphology of anaerobes, permitting a presumptive diagnosis and an initial guide to therapy. When culture results are subsequently known, more specific therapy may be selected. Since most hospital laboratories do not accurately test antimicrobial susceptibilities of anaerobic bacteria the clinician must often rely on published data concerning the specific anaerobe(s) identified (Martin *et al.*, 1972; Wilkins *et al.*, 1972; Sutter and Finegold, 1972).

Specimens which yield potential aerobic pathogens as well as anaerobes may present a therapeutic dilemma. In several such cases in the present series there was good response to therapy directed only against the anerobic isolates. However, if the concurrently recovered aerobe is resistant to the agent(s) selected and clinical response is suboptimal, or if the infection is fulminant, appropriate changes may be indicated.

Most of the anaerobic bacteria encountered in this review show good *in vitro* sensitivity to penicillin G (Martin *et al.*, 1972; Wilkins *et al.*, 1972; Sutter and Finegold, 1972) and extensive clinical data have established the efficacy of this agent. Ampicillin, cephaloridine and lincomycin have an anaerobic spectrum comparable to penicillin G (Martin *et al.*, 1972; Finegold *et al.*, 1967; Bartlett *et al.*, 1972). Other penicillins and cephalothin are less active *in vitro* (Martin *et al.*, 1972; Finegold *et al.*, 1970).

Bacteroides fragilis is relatively resistant to all of these agents. This organism was recovered in 21 of our cases. The therapeutic implications of this finding are inconclusive. It is conceivable that some of these cases would respond to the high doses of penicillin G often used. Even when very resistant *B. fragilis* strains are involved the patient might respond due to the activity of this agent against other components of infections containing mixtures of anaerobes. These possibilities could not be evaluated in the present series since alternative agents with established activity against this organism were selected in most of these 21 cases. In view of the serious nature of anaerobic pulmonary infections, and until these points are clarified, this latter approach is recommended.

Clindamycin shows excellent *in vitro* activity versus anaerobes including *B. fragilis* (Martin *et al.*, 1972; Bartlett *et al.*, 1972). Good clinical results, relative freedom from side effects, and ease of administration indicate that this may become the agent of choice for many anaerobic infections, including pleuropulmonary infections (Bartlett, 1972). Chloramphenicol is active against the entire spectrum of anaerobes and has been used extensively in extrapulmonary anaerobic infections with good results. Potential bone-marrow suppression and availability of alternative agents, however, has restricted its use in anaerobic pulmonary infections. Metronidazole also shows good *in vitro* activity against anaerobes; clinical experience with this agent is limited but results of initial trials appear promising (Tally *et al.*, 1972). Tetracycline and erythromycin are inconsistently active against anaerobes (Martin *et al.*, 1972) and should be used only when indicated by susceptibility data.

The timing, dosage and duration of antimicrobial therapy are important considerations. Patients with pneumonitis only will often respond well to short-term therapy. When parenchymal necrosis (abscess formation) has occurred the lesion is characterized by avascular tissue which is teeming with organisms. These cases often respond slowly to therapy and morbidity and mortality rates are high. Insufficient dosage or a delay in proper therapy may result in progressive pulmonary destruction, empyema formation, or distant dissemination, especially to the brain. Therefore early use of an appropriate agent is recommended. When fever and toxicity subside and x-ray improvement is noted, oral therapy may replace parenteral therapy. Antibiotics should be continued until the chest x-ray is clear or shows a small, stable residual. This may require several months and can often be accomplished on an outpatient basis. In some of our cases the premature discontinuation of therapy resulted in relapse; this emphasizes the need for extended treatment. Long-term follow-up is necessary in all cases to exclude an underlying lesion, particularly bronchogenic carcinoma.

Patients with anaerobic pulmonary infection without empyema can usually be treated medically. Surgical resection of a chronic pulmonary abscess which persists with symptoms following six weeks of therapy has sometimes been recommended (Waterman and Domm, 1954). In our experience, and that of others (Block *et al.*, 1969; Weiss, 1968), these lesions will eventually respond to intensive antimicrobial therapy for prolonged periods. Bronchoscopy may be useful in facilitating drainage of abscess cavities and in detecting underlying lesions such as bronchogenic carcinoma, foreign body or other cause of bronchial obstruction.

Adequate drainage of empyemas is crucial. When the fluid is thin and free-flowing this may be accomplished by needle aspiration. More frequently the fluid is thick or loculated and will require thoracotomy drainage. This was performed in 37 of the 41 empyema cases in this series. Although a closed drainage procedure was often performed first, 32 patients eventually required open drainage with rib resection.

REFERENCES

Allen, C.I., and Blackman, J.F.: Treatment of lung abscess with report of 100 consecutive cases. *J Thorac Surg, 6:*156, 1936.

Cato, E.P., Cummins, C.S., Holdeman, L.V., Johnson, J.L., Moore, W.E.C., Smibert, R.M., and Smith, L.DS.: *Outline of Clinical Methods in Anaerobic Bacteriology.* Blacksburg, Virginia Polytechnic Inst. and State Univ., 1970.

Attebery, H.R., and Finegold, S.M.: Combined screw-cap and rubber stopper closure for Hungate tubes (pre-reduced, anaerobically sterilized roll tubes and liquid media). *Appl Microbiol, 18:*558, 1969.

Bandt, P.D., Blank, N., and Castellino, R.A.: Needle diagnosis of pneumonitis. Value in high-risk patients. *JAMA, 220:*1578, 1972.

Barrett-Conner, E.: The nonvalue of sputum culture in the diagnosis of pneumococcal pneumonia. *Am Rev Respir Dis, 103:*845, 1971.

Bartlett, J.G., and Finegold, S.M.: Anaerobic pleuropulmonary disease. *Medicine, 51:*413, 1972.

Bartlett, J.G., Sutter, V.L., and Finegold, S.M.: Treatment of anaerobic infections with lincomycin and clindamycin. *N Engl J Med, 287:*1006, 1972.

Beerens, H., and Tahon-Castel, M.: *Infection Humaines à Bactéries Anaérobies Non-toxigènes.* Bruxelles, Presses Academiques Européennes, 1965.

Block, J.A., Wagley, P.F., and Fisher, M.A.: Delayed closure in lung abscess, a re-evaluation of the indications for surgery. *Johns Hopkins Med J, 125:*19, 1969.

Cecil, R.L.: *Textbook of Medicine.* Philadelphia, Saunders, 1929.

Cohen, J.: The bacteriology of abscess of the lung and methods for its study. *Arch Surg, 24:*171, 1932.

Finegold, S.M.: Isolation of Anaerobic Bacteria. In: Blair, J.E., Lenette, E.H., and Truant, J.P. (Eds.): *Manual of Clinical Microbiology.* Bethesda, Am Soc for Microbiology, 1970, ch. 32.

Finegold, S.M., Davis, A., and Miller, L.G.: Comparative effect of broad spectrum antibiotics on nonsporeforming anaerobes and normal bowel flora. *Ann NY Acad Sci, 145:*268, 1967.

Finegold, S.M., Sutter, V.L., and Sugihara, P.T.: Activity of antibiotics against anaerobic bacteria and human fecal flora. In: *Bact. Proc.— 1970.* Abstr. M269, Bethesda, Am Soc Microbiol, 1970, p. 114.

Guillemot, L., Halle, J.,- and Rist, E.: Recherches bacteriologiques et expérimentales sur les pleurésies putrides. *Arch Med Exper Pt D'Anat Path, 16:*571, 1904.

Hoeprich, P.D.: Etiologic diagnosis of lower respiratory tract infections. *Calif Med, 112:*1, 1970.

Kalinske, R.W., Parker, R.H., Brandt, D., and Hoeprich, P.: Diagnostic usefulness and safety of transtracheal aspiration. *N Engl J Med, 276:*604, 1967.

Kline, B.S., and Berger, S.S.: Pulmonary abscess and pulmonary gangrene. Analysis of ninety cases observed in ten years. *Arch Intern Med, 56:*753, 1935.

Lambert, A.V.S., and Miller, J.A.: Abscess of lung. *Arch Surg, 8:*446, 1924.

Lemierre, A.: On certain septicemias. *Lancet, 1:*701, 1936.

Martin, W.J., Gardner, M., and Washington, J.A.: *In vitro* antimicrobial susceptibility of anaerobic bacteria isolated from clinical specimens. *Antimicrob Agents Chemother, 1:*148, 1972.

McQuarrie, D.G., and Hall, W.H.: Actinomycosis of the lung and chest wall. *Surgery, 64:*905, 1968.

Pecora, D.V.: A comparison of transtracheal aspiration with other reliable methods in determining the bacterial flora of the lower respiratory tract. *N Engl J Med, 269:*664, 1963.

Pecora, D.V.: A method of screening uncontaminated tracheal secretions for bacterial examination. *J Thorac Surg, 37:*653, 1959.

Pilot, I., and Davis, D.J.: Studies in fusiform bacilli and spirochetes. IX. Their role in pulmonary abscess, gangrene and bronchiectasis. *Arch Intern Med, 34:*313, 1924.

Rosebury, T.: *Microorganisms Indigenous to Man.* New York, McGraw-Hill, 1962, pp. 314–331.

Schottmueller, H.: Allgernmeinen Krankenhaus Hamburg-Eppendorf. *Mitt Grenzt der Med und Chirg, 21:*450, 1910.

Smith, D.D., and Fekety, F.R.: Bacteroides empyema. *Ann Intern Med, 68:*1178, 1968.

Smith, D.T.: Experimental aspiratory abscess. *Arch Surg, 14:*231, 1927.

Smith, D.T.: Fusospirochaetal diseases of the lungs. *Tubercle, 9:*420, 1928.

Smith, D.T.: *Oral Spirochetes and Related Organisms in Fuso-Spirochetal Disease.* Baltimore, Williams and Wilkins, 1932.

Sutter, V.L., Attebery, H.R., Rosenblatt, J.E., Bricknell, K.S., and Finegold, S.M.: *Anaerobic Bacteriology Manual.* Univ. of Calif., Los Angeles Ext. Div., 1972.

Sutter, V.L., and Finegold, S.M.: Antibiotic disc susceptibility tests for rapid presumptive identification of gram-negative anaerobic bacilli, *Appl Microbiol, 21:*13, 1972.

Tally, F.P., Sutter, V.L., and Finegold, S.M.: Metronidazole versus anaerobes, *in vitro* data and initial clinical observations. *Calif Med, 117(6):*22, 1972.

Tillotson, J.R., and Lerner, A.M.: Bacteroides pneumonias. Characteristics of cases with empyema. *Ann Intern Med, 68:*308, 1967.

Thompson, L.: Bacteremia due to anaerobic gram-negative organisms (Bacteroides). *Mayo Clin Proc, 6:*372, 1931.

Varney, P.L.: Bacterial flora of abscesses of the lung. *Arch Surg, 19:*1602, 1920.

Warthin, T.A., and Busheuff, B.: Pulmonary actinomycosis. *Arch Intern Med, 101:*239, 1958.

Waterman, D.H., and Domm, S.E.: Changing trends in the treatment of lung abscess. *Dis Chest, 25:*40, 1954.

Weiss, W.: Delayed cavity closure in acute nonspecific primary lung abscess. *Am J Med Sci, 255:*313, 1968.

Wilkins, T.D., Holdeman, L.V., Abramson, I.J., and Moore, W.E.C.: Standardized single-disc method for antibiotic susceptibility testing of anaerobic bacteria. *Antimicrob Agents Chemother, 1:*451, 1972.

CHAPTER XXVI

INFECTIVE ENDOCARDITIS CAUSED BY ANAEROBIC BACTERIA

JOEL M. FELNER

ABSTRACT: *Findings are summarized for 48 cases of infective endocarditis due to anaerobic bacteria. The most common isolates were* Propionibacterium acnes *(15 cases) and* Bacteroides fragilis *(14); clostridia were found in 9. Clinical manifestations were similar to those in endocarditis due to facultative bacteria, and, as in the latter, the most common portal of entry was the oropharynx (15 cases). However, in endocarditis due to anaerobic bacteria more patients had gastrointestinal disease as a predisposing cause. Other differences were a lesser prevalence of preexisting heart disease and a greater incidence of emboli complicating the course of the anaerobic infection. Fourteen (29%) of the patients died. If a bactericidal antibiotic is not used in* B. fragilis *endocarditis, the prognosis is very poor; the best antibiotics currently available for treating this infection are clindamycin and carbenicillin.*

Anaerobic bacteria, like facultative bacteria, can cause endocarditis, but there are relatively few reports describing such cases. Therefore, bacteriologic and clinical data pertaining to 48 patients with endocarditis due to anaerobic bacteria are presented in this paper.

CLINICAL SERIES

The 48 cases were obtained from two sources: (1) patients from whom bacterial isolates had been submitted to the Center for Disease Control from 1963 to 1970 and whose clinical records were reviewed, and (2) patients admitted to the medical service at Grady Memorial Hospital in 1971 to 1972. All 48 had clinical courses compatible with endocarditis, and 44 had at least two positive blood cultures. In three patients with *Propionibacterium acnes* endocarditis and one with *Bacteroides fragilis* endocarditis, the organism was also recovered from the valvular vegetation either during surgery or post mortem. Confirmation of the diagnosis was obtained by autopsy findings in eight patients.

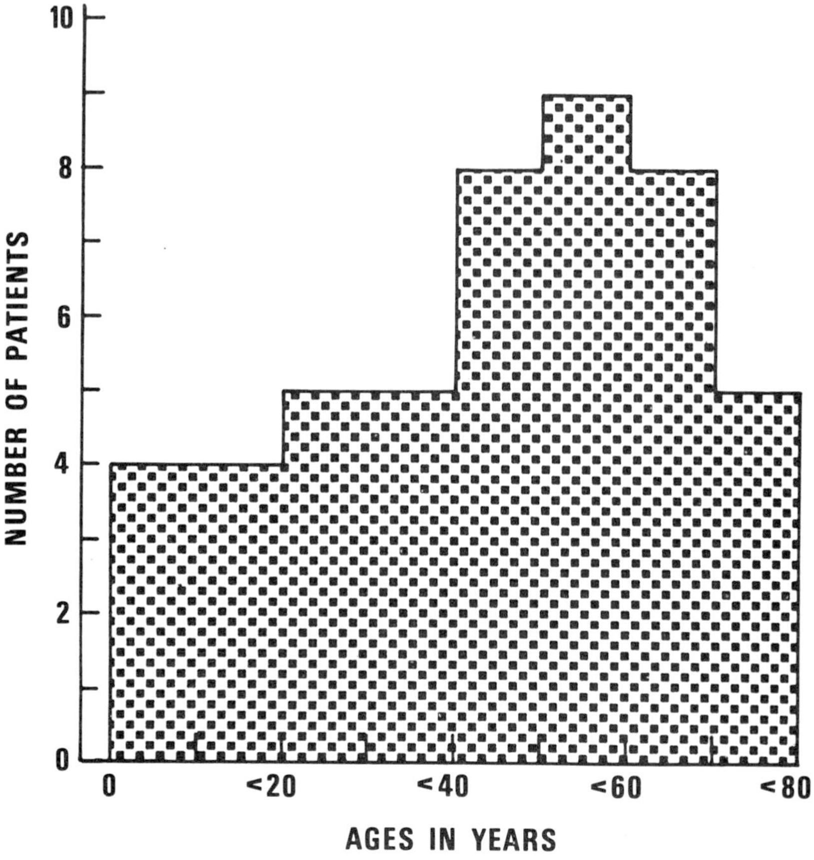

NUMBER OF PATIENTS

10
8
6
4
2
0

0 <20 <40 <60 <80

AGES IN YEARS

Figure XXVI-1. Age of patients with endocarditis due to anaerobic bacteria.

The series included 25 males and 23 females ranging in age from 5 to 78 years (median 45 years). Thirty (65%) were more than 39 years old (Fig. XXVI-1).

The organisms isolated are listed in Table XXVI-I. Bacteriologic methods were noted previously (Felner and Dowell, 1970). Probable portals of entry for the organisms are indicated in Table XXVI-II. Of the 15 patients with an oropharyngeal portal of entry, 13 had periodontal infections and two had sinusitis. Peptic ulcer, diverticulitis, and appendicitis were most common gastrointestinal conditions;

TABLE XXVI-I

ANAEROBIC ISOLATES IN 48 CASES OF ENDOCARDITIS

Organism	*No. Cases*
Nonsporulating anaerobes	
Gram-negative bacilli	
Bacteroides fragilis	14
B. melaninogenicus	1
Fusobacterium nucleatum	4
F. necrophorum	3
Gram-positive bacilli	
Propionibacterium acnes	15
Anaerobic cocci	
Peptostreptococcus intermedius	2
Clostridia	
Clostridium perfringens	3
C. sordellii	2
C. sporogenes	1
C. cochlearium	1
C. subterminale	1
C. septicum	1

in several cases abscesses had developed postoperatively. Genitourinary portals of entry included renal infections, septic abortion and pelvic inflammatory disease.

CLINICAL MANIFESTATIONS. All 48 patients were febrile during their hospital course; 16 (33%) developed splenomegaly; 11 (23%) petechiae; 5 Osler's nodes; 5 splinter hemorrhages; 3 subungual hematomas; 3 Roth spots; and 3 clubbing of the fingers. Anemia, the most common laboratory finding, occurred in 35 patients (73%) and leukocytosis in

TABLE XXVI-II

PROBABLE PORTAL OF ENTRY IN PATIENTS WITH ENDOCARDITIS

Organism	*Oro-pharynx*	*Lung*	*Gastro-intestinal Tract*	*Genito-urinary Tract*	*Skin*	*Unknown*
Bacteroides fragilis	3	1	6	2	2	0
B. melaninogenicus	1	0	0	0	0	0
Fusobacteria	5	1	1	0	0	0
Propionibacterium acnes	2	0	2	0	3	7
Clostridia	2	0	3	3	0	1
Peptostreptococci	1	0	0	1	0	0
Totals	15	2	12	6	5	8

23 (48%). Hematuria and proteinuria were also common laboratory findings.

Forty-six patients had audible heart murmurs; 37 of these were systolic only, 3 diastolic only and 6 both systolic and diastolic. In two patients without audible murmurs, endocarditis was confirmed at necropsy. Twenty-four patients (50%) had preexisting heart disease: rheumatic in 9, atherosclerotic in 9 and congenital in 6. Surgical procedures, prior antibiotic therapy, steroid administration and diabetes mellitus were other common predisposing factors.

The most common anaerobes isolated in this series were *P. acnes* and *B. fragilis*. Table XXVI-III compares some characteristics of cases from which these organisms were isolated.

COMPLICATIONS AND OUTCOME. Complications secondary to the cardiac problem were congestive heart failure, cardiogenic shock and arrhythmias, most commonly atrial fibrillation. Complications of bacteremia were metastatic abscesses, thrombophlebitis, endotoxic shock, empyema, peritonitis and fistulae. Forty-three episodes of emboli were documented in 31 patients: cerebral emboli in 12, renal in 9, pulmonary in 8, coronary arterial in 4, joint in 4, major arterial in 3 and hepatic in 3.

Fourteen patients (29%) died; all were more than 39 years old. Of the 7 patients with *B. fragilis* endocarditis who died,

TABLE XXVI-III

CHARACTERISTICS OF CASES OF ENDOCARDITIS DUE TO *B. FRAGILIS* AND *P. ACNES*

	P. acnes	*B. fragilis*
No. of cases	15	14
Age range in years	40 (8)*	40 (11)
Valve involved	aortic (8)	mitral (9)
Prior surgery	cardiovascular (5)	abdominal (6)
Complications	cardiovascular usually	cardiovascular or bacteremic
Deaths	5	7
Antibiotic(s) of choice	penicillin	clindamycin carbenicillin chloramphenicol

* No. patients.

5 were more than 60 years old, 3 had had recent abdominal surgery, and only 2 had received appropriate antibiotic therapy. Of the 5 with *P. acnes* endocarditis who died, only one was more than 60 years old and all had received appropriately high doses of penicillin, but 3 had had prosthetic aortic valves inserted. The other patients who died were a 5-year-old boy with tetralogy of Fallot and *Fusobacterium nucleatum* endocarditis who died following unsuccessful therapy for a cerebral abscess, and a patient with *Clostridium septicum* endocarditis who had in addition an occult spinal cord neoplasm (discovered at necropsy) which probably contributed to his demise.

ADDITIONAL CASE. Since the completion of this series, a 49th patient with infective endocarditis due to anaerobic bacteria has been treated on the medical service at Grady Memorial Hospital. A 33-year-old black housewife had a laparoscopic tubal ligation and 5 days later returned to the hospital with fever, chills, anemia and a new systolic heart murmur. Bacteroides bacteremia was demonstrated. Therapy with chloramphenicol was started, but when the patient's condition continued to deteriorate, high doses of intravenous clindamycin phosphate were substituted and an exploratory laparotomy was performed. This revealed a large pelvic abscess and an ovarian vein thrombosis. Despite multiple episodes of embolization, the patient has done well.

DISCUSSION

There are only a few published reports of endocarditis caused by anaerobic organisms other than anaerobic streptococci. As noted by Felner and Dowell (1970), of 1,046 patients included in four recently published series, only 14 (1.3%) had anaerobic bacteria isolated from the blood. However, 15 percent of the patients had "sterile" cultures, which leads to the suspicion that the actual percentage of anaerobic isolates might have been higher if adequate anaerobic culture procedures had been used.

The clinical manifestations of endocarditis due to anaerobic bacteria do not differ appreciably from those described for endocarditis due to facultative microorganisms. However, in the series reported here, the percentage of patients with emboli was higher and the percentage with preexisting heart disease was lower than the corresponding percentages reported for patients with endocarditis due to facultative organisms (Morgan and Bland, 1959; Vogler *et al.*, 1962; Pankey, 1961). Several factors may contribute to the higher frequency of embolization. In a series of 250 bacteremias due to bacteroides and fusobacteria (Felner and Dowell, 1971), thrombophlebitis was found in 20 percent and embolization in 30 percent of the patients. In patients with thrombophlebitis, early inflammatory masses may dislodge and result in renewed sepsis with subsequent development of endocarditis. Heparinase production by bacteroides and fusobacteria and inadequate or delayed antibiotic therapy may further contribute to septic embolization and renewed sepsis. In two series of infective endocarditis (Morgan and Bland, 1959; Vogler *et al.*, 1962), the overall frequency of embolization was approximately 35 percent. A much higher prevalence was found in the series reported here. Among 22 patients with endocarditis due to bacteroides or fusobacteria, 5 had thrombophlebitis and 16 (73%) had at least one documented episode of embolization. Heparinase activity of bacteroides and fusobacteria might account in part for this difference (Gesner and Jenkin, 1961; Bjornson *et al.*, 1970).

Among patients with endocarditis due to facultative bacteria (Morgan and Bland, 1959; Volger *et al.*, 1962), previously damaged heart valves were involved in approximately 75 percent of the patients. In the present series of endocarditis due to anaerobic bacteria, only 24 (50%) of the patients had clinical evidence of preexisting heart disease.

As in endocarditis due to facultative organisms, in this series the most common presumed portal of entry for the infecting organisms was the oropharynx. The majority of the normal oropharyngeal flora are composed of anaerobic bacteria, including fusobacteria and peptostreptococci, so this source is not unlikely. In the present series, unlike series of endocarditis cases due to facultative organisms, an apparent gastrointestinal focus of infection was frequently present early in the course of the illness. Bacteroides and clostridia account for most of the intestinal microbiota, and abdominal surgery may induce transient bacteremia and may compromise the blood supply to a loop of the intestine, resulting in hypoxia which enables the anaerobes to proliferate. In addition, small perforations may produce abscesses which may produce metastatic emboli and continued bacteremia.

Of the 15 patients in this series with *P. acnes* endocarditis, five had had cardiovascular surgical procedures, including prosthetic aortic valve insertion in three. Although diphtheroids rarely cause endocarditis, the incidence of *P. acnes* endocarditis is unusually high among patients who have had open heart surgery (Levin, 1966; Johnson *et al.*, 1968). This may be related to the resistance of this organism to the commonly used antibiotic prophylaxis and to contamination of sutures of synthetic graft material, which frequently proves refractory to all antibiotic treatment regimens.

The portal of entry for *P. acnes* was often hard to determine in this series. In patients who had undergone surgery the most likely source of *P. acnes* was the cardiopulmonary bypass machinery. However, since these organisms are normal skin inhabitants and are ubiquitous in nature, they may have been introduced by the surgeon or derived from the patient. Johnson and Kaye (1970) studied 52 patients with infections caused by aerobic or anaerobic diphtheroids. Thirty-one (60%) had endocarditis, and 13 of these had had cardiopulmonary bypass procedures; 11 had received prosthetic aortic valves. Of the 31 patients, 11 died despite appropriate antibiotics, and 6 of these 11 had had prosthetic aortic valve implantations. In another study of 9 patients who developed infective endocarditis after insertion of a valve prosthesis, 3 had diphtheroid endocarditis (Kaplan and Weinstein, 1969).

Even though many anaerobic bacteria are exquisitely sensitive to several antibiotics *in vitro*, endocarditis due to these organisms presents difficult therapeutic problems. Blood cultures may remain positive for many months despite apparently adequate treatment, and the organisms are often found in mixed cultures, e.g. peptostreptococci and *Bacteroides melaninogenicus*. Loculated abscesses, unless surgically drained, can provide a nidus for recurrent bacteremia—as frequently occurs with bacteroides and fusobacteria. Another problem is that recognition of diphtheroids, e.g. *P. acnes*, is difficult because these organisms are common contaminants of clinical cultures, frequently revert to L-forms, and can easily be confused with streptococci and listeria. Finally, when a bactericidal antibiotic is not employed for treatment, as often happens in *B. fragilis* endocarditis, the prognosis is extremely poor. Therefore, clindamycin phosphate and carbenicillin, given in high doses intravenously, are the best available antibiotics for treating *B. fragilis* endocarditis.

REFERENCES

Bjornson, H.S., Hill, E.O., and Atlemeier, W.A.: Role of L-forms, *Bacteroides* species, and *Sphaerophorus* species in acute and recurrent thromboembolic disease. *Bacteriol. Proc.—1970.* Washington, D.C., Am Soc Microbiol Abstract #M87, p. 87.

Felner, J.M., and Dowell, V.R., Jr.: Anaerobic bacterial endocarditis. *N Engl J Med, 283:*1188, 1970.

Felner, J.M., and Dowell, V.R., Jr.: "Bacteroides" bacteremia. *Am J Med, 50:*787, 1971.

Gesner, B.M., and Jenkin, C.R.: Production of heparinase by bacteroides. *J Bacteriol, 81:*595, 1961.

Johnson, W.D., Cobbs, C.G., and Arditi, L.I., *et al.:* Diphtheroid endocarditis after insertion of a prosthetic heart valve. *JAMA, 203:*919, 1968.

Johnson, W.D., and Kaye, D.: Serious infections caused by diphtheroids. *Ann NY Acad Sci, 174:*568, 1970.

Kaplan, K., and Weinstein, L.: Diphtheroid infections of man. *Ann Intern Med, 70:*919, 1969.

Levin, J.: Diphtheroid bacterial endocarditis after insertion of a Starr valve. *Ann Intern Med, 64:*396, 1966.

Morgan, W.L., and Bland, E.F.: Bacterial endocarditis in the antibiotic era. *Circulation, 19:*753, 1959.

Pankey, G.A.: Subacute bacterial endocarditis at the University of Minnesota Hospital, 1939 through 1959. *Ann Intern Med, 55:*550, 1961.

Vogler, W.R., Dorney, E.R., and Bridges, H.A.: Bacterial endocarditis. *Am J Med, 32:*910, 1962.

CHAPTER XXVII

A SURGEON'S VIEW OF THE IMPORTANCE OF ANAEROBIC BACTERIA IN MUSCULOSKELETAL INFECTIONS

HARVEY R. BERNARD

ABSTRACT: *Although infections caused by* Clostridium tetani *and* Clostridium perfringens *are obviously important and may be devastating if tardily recognized and improperly treated, they are unusual and should be preventable by good surgical care. The other anaerobic musculoskeletal infections are rarely lethal, and usually are relatively easily treated by simple surgical measures, with or without knowledge of the exact anaerobic flora. Until rapid methods are made available for anaerobe identification, the identification of anaerobes in musculoskeletal infection is of little practical value to the surgeon except in the rare case of a chronic nonhealing lesion.*

Anaerobic bacteria, recently "rediscovered," are now rising to popularity in the "procession of the pathogens" referred to by Professor R.E.O. Williams in his keynote address to the 1970 proceedings of the International Conference on Nosocomial Infections. Professor Williams emphasized the difficulties in determining the significance of changes in the prevalence of the various bacterial pathogens. Anaerobes are now and will be found associated with an increasing number of infectious lesions. Some of the increase in prevalence may relate to an increase in the number of diseases caused by anaerobic bacteria, but much is related to an increased awareness of the possibility of anaerobic growth and a resultant increase in the effort made to detect their presence. Since there is an important difference between prevalence in cultures and significant infections caused by anaerobic bacteria, working alone or synergistically, and since it is important that a reasonable perspective be achieved, I submit these comments.

The time afforded for preparation of this report precluded a prospective study, and I have drawn primarily upon our clinical experience with anaerobic infections of the musculoskeletal system at the Albany Medical Center over the last six years, a review of the autopsy reports for the last 24 months at the Albany Medical Hospital, and a study of the anaerobic bacteriology of surgical wounds associated with open colonic operations.

RECENT CLINICAL EXPERIENCE

Over the past six years we have been closely involved with the clinical management of 13 patients who have suffered severe musculoskeletal infections. Eight suffered acute severe myonecrosis as a consequence of infection with *Clostridium perfringens* as the dominant organism, usually in pure culture. Our laboratory has applied fluorescent antibody methods to the identification of *C. perfringens* in clinical infections. Seven of the eight infections involved the lower extremity. One invaded the muscular and fascial layers of the abdomen. All were treated by immediate and extensive debridement or excision of the involved muscle and fascial compartment. This meant amputation in each of the seven lower extremity wounds and involved excision of almost the entire right lower quadrant of the abdomen in the other. Five involved penetrating wounds resulting from automobile and farm accidents. Two occurred postoperatively in amputation stumps. The abdominal infection followed an operation for small intestinal obstruction from operative adhesions. All patients were treated with penicillin or cephalothin and were monitored carefully for evidence of respiratory failure. The surgical care of these patients proceeded parallel to the development of our Shock Unit and all were carefully monitored for deviations from normal in blood gas determinations. In each instance when this occurred normobaric oxygen therapy was utilized. No unusual measures have as yet been required in this regard and each of these patients so treated has lived and was discharged, healed, from the hospital.

Five additional patients treated at the beginning of this series are of interest. Two were infected with *Bacillus* species which resulted in confusion with *Clostridium perfringens* on initial Gram stain. Both underwent major amputations. In one the amputation was the consequence of the injury but in the other, the bacteriologic uncertainty contributed to the decision to perform a major amputation. Severe additional injuries were present in this patient and the patient eventually died as a consequence of many factors. A third patient, following compound fracture and dislocation of the knee, was found to be infected with 3 strains of clostridia, one of which was *C. perfringens.* Major amputation was eventually performed after extensive debridement because of unsatisfactory residual anatomical structures. The infection, however, was always relatively superficial and was controlled by local surgical measures. Not all gas which is found in tissue results from clostridial infections. The fourth and fifth patients exhibited rapidly progressing nonclostridial gas infections of the lower extremity. One resulted from perforation of a colonic diverticulum with dissection of intestinal gas along the iliopsoas muscle into the thigh. The other exhibited only a subcutaneous collection of gram-negative bacteria which infected a cutdown wound.

One infection with *Clostridium tetani* has been seen. This followed injury to a finger tip with a kitchen knife in a restaurant in a young Italian immigrant who had never been immunized against tetanus.

AUTOPSY SERIES

In the past two years approximately 50 percent of the deaths in our hospitals have come to autopsy, and we have studied autopsy records for this period to determine the likely frequency of the most important consequence of anaerobic musculoskeletal infection, that is, the death of the patient. The causes of death and other lesions existing at the time of death were tabulated in a search for lesions related to anaerobic bacterial infection in the musculoskeletal system. There were no deaths related to amputation or wounding of the extremities in association with spreading infection. One patient died of generalized sepsis following an operation on the lumbar spine. However, the pus from many wounds in this patient was never malodorous, and careful bacteriologic investigations for both anaerobic and aerobic flora disclosed only repeated isolation of *Micrococcus tetragenus*. In 950 autopsies 22 significant infections were associated with various disruptions, operative or primary, of the gastrointestinal tract (principally colonic). It is likely that a mixed bacterial flora was present, including anaerobic bacteria. Peritonitis was universally present. Although infection of the abdominal fascia and muscle occurred, no patient in this series died from invasive anaerobic infection of the extremities.

Over the same two-year period, our surgical bacteriology laboratory had the opportunity to study exudate from 75 perirectal abscesses, leg ulcers, infected diabetic feet and toes, *etc*. A wide variety of anaerobic bacteria has been recovered. There have also been many such contaminated wounds that we have not studied. However, despite the fact that the clinician usually treated his patients without exact knowledge of the anaerobic flora, therapeutic failure, at least as judged by the death of the patient, has not been a frequent consequence.

Our experience over the past two years is consistent with that of Nettles *et al.* (1969) at the Mayo Clinic, who noted only 11 musculoskeletal infections in 18 years caused by bacteroides. Their review indicated only nine previously published osseous infections involving bacteroides. Pearson and Harvey (1971) found 60 patients in whom bacteroides infections complicated orthopedic conditions. Bacteroides were recovered in pure culture in only three patients.

ANAEROBES AND OPERATIONS ON THE COLON

We have been interested in the role the anaerobic bacteria within the colon play in the frequency of infection following operations upon the gastrointestinal tract, especially operations on the colon. During a study of the effect of various antimicrobial preparations on the incidence of wound infections following such surgery, we identified and quantitated the anaerobic flora of the lumen of the intestine, the peritoneal cavity at some distance following completion of resection, and the superficial subcutaneous wound just prior to skin closure. Anaerobic bacteria were prevalent following colonic preparation with kanamycin. Bacteroides were cultivated from the intestinal lumen in all patients and *C. perfringens* in 30 percent. The findings indicated that contamination of the peritoneum and superficial wounds by various species of *Bacteroides* and *Clostridium* and other anaerobic bacteria is usual, and does not regularly lead to postoperative infection. We conclude that simple contamination by these bacteria of the peritoneal surface and the subcutaneous wound is of minor significance unless there is a continuing, uncontrolled leak of fecal content.

SUMMARY

Infections caused by *C. tetani* and *C. perfringens* are obviously important and may be devastating if tardily recognized and improperly treated. However, they are unusual and should be preventable by good surgical care. Musculoskeletal infections caused by a pure culture of anaerobic bacteria are decidedly unusual in our experience, and when a mixed flora is identified it is difficult to sort out the relative importance of these components. It is increasingly obvious that anaerobic bacteria are present, along with aerobes, in infections of the toes and feet, particularly of diabetics, and in infections associated with fistulae, sinuses and perforations of the gastrointestinal and genitourinary tract. However, except in cases in which infections are associated with peritonitis and abdominal wall sepsis, such infections are rarely lethal and usually are relatively easily treated by simple surgical measures, with or without knowledge of the exact anaerobic flora. Acute major skeletal osteomyelitis caused by anaerobic bacteria may occur, but must be rare; in our experience it has not been lethal. Secondary contamination of chronic osteomyelitis may or may not be significant, but has not caused fatality.

Cultivation and exact identification of many anaerobic bacteria require considerable time and cost. With the exception of clostridial infections, the anaerobic musculoskeletal infections are rarely lethal. Therefore, the identification of anaerobes in musculoskeletal infection will remain of little practical value to the surgeon, except in the rare case of a chronic nonhealing lesion, at least until more practical methods are made available for rapid identification of anaerobic bacteria.

REFERENCES

Nettles, J.L., Kelly, P.J., Martin, W.J., and Washington, J.A.: Musculoskeletal infections due to bacteroides. *J Bone Joint Surg,* [A] *51:*230, 1969.

Pearson, H.E., and Harvey, J.P. Jr.: Bacteroides infections in orthopedic conditions. *Surg Gynecol Obstet, 132:*876, 1971.

Williams, R.E.O.: Changing perspectives in hospital infection. In: Brachman, P.S., and Eickhoff, T.C. (Eds.): *Proceedings of the International Conference on Nosocomial Infections.* p. 1–10, Chicago, Ill., Amer. Hosp. Assoc., 1971.

CHAPTER XXVIII

URINARY TRACT INFECTIONS DUE TO ANAEROBIC BACTERIA

WILLIAM JEFFERY MARTIN
JOSEPH W. SEGURA

ABSTRACT: *The precise role of anaerobic bacteria in the urinary tract is not clear, although their association with significant urologic disease is undeniable. Case summaries are presented for ten patients who had anaerobes in urine specimens obtained by suprapubic bladder aspiration. Most had had multiple courses of therapy with antibiotics, but the significance of this association is conjectural. Anaerobes are probably not associated with uncomplicated cases of cystitis or asymptomatic bacteriuria in the absence of significant urologic disease, and previous investigators who concluded that routine anaerobic culture of urine is not warranted are certainly correct. The fact that anaerobes are associated with serious urologic problems indicates that anaerobic cultures of urine are needed in such patients.*

Anaerobic bacteria were first isolated from infections related to the urinary tract around the turn of the century by Albarran and Cottet (1898). The source of most of these isolations was frankly purulent drainage from periurethral abscesses, pyelonephrosis, renal abscesses, cases of frank gangrene, and cases of chronic infection. Few well-documented cases of anaerobic bacteriuria in the absence of associated purulent drainage have been described. In 1949 Beigelman and Rantz isolated *Bacteroides* species from urine in four cases. One isolate was associated with pyelonephritis and possible thrombophlebitis in a postpartum patient. In the other case in which a *Bacteroides* species was isolated in large numbers, right pyelonephritis with calculus disease was noted. In the other two cases the colony count was so low that contamination could not be ruled out. In the other published series up to 1964 there were no cases in which *Bacteroides* bacteriuria was demonstrated in a setting of uncomplicated cystitis or under circumstances such that contamination of the specimen could not be reasonably assumed.

The literature was reviewed by Finegold *et al.* in 1964,

when they reported 15 *Bacteroides* isolates which they accumulated over several years. Several patients in that study had nothing to suggest urinary tract infections, whereas others had chronic pyelonephritis and calculus disease. Until then, no effort had been made to determine the incidence of anaerobic bacteriuria in the urinary tract. These investigators cultured 100 randomly obtained urine specimens without taking any special anaerobic precautions and isolated 12 anaerobes from midstream specimens. Most of these were associated with low colony counts, although one was isolated in counts greater than 10^5 organisms/ml. Moreover, all 12 were associated with facultative bacteria.

In a series of analyses of routine specimens, Headington and Beyerlein (1966) recovered 158 anaerobes from 15,250 consecutive midstream specimens; again no particular effort was made to isolate anaerobic bacteria. More recently, Kuklinca and Gavan (1969) cultured 200 randomly collected urine samples and found anaerobes alone in 10; only one of these was associated with discernible urinary tract disease. Cases in which anaerobes might have been isolated in association with aerobes, however, were discarded from the study. Of the 200 samples in their study, 159 were discarded because of aerobic growth. Furthermore, no distinction was made between catheter and midstream samples. They concluded, correctly no doubt, that anaerobic cultures of routine urine specimens were not necessary.

In an effort to establish the anaerobic bacterial flora of urine from clinically normal persons, Holdeman, Cato and Moore (1972) examined 20 midstream specimens from men and 20 catheterized specimens from women; culturing was performed quantitatively in prereduced anaerobic brain-heart infusion agar (BHIA) and aerobic BHIA. Anaerobic and aerobic counts ranged from less than 10/ml to 1,200/ml for 19 specimens from women; one specimen had 10^7 facultative bacteria/ml. The counts in specimens from men did not exceed 10^3/ml. Anaerobic bacteria isolated by these workers were *Peptococcus variabilis*, *P. prevotii*, *Peptostreptococcus intermedius*, *Veillonella parvula*, *V. alcalescens*, *Peptococcus* sp., and *Propionibacterium* sp.

MAYO CLINIC STUDIES

In 1970, Segura *et al.* undertook studies to determine the minimal incidence of anaerobic bacteria in patients at the Mayo Clinic. Urine specimens were obtained for culture and for Gram stain on a routine basis by the midstream clean voided technique. In about 13 percent of the cases, significant numbers of bacteria were identified by Gram stain (defined as the presence of two or more bacteria per field under oil immersion) and then confirmed by culture (Segura *et al.*, 1972). Based on previous experience in this laboratory, such a positive Gram stain of 0.01 ml of uncentrifuged urine is almost always associated with colony counts of greater than 100,000/ml. On occasion, however, it was noted that certain of the bacteria observed on Gram stain did not grow in an aerobic environment. These observations suggested the presence of anaerobic bacteriuria in such cases. Inasmuch as microorganisms from the distal portion of the urethra and adjacent perineal and vulvar structures may contaminate the midstream specimen, confirmation of anaerobic bacteriuria by some other means was essential.

A convenient method of obtaining such specimens is by suprapubic bladder aspiration. This was emphasized by Finegold and associates (1964) and our study, as far as we know, is the first since then to utilize this technique in an attempt to determine the incidence of anaerobic bacteriuria. Suprapubic bladder aspirations were done on two groups of patients over a 5-month period in 1970: (1) patients whose aerobic culture of midstream urine specimens obtained by clean voided technique did not reveal organisms which were present in significant numbers on Gram stain of the same specimen; and (2) patients who required suprapubic bladder aspirations for other reasons, generally an inability to void in the lithotomy position; these patients served as a control group.

For suprapubic bladder aspiration, the patient is asked to report with a full bladder. The lower abdomen is prepared with alcohol and, at a point approximately two fingerbreadths above the symphysis pubis, the skin is infiltrated

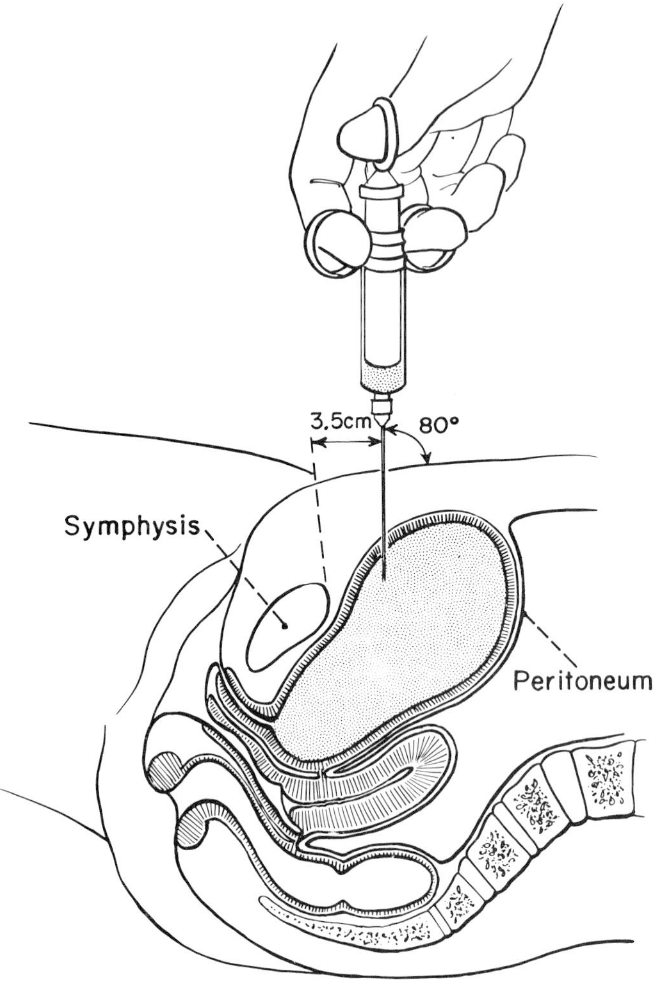

Figure XXVIII-1. Technique for suprapubic bladder aspiration.

with a local anesthetic. A long needle is then inserted directly into the bladder, at an angle of approximately 10° off the vertical (Fig. XXVII-1). The complication rate is almost zero. The only precaution necessary is that the patient must feel that the bladder is full. In women it frequently is difficult to palpate a full bladder, so one is dependent on the patient's report of a full bladder. Suprapubic aspirations can be done safely in pregnant patients, although the technique is somewhat different.

During the period of this study, a 1-ml sample of the aspirated urine was injected aseptically into prereduced sterilized "E" medium (Robin Laboratories, Division of Scott Laboratories, Chapel Hill, N.C.) with every effort being made to avoid any possible contamination with air. At the present time, however, we are inoculating the urine samples directly into an anaerobe specimen tube, such as that described by Attebery and Finegold (1969), rather than into a vial of "E" medium. Isolation and subsequent identification of anaerobic and aerobic bacteria have been described elsewhere (Segura *et al.*, 1972).

In the five months of the study, 5,781 midstream urine specimens were obtained by the clean voided technique for Gram stain and culture (Table XXVII-I). Bacteria were present in significant numbers on Gram stain in 795 specimens. Cultures of these specimens revealed aerobic bacteria in all but 25 specimens. Suprapubic bladder aspirations were done in 17 cases and anaerobic bacteria were confirmed in 10, for an incidence of anaerobic bacteriuria of 0.2 percent of the total patient population or of 1.2 percent of total patients with significant bacteriuria. Growth occurred as mixed infection with various aerobes and facultative bacteria. Members of

TABLE XXVIII-I

INCIDENCE OF ANAEROBIC BACTERIA IN URINE*

		Percentage of:		
Findings	*Total (5,781 Specimens)*	*Total*	*Positive Gram Stain*	*Positive Gram Stain and Negative Culture*
Positive Gram stain	795	13.8	...	...
Positive Gram stain and negative culture	25	0.4	3.2	...
Anaerobes by suprapubic bladder aspiration	10	0.2	1.3	40

* From Segura *et al.* (1972).

the genus *Bacteroides* were the most common anaerobes isolated (Segura *et al.*, 1972).

Suprapubic aspirations for miscellaneous reasons in 36 control patients were negative for anaerobic bacteria. In 13 of these patients, various aerobes were isolated. In all of the control patients, results with Gram stains agreed well with results obtained from culture.

Review of the histories of the 10 patients with anaerobes in the specimens obtained by suprapubic bladder aspiration revealed that all but one had significant urologic disease. These findings were comparable to those described in the earlier literature, suggesting that anaerobic bacteria are associated with significant disease in the urinary tract.

Case 1. The patient was a 61-year-old woman in whom bilateral renal calculi were first noted in 1952. Tuberculosis was suspected although cultures were negative at that time. There were recurrent urinary tract infections in the ensuing years until 1967 when a fistula spontaneously opened in the right flank. In 1969 tuberculosis was again strongly suspected, and therapy for this was started. A colocutaneous fistula was then demonstrated. This was resected although drainage persisted. *Bacteroides fragilis* was recovered in pure culture from her bladder.

Case 2. A 67-year-old man had a 6-year history of recurrent bladder infections. On evaluation, a nonfunctioning left kidney was discovered; markedly obstructive benign prostatic hypertrophy also was discovered. Suprapubic bladder aspiration revealed both *Bacteroides* and *Peptostreptococcus* species along with *Pseudomonas*.

Case 3. A 21-year-old woman, complaining primarily of condylomata acuminata, had a 3-year history of urinary tract infection which was responsive to antibiotics but also recurred readily when treatment was discontinued. Cystoscopic examination revealed an urethral diverticulum which was removed. Suprapubic bladder aspiration revealed *Bacteroides incommunis* (*B. fragilis* ss. *vulgatus*), *Staphylococcus aureus* and group D streptococcus.

Case 4. This 18-year-old woman had been born with spina bifida and sacral meningomyelocele. A left nephroureterectomy was performed in 1958. In 1966 the remaining ureteral stump was noted to be infected, and at this time a regimen of suppressive medication was begun. However, she had frequent urinary tract infections. The stump was subsequently removed at our institution. The organisms isolated were *B. fragilis* and *Peptococcus* species along with *Proteus morganii* and *Streptococcus faecalis*.

Case 5. The patient was a 20-year-old man who had been born with hypospadias and chordee. Surgical repair had been successful but persistent urethral stricture necessitated frequent sounding. He had recurrent urinary tract infections. Bladder aspiration resulted in culture of a *Peptococcus* species, *Escherichia coli*, and a *Citrobacter* species.

Case 6. A 56-year-old man with bilateral renal lithiasis and chronic pyelonephritis had chronic urinary tract infection despite a great variety of medications, and his urine frequently was of the consistency of pus. *Bacteroides* species, CDC group F2, was recovered along with *Proteus mirabilis* and a streptococcus of the viridans group.

Case 7. An 80-year-old man with carcinoma of the prostate had acute primary retention. He subsequently underwent transurethral resection; a 36-gm grade 3 adenocarcinoma of the prostate was removed. *Propionibacterium acnes* along with a species of *Corynebacterium* were recovered from his bladder.

Case 8. When this 41-year-old woman was a child, she had had multiple episodes of urinary tract infections. Many organisms were cultured and many antibiotics were used but infection always recurred. The diagnosis of leukoplakia of the bladder along with bladder calculi had been made on several occasions. More recently, chronic cystitis was noted on several cystoscopic examinations. She most likely had been refluxing for years since childhood. She had had a previous suprapubic cystostomy on two occasions and a vesicocutaneous fistula had been closed 22 years previously. *B. fragilis* and *B. incommunis* (*B. fragilis* ss. *vulgatus*) were recovered along with three different facultative bacteria from her bladder aspirate.

Case 9. A 29-year-old woman with chronic cystitis, bilateral reflux, and chronic atrophic pyelonephritis had had a previous left heminephrectomy because of an infected poorly functioning upper segment draining into an ectopic ureterocele on the floor of her urethra. She had had many urinary tract infections, and anaerobes had been recovered from her urine on culture of midstream specimens previously. Suprapubic bladder aspiration revealed *Bacteroides melaninogenicus* along with *E. coli.*

Case 10. This 61-year-old woman was the only patient whose urinary findings were minimal. She had had recurrent urinary tract infections but not in recent years; the diagnosis of cicatricial urethritis had been made on cystoscopic examination two years previously. Both *Peptococcus* and *Peptostreptococcus* species along with a viridans group streptococcus were recovered from her urine.

DISCUSSION

Urine itself is not a particularly good medium for the growth of anaerobic bacteria. Finegold *et al.* (1964) noted variable results in an attempt to grow certain anaerobes with urine as the sole nutrient medium. It seems unlikely that the anaerobes recovered in our series and the anaerobes recovered in most of the cases reported in the literature actually grow and thrive in urine itself. Probably most of these cases are associated with overt tissue infection or neoplasm of the urinary tract, where the oxygen tension can be assumed to be less than normal.

It has been known for years that urea-splitting facultative bacteria are associated with certain types of calculus disease of the urinary tract—specifically, magnesium ammonium phosphate or triple phosphate stones. There are at least two reports (Mencher and Leiter, 1938; Lazarus, 1944) of anaerobic bacteria in the urinary tract associated with renal calculi. Moreover, previous unpublished observations at the Mayo Clinic showed that anaerobes occasionally are found in association with infected renal calculi. In 62 instances of infected renal calculi, the urine, the renal pelvis, and the stone itself were cultured aerobically and anaerobically. Although no anaerobes were recovered, the usual urea-splitting facultative bacteria were recovered from these specimens.

REFERENCES

Albarran, J., and Cottet, J.: Note sur le rôle des microbes anaérobies dans les infections urinaires. *Assoc Fr Urol, 3:*83, 1898.

Attebery, H.R., and Finegold, S.M.: Combined screw-cap and rubber-stopper closure for Hungate tubes (pre-reduced anaerobically sterilized roll tubes and liquid media). *Appl Microbiol, 18:*558, 1969.

Beigelman, P.M., and Rantz, L.A.: Clinical significance of *Bacteroides. Arch Intern Med, 84:*605, 1949.

Finegold, S.M., Miller, L.G., Merrill, S.L., and Posnick, D.J.: Significance of anaerobic and capnophilic bacteria isolated from the urinary tract. In: Kass, E.H. (Ed.): *Progress in Pyelonephritis.* (Second International Symposium on Pyelonephritis, Boston, 1964.) Philadelphia, Davis, 1965, pp. 159–178.

Headington, J.T., and Beyerlein, B.: Anaerobic bacteria in routine urine culture. *J Clin Pathol, 19:*573, 1966.

Holdeman, L.V., Cato, E.P., and Moore, W.E.C.: Anaerobic bacterial flora of clinically normal urines and of the vagina of pregnant women. Abstr. #M4. Presented at 72nd Ann. Mtg. Am. Soc. for Microbiology, Philadelphia, April 23–28, 1972.

Kuklinca, A.G., and Gavan, T.L.: The culture of sterile urine for detection of anaerobic bacteria—not necessary for standard evaluation. *Cleve Clin Q, 36:*133, 1969.

Lazurus, J.A.: *Bacillus welchii* infections complicating surgical procedures upon the upper urinary tract. *J Urol, 51:*315, 1944.

Mencher, W.H., and Leiter, H.E.: Anaerobic infections following operations on the urinary tract. *Surg Gynecol Obstet, 66:*677, 1938.

Segura, J.W., Kelalis, P.P., Martin, W.J., and Smith, L.H.: Anaerobic bacteria in the urinary tract. *Mayo Clin Proc, 47:*30, 1972.

CHAPTER XXIX

SEPTIC ABORTION AND RELATED INFECTIONS OF PREGNANCY

EDWARD B. ROTHERAM, JR.

ABSTRACT: *Techniques which were used to isolate anaerobic bacteria in a study of septic abortion which have continued to prove useful include use of rabbit blood rather than sheep blood in agar, incubation of plates in an anaerobic jar, and a simplified method of quantitation. Of 87 patients with septic abortion, 56 had blood cultures; of the 34 positive cultures, 26 yielded anaerobic bacteria and 23 of these, anaerobes only. The most common isolates were anaerobic and capnophilic cocci and* Bacteroides *species. Aerobes were recovered from only 11 of the 34 positive blood cultures. Sixty-nine cervical exudates were cultured, and, although aerobes grew out almost as frequently as anaerobic and microaerophilic strains, they were much less apt to grow out in abundance. The results indicated that the nonclostridial anaerobes are the major invasive pathogens in septic abortion.*

In 1965, an unusual thing happened: an internist interested in infectious diseases was placed in control of the clinical microbiology laboratory in a hospital primarily devoted to obstetrics and gynecology. It was a marvelous opportunity to study mixed infection by indigenous microbes. Ultimately, the results of a study of septic abortion were published (Rotherman and Schick, 1969), since this was a common infectious entity well suited to statistical tabulation.

The logic behind our approach was as follows: (1) Many obstetrical and gynecological infections encountered in the hospital are not associated with a microbe of proven pathogenicity. (2) If *none* can be incriminated on the basis of known criminal behavior, then *all* the microbes present in clinical specimens must be suspect until proven innocent. (3) No microbial species should be absolved of guilt simply because it is difficult to grow or tedious to identify. (4) More than one microbial species may be actively involved in any

---

The author wishes to acknowledge the contributions of Suzanne F. Schick.

given infection. (5) Two qualities which might help to separate the pathogens from the innocent bystanders are the ability to proliferate in infected tissue and the ability to invade new territory.

EXPERIMENTAL METHODS

Isolation and Identification

An attempt was made to isolate as many microbial species as possible from the clinical specimens. This goal was approached by inoculating a variety of agar plates and incubating them under aerobic, capnophilic (candle jar) and anaerobic conditions. I have not detailed here the techniques we used in 1965 (Rotheram and Schick, 1969), for improvements are constantly being introduced. However, some general points seem to have withstood the test of time. Rabbit blood agar (as opposed to sheep blood) is a useful nonselective medium because it supports the growth of an amazingly wide variety of microbes and permits the eye to distinguish subtle differences in colonial morphology and hemolysis. An opaque medium such as chocolate blood agar renders such distinctions difficult and is best used for specific microorganisms such as *Neisseria gonorrhoeae.* The large surface area provided by the standard petri dish permits good separation of colonies which can be easily inspected and manipulated. The simultaneous inspection of three or four such plates, differing only in selectivity or conditions of incubation, can yield a great deal of information very quickly. I have been reluctant to abandon these advantages for others, such as those offered by prereduced media in tubes.

Quantitation of Microbial Growth

Standardized streaking of each primary plate permits a valid quantitation of the growth of each microbial species recovered from the specimen. As shown schematically in Figure XXIX-1, a primary streak is made by rolling the swab across a portion of the agar surface. Microorganisms are then moved out from the primary streak with a straight wire to form secondary

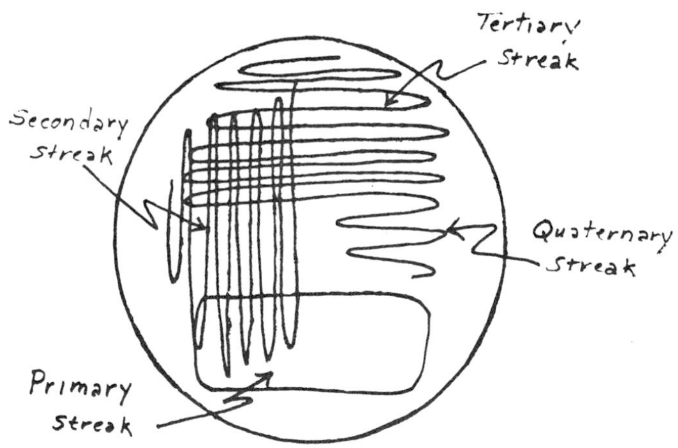

Figure XXIX-1. Streaking of primary plate for quantitation.

and tertiary streaks. A fourth streak is necessary to assure separation of colonies in heavily inoculated plates.

The growth of an individual microbial species is reported as follows:

Very light —Colonies are readily counted in the primary streak.
Light —Colonies are numerous in the primary streak but readily counted in the secondary streak.
Moderate —Colonies are readily counted in the tertiary streak.
Abundant —Colonies are readily countable only in the quaternary streak.

Figure XXIX-2 illustrates the results of an experiment which we still do periodically in the clinical laboratory. When serial ten-fold dilutions of an overnight culture are plated out as described above:

Very light growth indicates 10^3 bacteria per ml or less.
Light growth indicates from 10^4 to 10^5 bacteria per ml.
Moderate growth indicates about 10^6 bacteria per ml.
Abundant growth indicates 10^7 bacteria per ml or more.

Obviously, certain factors such as fastidious growth requirements, antibotic therapy, or clumping within phagocytes, may

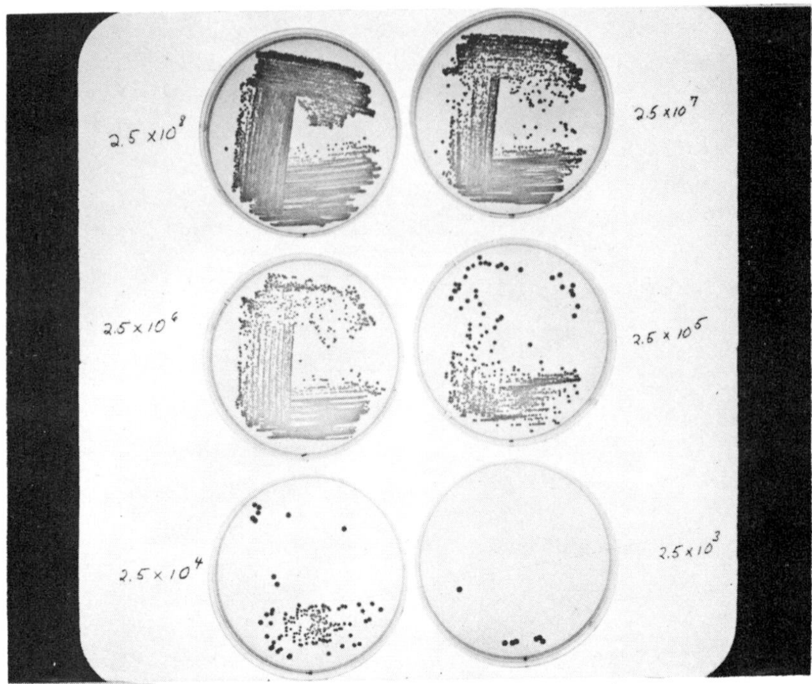

Figure XXIX-2. Growth patterns obtained from quantitative plating of serial ten-fold dilutions of a broth culture of *E. coli*.

invalidate this quantitation method, and the Gram stain serves as an independent and reliable check on quantitation which should *never* be overlooked in the clinical laboratory. I wish to stress how helpful such simple quantitation can be in evaluating mixed infection. More elegant techniques which involve liquefaction of the specimen, serial dilutions, and pour plates might destroy fastidious organisms and seriously bias the results in favor of the more robust species such as *Escherichia coli*.

Both rabbit blood agar plates in Figure XXIX-3 were inoculated with the same specimen and streaked in the standard

>>

Figure XXIX-3. (A, B) Cervical exudate yielded a light growth of *E. coli* and a moderate growth of capnophilic streptococcus on Plate A (top) incubated in a candle jar. An abundant growth of four strictly anaerobic species was obtained on Plate B (bottom) incubated in an anaerobic jar.

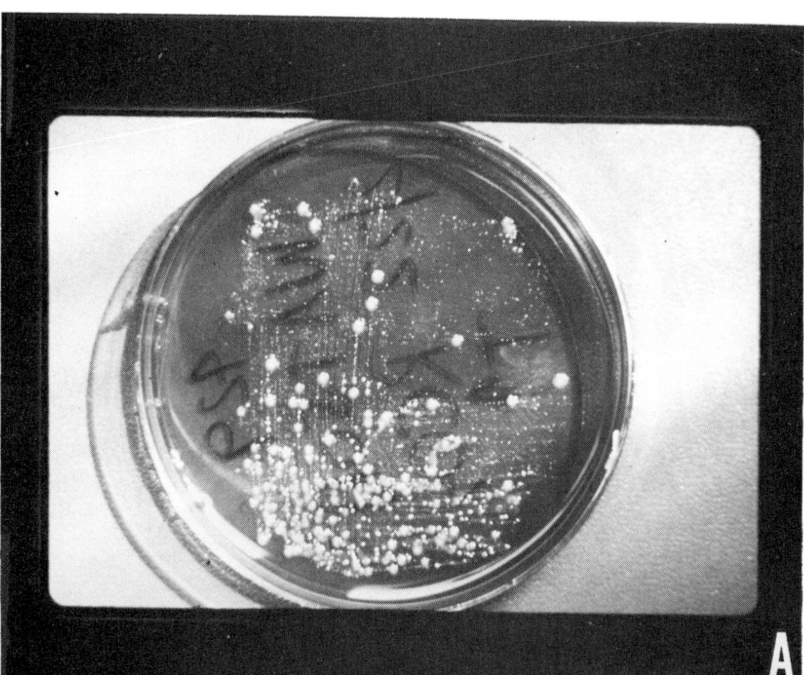
A

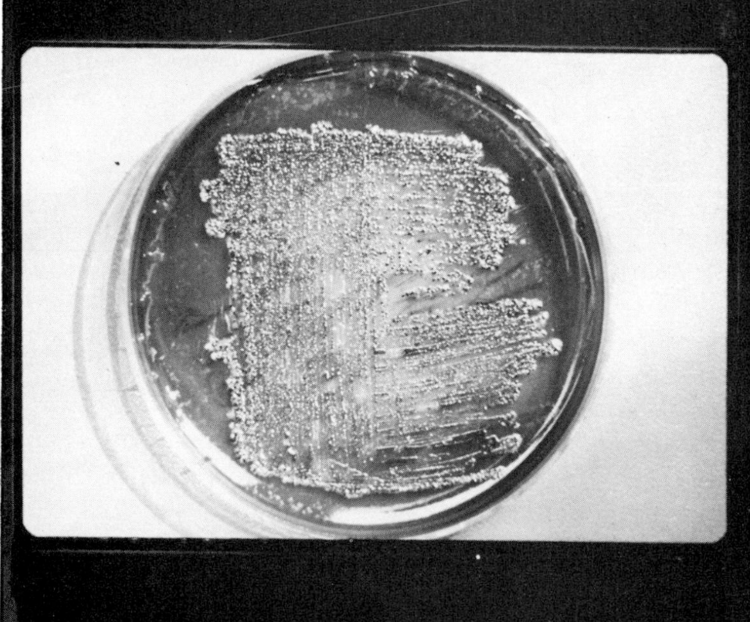
B

fashion. Plate A, incubated in a candle jar, shows a light growth of *E. coli* and a moderate growth of microaerophilic (capnophilic) streptococci. Plate B, incubated in an anaerobic jar, shows an abundant growth of four strictly anaerobic species.

Demonstration of Invasiveness

Invasiveness is perhaps best demonstrated by bacteremia. In this study all blood culture bottles were subcultured aerobically and anaerobically after appropriate preliminary incubation. An anaerobic organism was not assumed to account for all the bacteria seen on Gram stain of a positive broth. The possibility of a mixed bacteremia was always considered.

RESULTS OF THE STUDY OF SEPTIC ABORTION

Blood Cultures

Of 87 patients with septic abortion, 56 had their blood cultured on admission and 34 (60.7%) had bacteremia. This incidence of bacteremia was much higher than that reported in any previous series. The difference is largely explained by the fact that anaerobic and capnophilic bacteria predominated in our series but had appeared rarely in previous series. As detailed in Table XXIX-I, anaerobes and capnophiles were isolated 47 times, whereas aerobes were isolated only 12 times. A total of 59 organisms were isolated from 34 positive blood cultures; half of the positive cultures contained two or more strains.

In our discussion, capnophiles and anaerobes are considered together under the term "anaerobes." This is reasonable, since capnophiles are likely to be missed in laboratories not attuned to the recovery of anaerobes. Table XXIX-II shows that anaerobes were recovered from 26 of the 34 positive blood cultures (76.5%) and that they were the *only* type of bacteria recovered from 23. If these 23 cultures are excluded, our incidence of positive blood cultures would drop from 61 to 19 percent (11 of 56), a figure very

TABLE XXIX-I

BACTERIA ISOLATED FROM THE BLOOD OF 34 PATIENTS

Bacteria		*No. of Isolations*
Strictly anaerobic		
Streptococcus sp.		16
Bacteroides fragilis		9
Bacteroides sp.		4
Bacteroides melaninogenicus		3
Micrococcus sp.		2
	Total	34
Capnophilic (candle jar)		
Streptococcus sp.		7
Hemophilus vaginalis		5
Actinomyces sp.		1
	Total	13
Aerobic or facultative		
Escherichia coli		7
Group B streptococcus		2
Group D streptococcus		1
Pseudomonas aeruginosa		1
Diphtheroid		1
	Total	12

TABLE XXIX-II

ANAEROBIC BACTERIA IN BLOOD CULTURES

	Number	*%*
Cultures positive for bacteria:	34	
Cultures yielding anaerobes:	26	
Percentage yielding anaerobes:		76.5
Cultures yielding anaerobes only:	23	
Three species recovered in 6		
Two species recovered in 8		
One species recovered in 9		

near the highest incidence previously reported in the literature (Neuwirth and Friedman, 1963). Anaerobes also accounted for the high incidence of mixed bacteremia in our series (Table XXIX-II).

Table XXIX-III gives similar data for aerobic isolates, which were recovered from only 11 of 34 positive cultures (32.4%). Two aerobes were recovered together only once but an aerobe was recovered with anaerobes three times. Of particular interest, *E. coli* was recovered with *Bacteroides fragilis* and *B. melaninogenicus* in one culture, and with an anaerobic streptococcus and a *Bacteroides* species in another.

TABLE XXIX-III

AEROBIC BACTERIA IN BLOOD CULTURES

	Number	%
Patients with cultures on admission:	56	
Cultures positive for bacteria:	34	
Cultures yielding aerobes:	11	
% of total cultures yielding aerobes:		19.6
% of positive cultures yielding aerobes:		32.4
Cultures yielding aerobes only:	8	
Two species recovered in 1		

Cervical Cultures

Sixty-nine cervical exudates were cultured; in six, aerobic cultures only were performed. Anaerobic species were found frequently, but certain aerobes were recovered almost as frequently (Table XXIX-IV). The striking difference lies in the frequency with which the various microbes were isolated *in abundance.* The anaerobes grew out in abundance four times as often as the aerobes (81 times versus 21 times). Clearly, intrauterine conditions favor the proliferation of capnophiles and strict anaerobes. When the individual cultures were carefully analyzed, only six appeared to be documented instances of predominantly aerobic infection, all attributable to *E. coli.* And in one of these cases, anaerobes together with *E. coli* were recovered from blood, indicating that for some reason anaerobes were missed in the cervical exudate.

The results indicate that the same microbes which proliferate in the uterine cavity are most frequently isolated from the blood, and therefore establish nonclostridial anaerobes as the major invasive pathogens in septic abortion.

DISCUSSION

Septic abortion provides an excellent model for studying mixed infection by indigenous bacteria. An avascular and poorly defended space is created, filled with nutritious material, and inoculated with a variety of microbes belonging to the normal flora of an adjacent region. In septic abortion, the space is the uterine cavity; the material, devitalized fetal tissues; the flora, vaginal. But other newly created avascular

TABLE XXIX-IV

ORGANISMS ISOLATED FROM THE CERVIX OF 69 PATIENTS

Organism	*No. of Isolations*	*Isolations in Abundance*
Strictly anaerobic		
Bacteroides fragilis	35	18
Streptococcus sp.	33	20
Bacteroides melaninogenicus	22	11
Bacteroides sp.	13	7
Micrococcus sp.	4	3
Diphtheroid	2	1
Clostridium perfringens	2	1
		Total 61
Capnophilic (candle jar)		
Hemophilus vaginalis	20	9
Streptococcus sp.	17	6
Lactobacillus sp.	6	3
Actinomyces sp.	1	0
Diplococcus pneumoniae	1	0
Neisseria gonorrhoeae	4	2
		Total 20
Aerobic or facultative		
Coagulase-negative staphylococcus	27	2
Escherichia coli	25	6
Diphtheroid	24	4
Group B streptococcus	11	3
Streptococcus viridans	11	2
Group D streptococcus	5	1
Klebsiella pneumoniae	5	1
Enterobacter sp.	4	0
Candida sp.	4	0
Proteus mirabilis	3	1
Staphylococcus aureus (coagulase-positive)	2	0
Diplococcus pneumoniae	1	0
Proteus vulgaris	1	0
Bacterium anitratum	1	1
Mima polymorpha	1	0
Aspergillus fumigatus	1	0
		Total 21

spaces might be the lumen of an occluded Fallopian tube, a thrombosed vein, a hematoma dissecting along fascial planes, or fluid loculated in the peritoneal cavity. The nutritious material might be liquid blood, clot, or feces. The flora might derive from the bowel or the mouth rather than the vagina. Nevertheless, subsequent events are often similar. Certain microbial species among those inoculated proliferate. Because of the low oxidation-reduction potential of the tissue in these areas, the bacteria which proliferate are most often anaerobic. Unlike certain clostridial species, most of these anaerobes are not equipped with obvious offensive weapons.

Still they can invade, either locally by creating more spaces (hemorrhage and exudates) or, more importantly, by inducing *septic thrombophlebitis* (Tynes and Frommeyer, 1962; Reid *et al.*, 1945). A "pipeline" is thus established whereby the mixed infection within a space that is avascular can, paradoxically, gain access to the blood stream. Usually this thrombophlebitis seals off the infection and the bacteremia is transient. Rarely, invasion proceeds, more spaces are created, larger and larger veins are involved, the bacteremia persists, and septic emboli create metastatic abscesses whose population is also often polymicrobial (Sabbaj *et al.*, 1972).

Occasionally, usually avirulent aerobes (such as group D or group B streptococci) playing a bystander role in the pelvic infection may be swept into the blood stream to confuse the issue. At other times, weakly virulent aerobes or facultative organisms such as *E. coli* may produce serious complications through their endotoxins. Rarely, aerobes such as group A streptococci or coagulase-positive staphylococci may prove overwhelmingly invasive. Nevertheless, nonclostridial anaerobes are likely to be dominating the local infection.

REFERENCES

Neuwirth, R.S., and Friedman, E.A.: Septic abortion, changing concept of management. *Am J Obstet Gynecol, 85:*24, 1963.

Reid, J.D., Snider, G.E., and Toone, E.C., et al.: Anaerobic septicemia. *Am J Med Sci, 209:*296, 1945.

Rotheram, E.B., Jr., and Schick, S.F.: Nonclostridial anaerobic bacteria in septic abortion. *Am J Med, 46:*80, 1969.

Sabbaj, J., Sutter, V.L., and Finegold, S.M.: Anaerobic pyogenic liver abscess. *Ann Intern Med, 77:*629, 1972.

Tynes, B.S., and Frommeyer, W.B.: *Bacteroides* septicemia: cultural, clinical, and therapeutic features in a series of twenty-five patients. *Ann Intern Med, 56:*12, 1962.

CHAPTER XXX

ANAEROBIC BACTERIA IN INFECTIONS OF THE FEMALE GENITAL TRACT

ROBERT M. SWENSON

ABSTRACT: *Between January 1971 and June 1972, 91 hospitalized patients were entered into a systematic study of the bacteriology of infections of the female genital tract. Specimens were collected by methods designed to maintain anaerobic conditions, and culture and isolation procedures were carried out in an anaerobic chamber. A total of 141 anaerobic isolates were recovered from 67 of the 91 cases, and in 34 cases, only anaerobes were isolated. The most common isolates were* Bacteroides *species, which comprised 43 percent of the isolates, and anaerobic gram-positive cocci. Facultative bacteria were isolated in 57 of the 91 cases. The most frequent isolates were* Escherichia coli *and streptococci. The findings indicated that, with the exceptions of gonococcal infections and wound infections following "clean" surgical procedures, the majority of infections related to the female genital tract involve nonsporeforming anaerobic bacteria.*

There has recently been an increasing interest in the role of anaerobic bacteria in a wide variety of human infections. Previous studies have suggested that anaerobes may often be involved in infections of the female genital tract. Rotheram and Schick (1969) found anaerobic species in blood cultures from 34 of 56 cases of septic abortion. The findings in studies by Hall *et al.* (1967) indicated that nonsporeforming anaerobic bacteria were frequently involved in postoperative pelvic infections. However, to date there has not been a systematic prospective study of infections of the female genital tract.

In early 1969 we at Temple University Hospital set up a special anaerobe laboratory to facilitate the isolation and identification of anaerobic bacteria from clinical specimens. Studies at that time suggested that many infections of the female genital tract might be due to anaerobic bacteria. Thus, in January 1971, we undertook a systematic study of the bacteriology of such infections in hospitalized patients.

MATERIALS AND METHODS

Patients studied were inpatients on the Obstetrical or Gynecological services at Temple University Hospital during the period from January 1971 through June 1972. Cases were included in the present study only if the patient had received no antibiotic therapy within the prior seven days or received antibiotics for only 24 hours or less. In most instances specimens were collected by aspiration from a closed cavity using an 18-gauge needle and syringe. Exceptions to this were endometrial cultures, which were obtained transcervically, and some wound infections in which pus was merely aspirated from the open wound.

After collection, specimens were immediately injected into an oxygen-free tube for transportation to the special anaerobe laboratory. There, specimens were passed into an anaerobic chamber where all further manipulations were carried out. Anaerobic bacteria were identified by colonial morphology, reaction to Gram stain, growth on selective media, and gas-liquid chromatography of fermentation products. The criteria used were those outlined by the Anaerobe Laboratory at Virginia Polytechnic Institute (Holdeman and Moore, 1972). Facultative bacteria were isolated and identified using standard techniques (Blair *et al.*, 1970).

RESULTS

During the period of study a total of 91 patients fulfilled the criteria outlined for inclusion in the study. The sites of infection in these patients are shown in Table XXX-I. Ten cases of endomyometritis resulted from incomplete septic abortion while four cases followed prolonged rupture of fetal membranes. The cases listed as peritonitis represent 14 instances of recurrent pelvic inflammatory disease with marked signs of peritoneal irritation in which the specimen was obtained by culdocentesis. The other two cases followed rupture of a pelvic abscess. All of the specimens from pelvic abcesses were obtained at laparotomy to drain the lesion. All cases of vaginal cuff abscess followed abdominal hysterectomy and all wound infections followed operations on the female genital tract.

Anaerobic bacteria were isolated from 67 of 91 infections (74%). In 43 of 67 cases (64%) multiple species of anaerobic

TABLE XXX-I

ANAEROBIC BACTERIA IN INFECTIONS OF THE FEMALE GENITAL TRACT

Infection	*Total No. Cases*	*No. Cases Yielding Anaerobes*	*Multiple Anaerobic Species*	*Anaerobes Only—No Aerobes*
Endomyometritis	14	13	7	6
Salpingitis, peritonitis*	16	10	6	4
Pelvic abscess	10	8	6	5
Vaginal cuff abscess following abdominal hysterectomy	21	16	13	11
Bartholin's abscess	15	10	7	4
Wound infection†	15	10	4	4
Total	91	67	43	34

* Includes cases of peritonitis secondary to salpingitis. Specimens obtained by culdocentesis.

† Following surgical procedure on female genital tract.

bacteria were found. Finally, in 34 of 91 instances (37%) only anaerobic bacteria were isolated. In general, the results were similar regardless of the site of infection.

Eight patients with pelvic bleeding, e.g. ectopic pregnancy, were also studied. In all instances no bacteria were isolated from culdocentesis specimens from these patients. Similarly, specimens obtained from four uninfected Bartholin's cysts were sterile.

A summary of the anaerobic isolates is presented in Table XXX-II. There were a total of 141 anaerobic isolates from these 91 infections. *Bacteroides* species were most commonly found, making up 43% of the total. *Bacteroides melaninogenicus* and *Bacteroides fragilis* were the most frequently isolated of the *Bacteroides* species.

There were 21 isolates of *Peptococcus* species. *Peptococcus prevotii* was found most frequently, while *Peptococcus magnus* was isolated on only four occasions. *Peptostreptococcus species* were also frequently isolated, with *Peptostreptococcus anaerobius* and *Peptostreptococcus intermedius* being found most commonly.

Anaerobic gram-positive cocci which could not be definitely placed as either *Peptococcus* or *Peptostreptococcus* species were designated as anaerobic streptococci. There

TABLE XXX-II

ANAEROBIC BACTERIA ISOLATED FROM 91 INFECTIONS OF THE FEMALE GENITAL TRACT

Bacteroides melaninogenicus	19	
Bacteroides fragilis	15	
Bacteroides spp.	27	
		61
Peptococcus prevotii	10	
Peptococcus magnus	4	
Peptococcus spp.	7	
		21
Peptostreptococcus anaerobius	11	
Peptostreptococcus intermedius	7	
Peptostreptococcus spp.	1	
		19
Anaerobic streptococcus	9	
Fusobacterium spp.	9	
Eubacterium spp.	9	
Clostridium spp.	6	
Others	7	
		40
Total Isolates		141

were nine isolates of such organisms. *Eubacterium* and *Fusobacterium* species were each isolated nine times. However, *Clostridium* species were found on only six occasions.

Facultative bacteria were found in 57 of the 91 cases (63%). There was a total of 66 such isolates (Table XXX-III). Twenty-two (33%) were *Escherichia coli* and an equal number *Streptococcus* species. Of the latter, four strains were group A and five strains group D. *Klebsiella* and *Proteus* species were isolated significantly less frequently. *Neisseria gonorrhoeae* and *Staphylococcus aureus* were found on only four occasions.

DISCUSSION

Anaerobic bacteria make up a large portion of the normal vaginal flora (Swenson *et al.*, unpublished data). Therefore, it is not surprising that these organisms are frequently involved in infections of the female genital tract, as indicated by the data presented here. In the majority of cases these

TABLE XXX-III

FACULTATIVE BACTERIA ISOLATED FROM 91 INFECTIONS OF THE FEMALE GENITAL TRACT

Escherichia coli		22
Streptococcus spp. (total)		22
Group A	4	
Group D	5	
Proteus spp.		7
Klebsiella spp.		5
Neisseria gonorrheae		4
Staphylococcus aureus		4
Hemophilus vaginalis		1
Pseudomonas spp.		1
Total		66

were mixed infections involving multiple anaerobic species and often other facultative bacteria were found. The high incidence of anaerobic bacteria in this study is explainable by the fact that great care was taken to ensure anaerobic collection of specimens and special procedures were used for the isolation and identification of anaerobic isolates. Without such procedures, many anaerobes would not have been isolated (Vargo *et al.*, 1971). Since such sensitive techniques were employed, it is significant that culdocentesis specimens from noninfected patients and specimens from Bartholin's cysts were found to be sterile. These findings lend support to the idea that the anaerobes isolated from other cases were not merely contaminants from the normal anaerobic flora of the vagina.

Bacteroides species were the most frequently encountered anaerobes, while the clostridia were only infrequently found. These findings have significant implications since the majority of isolates of *B. fragilis* are resistant to penicillin (Martin *et al.*, 1972), an antibiotic commonly employed for infections of the female genital tract.

The facultative bacteria isolated were predominantly gram-negative bacilli and *Streptococcus* species. The infrequent isolation of *N. gonorrhoeae* is explainable by the fact that this study was confined to hospitalized patients, and at our institution the majority of patients with gonococcal infections are treated as outpatients.

The findings indicate that, with the exceptions of gonococcal infections and wound infections following "clean" surgical procedures, e.g. removal of an ovarian cyst, the majority of infections related to the female genital tract involve nonsporeforming anaerobic bacteria.

REFERENCES

Blair, J.E., Lennette, E.H., and Truant, J.P.: *Manual of Clinical Microbiology*. Baltimore, Williams and Wilkins, 1970.

Hall, W.L., Sobel, A.I., Jones, C.T., and Parker, R.T.: Anaerobic postoperative pelvic infections. *Obstet Gynecol, 30:*1, 1967.

Cato, E.P., Cummins, C.S., Holdeman, L.V., Johnson, J.L., Moore, W.E.C., Smibert, R.M., and Smith, L.DS.: *Outline of Clinical Methods in Anaerobic Bacteriology*. Blacksburg, Virginia Polytechnic Institute and State University, 1970.

Martin, W.J., Gardner, M., and Washington, J.A.: *In vitro* antimicrobial susceptibility of anaerobic bacteria isolated from clinical specimens. *Antimicrob Agents Chemother, 1:*148, 1972.

Rotheram, E.B., and Schick, S.F.: Nonclostridial anaerobic bacteria in septic abortion. *Am J Med, 46:*80, 1969.

Swenson, R.M., Michaelson, T.C., and Spaulding, E.H.: Unpublished observations.

Vargo, V., Michaelson, T.C., Spaulding, E.H., Vitagliano, R., Swenson, R.M., and Forsch, E.: Comparison of a prereduced anaerobic method and GasPak for isolating anaerobic bacteria. In: *Bact Proc* Ann Arbor, Soc for Microbiol, 1971, p. 109.

PART V

DISEASE SYNDROMES (CONTINUED) AND *IN VITRO* ANTIBIOTIC SUSCEPTIBILITY TESTING

CHAPTER XXXI

LIVER ABSCESS: THE ETIOLOGIC ROLE OF ANAEROBIC BACTERIA

W.A. Altemeier

Abstract: *A 17-year study of 83 patients with liver abscess has emphasized (1) the crucial importance of surgical drainage, since mortality was 100 percent when the abscess was not drained, and (2) increasingly strong evidence of the role of anaerobic bacteria in the etiology of the disease. The majority of anaerobic isolates were* Bacteroides, Sphaerophorus, Fusobacterium *and* Peptostreptococcus *species. All of the last 12 patients in the series with pyogenic liver abscesses had anaerobic isolates, and anaerobes were the only isolates in 8 of these 12.*

The capability of diagnosing and localizing abscesses 2 cm or larger has been greatly increased during the past six years by use of radioisotopic liver scanning technics. More recently, hepatic arteriography has also been used effectively. The earlier and more definitive diagnosis made possible by these technics has permitted life-saving surgical drainage in more patients.

Recent clinical and laboratory experience has revealed that anaerobes play an important role in the etiology of liver abscess (Altemeier, 1970, 1971; Sherman and Robbins, 1960; St. John *et al.*, 1942; Block *et al.*, 1964). This role has become increasingly apparent during a study of 83 patients with liver abscess seen in four hospitals between 1955 and 1972. The study was undertaken to obtain a better understanding of the incidence, etiology, pathogenesis, diagnostic problems, effectiveness of therapy, morbidity and mortality of this serious disease.

In our early experience, hepatic abscess was a life-threatening disease characterized by obscure onset, lack of prominent localizing signs, association with other more dominant diseases, prolonged diagnostic delays or failures, and a high mortality rate (Altemeier, 1970, 1971; Sherman and Robbins, 1960; St. John *et al.*, 1942; Block *et al.*, 1964; Ochsner *et al.*, 1938; Eliason, 1926; McFadzean *et al.*, 1953; Cronin, 1953; Ryes and Reyes, 1969; Ostermiller and Carter, 1967;

This work supported in part by U.S.P.H.S. Grant 5-PO1-GM-15428 and U.S. Army Contract DA-49-193-MD-2531.

Berke and Pecora, 1966; Warren and Hardy, 1968; Price *et al*, 1967; Grant *et al.*, 1969). Antibiotic therapy, while a valuable adjunctive treatment, did not prevent death when these abscesses were not drained. More recently, earlier clinical diagnosis, more accurate bacteriologic diagnosis, and improved localization of hepatic abscesses have made reduced mortality possible (Price *et al.*, 1967; Grant *et al.*, 1969; Gaisford and Mark, 1969; Schuman *et al.*, 1964; Gollin *et al.*, 1964; Shingleton *et al.*, 1966; Gottschalk, 1967; McCarthy *et al.*, 1967).

Some of the results of our 17-year study have been described previously (Altemeier, 1970, 1971).

PATIENTS AND DIAGNOSTIC CHARACTERISTICS

All patients with documented hepatic abscess seen during the 17 years from January 1955 through October 1972 were included in a study of the case material from four hospitals of the University of Cincinnati Hospital group: Cincinnati General Hospital, Cincinnati Veterans Administration Hospital, Christian R. Holmes Hospital, and Cincinnati Childrens' Hospital. A total of 83 cases were studied, 74 being classified as pyogenic, 6 amebic, 2 fungal, and 1 echinococcal.

For the purposes of comparison and discussion of the incidence, etiology, diagnostic problems, types and effectiveness of treatment, morbidity, and mortality, this study was divided into four periods (Table XXXI-I). Included in the total series were four amebic abscesses in the period 1955 to 1959 and two in the period of 1970 to 1972. All of these

TABLE XXXI-I

ETIOLOGY OF LIVER ABSCESS IN 83 CASES

Period	*Pyogenic*	*Amebic*	*Fungal*	*Echinococcal*	*Total*
1955–59	22	0	1	0	23
1960–64	19	0	0	0	19
1965–69	17	4	1	1	23
1970–72	16	2	0	0	18
Totals	74	6	2	1	83

TABLE XXXI-II

ETIOLOGY OF LIVER ABSCESSES AMENABLE TO SURGICAL INTERVENTION IN 51 CASES

Period	*Pyogenic*	*Amebic*	*Fungal*	*Echinococcal*	*Total*
1955–59	9	0	1	0	10
1960–64	11	0	0	0	11
1965–69	12	4	1	1	18
1970–72	10	2	0	0	12
Totals	42	6	2	1	51

were found in returning veterans from Vietnam. There were two fungal abscesses and one echinococcal multiple cyst with infection. Further subdivision of the patients into two groups was done according to whether or not surgical intervention with drainage of the abscesses was or would have been beneficial. The liver abscesses in 51 of the patients were amenable to surgical drainage, and were 2.0 cm in diameter or greater and were noted to be either solitary and discrete, or confluent (Table XXXI-II). The largest abscess, which contained 1200 cc of pus, measured 15 cm in diameter.

AGE, SEX AND RACE. In the total series of 83 patients, there were 58 males and 25 females. Fifty-one were white and 32 were black. The average age was 47.5 years, the youngest patient being 12 days and the oldest 90 years. The majority of the patients were in the first, fifth, sixth, seventh, and eighth decades of life.

ABSCESS LOCATION. Of the 83 cases, 39 involved a single lobe, while 44 were bilateral and multiple. Of the 51 patients with abscesses considered amenable to surgery, 39 had single abscesses and 30 of these were located in the right lobe (Table XXXI-III). Among the last cases seen was one of special interest, since it was characterized by an irregular coalesced abscess in the right lobe and multiple smaller abscesses 1 to 3 mm in diameter throughout both lobes. Of the 44 patients with bilateral and multiple abscesses, 32 fell into the nonoperable group.

SIGNS, SYMPTOMS AND DIAGNOSTIC PROCEDURES. The clinical features noted in the 51 patients with hepatic abscesses

TABLE XXXI-III

LOCATION OF LIVER ABSCESSES AMENABLE TO SURGICAL INTERVENTION IN 51 CASES

Type	*Right Lobe*	*Left Lobe*	*Bilateral*	*Total*
Single	27	4	0	31
Multiple	3	5	12	20
Totals	30	9	12	51

amenable to surgery are listed in Table XXXI-IV; 98 percent of the patients had fever, and 53 percent had spikes as high as 104°, 105°, or 106°F. Abdominal pain was present in only 67 percent, which was somewhat surprising in view of the large size of some of the abscesses (≥ 1000 cc). Chills of the hard, bed-shaking variety were observed in 25 percent. The x-ray findings in these patients have been of relatively little help. As a rule there was some elevation of the right diaphragm, but few or no signs in the pulmonary area above the diaphragm. Elevation and immobilization of the diaphragm have been noted, but have not been constant. Marked diaphragmatic elevation was noted with the perforation of large hepatic abscesses into the subphrenic space, and in these instances the responsible organisms were *Bacteroides* and *Peptostreptococcus* species. In the past three years, liver scan has been performed regularly in patients

TABLE XXXI-IV

CLINICAL FEATURES IN 51 CASES OF LIVER ABSCESS AMENABLE TO SURGERY

Signs and Symptoms	*No. Patients (%)*
Temperature elevation	49 (98%)
Abdominal pain	34 (67%)
Anorexia	30 (59%)
Hepatomegaly	28 (55%)
Abdominal tenderness (right upper quadrant)	27 (53%)
Jaundice	23 (45%)
Pulmonary signs and symptoms	14 (27%)
Chills	13 (25%)
Vomiting	10 (19%)

suspected of having a liver abscess because of a high spiking fever, the absence of any localizing signs, and an elevation of the right diaphragm. With the various radioisotopic techniques available it has been possible to identify a space-occupying lesion in those instances where the abscess was greater than 2 cm in diameter (Fig. XXXI-1). Scanning procedures have been a great help to us in identifying and locating abscesses. The majority of the scans were performed with the sulfur colloid, techneticum sulfide Tc 99m, and Photogamma III scintillation cameras.

Selective arteriography has also been employed in all of the most recent cases. It has been of value in confirming moderate- and large-sized defects demonstrated by liver scans. Such areas suspected of representing an abscess were proven to be avascular by the selective arteriogram (Fig. XXXI-2).

ASSOCIATED DISEASES. Of interest also in the 51 patients with liver abscesses amenable to surgical drainage were the various associated diseases (Table XXXI-V). Only 22 of the 51 had associated biliary tract diseases; 15 were inflammatory, and seven were malignant. Intrahepatic diseases present in nine were polycystic disease of the liver in three, metastatic carcinoma with necrosis and secondary infection in five, and tuberculous involvement in one. Extrahepatic biliary disease was found in 20 patients, and these were largely inflammatory diseases involving the abdominal alimentary tract.

Microbial Etiology

Bacteriological data were available in 31 of the 42 cases of pyogenic abscess amenable to surgery, and the organisms are listed in Table XXXI-VI. More sophisticated anaerobic culture technics in the past 8 years have led to a greatly increased frequency of isolation of anaerobic bacteria. The most common isolates were members of the family Bacteroidaceae. In the last 12 cases of pyogenic abscess seen by the author between January 1970 and November 1972, we get a clearer picture of the importance of the anaerobes

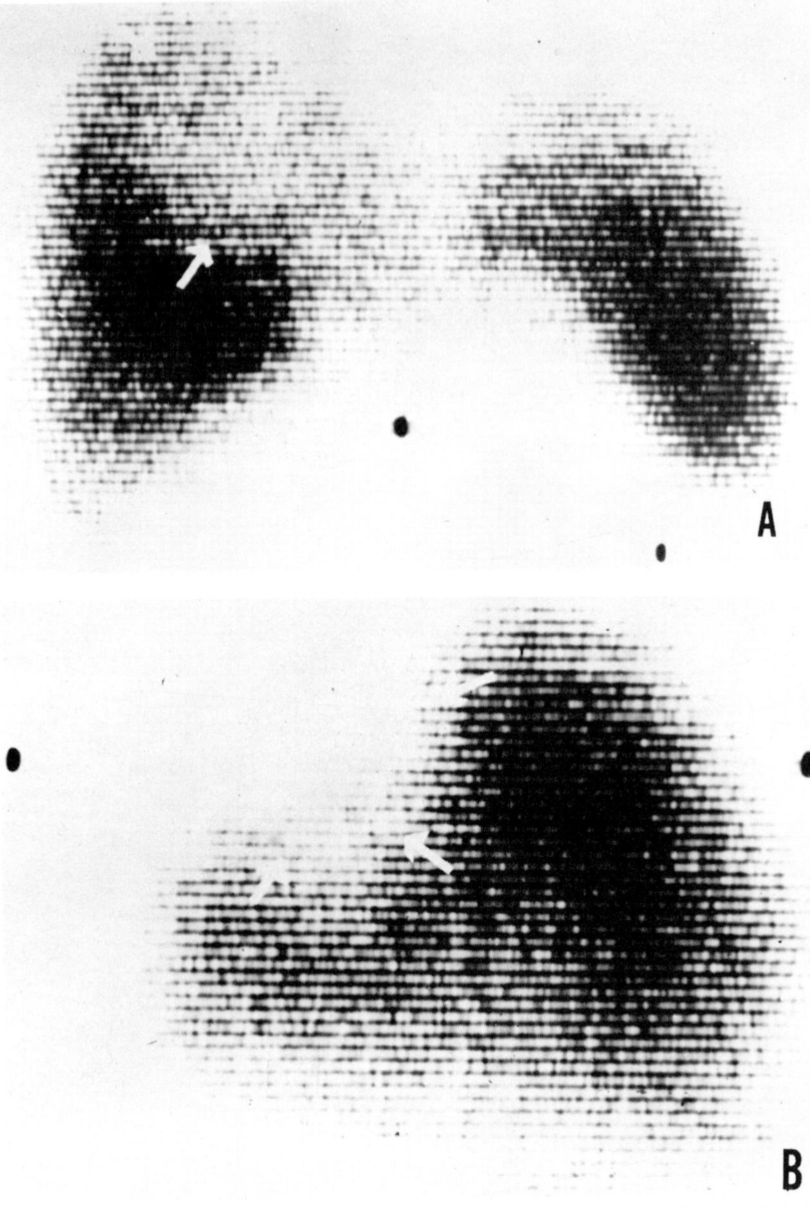

Figure XXXI-1. (A, B) Liver scan posterior-anterior (A) and lateral (B) showing presence and location of a large lesion in the superior and posterior aspects of the right lobe as indicated by arrows. (Reprinted from *Del Med J*, *43*:327, 1971).

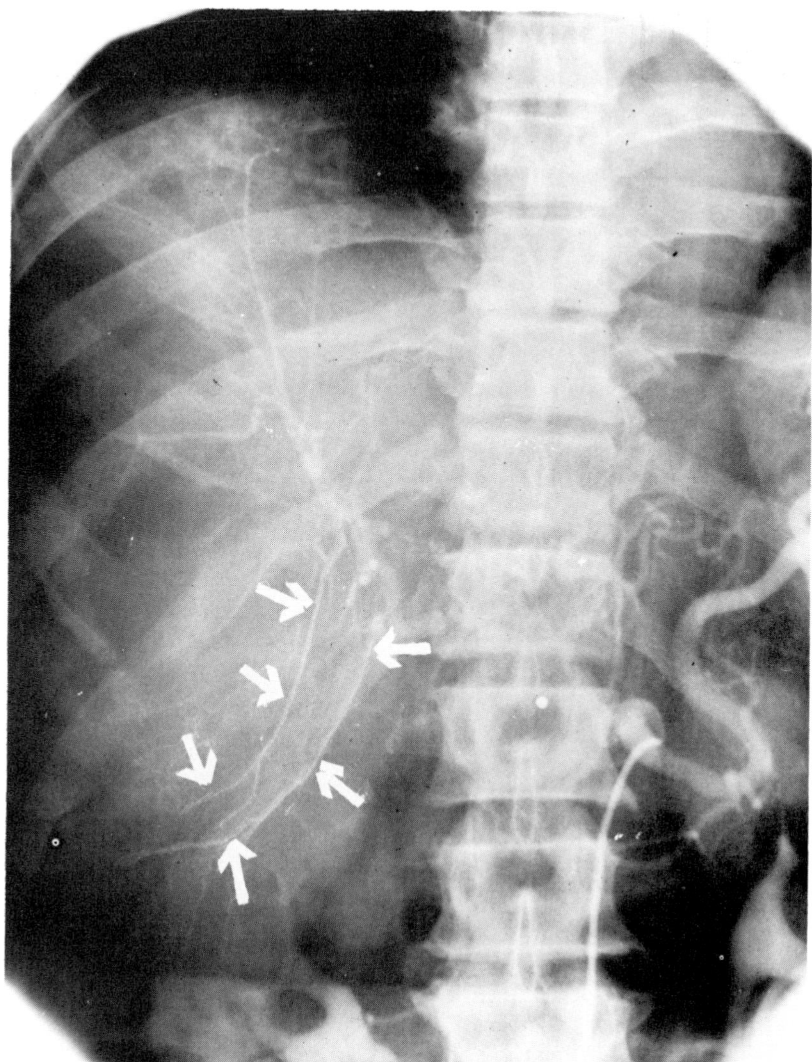

Figure XXXI-2. Arteriogram with selective catheterization of celiac presence and location of an avascular area in right lobe of liver. Note curved branches of hepatic artery (arrow) stretched around the area. (Reprinted from *Arch Surg*, *101*:258, 1970. Copyright 1970, Amer. Med. Assoc.).

(Table XXXI-VII), for anaerobes were present in all 12 and were the only isolates in 8. *Bacteroides* species were cultured in 8 of the 12.

TABLE XXXI-V

DISEASES ASSOCIATED WITH LIVER ABSCESSES AMENABLE TO SURGERY IN 51 CASES

Disease		*No. Cases*
Biliary Tract Disease		22
Inflammatory disease	15	
Malignant disease	7	
Intrahepatic Disease		9
Polycystic disease of liver	3	
Metastatic carcinoma	5	
Tuberculous involvement	1	
Extrahepatic extrabiliary disease		20
Pneumonia	2	
Peritonitis	3	
Probable preexisting appendicitis	1	
Chronic pancreatitis	1	
Diverticulitis with abscess	3	
Subphrenic and interloop abscesses	1	
Regional enteritis	3	
Ulcerative proctitis	1	
Acute hemorrhagic pancreatitis	1	
Pneumonia, meningitis, septicemia	1	
Urinary tract infection	1	
Unknown	2	
Total		51

TABLE XXXI-VI

CULTURE RESULTS IN 31 CASES OF PYOGENIC LIVER ABSCESS AMENABLE TO SURGERY

Aerobes		*Anaerobes*	
Escherichia coli	11 (35%)	*Fusobacterium* sp.	3 (10%)
Streptococcus viridans	6 (20%)	*Bacteroides* sp.	9 (30%)
Pseudomonas sp.	4 (13%)	*Sphaerophorus* sp.	7 (23%)
Aerobacter aerogenes	1 (3%)	*Peptostreptococcus*	8 (26%)
		Clostridium perfringens	1 (3%)
		Diphtheroids	2 (6%)

TABLE XXXI-VII

CULTURE RESULTS IN 12 CONSECUTIVE CASES OF PYOGENIC LIVER ABSCESSES (JANUARY 1970–NOVEMBER 1972)

Aerobes (4 cases)		*Anaerobes (12 cases; 8, anaerobes only)*	
E. coli	4 (33%)	*Bacteroides* sp.	8 (66%)
S. viridans	2 (17%)	*Peptostreptococcus* sp.	6 (50%)
Pseudomonas sp.	2 (17%)	*Sphaerophorus* sp.	4 (33%)
Aerobacter aerogenes	1 (8%)	*Fusobacterium* sp.	2 (17%)
		C. perfringens	1 (8%)
		Diphtheroids	1 (8%)

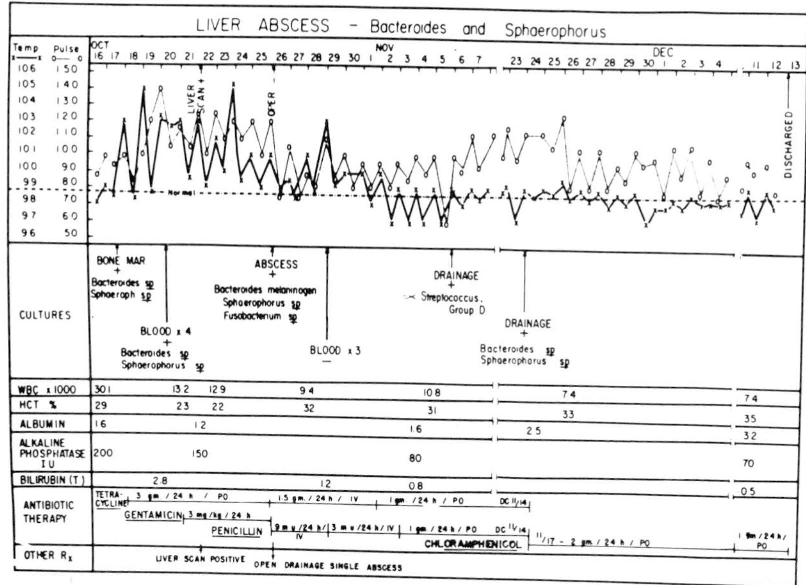

Figure XXXI-3. Septic course of 44-year-old male patient with obscure liver and eight-week history. Liver scan revealed large defect in lateral aspect of right lobe. (Reprinted from *Del Med J*, *43:*327, 1971).

Fig. XXXI-3 illustrates the clinical findings in a patient with a liver abscess from which *Bacteroides* and *Sphaerophorus* species were isolated repeatedly.

TREATMENT AND RESULTS

Review of these cases has emphasized the fact that mortality of liver abscess is 100 percent when undiagnosed and therefore undrained surgically. Definitive and earlier diagnosis of liver abscess has permitted life-saving surgical drainage. As shown in Table XXXI-VIII, there were only two deaths among 33 cases in which a diagnosis was made and surgical drainage instituted. One of the deaths resulted from pulmonary embolism in the postoperative period. In the other, which occurred in a case seen in 1972, autopsy showed a massive infarct of the liver with a severe mixed infection with *Bacteroides* species, *Peptostreptococcus* species and *Clostridium perfringens*, which could not be controlled.

TABLE XXXI-VIII

MORTALITY AND SURGICAL DRAINAGE AMONG 57 CASES OF LIVER ABSCESS

Period	*No. Cases*	*Number Diagnosed (%)*	*Number Drained*	*Mortality in Drained Cases*	*Mortality with No Drainage*	*Overall Mortality*
1955–59	10	2 (20%)	2	0%	100%	80%
1960–64	11	2 (18%)	2	0%	100%	82%
1965–69	18	14 (78%)	14*	7%†	100%	28%
1970–72	18	15 (93%)	15	6%§	100%	17%

* One patient spontaneously decompressed his abscess.
† One patient died with a pulmonary embolism in the postoperative period.
§ One diabetic patient died of massive infarct of liver with severe *C. perfringens* infection.

The surgical approach in all but one case was transabdominal; the procedure has been described previously (Altemeier, 1970). Antibacterial therapy was used in all cases preoperatively, intraoperatively, and postoperatively. When possible the selection of the antibacterial agent(s) was made preoperatively on the basis of cultures and sensitivity tests, usually made on the microorganisms isolated from the blood stream. When such cultures were not available preoperatively, the selection of the antibiotic agents was made on the basis of presumptive etiology, pending more specific information provided by examination of gram-stained preparations of the pus at the time of operation, or later by the results of the cultures and sensitivity tests.

Antimicrobial therapy was extended postoperatively for variable periods of time depending upon the size of the abscess, the association of septicemia, presence of other smaller abscesses, and the patient's postoperative course. In most instances the duration of therapy exceeded 14 days, and occasionally was extended for as long as 10 weeks.

Fig. XXXI-3 illustrates the clinical course and response to surgical drainage and adjunctive antimicrobial therapy in one patient.

COMMENTS

The findings in this 17-year study indicate the advisability of liver scanning, hepatic arteriography and anaerobic as well as aerobic cultures in patients with signs of general sepsis but without localizing signs or symptoms. Early diagnosis permits surgical drainage, which is crucial to recovery, under planned and safer conditions. Adjunctive antimicrobial therapy should be based on results of susceptibility tests. When therapy must be started on the basis of presumptive bacterial etiology, the likelihood of the presence of anaerobic bacteria, alone or with facultative organisms, should be taken into consideration.

REFERENCES

Altemeier, W.A., Schowengerdt, C.G., and Whiteley, D.H.: Abscesses of the liver. *Arch Surg, 101:*258, 1970.

Altemeier, W.A.: Recent trends in the management of hepatic abscess. *Del Med J, 43:*327, 1971.

Berke, J., and Pecora, C.: Diagnostic problems of pyogenic hepatic abscess. *Am J Surg, 111:*678, 1966.

Block, M.A., Schuman, B.M., Eyler, W.R., Truant, J.P., and DuSault, L.A.: Surgery of liver abscesses. *Arch Surg, 88:*602, 1964.

Cronin, K.: Pyogenic abscess of the liver. *Gut, 2:*53, 1953.

Eliason, E.L.: Pylephlebitis and liver abscess following appendicitis. *Surg Gynecol Obstet, 42:*510, 1926.

Gaisford, W.E., and Mark, J.B.D.: Surgical management of hepatic abscess. *Am J Surg, 118:*317, 1969.

Gollin, F.F., Sims, J.L., and Cameron, J.R.: Liver scanning and liver function tests—a comparative study. *JAMA, 187:*11, 1964.

Gottschalk, A.: Liver scanning. *JAMA, 200:*630, 1967.

Grant, R.N., Morgan, L.R., and Cohen, A.: Hepatic abscesses. *Am J Surg, 118:*15, 1969.

McCarthy, C.F., Read, A.E.A., and Ross, F.G.M., *et al.:* Ultrasonic scanning of the liver. *Q J Med, 36:*517, 1967.

McFadzean, J.S., Chang, K.P.S., and Wong, C.C.: Solitary pyogenic abscess of the liver treated by closed aspiration and antibiotics: A report of 14 consecutive cases with recovery. *Br J Surg, 41:*141, 1953.

Ochsner, A., DeBakey, M., and Murray, S.: Pyogenic abscess of the liver: II. An analysis of 47 cases with review of the literature. *Am J Surg, 40:*292, 1938.

Ostermiller, W., and Carter, R.: Hepatic abscess: current concepts in diagnosis and management. *Arch Surg, 94:*353, 1967.

Price, J.E., Joseph, W.L., and Mulder, D.G.: Diagnosis and treatment of intrahepatic abscess. *Am Surg, 33:*820, 1967.

Ryes, A.I., and Reyes, D.A.: Hepatic abscess: an analysis of 86 cases. *Int Surg, 52:*1173, 1969.

Schuman, B.M., Block, M.A., Eyler, W.R., and DuSault, L.A.: Liver abscess: Rose Bengal I-131 hepatic photoscan in diagnosis and management. *JAMA, 187:*708, 1964.

Sherman, J.D., and Robbins, S.L.: Changing trends in the casuistics of hepatic abscess. *Am J Med, 28:*943, 1960.

Shingleton, W.W., Taylor, L.A., and Pircher, F.J.: Radioisotope photoscan of the liver in differential diagnosis of upper abdominal disease: Review of 232 cases. *Ann Surg, 163:*685, 1966.

St. John, F.B., Pulaski, E.J., and Ferrer, J.M.: Primary abscess of the liver due to anaerobic nonhemolytic streptococcus. *Ann Surg, 116:*217, 1942.

Warren, K.W., and Hardy, K.J.: Pyogenic hepatic abscess. *Arch Surg, 97:*40, 1968.

CHAPTER XXXII

ANAEROBIC MICROORGANISMS IN INTRAABDOMINAL INFECTIONS

SHERWOOD L. GORBACH, HARAGOPAL THADEPALLI AND JEANNETTE NORSEN

ABSTRACT: *The application of rigorous techniques for culturing fastidious anaerobes has produced a new understanding of the microflora of intra-abdominal infections. Of 43 cases of intraabdominal sepsis following abdominal surgery, 40 (93%) yielded anaerobic bacteria from wound cultures. The microbial flora contained mixed anaerobes and aerobes in 33, anaerobes alone in 7, and only aerobic or facultative bacteria in 3. Most cultures yielded polymicrobial growth (up to 13 species). Bacteroides and clostridia were the most common anaerobic microorganisms, while* Escherichia coli *was the most frequently isolated facultative organism. Good results from treatment of mixed anaerobic-aerobic infections with an agent effective only against the anaerobes strengthens the evidence for an important pathogenic role for the anaerobes in intraabdominal infections.*

Our understanding of the microorganisms involved in intraabdominal infections has undergone a radical change in recent years. Previous studies had suggested that the offending bacteria were usually aerobic or facultative enteric bacilli (Altemeier, 1944, 1972; Sanchez *et al.*, 1958; Barnes *et al.*, 1959; Karl *et al.*, 1966; Quick and Bogan, 1968; Robson and Heggers, 1969; Polk and Lopez-Mayor, 1969; Konvolinka and Olearczyk, 1972). Medical students are still taught to recognize the odor and consistency of "*E. coli* pus."

Antibiotic therapy for abdominal infections has reflected this conception of the microflora of such wounds; the most popular drugs in surgical units have been the penicillins, cephalosporins and aminoglycoside antibiotics.

The recent availability of sophisticated anaerobic techniques (Hungate, 1950; Moore *et al.*, 1969) to culture clinical specimens has prompted us to take a fresh look at the microorganisms involved in intraabdominal infections. We are reporting our experiences with complicated intraabdominal infections that were refractory to conventional therapy. Our results suggest that anaerobic microorganisms play a significant role in the pathogenesis of intraabdominal sepsis.

MATERIALS AND METHODS

PATIENTS. This series consists of 46 patients who had been admitted to the Department of Surgery of the Cook County Hospital in Chicago. They were subsequently seen in consultation by a member of the Infectious Disease Service for intraabdominal sepsis. These were all adults, ages 20 to 70 years. Because of the predominance of traumatic injuries, the majority of patients were under 40 years of age.

MICROBIOLOGIC METHODS. Specimens from the wounds and drainage sites were obtained at the bedside by a member of the Infectious Disease Service. A gassed-out CO_2 tube was used with two sterile swabs packed in another gassed-out tube. Specimens were brought immediately to the laboratory, and were placed on freshly prepared media for anaerobic culture. In addition, a swab was inoculated into a prereduced, chopped meat glucose broth. Isolates were identified according to the anaerobic manual of the Virginia Polytechnic Institute (Holdeman and Moore, 1972).

RESULTS

INTRAABDOMINAL INFECTIONS. Forty-six patients were referred to the Infectious Disease Service for consultation. These included 32 patients with intraabdominal abscess, ten with generalized peritonitis, and four with miscellaneous infections such as retroperitoneal and prostatic abscesses and infected aortic graft site. All these patients had undergone prior abdominal surgery and had received postoperative antibiotics for signs of infections. The infectious disease consultation was requested because of persisting infection.

Predisposing conditions included trauma, generally due to gunshot or to knife injuries, in 22 patients; carcinoma of the colon, pancreas or kidney in seven patients; intestinal surgery in seven patients; perforated appendix in four; cirrhosis with spontaneous peritonitis in three, and peritoneal dialysis in three.

Forty-three specimens were collected directly from purulent material or surgical drains. Three patients had positive blood cultures but no abdominal site available for culture. The microbial flora of 33 of the 43 specimens of purulent material contained mixed aerobes and anaerobes. An exclusively anaerobic flora was present in seven patients, while three patients had only aerobic or facultative bacteria. Thus, anaerobic bacteria could be cultured from 40 of 43 wounds (93%) in patients with severe intraabdominal infections.

Polymicrobial growth was present in most infections with an average number of five microbial isolates per specimen. This included an average of two aerobes and three anaerobes. The range was from one to 13 different species of microorganisms isolated from a single sample.

Of the aerobic and facultative microorganisms, *Escherichia coli* led the list (Table XXXII-I). Other gram-negative enteric organisms such as *Proteus*, *Klebsiella*, *Pseudomonas* and *Enterobacter* species were also isolated with relative frequency. Although many specimens contained *Staphylococcus epidermidis*, only one contained the more virulent species, *Staphylococcus aureus*. This represents a dramatic change in the ecology of wound infections from a previous era when *S. aureus* was the scourge of surgical wards.

A wide variety of anaerobic species were cultured from these patients (Table XXXII-II). Bacteroides was the most common isolate, recovered on 36 occasions; *B. fragilis* represented 28 of these strains, while the other species were

TABLE XXXII-I

AEROBIC AND FACULTATIVE MICROORGANISMS ISOLATED FROM 46 INTRAABDOMINAL INFECTIONS

Escherichia coli	28
Staphylococcus epidermidis	15
Proteus	10
Klebsiella	9
Pseudomonas	8
Enterobacter	8
Candida	4
Streptococcus	3
Enterococcus	2
Staphylococcus aureus	1

TABLE XXXII-II

ANAEROBIC MICROORGANISMS ISOLATED FROM 43 INTRAABDOMINAL WOUNDS

Gram-negative nonsporing rods		Clostridia	31
Bacteroides	36	*C. ramosum* (3)	
		C. perfringens (3)	
B. fragilis (28)		*C. butyricum* (2)	
B. melaninogenicus (4)		*C. bifermentans* (2)	
B. clostridiiformis (2)		*C. sporogenes* (2)	
B. biacutus (1)		*C. innocuum* (2)	
Bacteroides species (1)		*C. sordellii* (1)	
		C. fallax (1)	
Fusobacterium	6	*C. beijerinckii* (1)	
		C. barati (1)	
F. varium (1)		*C. difficile* (1)	
F. necrophorum (1)		*C. ghoni* (1)	
Fusobacterium species (4)		*C. sphenoides* (1)	
		Clostridium species (10)	
Cocci			
		Gram-positive nonsporing rods	
Peptostreptococci	11		
		Eubacterium	11
Peptococcus	3		
		E. lentum (6)	
P. magnus (2)		*E. filamentosum* (2)	
P. prevotii (1)		*E. multiforme** (1)	
		E. limosum (1)	
Veillonella	1	*Eubacterium* species (1)	
		Propionibacterium	2
		Lactobacillus catenaforme	2
		Bifidobacterium	1

* Now classified as *Clostridium ramosum.*

present less frequently. Fusobacteria were isolated on six occasions, and four of these strains could not be further speciated.

Clostridial species were isolated nearly as frequently as bacteroides, i.e. from 31 specimens. These included thirteen known species and ten strains that could not be further identified. It is interesting to note that *C. perfringens* was isolated in only three patients. Neither these patients nor others associated with histotoxic strains of clostridia had clinical signs of gas gangrene.

Anaerobic cocci were present in 14 patients, with a preponderance of peptostreptococci. Peptococci and veillonella were isolated in three and one patients, respectively.

Gram-positive, nonsporing rods were recovered in 16 patients. The most common isolates were eubacteria. These gram-positive anaerobic rods were always present in mixed culture, usually associated with a Bacteroides species.

Positive blood cultures were found in 13 (28%) of the 46 patients with intraabdominal infections (Table XXXII-III). Multiple blood cultures were obtained from these patients, and these isolates represented consistently positive cultures within a specific period of time. Eleven had exclusively anaerobic growth; two had mixed aerobic and anaerobic growth, but there were none with exclusively aerobes. *Bacteroides fragilis* was the predominant organism, found in eight of the 13 positive cases. Two of these patients had a mixture of *B. fragilis* and *Klebsiella* species. Clostridial septicemia was noted in three patients; one patient had two strains of clostridia, *C. perfringens* and *C. tertium*, in his blood on repeated occasions. *Bacteroides melaninogenicus* and fusobacteria were each present in one patient.

DISCUSSION

The application of rigorous techniques for culturing fastidious anaerobes in clinical material has produced a new understanding of the microflora of intraabdominal infections. It has been known for several years that the predominant microflora of the gastrointestinal tract is anaerobic, principally bacteroides, peptostreptococci and clostridia (VanHoute and Gibbons, 1966; Drasar, 1967; Floch, 1968, 1970; Finegold,

TABLE XXXII-III

MICROORGANISMS IN 13 SEPTICEMIAS ASSOCIATED WITH INTRAABDOMINAL INFECTIONS*

Bacteroides fragilis	8
Bacteroides melaninogenicus	1
Clostridium perfringens	2
Clostridium tertium	1
Clostridium difficile	1
Fusobacterium nucleatum (*F. fusiforme*)	1
Klebsiella	2

* Two patients had *B. fragilis* and klebsiella; one patient had 2 strains of clostridia.

1969; Gorbach, 1971). In this respect, it is curious that a discrepancy has existed between the high frequency of anaerobes in normal intestinal contents and the relatively uncommon recovery of these forms in infections associated with intestinal perforation (Altemeier, 1944, 1972; Sanchez *et al.*, 1958; Barnes *et al.*, 1959; Karl *et al.*, 1966; Quick and Bogan, 1968; Robson and Heggers, 1969; Polk and Lopez-Mayor, 1969; Konvolinka and Olearczyk, 1972). This paradox appears to have been solved by utilizing the newer anaerobic methods in culturing these infections. In our study of 46 patients with intraabdominal sepsis, all but three had anaerobic bacteria associated with their disease. Indeed, there were more patients with an exclusively anaerobic flora than there were with an exclusively aerobic or facultative flora. Furthermore, we found no cases of "sterile abscess," the former refuge of bacteriologic ineptitude.

The recognition of anaerobes in abdominal wounds has important implications for antimicrobial therapy. On the basis of *in vitro* (Ingham *et al.*, 1968; Sutter and Finegold, 1971; Martin *et al.*, 1972; Kislak, 1972; Nastro and Finegold, 1972; Wilkins *et al.*, 1972) and *in vivo* (Tracy *et al.*, 1972; Bartlett *et al.*, 1972; Thadepalli and Gorbach, 1972) data from several laboratories, it would appear that certain drugs are acceptable for treatment of anaerobic infections, whereas other antibiotics are unacceptable. In the acceptable group, chloramphenicol and clindamycin are the most satisfactory agents. On the other hand, there is good evidence that the penicillins, cephalosporins and aminoglycoside antibiotics have relatively poor activity against *Bacteroides fragilis*, the most frequent isolate in intraabdominal infections. The new information on antibiotic susceptibilities combined with the data on anaerobic bacteriology should cause a reassessment of the use of antibiotics in abdominal infections.

The microbial flora of an intraabdominal abscess is often a complex ecosystem. As many as 13 distinct species of bacteria can be isolated from infected sites. This raises the question of which organism(s) is critical to the pathogenesis of this septic complication. Does the presence of multiple organisms mean that there are necessarily multiple pathogens? Is it necessary to treat all cultivable microorganisms that are potentially pathogenic?

An approach to answering these questions is the use of an antibiotic with a relatively limited spectrum in treating patients with mixed infections. Clindamycin fulfills this criterion in that it has no activity against most coliforms and pseudomonads (McGehee *et al.*, 1968) but is highly effective against a wide range of anaerobic bacteria (Ingham *et al.*, 1968; Sutter and Finegold, 1971; Martin *et al.*, 1972; Kislak, 1972; Nastro and Finegold, 1972; Wilkins *et al.*, 1972). We recently reported our results in treating 12 patients who had mixed aerobic and anaerobic intraabdominal infections with clindamycin alone (Thadepalli and Gorbach, 1972). The aerobic bacteria included *E. coli,* proteus, klebsiella or pseudomonas; an average of two of these organisms were present in each case, in addition to bacteroides, clostridia or peptostreptococci. The clinical results in these 12 patients were entirely satisfactory, with decreased drainage from the wound, a lowering of fever and reduction of systemic toxicity. All patients were essentially cured of intraabdominal sepsis by this single agent, despite the continued growth of the aerobic and facultative bacteria in the wound cultures.

Several lines of evidence have implicated anaerobic bacteria in the pathogenesis of intraabdominal abscess. However, while our advanced technology has revealed a wide variety of microbial components in such infections, it has at the same time raised new queries about the relevance of the findings. We may no longer be teaching our students about "*E. coli* pus", but there is no comfort in promulgating what is basically the same old fiction under a different set of titles. Clearly, our understanding of the true role of anaerobes in infectious processes is at an early, and as yet unresolved, stage.

REFERENCES

Altemeier, W.A.: Bacteriology of traumatic wounds. *JAMA, 124:*413, 1944.

Altemeier, W.A.: The significance of infection in trauma. *Bull Am Coll Surg, 57:*7, 1972.

Barnes, B.A., Benringer, G.E., and Wheelock, F.C., *et al.*: Surgical sepsis. *N Engl J Med, 261:*1351, 1959.

Bartlett, J.G., Sutter, V.L., and Finegold, S.M.: Treatment of anaerobic infections with lincomycin and clindamycin. *N Engl J Med, 287:*1006, 1972.

Drasar, B.S.: Cultivation of anaerobic intestinal bacteria. *J Pathol Bact, 94:*417, 1967.

Finegold, S.M.: Intestinal bacteria—the role they play in normal physiology, pathologic physiology and infection. *Calif Med, 110:*455, 1969.

Floch, M.H., Gershengoren, W., and Freedman, L.R.: Methods for the quantitative study of the aerobic and anaerobic intestinal bacterial flora of man. *Yale J Biol Med, 41:*50, 1968.

Floch, M.H., Gorbach, S.L., and Ludkey, T.D., Eds.: Intestinal microflora. Part I, Part II. *Am J Clin Nutr, 23:*1425, 1970.

Gorbach, S.L.: Intestinal microflora. *Gastroenterology, 69:*1110, 1971.

Holdeman, L.V., and Moore, W.E.C.: *Anaerobe Laboratory Manual.* Blacksburg, Virginia Polytechnic Institute and State Univ., 1972.

Hungate, R.E.: The anaerobic, mesophilic, cellulolytic bacteria. *Bact Rev, 14:*1, 1950.

Ingham, H.R., Selkon, J.B., and Codd, A.A., *et al.*: A study *in vitro* of the sensitivity to antibiotics of *Bacteroides fragilis. J Clin Pathol, 21:*432, 1968.

Karl, R.C., Mertz, J.J., and Veith, F.J.: Prophylactic antimicrobial drugs in surgery. *N Engl J Med, 275:*305, 1966.

Kislak, J.W.: The susceptibility of *Bacteroides fragilis* to 24 antibiotics. *J Infect Dis, 125:*295, 1972.

Konvolinka, C.W., and Olearczyk, A.: *Subphrenic Abscess, Current Problems in Surgery.* Chicago, Year Book Med. Pub. Inc., Jan, 1972.

Martin, W.J., Gardner, M., and Washington, J.A., II: *In vitro* antimicrobial susceptibility of anaerobic bacteria isolated from clinical specimens. *Antimicrob Agents Chemother, 1:*148, 1972.

McGehee, R.F., Smith, C.B., and Wilcox, C., *et al.*: Comparative studies of antibacterial activity *in vitro* and absorption and excretion of lincomycin and clindamycin. *Am J Med Sci, 256:*279, 1968.

Moore, W.E.C., Cato, E.P., and Holdeman, L.V.: Anaerobic bacteria of the gastrointestinal flora and their occurrence in clinical infection. *J Infect Dis, 119:*641, 1969.

Nastro, L.J., and Finegold, S.M.: Bactericidal activity of five antimicrobial agents against *Bacteroides fragilis. J Infect Dis, 126:*104, 1972.

Polk, H.C., and Lopez-Mayor, J.F.: Post-operative wound infection: A prospective study of determinant factors and prevention. *Surgery, 66:*97, 1969.

Robson, M.C., and Heggers, J.P.: Surgical infections. I. Single bacterial species or polymicrobiotic in origin. *Surgery, 65:*608, 1969.

Sanchez, U.R., Fernand, E., and Rousselot, L.M: Complication rate in general surgical cases. *N Engl J Med, 259:*1045, 1958.

Sutter, V.L., and Finegold, S.M.: Antibiotic disc susceptibility tests for rapid presumptive identification of gram-negative anaerobic bacilli. *Appl Microbiol, 21:*13, 1971.

Thadepalli, H., and Gorbach, S.L.: Clindamycin in the treatment of pure and mixed anaerobic infections. Abstr. #135, p. 71. Paper presented at the 12th Interscience Conference on Antimicrobial Agents and Chemotherapy, Atlantic City, Sept. 26–29, 1972.

Tracy, O., Gordon, A.M., and Moran, F.: Lincomycin in the treatment of bacteroides infections. *Br Med J, 1:*280, 1972.

Quick, C.A., and Bogan, T.D.: Gram-negative rods and surgical wound infections. *Lancet, 1:*163, 1968.

van Houte, J., and Gibbons, R.J.: Studies of the cultivable flora of normal human feces. *Antonie V. Leeuwenhoek, 32:*212, 1966.

Wilkins, T.D., Holdeman, L., and Abramson, I.J., *et al.*: Standardized single-disc method for antibiotic susceptibility testing of anaerobic bacteria. *Antimicrob Agents Chemother, 1:*451, 1972.

CHAPTER XXXIII

DENTAL INFECTIONS

WALTER J. LOESCHE

ABSTRACT: *Dental plaque is a unique ecosystem which colonizes a nonrenewable body surface. Plaque is comprised mostly of bacteria capable of anaerobic growth. The exact microbial composition of plaque is dependent upon several selection factors among which are oxygen sensitivity, adherence and nutrient supply. Certain plaques predispose to caries on the tooth surface. Other plaques are associated with gingival inflammation in the approximating soft tissue. Some aspects of these two important dental diseases may be due to the presence of specific bacteria such as* Streptococcus mutans, Actinomyces viscosus *or* Actinomyces naeslundii *in the plaque.*

Mammalian surfaces are colonized by considerable numbers of microbes (Rosebury, 1962). In the oral cavity the resident flora has long been suspected of contributing to dental caries, periodontal disease, certain bacteremias and as posing a serious threat to health when the integrity of the host tissues is compromised by malnutrition or irradiation. The actual role which the oral flora plays in these pathological conditions has long been obscured because the sampling procedures and culturing techniques did not take into account the anaerobic nature and diversity of the oral environment. Only recently has there come the realization that the oral flora is essentially anaerobic and that distinct microbial ecosystems exist on the various oral structures. In this essay the microbes resident in one such ecosystem, the dental plaque, will be described and their contribution to dental and systemic infections discussed.

Dental plaque is the term given to the aggregation of bacteria, salivary glycoproteins and inorganic salts which adheres to the tooth surface. If the plaque is located above the gingival margin, it is called supragingival plaque, whereas if it is below the gingival margin it is called subgingival plaque. This distinction is necessary, as the resident flora of each plaque differs and is associated with different dental diseases. The plaque is unique in that it is attached to a nonrenewing tooth surface, and therefore will accumulate if it is not removed by mastication or mechanical brushing. Thus in certain oral disease situations, about 100 mg or more wet weight, or close to 10^{11} bacteria, are concentrated on the tooth surfaces in direct contact with the mucosal epithelial cells (Socransky *et al.*, 1963). Accumulations of this magnitude presumably affect the underlying host tissue as the gingival surfaces are permeable to microbial byproducts and may be vulnerable to microbial invasion. The microbial load on epithelial surfaces elsewhere in the body is markedly lower due to shedding of the mucosal cell.

ANAEROBIC NATURE OF ORAL CAVITY

The oral cavity is not considered an anaerobic environment because of its easy access to atmospheric oxygen. Yet when quantitative culturing procedures were used to isolate plaque bacteria, only 20 percent of the microscopic count could be recovered in the viable count, despite the use of anaerobic jars (Socransky *et al.*, 1963; Gibbons *et al.*, 1963, 1964c, Gilmour and Poole, 1970). Among the explanations given for the low recoveries was that the manipulation of the sample under aerobic conditions killed many sensitive anaerobes. Subsequent studies have shown that if gingival plaque is collected and manipulated under systems of complete anaerobiosis, i.e. roll tube (Gordon *et al.*, 1971) or anaerobic chamber (Aranki *et al.*, 1969), the recoveries of bacteria are tripled over those obtained with anaerobic jars. Apparently some oral anaerobes, such as the spirochetes, are killed by short exposures to oxygen, a situation quite similar to that found with rumen anaerobes (Loesche, 1969). The spirochetes are unusual in that their *in vitro* propagation requires that the media Eh be poised at −180mv (Socransky *et al.*, 1964). The recovery of this strict anaerobe from subgingival plaque argued that the crevice site must be anaerobic. The Eh of periodontal pockets was found to be about −50mv by direct Eh electrode measurements (Kenney and Ash, 1969) and that of plaque as low as −200 mv (Onisi *et al.*, 1960). Also in two subjects an Eh electrode was maintained *in situ* on an artificial tooth and plaque was allowed to accumulate over the electrodes. After four to seven days of undisturbed growth, the plaque Eh was in the range of -110 to -140mv (Kenney and Ash, 1969). When Eh indicators such as benzyl viologen (midpoint potential -359mv) are applied directly *in vivo* to plaque, discrete sections of the plaque near the gingiva crevice are capable of reducing the dye (Socransky and Manganiello, 1971). In other studies the pO_2 of the oral cavity when the lips are together varied from 16 percent over the anterior tongue to less than 1 percent in the buccal fold (Eskow and Loesche, 1971). The low pO_2 readings obtained in the buccal fold probably reflect that under normal conditions the cheek mucosa approximates the tooth and gingival surfaces, thereby minimizing gas exchange between this site and the space over the tongue.

Each of these observations, i.e. the need for complete anaerobiosis to grow the majority of the plaque bacteria, the *in vitro* Eh requirement of the spirochete, the *in vivo* plaque Eh and the pO_2 of the buccal fold, collectively demonstrate that the plaque represents an extremely anaerobic ecosystem.

ECOLOGICAL NICHES

Many investigations of the oral flora were performed on saliva, partly because of the ease with which the sample could be obtained, and partly because it was assumed that the oral flora and the salivary flora were synonymous. These studies demonstrated the complexity of the oral flora, but they were not interpreted as indicating that this complexity might be due to the washoff of bacteria from several diverse ecosystems within the oral cavity. Later when dental plaque was examined, Krasse (1954) and Gibbons *et al.* (1964a) noted the virtual absence of *Streptococcus salivarius* in the plaque, despite the fact that this organism was the most prominent streptococcus in saliva. Similarly, spirochetes and *Bacteroides melaninogenicus* could be recovered from subgingival plaque but rarely from saliva (Gibbons *et al.*, 1964b). Subsequent quantitative cultural studies of various sites in the oral cavity, as well as the same sites in different clinical situations, have consistently demonstrated a localization of various microbes to certain anatomic areas or disease situations. *S. salivarius* colonizes primarily the tongue and cheek (Gibbons and van Houte, 1971). As these surfaces are the largest in the oral cavity, their sheddings make the greatest contribution to the saliva, which accounts for the prominance of *S. salivarius* in saliva. The question of what selection factors determine the various ecosystems merits some attention. Thus far, three ecological determinants have been identified: anaerobiosis, adherence and nutrition.

Anaerobiosis

Clearly the oxygen tension and Eh of various anatomic areas should select out or discriminate against certain of the organisms which gain access to the oral cavity. The teeth and associated gingival crevice area are necessary for growth of the spirochetes. When teeth are absent, as in the edentulous mouth, spirochetes are not found (Lewkowicz, 1901; Brailovsky-Lounkevitch, 1915; Kostecka, 1924; Rosenthal and Gootzeit, 1942). This would seem to indicate that only in the crevice area can the spirochetes find the optimal growth conditions, among which would be a low Eh. Support for the importance of Eh in the localization of strict anaerobes on mucous membranes is provided by germfree animal studies. Anaerobes do not establish themselves in pure cultures in germfree animals, presumably because the Eh potentials of the mucosal surfaces are too high (Loesche, 1969; Wostmann and Bruckner-Kardoss, 1966). Prior establishment of a facultative organism usually permits subsequent colonization by anaerobes (Syed *et al.*, 1970) although important exceptions occur. Thus, *B. melaninogenicus* and the oral spirochetes could not be established in germfree mice even after the animals were contaminated with a variety of facultative organisms (Gibbons *et al.*, 1964b). Other determinants in addition to anaerobiosis must contribute to the ecosystems found in the oral cavity. Two of these, i.e. adherence and nutritional localization, can be documented.

Adherence

Previously the localization of *S. salivarius* to the tongue was mentioned as the reason for its numerical prominance in saliva. The corollary to this observation was the absence of this organism on the tooth surface. Van Houte and coworkers (1970, 1971) have studied the proportional distribution of *S. salivarius, S. sanguis* and *S. mutans* on various oral structures after a dental prophylaxis or after an exposure to a mouthrinse containing streptomycin-labeled strains of these species. These streptococci were selected because *S. salivarius*, is normally dominant on the tongue (Gordon and Gibbons, 1966); *S. sanguis* in the dental plaque (Carlsson, 1965); and *S. mutans* in the carious tooth (Krasse *et al.*, 1968; Littleton *et al.*, 1968; Shklair *et al.*, 1972); and because colonies of these organisms can be readily identified on Mitis Salivarius agar (Jordan *et al.*, 1968). Their results showed a remarkable selectivity of these organisms to the sites where they normally are found. *S. salivarius* comprised only a small percentage of the facultative streptococci recovered from the tooth surfaces 30 to 120 minutes after a prophylaxis, despite the fact that in the bathing saliva the mean percentage of this organism was ten times as high. *S. sanguis*, which comprised about 15 percent of the salivary streptococci, adsorbed to the tooth surface and accounted for about 55 percent of the streptococci found on the tooth surface after 120 minutes. When antibiotic-labeled strains were given in a mouthrinse, *S. salivarius* did not adhere to the clean tooth surface, or dental plaque, but did attach to the tongue and to a lesser extent the vestibular mucosa. *S. sanguis* attached to the clean tooth surface, dental plaque, vestibular mucosa and to a lesser extent the tongue. *S. mutans* did not adhere to the soft tissues. From these experiments it would appear that *S. salivarius* and *S. mutans* have receptor or attachment sites which can distinguish between tooth and soft tissues. These attachment mechanisms help determine the *in vivo* localization of these organisms and may in the case of *S. mutans* account for its role in dental caries. A more general role for microbial surface attachment has been postulated by Gibbons and coworkers in regard to *Corynebacterium diphtheriae* and *S. pyogenes* infections (Gibbons and van Houte, 1971; Ellen and Gibbons, 1972).

Nutritional Requirements

The distribution of certain oral organisms may be related to the availability of required nutrients only in specific sites. *B. melaninogenicus* has a requirement for hemin (Evans, 1951) and vitamin K (Lev, 1958; Gibbons and Macdonald, 1960) and *Treponema microdentium* shows a requirement for α-2 globulin (Socransky and Hubersak, 1967). These compounds would be present in serum. The gingival crevice is bathed by a serum transudate (Brill, 1959) which might contain these nutrients and thereby explain the localization of these anaerobes in this site. If plaque accumulates to the extent that gingivitis results, the serum transudate increases and bleeding occurs, a situation that would enhance the supply of hemin, vitamin K and α-2 globulin in the crevice area. In plaques associated with gingivitis the levels of *B. melaninogenicus* (Loesche et al., 1972) and spirochetes (Socransky *et al.*, 1963) tend to be elevated.

Other members of the oral flora are dependent on microbial fermentation or other by-products for essential nutrients. Lactate is catabolized by *Veillonella* species (Rogosa, 1964) and formate and/or hydrogen is utilized by *Vibrio sputorum* (Loesche, 1968). Several oral species produce vitamin K which can be used by *B. melaninogenicus* (Gibbons and Engle, 1964). Some of the spirochetes' nutrient requirements, i.e. spermine and isobutyrate, are provided by coinhabiting fusobacteria and diphtheroids (Socransky *et al.*, 1964). Thus some oral species are microbe dependent and their presence in the oral cavity would be secondary to colonization by the nutrient-producing bacteria (Loesche, 1968).

The main determinant of the type of flora found in plaque, however, is the mammalian diet. Frequent ingestion of a readily metabolizable substrate, such as sucrose, inevitably leads to selection of a cariogenic flora, whereas low-sucrose diets are associated with low caries incidence. Epidemiologic and clinical evidence documents that the shift from a low- or no-sucrose diet to a high-sucrose diet caused caries among Eskimos (Waugh and Waugh, 1940), inhabitants of a remote Pacific island, Trista da Cunha (Sognnaes, 1954; Holloway *et al.*, 1963), and among the inmates of a mental institution in Sweden (Gustafsson *et al.*, 1954). Conversely, a shift away from high-sucrose diets as occurred in war-torn Europe was associated with a decline in caries (Torerud, 1949; Hardwick, 1960). Also, populations which ingest low-sucrose diets as a way of life, such as residents of underdeveloped countries or mentally institutionalized individuals, tend to have a low caries incidence (Bibby, 1970). In these latter individuals, oral hygiene is almost nonexistent, leading to massive plaque accumulations which invariably are associated with periodontal breakdown.

DENTAL CARIES AND PERIODONTAL DISEASE AS INFECTIONS

The complexity of the oral flora and the inability to distinguish differences between the plaque of carious and noncarious individuals in the past led to acceptance of the idea that caries was due to a nonspecific overgrowth on the teeth of acid-producing organisms. Likewise, periodontal disease was considered due to a nonspecific overgrowth of bacteria which released irritants such as acid, enzymes, antigens and endotoxin into the gingival tissues. The difficulty of studying these chronic nonspecific infections in humans led investigators to use animal models for the study of caries and periodontal disease. Genetic dietary and bacterial components were readily identified. The essentiality of bacteria was demonstrated by studies in germfree rats in which genetically caries-susceptible rats fed a cariogenic sucrose diet did not develop caries (Orland *et al.*, 1954; Fitzgerald *et al.*, 1960). However, the role of bacteria was thought to be nonspecific until 1960, when Keyes demonstrated the transmissibility of caries in hamsters. Subsequently a streptococcus was isolated which by itself could cause destruction of teeth, provided it was established in the oral cavity (Fitzgerald and Keyes,

1960; Krasse, 1965); this streptococcus is now designated as *Streptococcus mutans.* Jordan and Keyes (1964) then went on to demonstrate that a form of periodontal disease was also transmissible in hamsters and identified a new species, *Actionomyces viscosus*, as the etiologic agent (Howell *et al.*, 1965). These findings have led to studies seeking to implicate these or similar microbes in human dental disease.

Dental Caries

Caries can occur on different aspects of the tooth such as in pit and fissures, on the smooth surfaces of the crown, and on the root surfaces. In each case, the loss of tooth substance presumably occurs as a result of solubilization of tooth mineral by an acid solution. However, such a unitary hypothesis may have multiple variations, some of which reside in the morphology and composition of the tooth. Thus, genetic and developmental influences control the dimensions of the pits and fissures, and their size in turn determines the microbial load retained by them. In smooth surface and root surface caries, microbial retention is associated with plaque-forming bacteria. Smooth surface caries requires the dissolution of enamel, the hardest tissue in the body, which is 99 percent inorganic and 1 percent organic. Root surface caries occurs below the enamel in the cementum and dentine, which is 70 percent inorganic and 30 percent organic. In the animal models, different microbes exhibited a predilection for one or more of these surfaces, which prompted Keyes (1968) to speculate that the events leading to caries may be different for each surface. These microbes and the mechanisms by which they contribute to caries will now be discussed.

Streptococci

Clark (1924) and Maclean (1927) described *S. mutans* as the etiologic agent of caries approximately 50 years ago. Their data consisted mainly of the isolation of this organism from the carious lesion and provided no evidence of animal infectivity. The statistical association which they demonstrated between *S. mutans* and caries was no more convincing than the concurrent studies at that time associating lactobacillus with caries (Enright *et al.*, 1932; Bunting and Palmerlee, 1925). Inasmuch as a simple diagnostic medium for the identification of lactobacilli emerged first (Hadley, 1933), most of the microbiologic research in caries subsequently turned away from the streptococci and towards the lactobacilli. Eventually, when quantitative culturing procedures were introduced, streptococci were found to outnumber the lactobacilli by a factor of 10,000 to 1 (Stralfors, 1950; Socransky and Manganiello, 1971). But the concept of specificity was not revived until the demonstration of transmissibility of caries in the hamster (Keyes, 1960), and the observation that only certain streptococci caused caries in the germfree rat (Fitzgerald, 1968). Investigations of this hamster infection revealed the essentiality of sucrose for the establishment of the cariogenic streptococcus in the oral cavity (Krasse, 1965). Gibbons *et al.*, (1966) showed that sucrose was converted by this streptococcus into a variety of extracellular polysaccharides, some of which permitted these organisms to adhere to solid surfaces. These polysaccharides subsequently were shown to be present in plaque (Gibbons and Banghart, 1967; Critchley *et al.*, 1967) and were thought to provide the mechanisms by which these and other bacteria adhered to the tooth surface. The initial chemical analysis of the polysaccharides revealed dextrans and levans (Gibbons and Banghart, 1967; Critchley *et al.*, 1967; Guggenheim and Schroeder, 1967), but now it appears that the main insoluble polysaccharide is a unique α1–3 glucan, which has been given the name mutan (Hotz *et al.*, 1972). These findings encouraged taxonomic studies of the oral streptococci which resulted in the recognition of several distinct species, one of which contained all of the cariogenic streptococci (Carlsson, 1968; Edwardsson, 1968; Guggenheim, 1968). These cariogenic streptococci were unique in their ability to ferment mannitol and sorbitol, and as these characteristics were used by Clarke in 1924 to identify *S. mutans*, this designation has been retained for the cariogenic streptococci. The precise taxonomic status of *S. mutans*, however, remains undecided, because DNA analysis reveals three distinct species among this seemingly homogeneous group of cariogenic organisms (Coykendall, 1971).

S. mutans has been isolated from humans on a world-wide basis (Krasse *et al*, 1968; Jordan *et al.*, 1969; Zinner *et al.*, 1965; Gibbons and Loesche, 1967; deStoppelaar *et al.*, 1969; Schamschula and Barnes, 1970; Bratthall, 1972). The correlation between *S. mutans* and total caries experience, i.e. the standard decayed, missing and filled teeth index (DMF teeth) used to document caries morbidity, has not been good (Jordan *et al.*, 1969). However, as Krasse *et al.* (1968) have indicated, significant associations between *S. mutans* and active caries can usually be found. *S. mutans* accounted for 17 percent of the cultivable isolates from carious teeth but only 1.5 percent of the isolates from noncarious teeth (Loesche *et al.*, 1973). *S. mutans* was detected in 26 of 27 preschool children diagnosed as having rampant caries and averaged 22 percent of the cultivable flora. This association of *S. mutans* with human caries may indicate causation or a secondary phenomenon related to the carious lesion. Long term clinical studies of one to two year's duration are now needed to demonstrate that the presence of certain levels of *S. mutans* on a tooth surface will inevitably lead to caries of that surface.

Actinomyces

Keyes and Jordan have described animal infections caused by a plaque-forming bacillus which resulted in extensive bone loss and root surface caries (Jordan and Keyes, 1964; Keyes and Jordan, 1964; Jordan *et al.*, 1969). The organism possessed a catalase and was initially described as *Odontomyces viscosus* (Howell *et al.*, 1965), but its similarity to the *Actinomyces* has led to its redesignation as *Actinomyces viscosus* (Pine, 1970). The plaque formed by this organism is starch dependent and thus differs from the sucrose dependent *S. mutans* plaque. Only preliminary information is available concerning the significance of this organism in human root surface caries. This can be attributed to the lack of a selective medium for this organism, and the inability to distinguish by simple tests *Actinomyces viscosus* from the other

Actinomyces species present in dental plaque. Recent reports demonstrate that this organism as well as other *Actinomyces* can be isolated from root surface decay found in teeth extracted from mentally retarded individuals (Jordan and Hammond, 1972). Also *A. viscosus* was recovered from plaque associated with extensive root decay in New Guinea tribesmen (Schamschula *et al.*, 1972). Both of these populations tend to ingest a high percentage of carbohydrate calories as starch and are low in sucrose consumption. The data suggest that *Actionomyces viscosus* may be an important contributor to root surface decay.

Lactobacillus

Lactobacillus species have long been associated with caries (Enright *et al.*, 1932) but whether this is a cause and effect relationship has never been satisfactorily demonstrated. Lactobacilli were suspected because of their aciduric characteristics which would enable them to survive in the acid pHs necessary to dissolve enamel, i.e. pH<5.1. Simple selective primary isolation media have facilitated their detection (Hadley, 1933). Salivary lactobacillus counts have been used to diagnose human caries (Burnett and Sherp, 1968) and to monitor the efficacy of a restrictive carbohydrate diet (Jay, 1947). Lactobacilli are not numerically important in nondisease-associated human plaque, being outnumbered by the streptococci by about 10,000 to 1 (Gibbons *et al.*, 1964; Stralfors, 1950). However, in carious dentine *Lactobacillus caseii* may be the dominant isolate (Shovlin and Gillis, 1969) or at least be as numerous as the various *Streptococcus* species (Loesche and Syed, 1973). In germfree rats lactobacillus species will cause pit and fissure caries (Fitzgerald *et al.*, 1966; Rosen *et al.*, 1968) which do not appear to be sucrose dependent.

These observations can be explained by ecological considerations. Normally the lactobacilli are in low numbers in the oral cavity because they cannot adhere to the dental structures (van Houte *et al.*, 1972). Perhaps they reside in sites such as the pit and fissures or in interproximal areas where retention is mechanical (Socransky and Manganiello, 1971). If a cariogenic plaque exists on these sites, the low pH produced could act to select out the aciduric flora. As decay progresses through the enamel and spreads into the dentine, the lesion is almost a closed system, far removed from any buffering influence of the saliva. In this acid environment, the lactobacilli apparently thrive. This focus of infection would be adequate to seed the saliva, thereby accounting for the elevated lactobacilli counts associated with caries. In this context, lactobacilli would be secondary pathogens to the plague-forming organisms such as *S. mutans* on the smooth surfaces, but could well be the primary pathogens in pit and fissure caries.

Periodontal Disease

If debridement of the tooth surfaces by various mechanical devices is not performed routinely, plaque will accumulate and lead first to gingivitis and eventually to periodontal disease and tooth loss. A considerable amount of clinical evidence can be found to support this statement (Glickman, 1971; Loe *et al.*, 1965). Dental surveys of various underdeveloped countries, where toothbrushes are not used, note the early onset of periodontal disease (Russell, 1966). Residents of mental institutions, who are unable to practice oral hygiene procedures, also have periodontal destruction at an early age (Cutress, 1971). Socransky *et al.* (1963) noted that 10 to 20 times more plaque could be removed from periodontal patients than from nonperiodontal patients. Loe and associates have shown that after two to three weeks without oral hygiene procedures, human volunteers will exhibit clinical gingivitis (Loe *et al.*, 1965; Loe, 1971). If plaque accumulation was suppressed by twice daily rinsing with a chlorhexidine mouthwash, no gingivitis occurred (Loe and Schiott, 1970). A broad spectrum antibacterial agent was necessary, as previous clinical trials with vancomycin (gram-positive spectrum) and polymyxin (gram-negative spectrum) were not successful in preventing plaque accumulation and gingivitis (Jensen *et al.*, 1968).

These observations have been interpreted as indicating that a certain bulk of plaque in contact with gingival crevice epithelium will elicit an inflammatory response in the underlying tissue. Direct microbial invasion is not observed, with the exception of acute necrotizing ulcerative gingivitis, noma and possibly during the bacteremias which are of common occurrence following dental procedures (Bender *et al.*, 1963). Microbial byproducts ranging from low molecular weight organic acids (Sharawy and Socransky, 1967), H_2S (Rizzo, 1967), NH_4 (Macdonald and Gibbons, 1962) and amines (Bahn, 1970) to high molecular weight antigens (Wittwer *et al.*, 1969), enzymes (McDougall, 1971) and endotoxin (Schwartz *et al.*, 1972) can enter the tissue. These compounds probably penetrate the gingival tissue at all times. When plaque accumulations are minimal, the deleterious effects of these irritants are difficult to detect, perhaps due to such host factors as mucosal cell sloughing, good vascularization of the gingival tissues and the cleansing action of the gingival crevice fluid. But with increased plaque mass, the increased amounts of noxious products which enter the tissue presumably trigger additional defense mechanisms. If the microbial molecule entering the tissue is large enough to be antigenic, then an antigen-antibody complex which activates the complement system could occur (Mergenhagen *et al.*, 1970). This would attract polymorphonuclear leukocytes (PMN's) to the tissue, and the subsequent release of their lysosomal enzymes could result in a net loss of tooth-supporting tissue (Taichman *et al.*, 1966). One much studied antigen in this regard is endotoxin, which appears capable of penetrating healthy gingiva (Schwartz *et al.*, 1972) as well as ulcerated gingiva (Rizzo, 1970). Endotoxin could activate the complement system by forming an antigen-antibody complex, or it could activate the C'3 component of complement directly (Mergenhagen *et al.*, 1970). The plaque also contains chemotactic factors which would attract the PMN's to the gingival area (Tempel *et al.*, 1970). *Actinomyces viscosus* secretes a peptide with molecular weight of 10 to 15,000 which is directly chemotactic for PMN's *in vitro* in a Boyden Chamber. Other investigations suggest that the important inflammatory cells in gingivitis are mononuclear (Lehner, 1972). Lymphocytes from peripheral blood from periodontal patients showed blast transformations when stimulated with plaque or with antigens derived from pure cultures of certain plaque bacteria such as *Veillonella alcalescens* or *Actinomyces viscosus* (Ivanyi, 1972), but did not respond to antigens obtained from intestinal bacteria. Furthermore, lymphocytes from nonperiodontal patients did not respond to any of the bacterial antigens. This and other tests for cellular antibodies, such as cytotoxicity of lymphocytes against unrelated target cells and the migration inhibition of lymphocyte culture supernatants on guinea pig macrophages, indicate that the small amounts of bacteria resident on the tooth surfaces apparently can elicit a systemic response of cellular antibodies (Ivanyi *et al.*, 1972).

Another defense mechanism of the host could be the detoxification of plaque by calculus formation. Most dentists consider calculus as a contributor to periodontal disease, but Keyes (1970) suggests that the biological significance of calculus is to embed and thereby eliminate irritating bacteria.

All of the defense mechanisms cited are directed towards the protection of the host from microbial penetration at a vulnerable soft tissue-hard tissue interface. The ultimate result of this protection would be tooth loss, which is a small price to pay for systemic health.

The available evidence suggests that plaque mass is the important parameter in gingivitis and periodontal disease. It is generally assumed that increased plaque bulk results from poor oral hygiene and does not involve the presence of specific bacteria. However, *Actinomyces viscosus* can cause a transmissible periodontal disease in hamsters which is associated with plaque formation at the gingival margin (Jordan and Keyes, 1964). Also, a periodontal syndrome occurring in the rice rat could be eliminated by penicillin (Mulvihill *et al.*, 1967) and was subsequently shown to be transmissible by gram-positive rods (Dick *et al.*, 1967). Human isolates of *S. mutans* (Gibbons *et al.*, 1966) and *Actinomyces naeslundii* (Socransky *et al.*, 1970) will cause extensive periodontal destruction in germfree rats. The mechanism for pathophysiology is not bacterial invasion of the tissue, and therefore probably resides either in the ability of these bacteria to elaborate products which are more antigenic, more chemotoxic *etc*, or in their ability to form more plaque, thereby increasing the overall levels of noxious products. Thus far, all the animal periodontal disease syndromes are associated with plaque formation. However, antigens from *Actinomyces viscosus* are several times more active than veillonella antigens in eliciting blast transformations. In addition, sensitization to actinomyces antigens can be demonstrated in the skin of periodontal patients (Nisengard and Beutner, 1970). Collectively these findings suggest that some microbial specificity may be involved in periodontal disease. If so, then treatment approaches aimed at the elimination of these bacteria from plaque might offer some hope for control of this disease in the future.

SYSTEMIC INFECTIONS

Bacteremia

The mucous membranes form a physical barrier preventing the invasion of microbes into the host tissue. In addition, there would appear to be a physiochemical barrier, as the host tissue is aerobic with an Eh in the vicinity of +50 to +200 mv (Hewitt, 1950), whereas the plaque (Kenney and Ash, 1969; Onisi *et al.*, 1960) and intestinal contents (Wostmann and Bruckner-Kardoss, 1966) are anaerobic with an Eh of −50 to −200 mv. Thus, if anaerobic bacteria were to enter healthy aerobic tissue, it is unlikely that they could propagate and cause infection. A somewhat analogous situation is the inability to establish strict anaerobes on the mucous membranes of germfree rats when the Eh is positive (Syed *et al.*, 1970). In the oral cavity plaque organisms gain access to the tissue via the physical manipulation of the teeth and periodontal tissue during extractions and periodontal surgery (O'Kell and Elliot, 1935), carious pulpal exposure, scaling procedures (Rogosa *et al.*, 1960), dental prophylaxis (Winslow and Kobernick, 1960), and even mastication (Cobe, 1954; Round *et al.*, 1936). Usually a transient bacteremia of 5 to

20 minutes duration occurs and its significance in terms of systemic disease is usually ignored (Martin, 1967). It is unlikely that the anaerobic members of this flora, if introduced in small numbers, could survive and cause disease. However, the facultative streptococci can survive in aerobic tissues and thus could seed the body, especially sites such as heart valves, traumatized tissue and traumatized teeth. Alpha-hemolytic streptococci form the single most numerous group of bacteria associated with subacute bacterial endocarditis (SBE) (Wilson and Miles, 1964). Approximately one third of the streptococcal isolates are *Streptococcus sanguis* (van Houte *et al.*, 1971b), a dextran-forming organism which colonizes the tooth surface, the cheek, the tongue and the vestibular mucosa (van Houte *et al.*, 1971a). The early attempts to determine the reservoir of *S. sanguis* on the mucous membranes were unsuccessful (Hehre, 1948) until Carlsson (1965) demonstrated that it is the most numerous of the streptococci in dental plaque. Subsequently van Houte *et al.* (1971b) examined dental plaque and feces of the same individual for this organism. They found high concentrations of *S. sanguis* in plaque and a relative absence of it in feces, and concluded that the mouth probably is the likely portal of entry of this organism in SBE.

Gibbons and Banghart (1968) showed that a streptococcus isolated from the blood of a patient with SBE caused caries in germfree rats. This suggested that bacteria could gain access to the body via extension of the carious lesion into the pulp. De Stoppelaar (1971) examined the occurrence of *S. mutans* and *S. sanguis* in the blood of endocarditis patients. He identified 35 strains as *S. mutans* and 208 strains as *S. sanguis*. The strains were serologically identical to *S. mutans* and *S. sanguis* strains isolated from plaque.

These data point to the dental plaque as the main reservoir of these streptococci which are so prominent in SBE. If so, then the contribution of the plaque is way out of proportion to its surface area and mass, when compared to the surface area of the intestinal tract and the mass of the intestinal contents. Thus, entry of these organisms to the blood stream must be dependent on peculiarities unique to the oral cavity and dental plaque. Prominent among these would seem to be dental extractions and periodontal procedures. However, it may be that the accumulation of a certain mass of plaque on the nonshedding tooth surface, in contact with gingival or cheek epithelium may be too much of a microbial challenge for any tissue to effectively resist. In situations where the integrity of the epithelium is impaired via nutritional or immunological deficiencies, plaque accumulations may become frankly pathogenic. This occurs in malnourished children with cancrum oris and in radiation-treated individuals who experience osteoradial necrosis and rampant caries.

Cancrum Oris (Noma)

The bacterial load on the mucosal surfaces is presumably regulated by epithelial cell renewal and the antibacterial factors present in the bathing secretions. When the level of tissue metabolism is decreased to the point where epithelial cells do not slough as often, and the secretions are diminished, the bacterial loads on these surfaces should increase. In this situation, usually associated with protein malnutrition, the normal flora can figure importantly in infections (Scrimshaw *et al.*, 1968). In the oral cavity the most common infection is cancrum oris, an oro-facial gangrenous lesion often associated with a fuso-spirochetal flora (Goldberg, 1966). Most investigators of this lesion have commented upon the poor oral hygiene resulting in heavy plaque growth on the involved teeth and gingiva (Emslie, 1963; Malberger, 1967). Enwonwu (1972) examined for the presence of this lesion in Nigerian children who were well nourished, malnourished or hospitalized with kwashiorkor. No cases of cancrum oris were found in the well-nourished children, whereas 15 percent of the malnourished and 27 percent of the kwashiorkor patients exhibited the lesion. He attributed the high incidence of the lesion in these individuals to their inadequate diet. In certain cases, lesions of the buccal mucosa and ulcerations of the mucosa of the adjacent cheek were present simultaneously. This would suggest that the plaque bacteria spread from the tooth surface into the approximating cheek mucosa. This invasion of the cheek mucosa may well have been preceded by an increased permeability of the tissue to microbial byproducts, a finding common in protein calorie malnutrition (Goldberg *et al.*, 1956). This in turn could lead to tissue damage, ulceration, and subsequent bacterial invasion.

Cancrum oris may be an advanced stage of the necrotizing ulcerative gingivitis (NUG) seen in young, well-nourished adults. In NUG an association with emotional stress has been noted (Goldhaber and Giddon, 1964; Rao *et al.*, 1968). It has been speculated that because of the stress, elevated levels of cortisone would exert an anti-inflammatory effect on the gingival tissues. The cortisone would presumably decrease protein turnover and mitotic activity, thereby increasing the susceptibility of the gingival epithelial cells to plaque irritants. If the plaque mass is great, due to poor oral hygiene, then tissue breakdown and microbial invasion might occur. In the otherwise well nourished, the bacteria invade a vulnerable gingival papilla, but in the malnourished, the bacterial penetration involves the cheek, buccal mucosa and gingiva, giving rise to cancrum oris (Enwonwu, 1972). The severity of the disease in the malnourished reflects their reduced protein intake, the state of continuous stress accompanied by adrenal hyperfunction (Frank *et al.*, 1965) and the absence of oral hygiene procedures.

Rampant Caries

Irradiation treatment of oral cancer invariably results in a rapid development of extensive caries associated with a pronounced salivary gland atrophy, characterized by acinar degeneration, adiposis, and fibrosis (Llory *et al.*, 1972). The bacterial changes which occur with this rather predictable development of caries have only recently been investigated. Llory *et al.* (1972) noted that plaque accumulates on the teeth following irradiation. Important proportional increases in *S. mutans*, yeast, and actinomyces species were detected in the plaque. The most impressive was the increase of *S. mutans* from 0.6 percent of the plaque facultative streptococci prior to treatment to 44 percent after treatment. Given the unique involvement of this organism in animal caries, the data obtained would suggest that the caries observed was due to a *S. mutans* infection. If so, then this unfortunate medical situation might provide the human model needed to demonstrate the specific infectious nature of at least some forms of dental caries in the human.

REFERENCES

Aranki, A., Syed, S.A., Kenney, E.B., and Freter, R.: Isolation of anaerobic bacteria from human gingiva and mouse cecum by means of a simplified glove box procedure. *Appl Microbiol, 17:*568, 1969.

Bahn, A.N.: Microbial potential in the etiology of periodontal disease. *J Periodontol, 41:*603, 1970.

Bender, I.B., Seltzer, S., Tashman, S., and Meloff, G.: Dental procedures in patients with rheumatic heart disease. *Oral Surg, 16:*466, 1963.

Bibby, B.G.: Inferences from naturally occurring variations in caries prevalence. *J Dent Res, 49:*1194, 1970.

Brailovsky-Lounkevitch, Z.A.: Contribution a l'étude de la flore microbienne habituelle de la bouche normale (Nouveau-nés, enfants, adultes). *Ann Inst Pasteur, 29:*379, 1915.

Bratthall, D.: Demonstration of *Streptococcus mutans* strains in some selected areas of the world. *Odont Revy, 23:*401, 1972.

Brill, N.: Effect of chewing on flow of tissue fluid into human gingival pockets. *Acta Odont Scand, 17:*277, 1959.

Bunting, R.W., and Palmerlee, F.: The role of *Bacillus acidophilus* in dental caries. *J Am Dent Assoc, 12:*381, 1925.

Burnett, G.W., and Sherp, H.W.: *Oral Microbiology and Infectious Diseases*, 3rd ed. Baltimore, Williams and Wilkins, 1968.

Carlsson, J.: A numerical toxonomic study of human oral streptococci. *Odont Revy, 19:*137, 1968.

Carlsson, J.: Zooglea-forming streptococci resembling *Streptococcus sanguis* isolated from dental plaque in man. *Odont Revy, 16:*348, 1965.

Clarke, J.K.: On the bacterial factor in the aetiology of dental caries. *Br J Exp Pathol, 5:*141, 1924.

Cobe, M.H.: Transitory bacteremia. *Oral Surg Oral Med Oral Pathol, 7:*609, 1954.

Coykendall, A.L.: Genetic heterogeneity in *Streptococcus mutans. J Bacteriol, 106:*192, 1971.

Critchley, P., Wood, J.M., Saxton, C.A., and Leach, S.A.: The polymerization of dietary sugars by dental plaque. *Caries Res, 1:*112, 1967.

Cutress, T.W.: Periodontal disease and oral hygiene in trisomy 21. *Arch Oral Biol, 16:*1345, 1971.

de Stoppelaar, J.P.: *Streptococcus mutans, Streptococcus sanguis and Dental Caries.* Holland, Dept. of Preventive Dentistry, Univ. of Utrecht, 1971.

de Stoppelaar, J.D., van Houte, J., and Backer-Dirks, O.: The relationship between extracellular polysaccharide-producing streptococci and smooth surface caries in 13-year-old children. *Caries Res, 3:*190, 1969.

Dick, D.S., Shaw, J., and Socransky, S.S.: Further studies on the microbial agent or agents responsible for the periodontal syndrome in the rice rats. *Arch Oral Biol, 13:*215, 1967.

Edwardsson, S.: Characteristics of caries-inducing human streptococci resembling *Streptococcus mutans. Arch Oral Biol, 13:*637, 1968.

Ellen, R., and Gibbons, R.J.: M protein-associated adherence of *Streptococcus pyogenes* to epithelial surfaces: prerequisite for virulence. *Infect and Immun, 5:*826, 1972.

Emslie, R.D.: Cancrum oris. *Dent Pract, 13:*481, 1963.

Enright, J.J., Friesell, H.E., and Trescher, M.D.: Studies on the cause and nature of dental caries. *J Dent Res, 12:*759, 1932.

Enwonwu, C.O.: Epidemiological and biochemical studies of necrotizing ulcerative gingivitis and noma (cancrum oris) in Nigerian children. *Arch Oral Biol, 17:*1357, 1972.

Eskow, R.N., and Loesche, W.J.: Oxygen tensions in the human oral cavity. *Arch Oral Biol, 16:*1127, 1971.

Evans, R.J.: Haematin as a growth factor for a strict anaerobe *Fusiformes melaninogenicus. Proc Soc Gen Microbiol, 5:*29, 1951.

Fitzgerald, R.J.: Dental caries research in gnotobiotic animals. *Caries Res, 2:*139, 1968.

Fitzgerald, R.J., Jordan, H.V., and Archard, H.O.: Dental caries in gnotobiotic rats infected with a variety of *Lactobacillus acidophilus. Arch Oral Biol, 11:*473, 1966.

Fitzgerald, R.J., Jordan, H.V., and Stanley, H.R.: Experimental caries and gingival pathologic changes in the gnotobiotic rat. *J Dent Res, 39:*923, 1960.

Fitzgerald, R.J., and Keyes, P.H.: Demonstration of the etiologic role of streptococci in experimental caries in the hamster. *J Am Dent Assoc, 61:*9, 1960.

Frank, R.M., Herdly, J., and Phillippe, E.: Acquired dental defects and salivary gland lesions after irradiation for carcinomas. *J Am Dent Assoc, 70:*868, 1965.

Gibbons, R.J., and Banghart, S.: Induction of dental caries in gnotobiotic rats with a levan-forming streptococcus and a streptococcus isolated from subacute bacterial endocarditis. *Arch Oral Biol, 13:*297, 1968.

Gibbons, R.J., and Banghart, S.: Synthesis of extracellular dextran by cariogenic bacteria and its presence in human dental plaque. *Arch Oral Biol, 12:*11, 1967.

Gibbons, R.J., Berman, K.S., Knoettner, P., and Kapsimalis, B.: Dental caries and alveolar bone loss in gnotobiotic rats infected with capsule forming streptococci of human origin. *Arch Oral Biol, 11:*549, 1966.

Gibbons, R.J., and Engle, L.P.: Vitamin K compounds in bacteria that are obligate anaerobes. *Science, 146:*1307, 1964.

Gibbons, R.J., Kapsimalis, B., and Socransky, S.S.: The source of salivary bacteria. *Arch Oral Biol, 9:*101, 1964a.

Gibbons, R.J., and Loesche, W.J.: Isolation of cariogenic streptococci from Guatemalan children. *Arch Oral Biol, 12:*1013, 1967.

Gibbons, R.J., and Macdonald, J.B.: Hemin and vitamin K compounds as required factors for the cultivation of certain strains of *B. melaninogenicus. J Bacteriol, 80:*164, 1960.

Gibbons, R.J., Socransky, S.S., De Araujo, W.C., and van Houte, J.: Studies of the predominant cultivable microbiota of dental plaque. *Arch Oral Biol, 9:*365, 1964c.

Gibbons, R.J., Socransky, S.S., and Kapsimalis, B.: Establishment of human indigenous bacteria in germfree mice. *J Bacteriol, 88:*1316, 1964b.

Gibbons, R.J., Socransky, S.S., Sawyer, S., Kapsimalis, B., and Macdonald, J.B.: The microbiota of the gingival crevice area of man. II. The predominant cultivable organisms. *Arch Oral Biol, 8:*281, 1963.

Gibbons, R.J., and van Houte, J.: Selective bacterial adherence to oral epithelial surfaces and its role as an ecological determinant. *Infect Immun, 3:*567, 1971.

Gilmour, M.N., and Poole, A.E.: Growth stimulation of the mixed microbial flora of human dental plaques by haemin. *Arch Oral Biol, 15:*1343, 1970.

Glickman, I.: Periodontal disease. *N Engl J Med, 284:*1071, 1971.

Goldberg, H.J.V.: Acute necrotizing ulcerative gingivitis. *J Oral Ther Pharm, 2:*415, 1966.

Goldberg, H., Ambinder, W.J., Cooper, L., and Abrams, L.A.: Emotional status of patients with acute gingivitis. *NY State Dent J, 22:*308, 1956.

Goldhaber, P., and Giddon, D.B.: Present concepts concerning the etiology of acute necrotizing ulcerative gingivitis. *Int Dent J Lond, 14:*468, 1964.

Gordon, D.F., and Gibbons, R.J.: Studies on the predominant cultivable microorganisms from the human tongue. *Arch Oral Biol, 11:*627, 1966.

Gordon, D.F., Stutman, M., and Loesche, W.J.: Improved isolation of anaerobic bacteria from the gingival crevice area of man. *Appl Microbiol, 21:*1046, 1971.

Guggenheim, B.: Streptococci of dental plaques. *Caries Res, 2:*147, 1968.

Guggenheim, B., and Schroeder, H.E.: Biochemical and morphological aspects of extracellular polysaccharides produced by cariogenic streptococci. *Helv Odont Acta, 11:*131, 1967.

Gustafsson, B.E., Quensel, C.E., Lanke, L., Lundquist, C., Grahnen, H., Bonow, B.E., and Krasse, B.: The Vipeholm dental caries study: effect of different levels of carbohydrate intake on caries activity in 436 individuals observed for five years. *Acta Odont Scand, 11:*232, 1954.

Hadley, F.P.: A quantitative method for estimating *Bacillus acidophilus* in saliva. *J Dent Res, 13:*415, 1933.

Hardwick, J.L.: The incidence and distribution of caries throughout the ages in relation to the Englishman's diet. *Br Dent J, 108:*9, 1960.

Hehre, E.J.: Dextran-forming streptococci from the blood in subacute endocarditis and from the throats of healthy persons. *Bull NY Acad Med, 24:*543, 1948.

Hewitt, L.F.: *Oxidation-Reduction Potentials in Bacteriology and Biochemistry,* 6th ed. Baltimore, Williams and Wilkins, 1950.

Holloway, P.J., James, P.M.C., and Slack, G.L.: Dental disease in Tristan da Cunha. *Br Dent J, 115:*19, 1963.

Hotz, P., Guggenheim, B., and Schmid, R.: Carbohydrates in pooled dental plaque. *Caries Res, 6:*103, 1972.

Howell, A., Jr., Jordan, H.V., and Georg, L.K., *et al.: Odontomyces viscosus,* gen. nov., spec. nov., a filamentous microorganism isolated from periodontal plaque in hamsters. *Sabouraudia, 4:*65, 1965.

Ivanyi, L.: *Conference on Dental Plaque.* Seattle, Univ. of Washington School of Dentistry, 1972.

Ivanyi, L., Wilton, J.M., and Lehner, T.: Cell-mediated immunity in periodontal disease—cytotoxicity, migration inhibition and lymphocyte transformation studies. *Immunology, 22:*141, 1972.

Jay, P.: The reduction of oral *Lactobacillus acidophilus* counts by the periodic restriction of carbohydrates. *Am J Orthod Oral Surg, 33:*162, 1947.

Jensen, S.B., Loe, H., Schiott, C.R., and Theilade, E.: Experimental gingivitis in man. 4. Vancomycin induced changes in bacterial plaque composition as related to development of gingival inflammation. *J Periodont Res, 3:*284, 1968.

Jordan, H.V., Englander, H.R., and Lim, S.: Potentially cariogenic streptococci in selected population groups in the western hemisphere. *J Am Dent Assoc, 78:*1331, 1969.

Jordan, H.V., and Hammond, B.F.: Filamentous bacteria isolated from human root surface caries. *Arch Oral Biol, 17:*1333, 1972.

Jordan, H.V., and Keyes, P.H.: Aerobic, gram-positive, filamentous bacteria as etiologic agents of experimental periodontal disease in hamsters. *Arch Oral Biol, 9:*401, 1964.

Jordan, H.V., Keyes, P.H., and Lim, S.: Plaque formation and implantation of *Odontomyces viscosus* in hamsters fed different carbohydrates. *J Dent Res, 48:*824, 1969.

Jordan, H.V., Krasse, B., and Moller, A.: A method of sampling human dental plaque for certain "caries-inducing" streptococci. *Arch Oral Biol, 13:*919, 1968.

Kenney, E.B., and Ash, M.M., Jr.: Oxidation-reduction potential of developing plaque, periodontal pockets and gingival sulci. *J Periodontol 40:*630, 1969.

Keyes, P.H.: Are periodontal pathoses caused by bacterial infections on cervicoradicular surfaces of teeth? *J Dent Res, 49:*223, 1970.

Keyes, P.: Infections and transmissible nature of experimental dental caries. *Arch Oral Biol, 1:*304, 1960.

Keyes, P.H.: Research in dental caries. *J Am Dent Assoc, 76:*1357, 1968.

Keyes, P.H., and Jordan, H.V.: Periodontal lesions in the Syrian hamster. III. Findings related to an infectious and transmissible component. *Arch Oral Biol, 9:*377, 1964.

Kostecka, F.: Relation of the teeth to the normal development of microbial flora in the oral cavity. *Dental Cosmos, 66:*927, 1924.

Krasse, B.: The effect of caries-inducing streptococci in hamsters fed diets with sucrose or glucose. *Arch Oral Biol, 10:*223, 1965.

Krasse, B.: The proportional distribution of *Streptococcus salivarius* and other streptococci in various parts of the mouth. *Odont Revy, 5:*203, 1954.

Krasse, B., Jordan, H.V., Edwardsson, S., and Trell, L.: The occurrence of certain "caries-inducing" streptococci in human dental plaque material with special reference to frequency and activity of caries. *Arch Oral Biol, 13:*911, 1968.

Lehner, T.: Cell-mediated immune responses in oral disease. A review. *J Oral Pathol, 1:*39, 1972.

Lev, M.: Apparent requirement for vitamin K of rumen strains of *Fusiformes nigrescens. Nature, 181:*203, 1958.

Lewkowicz, X.: Recherches sur la flore microbienne de la bouche des nourrissons. *Arch Med Exp Anat Pathol, 13:*633, 1901.

Littleton, N.W., Kakehashi, S., and Fitzgerald, R.J.: Recovery of specific "caries-inducing" streptococci from carious lesions in the teeth of children. *Arch Oral Biol, 15:*461, 1970.

Llory, H., Dammron, A., Gioanni, M., and Frank, R.M.: Some population changes in oral anaerobic microorganisms, *Streptococcus mutans* and yeasts following irradiation of the salivary glands. *Caries Res, 6:*298, 1972.

Loe, H.: Human research model for the production and prevention of gingivitis. *J Dent Res, 50:*256, 1971.

Loe, H., and Schiott, C.R.: The effect of mouthrinses and topical application of chlorhexidine on the development of dental plaque and gingivitis in man. *J Periodont Res, 5:*79, 1970.

Loe, H., Theilade, E., and Jensen, S.B.: Experimental gingivitis in man. *J Periodont, 36:*177, 1965.

Loesche, W.J.: Effect of bacterial contamination on cecal size and cecal contents of gnotobiotic rodents. *J Bacteriol, 99:*520, 1969.

Loesche, W.J.: Importance of nutrition in gingival crevice microbial ecology. *Periodontics, 6:*245, 1968.

Loesche, W.J.: Oxygen sensitivity of various anaerobic bacteria. *Appl Microbiol, 18:*723, 1969.

Loesche, W.J., Hockett, R.H., and Syed, S.A.: The predominant cultivable flora of tooth surface plaque removed from institutionalized subjects. *Arch Oral Biol, 17:*1311, 1972.

Loesche, W.J., and Syed, S.A.: The predominant cultivable flora of carious plaque and carious dentine. *Caries Res, 7:*201, 1973.

Loesche, W.J., Walenga, A., and Loos, P.: Recovery of *Streptococcus mutans* and *Streptococcus sanguis* from a dental explorer after clinical examination of single teeth. *Arch Oral Biol, 18:*571, 1972.

Macdonald, J.B., and Gibbons, R.J.: The relationship of indigenous bacteria to periodontal disease. *J Dent Res, 41:*320, 1962.

Maclean, I.H.: The bacteriology of dental caries. *Br Dent J, 48:*579, 1927.

Malberger, E.: Acute infectious oral necrosis among young children in the Gambia, West Africa. *J Periodont Res, 2:*154, 1967.

Martin, W.J.: Bacteremia: Common pathogens and methods of management. *Journal-Lancet, 87:*439, 1967.

McDougall, W.A.: Penetration pathways of a topically applied foreign protein into rat gingiva. *J Periodont Res, 6:*89, 1971.

Mergenhagen, S.E., Tempel, T.R., and Snyderman, R.: Immunologic reactions and periodontal inflammation. *J Dent Res, 49:*256, 1970.

Mulvihill, J.E., Susi, F.R., Shaw, J.H., and Goldhaber, P.: Histological studies of the periodontal syndrome in rice rats and the effects of penicillin. *Arch Oral Biol, 12:*733, 1967.

Nisengard, R.J., and Beutner, E.H.: Immunologic studies of periodontal disease. V. IGG type antibodies and skin test responses to Actinomyces and mixed oral flora. *J Periodont, 41:*149, 1970.

O'Kell, C.C., and Elliot, S.D.: Bacteremia and oral sepsis with special reference to the etiology of subacute endocarditis. *Lancet, 2:*869, 1935.

Onisi, J., Kondo, W., Hoiuchi, I., and Uchiyama, Y.: Preliminary report on the oxidation-reduction potential obtained on surfaces of gingiva and tongue and in interdental space. *Bull Tokyo Med Dent Univ, 7:*161, 1960.

Orland, F.J., Blayney, J.R., Harrison, R.W., Reyniers, J.A., Trexler, P.C., Wagner, M., Gordon, H.A., and Luckey, T.D.: Use of germfree animal technic in the study of experimental dental caries. I. Basic observations on rats reared free of all microorganisms. *J Dent Res, 33:*147, 1954.

Pine, L.: Classification and phylogenetic relationship of microaerophilic actinomycetes. *Int J Syst Bacteriol, 20:*445, 1970.

Rao, K.S.J., Srikantia, S.G., and Gopalan, C.: Plasma cortisol levels in protein-calorie malnutrition. *Arch Dis Child, 43:*365, 1968.

Rizzo, A.A.: Histologic and immunologic evaluation of antigen penetration into oral tissues after topical application. *J Periodont, 41:*210, 1970.

Rizzo, A.A.: The possible role of hydrogen sulfide in human periodontal disease. I. Hydrogen sulfide production in periodontal pockets. *Periodontics, 5:*233, 1967.

Rogosa, M.: The genus *Veillonella*. I. General cultural, ecological and biochemical considerations. *J Bacteriol, 87:*162, 1964.

Rogosa, M., Hampp, E.G., Nevin, T.A., Wagner, H.N., Driscoll, E.J., and Baer, P.N.: Blood sampling and sampling and cultural studies in the detection of postoperative bacteremias. *J Am Dent Assoc, 60:*171, 1960.

Rosebury, T.: *Microorganisms Indigenous to Man.* New York, McGraw-Hill, 1962.

Rosen, S., Lenney, W.S., and O'Malley, J.E.: Dental caries in gnotobiotic rats inoculated with *Lactobacillus casei*. *J Dent Res, 47:*358, 1968.

Rosenthal, S.L., and Gootzeit, E.H.: The incidence of *B. fusiformes* and spirochetes in the edentulous mouth. *J Dent Res, 21:*373, 1942.

Round, H., Kirkpatrick, H.J.R., and Hails, C.G.: Further investigation on bacteriological infections of the mouth. *Proc R Soc Med, 29:*1552, 1936.

Russell, A.L.: World epidemiology and oral health. In: Kreshover, S.J., and McClure, J.J. (Eds.): *Environmental Variables in Oral Disease.* Washington, Am. Assoc. Advance. Sci., #81, 1966.

Schamschula, R.G., and Barnes, D.E.: A study of the streptococcal flora of plaque in caries-free and caries-active primitive peoples. *Aust Dent J, 15:*377, 1970.

Schamschula, R.G., Keyes, P.H., and Horrabrook, R.W.: Root surface caries in Lufa, New Guinea. I. Clinical observations. *J Am Dent Assoc, 85:*603, 1972.

Schwartz, J., Stinson, F.L., and Parker, R.B.: The passage of tritiated bacterial endotoxin across intact gingival crevicular epithelium. *J Periodont, 43:*270, 1972.

Scrimshaw, N.S., Taylor, C.E., and Gordon, J.E.: *Interactions of Nutrition and Infection.* Geneva, WHO Monograph Ser. 57, 1968.

Sharawy, A.M., and Socransky, S.S.: Effect of human streptococcus strain GS-5 on caries and alveolar bone loss in conventional mice and rats. *J Dent Res, 46:*1385, 1967.

Shklair, I.L., Keene, H.J., and Simonson, L.G.: Distribution and frequency of *Streptococcus mutans* in caries-active individuals. *J Dent Res, 51:*882, 1972.

Shovlin, F.E., and Gillis, R.E.: Biochemical and antigenic studies of lactobacilli isolated from deep dentinal caries. I. Biochemical aspects. *J Dent Res, 48:*356, 1969.

Sognnaes, R.F.: *Oral Health Survey of Tristan da Cunha.* Oslo, Det Norshe Videnskaps-Akademi 1, 1954.

Socransky, S.S., Gibbons, R.J., Dale, A.C., Bortnick, L., Rosenthal, E., and Macdonald, J.T.: The microbiota of the gingival crevice of man. I. Total microscopic and viable counts of specific organisms. *Arch Oral Biol, 8:*275, 1963.

Socransky, S.S., and Hubersak, C.: Replacement of ascitic fluid or rabbit serum requirement of *Treponema dentium* by α-globulin. *J Bacteriol, 94:*1795, 1967.

Socransky, S.S., Hubersak, C., and Propas, D.: Induction of periodontal destruction in gnotobiotic rats by a human oral strain of *Actinomyces naeslundii. Arch Oral Biol, 15:*993, 1970.

Socransky, S.S., Loesche, W.J., Hubersak, C., and Macdonald, J.B.: Dependency of *Treponema microdentium* on other oral organisms for isobutyrate, polyamines and a controlled oxidation-reduction potential. *J Bacteriol, 88:*200, 1964.

Socransky, S.S., and Manganiello, S.D.: The oral microbiota of man from birth to senility. *J Periodontol, 42:*485, 1971.

Stralfors, A.: Investigations into the bacterial chemistry of dental plaques. *Odontol Tidskr, 58:*155, 1950.

Syed, S.A., Abrams, G.D., and Freter, R.: Efficiency of various intestinal bacteria in assuming normal functions of enteric flora after association with germ-free mice. *Infect Immun, 2:*376, 1970.

Taichman, N.S., Freedman, H.L., and Uriuhara, T.: Inflammation and tissue injury. I. The response to intradermal injections of human dentogingival plaque in normal and leukopenic rabbits. *Arch Oral Biol, 11:*1385, 1966.

Tempel, T.R., Snyderman, R., and Jordan, H.V., *et al.:* Factors from saliva and oral bacteria. Chemotactic for polymorphonuclear leukocytes—their possible role in gingival inflammation. *J Periodont, 41:*71, 1970.

Torerud, G.: Dental caries in Norwegian children during and after the last World War. A preliminary report. *Proc R Soc Med, 42:*249, 1949.

van Houte, J., Gibbons, R.J., and Banghart, S.B.: Adherence as a determinant of the presence of *Streptococcus salivarius* and *Streptococcus sanguis* on the human tooth surface. *Arch Oral Biol, 15:*1025, 1970.

van Houte, J., Gibbons, R.J., Liljemark, W.F., and Pulkkinen, A.J.: Ecology of human oral lactobacilli. Abstr. #237. Paper presented at Int. Assoc. Dent. Res., Las Vegas, 1972.

van Houte, J., Gibbons, R.J., and Pulkkinen, A.J.: Adherence as an ecological determinant for streptococci in the human mouth. *Arch Oral Biol, 16:*1131, 1971a.

van Houte, J., Jordan, H.V., and Bellack, S.: Proportions of *Streptococcus sanguis*, an organism associated with subacute bacterial endocarditis in human feces and dental plaque. *Infect Immun, 4:*658, 1971b.

Waugh, D.B., and Waugh, L.: Effects of natural and refined sugars in oral lactobacilli and caries among primitive Eskimos. *Am J Dis Child, 59:*483, 1940.

Wilson, G.S., and Miles, A.A.: *Topley and Wilson's Principles of Bacteriology and Immunology.* Baltimore, Williams and Wilkins, 1964.

Winslow, M., and Kobernick, S.D.: Bacteremia after prophylaxis. *J Am Dent Assoc, 61:*69, 1960.

Wittwer, J.W., Toto, P.D., and Dickler, E.H.: *Streptococcus mitis* antigen in inflamed gingiva. *J Periodont, 40:*639, 1969.

Wostmann, B.S., and Bruckner-Dardoss, E.: Oxidation-reduction potentials in cecal contents of germfree and conventional rats. *Proc Soc Exp Biol Med, 121:*1111, 1966.

Zinner, D., Jablon, J., Aran, A., and Saslaw, M.: Experimental caries induced in animals by streptococci of human origin. *Proc Soc Exp Biol Med, 18:*766, 1965.

CHAPTER XXXIV

DERMATOLOGIC ANAEROBIC INFECTIONS (INCLUDING ACNE)

S. Madli Puhvel and Ronald M. Reisner

Abstract: *Dermatologic infections due to anaerobic bacteria as classically defined are relatively uncommon. In addition to* Clostridium *species, the pathogens include species of* Peptococcus, Peptosptreptococcus, *and* Bacteroides, *which are apt to act in synergism with aerobic pathogens. The anaerobic diphtheroid,* Propionibacterium acnes, *is directly involved in the pathogenesis of the almost universal dermatologic syndrome of acne vulgaris. Free fatty acids in sebum apparently instigate the inflammatory response in acne patients, and lipolysis of triglycerides in sebum to free fatty acids depends on bacterial lipases, especially from* P. acnes. *This organism also appears to be involved in a hyperkeratinization process in the sebaceous ducts in acne patients. The success of low-dose tetracycline therapy in acne appears to be due to the drug's direct antibacterial effect on* P. acnes.

The anaerobes associated with infections of the skin are all members of the indigenous microflora of the gastrointestinal tract, oral cavity or skin of man. When they produce dermatologic lesions, their invasiveness is often related to some alteration in the host's defense mechanism. These include underlying systemic illness, such as diabetes mellitus, atherosclerosis or connective tissue disease, and traumatic injury, which provides a suitable environment of devitalized tissue in which the anaerobic organisms can survive and proliferate. In rare instances a combination of anaerobic bacteria acting in synergism with aerobic bacteria can cause seemingly spontaneous ulcerative lesions in normal skin (Yelderman and Weaver, 1969), but generally the lesions follow some form of trauma such as surgery, mechanical injury, inoculation, or introduction of foreign bodies into the skin.

Anaerobes can also superinfect underlying dermatologic lesions, such as pilonidal and sebaceous cysts or decubitus ulcers. The superinfecting organisms can aggravate and modify the preexisting lesion, but by themselves cannot initiate the lesion in undamaged skin. Often in such cases a combination of anaerobic organisms together with aerobic organisms may be involved, and the actual role of the anaerobe may be difficult to define.

Anaerobic skin infections are rarities when compared in number to aerobic bacterial infections such as furuncles, carbuncles, impetigo and erysipelas, and to superficial fungal infections, which together form the bulk of cutaneous infections. However, in addition to the classically accepted dermatologic infections produced by anaerobic organisms, acne vulgaris should be included as a dermatologic disease of multifactorial etiology in which the anaerobic skin diphtheroid, *Propionibacterium acnes* (formerly *Corynebacterium acnes*), is one important pathogenic factor. It would be misleading to term acne vulgaris a bacterial infection, but recent research indicates that the ubiquitous *P. acnes* may be of prime significance in the etiology of this universal dermatologic syndrome.

In this paper we summarize briefly the dermatologic syndromes characteristically caused by anaerobic organisms, and refer the reader to more comprehensive reviews for detailed discussion of some specific clinical syndromes. Many of the existing classifications have evolved from descriptions of clinical entities as viewed by different observers widely separated in time and place. This has resulted in many synonymous and cumbersome descriptive names. To provide a better working tool, a functional classification should be developed based on such features as depth of tissue involvement (skin, muscle, *etc*); area and extent of involvement; special identifying features (crepitation, gangrene, *etc*); and specific organisms isolated. In this review the infections are discussed according to the organisms involved, although in some instances more than one genus may be involved in similar dermatologic syndromes.

PEPTOCOCCUS (ANAEROBIC STAPHYLOCOCCUS)

The anaerobic staphylococci are part of the indigenous flora of the oral cavity, skin, gastrointestinal tract and genitourinary system (Thomas and Hare, 1954). Their role in clinical infection has not been clearly established, but they may act in synergism with other commensals and produce infection in circumstances in which the host's resistance has been lowered. They have been isolated from infected sebaceous cysts (Stokes, 1958), wound infections (Finegold *et al.*, 1968), and from ischemic foot and leg ulcers and skin infections such as hidradenitis suppurativa (Pien *et al.*, 1972).

Pien and coworkers (1972) found that of 85 strains of gram-positive anaerobic cocci isolated during a four-week period from 70 patients, all but nine appeared on the basis of morphology to belong to the genus *Peptococcus*. About 90 percent of the 85 anaerobic gram-positive cocci were isolated from surgical wounds (42), ischemic ulcers of the leg and foot (19) or superficial skin infections (16). An average of three other bacteria were isolated in association with the anaerobic cocci (range of zero to nine). The majority of these were aerobic bacteria. In 36 cases other anaerobes were isolated with the anaerobic gram-positive cocci. Most were *Bacteroides* species, although *Clostridium perfringens* was isolated in five cases. The authors concluded that *Peptococcus* species may act as opportunists in cases of decreased resistance, and can, in synergism with other organisms such as *Bacteroides* species, produce necrosis.

PEPTOSTREPTOCOCCUS (ANAEROBIC STREPTOCOCCUS)

Peptostreptococci are found less frequently than peptococci in clinical specimens. Their frequent association with *Bacteroides* species and with *Staphylococcus aureus* suggests that they act in synergism with these organisms in producing infection. As with the anaerobic staphylococci, peptostreptococci have been recovered from wound infections, infected pilonidal and sebaceous cysts, skin ulcers and subcutaneous abscesses, particularly in the perirectal and inguinal areas (Bornstein *et al.*, 1964).

In addition, anaerobic streptococci have been implicated in several distinct dermatologic syndromes which usually have resulted from trauma or wound infections. While these have been described as specific syndromes there may be a considerable overlap of findings and in some cases the distinction may be primarily of bacteriological rather than clinical significance.

Chronic Burrowing Ulcer

This is a syndrome discussed more frequently in the older literature. The lesion begins as a small superficial ulcer following trauma or surgery. This usually enlarges in size very slowly and can persist for months or years. Typically it is not associated with pain. The cardinal characteristic of this type of ulcer is its markedly undermined border, along with the tendency to extend insidiously until large ulcerative plaques are formed. Generally, the lesion does not spread in depth beyond the skin and subcutaneous tissue. The borders of the burrowing ulcers are slightly raised, edematous, and boggy and usually have a distinct purple color. Beyond this border there is an erythematous area which fades into normal skin. Anaerobic streptococci may be cultured from the undermined borders of the ulcer. Often the ulcer may be associated with ulcerative colitis, although other underlying diseases, such as empyema, may be present.

Treatment consists of appropriate vigorous antibiotic therapy in addition to debridement. The local application of activated zinc peroxide compresses was recommended by Meleney and Johnson in 1937 and is still considered the treatment of choice by some (Yelderman and Weaver, 1969).

Progressive Synergistic Bacterial Gangrene

This is a progressive necrotizing infection of the skin and subcutaneous tissues. Meleney was the first to document its synergistic nature (Brewer and Meleney, 1926; Meleney, 1931). Anaerobic streptococci together with hemolytic S. *aureus* were found in the classical infection. Neither organism alone could produce lesions in the experimental animal, but together they formed a necrotic dermal ulcer. Cases of progressive bacterial synergistic gangrene involving gram-negative bacilli acting in synergism with anaerobic streptococci have also been recorded (Smith, 1956).

In the classic syndrome described by Meleney, the lesion begins as a small painful area of edema with a deep red or purple color. This breaks down, leaving a necrotic center which can be shaggy gray-black in color. Surrounding it is a rim of gangrenous skin encircled by a zone of purple erythema which blends into the surrounding edematous skin.

Meleney stressed that the anaerobic streptococci were found in the peripheral erythematous zone, but not in the central necrotic area. Aerobic organisms could usually be recovered from the central shaggy area. Progressive synergistic bacterial gangrene usually is a surgical complication, although it may also superinfect an area of previously diseased skin. An underlying systemic disease greatly predisposes to its development.

Specific antibiotics and appropriate surgical debridement are the mainstays of therapy. The sensitivity of all organisms recovered from the wound should be determined, and an appropriate combination of antibiotics selected (Pillsbury *et al.*, 1956).

Synergistic Necrotizing Cellulitis

This is a frequently lethal cellulitis caused by anaerobic streptococci and/or *Bacteroides* species in synergism with aerobic gram-negative rods (Stone and Martin, 1972). The initial lesion is often a small skin ulcer draining thin reddish-brown, foul-smelling fluid. The ulcer is usually surrounded by areas of necrotic skin. Gaseous crepitation may or may not be present in adjacent tissues. Unlike clostridial cellulitis, the areas involved in synergistic cellulitis are extremely painful and tender. There is widespread infection of the muscles and fascia. The patients usually exhibit symptoms of generalized toxemia.

Reviewing their 63 cases of synergistic necrotizing cellulitis, Stone and Martin (1972) found that 94 percent of the infections were located within the perianal area and the lower extremities; 38 percent started as perirectal abscesses. Underlying metabolic illness was a frequent finding: 75 percent of the patients had diabetes mellitus and 53 percent cardiovascular and/or renal disease. More than 80 percent showed extreme states of altered nutrition, either marked obesity or significant wasting. The overall mortality rate was 76 percent. The best chance for survival was in those patients in whom amputation of the infected area was possible. In others, massive systemic antibiotic therapy with radical debridement of infected tissue was the only alternative means of treatment. The mortality rate was higher in patients with underlying renal disease or diabetes. Bacteriologically, 80 percent of the wounds contained anaerobic streptococci, bacteroides or both, together with gram-negative aerobic rods, primarily *Klebsiella-Aerobacter* or *Proteus* species.

Stone and Martin's article presents an excellent discussion of the differential diagnosis between synergistic necrotizing cellulitis and the other streptococcal necrotizing infections of the skin. In essence, necrotizing cellulitis differs from anaerobic streptococcal myonecrosis mainly in that the latter, like clostridial myonecrosis, is primarily an infection of the muscles and deeper tissues. Also, in anaerobic streptococcal myonecrosis, the wound and draining fluids are usually purulent.

Synergistic necrotizing cellulitis differs from progressive synergistic bacterial gangrene mainly in the organisms involved. Classically, anaerobic streptococci act in synergism with *S. aureus* in Meleney's syndrome and with gram-negative rods in necrotizing cellulitis. Also, progressive synergistic bacterial gangrene is a much more indolent process. It primarily involves skin and subcutaneous tissues, with subsequent necrosis of the deep enveloping fascia only as a late development. Very rarely is muscle destruction present. Progressive synergistic gangrene is much more responsive to treatment with antibiotics and to conservative (as opposed to radical) debridement.

For this last reason, it is important to always distinguish between synergistic gangrene and synergistic cellulitis. In cases where anaerobic streptococci and/or *Bacteroides* species are found together with gram-negative rods in the wound or draining fluids, the diagnosis of synergistic necrotizing cellulitis is most likely. Where only gram-positive cocci are isolated, and these are anaerobic streptococci and aerobic *S. aureus*, progressive synergistic bacterial gangrene is the likely diagnosis.

Anaerobic Streptococcal Myonecrosis

This is a myonecrosis or gangrene of the muscles which is produced primarily by anaerobic streptococci. MacLennan (1962) has presented an excellent review of the symptoms and differential diagnosis of the various types of myonecrosis in man. Streptococcal myonecrosis or myositis resembles a subacute clostridial myonecrosis. In onset it is slower than the commonest form of clostridial gas gangrene. Usually incubation is three to four days. Swelling, edema, and a purulent discharge from the wound are presenting signs. Pain comes later than in clostridial gangrene, but may become just as intense. The infected muscles are pale and soft at first, and later bright red. The discharge is seropurulent and has a sour odor. There is no generalized toxemia. Gram stains of the seropurulent discharge from gangrenous lesions will reveal gram-positive cocci and no gram-positive bacilli. Treatment consists of radical debridement of devitalized tissues and vigorous appropriate antibiotic therapy.

CLOSTRIDIA

The clostridial infections involving skin characteristically follow trauma and represent wound infections. These infections have been reviewed at length by Willis (1969), MacLennan (1962), Bornstein *et al.* (1964), Altemeier and Fullen (1971) and others, as well as at this conference by Willis and Brummelkamp.

BACTEROIDES

The major species of this genus of anaerobic gram-negative nonsporeforming rods which are associated with human infection are *B. fragilis* and *B. melaninogenicus*.

In feces, *Bacteroides* species outnumber all other anaerobic and facultative anaerobic bacteria. This may be one reason why *Bacteroides* species are the most common anaerobic bacteria isolated from skin infections such as infected wounds and ulcers, pilonidal cysts and perirectal abscesses.

Usually *Bacteroides* species are recovered in combination with other organisms, particularly with anaerobic streptococci and aerobic gram-negative rods (Bornstein *et al.*, 1964). The isolation of a bacteroides as the alternate anaerobic organism found in synergistic necrotizing cellulitis has already been discussed under the section discussing infections caused by anaerobic streptococci. Generally, it is believed that *Bacteroides* species, like most of the other anaerobic commensals, can act as pathogenic opportunists in cases of decreased resistance, and can, particularly in synergism with other organisms, produce genuine infections and necrosis.

ACTINOMYCES

Actinomycosis is discussed elsewhere in this volume (see Lerner, page 571) and our discussion will be limited to a description of the skin involvement in this disease. The skin is involved in actinomycosis usually only as a result of the formation of draining abscesses to the surface from the underlying subcutaneous lesions. Such abscesses are particularly common in cases of cervicofacial actinomycosis which frequently follows extraction of infected teeth. The mandible is commonly involved near the angle of the jaw, and the maxillary sinus is also particularly susceptible. One of the initial symptoms of infection is the development of an extremely hard indurated area beneath the skin immediately over the site of infection. This may, in its initial stages, simulate an infected sebaceous cyst in appearance. Soon the area becomes irregularly swollen, abscesses develop and finally rupture or are incised. Sinus tracts extending from the deeper tissues can persist for months draining serosanguinous or purulent fluid which contains the diagnostic sulfur granules consisting of tangled hyphal forms of *Actinomyces israelii* or other *Actinomyces* and *Arachnia* species.

Any part of the body may be infected, either by direct extension, hematogenous spread, or even by direct inoculation through the skin. This latter form of infection is most common in the foot or leg. In such cases, ulceration of the skin is the initial lesion followed by deeper and deeper penetration of the infection into the tissues. This form of actinomycosis is also referred to as "Madura foot", "maduramycosis" or "mycetoma". Maduramycosis is more commonly caused by aerobic organisms such as *Nocardia* but can also be caused by *A. israelii* (Wilson and Plunkett, 1965).

ANAEROBIC DIPHTHEROIDS AND ACNE VULGARIS

Anaerobic gram-positive diphtheriods are common isolates from clinical specimens but only rarely has clinical significance been ascribed to their isolation (Kimbrell, 1964; Ballantine and Shealy, 1959; Waitzkin, 1969).

The commonest anaerobic member of the normal flora of human skin is *Propionibacterium acnes*. It outnumbers all other bacteria in the areas of skin rich in sebaceous glands. These organisms proliferate within the depths of the pilary canals of the skin close to the opening of the sebaceous ducts. *P. acnes* is also the predominant anaerobe in the anterior nares (Watson *et al.*, 1962) and on the scalp of man. Recent studies have demonstrated that *P. acnes* is directly involved in the pathogenesis of the almost universal dermatologic syndrome of acne vulgaris.

Acne vulgaris is a disease of the sebaceous follicles in man. The primary acne site is the face and to a lesser extent the back, chest, and shoulders. The disease is characterized by five basic types of lesions, the comedone (open and closed), papule, pustule, nodule and cyst. While any one lesion may dominate, usually several types are present simultaneously. The comedone represents the noninflammatory phase of the disease. The other lesions are associated with inflammation with or without pus formation.

In order to fully understand the pathogenesis of acne, an appreciation of the anatomy of the pilosebaceous follicles in skin is necessary. Pilosebaceous follicles are present in skin in all areas of the body, except on the soles of the feet and the palms of the hand. However, they are most numerous and largest in size on the face, the upper back, the chest and shoulders. In these areas the glands may exist independently with ducts leading directly to the skin surface, or they may be associated with rudimentary hairs, in which case the sebaceous duct may open into widely dilated pilary canals. This peculiar structure with its rudimentary hair, dilated pilary canal and huge sebaceous glands, has been called the sebaceous follicle and is the primary site of the acne lesion.

Sebaceous glands are holocrine glands. The epithelial cells lining the basement membrane of the gland differentiate into mature lipid forming cells which increase in lipid content as they migrate away from the basement membrane. Final rupture of the cell releases lipid-rich cellular contents into the sebaceous duct, and a combination of capillary action and pressure from accumulating sebum keeps a continual flow of lipid passing through the sebaceous duct.

Acne is a disease which begins during or after puberty. Normally, after the neonatal period and before puberty the sebaceous glands are small and lipid production is minimal. The increase in the size and activity of the sebaceous gland at puberty can be directly correlated with increases in hormonal levels at that time. The prime and possibly the sole direct stimulus for the development of sebaceous glands is androgen (Strauss *et al.*, 1962). There is no doubt that androgen metabolism in some as yet not fully elucidated manner has a direct bearing on the development of acne. Eunuchs have smooth skin and do not develop acne (Hamilton, 1941). Females with excess androgen-producing syndromes have an unusually high frequency of severe cystic acne (Pillsbury *et al.*, 1956).

The prime response of the sebaceous glands to androgen stimulation is an increase in sebum production. It has been established that excessive sebum production is a characteristic common to patients with acne. Highest levels of sebaceous gland activity tend to occur in those with the most severe acne (Pochi and Strauss, 1964). However, even though on the average, patients with acne produce more sebum than normals, there is a wide overlap of sebum production levels in both groups, so that quantitative differences in sebum production alone cannot account for the development of acne. Gross seborrhea may be present without acne developing.

Androgens may also affect the sebaceous gland by altering the quantity or quality of keratins produced by the epithelium lining the sebaceous ducts and follicular canals. At the present time, however, no proof for this has been presented.

The composition of sebum from patients with acne appears to be very similar to that of sebum from nonacne controls. No obvious qualitative differences in lipid production in the skin has been found (Boughton *et al.*, 1959).

Studies on the inflammatory effect of sebum have demonstrated that injection of sebum into skin produces an intense lymphocytic infiltrate very similar to that seen in acne lesions (Strauss and Pochi, 1965). Chemical separation of sebum into different components and injection of the separated components into skin, has demonstrated that the free fatty acid fraction of sebum is responsible for producing the lymphocytic response. Sebum from which the free fatty acids have been removed no longer induces the characteristic severe inflammation. Thus, free fatty acids of sebum are thought to be important instigators of the inflammatory response in acne patients.

The source of the free fatty acids in sebum is related to the presence of *P. acnes* in the sebaceous follicles. Lipid analysis of intact sebaceous glands has shown that very little free fatty acid is present in sebum as it is first released by sebaceous cells. All fatty acids exist in the esterified, or triglyceride form in newly formed sebum (Kellum, 1967). However, in sebum collected from the skin surface, approximately half the fatty acids exist in the free form. Thus, triglycerides of pure sebum appear to undergo considerable lipolysis during passage from the sebaceous gland through the follicular duct to the surface of the skin.

The major source of active lipases in the follicular duct appears to be bacterial in origin. Furthermore, even though all three groups of organisms which form the resident flora of the sebaceous ducts, *S. epidermidis, Pityrosporum orbiculare,* and *P. acnes* have been shown in *in vitro* studies to produce lipases capable of hydrolyzing the triglycerides found in sebum to release free fatty acids, selective suppression of each of the three groups of organisms by specific antimicrobial therapy results in a reduction in the proportion of free fatty acids in sebum only when the density of *P. acnes* in skin is reduced (Marples *et al.*, 1971).

Clinically the most successful innovation in the treatment of acne during the past decade has been the use of antibiotics. The antibiotic of choice has been tetracycline and doses of 250 mg or less of tetracycline a day have been found to be sufficient to adequately control the disease in some patients. Administration of tetracycline has been correlated with a reduction in free fatty acids in skin surface lipids, a reduction in the density of *P. acnes* on the skin and clinical improvement in acne (Marples *et al.*, 1971; Freinkel *et al.*, 1965).

In addition to tetracycline, *P. acnes* is sensitive to a wide range of antibiotics *in vitro*. Yet some of the antibiotics to which it is sensitive *in vitro*, such as penicillin, are not effective *in vivo* in the treatment of acne. It was primarily for this reason that it was thought that the explanation of the effectiveness of tetracyclines in acne had to be related to undetermined metabolic effects rather than to direct antibacterial activity. It was speculated that the effectiveness of tetracylines as lipase inhibitors might be significant in directly reducing the proportion of irritating free fatty acids in the sebaceous follicles (Shalita and Wheatley, 1970). However, recent studies have demonstrated that the most plausible explanation of the effectiveness of tetracycline in acne is still its direct antibacterial effect on *P. acnes*. The levels of tetracycline necessary to produce significant inhibition of *P. acnes* lipase *in vitro* are far in excess of the levels of tetracycline required to inhibit the growth of *P. acnes in vitro* (Puhvel and Reisner, 1972).

The difference in the effectiveness of various antibiotics in improving acne clinically appears to be related to the route by which each reaches the sebaceous duct. Tetracycline probably is incorporated directly into the sebaceous gland and reaches the skin through sebum, thus gaining direct access to the organisms in the sebaceous follicles (Marples and Kligman, 1971). There has also been ample evidence demonstrating that tetracycline has the characteristic of concentrating in areas of inflammation in the skin (Cullen and Crounse, 1965), and it presumably does so also in lesions of acne. Therefore, even low doses of tetracycline could produce high accumulative levels in areas of skin involved in acne.

Even though the so-called free fatty acid theory of acne etiology is tempting to accept, it has left many questions unanswered. The most obvious is, if all individuals harbor *P. acnes* in their follicles, and the sebum from patients with acne is similar in composition to controls without acne, then why do some individuals develop acne, while others do not? Obviously, some missing factor remains to be identified.

The pathology of the acne lesion begins with hyperkeratinization of the follicular duct. As the follicular duct becomes hyperkeratinized, keratin, sebum, bacteria and, in some cases, hair remnants, plug the duct. Either of two possibilities may occur. The lesion can remain in a noninflammatory state or it can progress to an inflammatory lesion.

In the open comedone, the follicular wall gradually thins out as it becomes distended by the central plug, the associated sebaceous gland atrophies and the lesion may stay in this state indefinitely. Melanin from melanocytes trapped in the plug produce a pigment which oxidizes to a black-gray color at the exposed surface of the plug to give the typical appearance which gives the lesion the common name "blackhead" (Blair and Lewis, 1970).

The closed comedone, which has only a microscopic communication to the surface, may progress into an inflammatory lesion. The thinned wall of the follicle ruptures as the sebaceous duct fills with keratin, lipid and bacteria, and the follicular content is spilled into the dermis. Occasionally, similar disruption of the follicular wall may occur in apparently unobstructed sebaceous follicles. An intense lymphocytic infiltrate develops. Although *P. acnes* is usually not seen in the extrafollicular infiltrate, immunofluorescence has demonstrated *P. acnes* antigens in this location (Imamura *et al.*, 1969).

The cause of the initial hyperkeratinization in acne is the key which will probably eventually solve the complex etiology of this disease. Very recent electron microscopic studies (Knutsen, unpublished data) have demonstrated that in normal follicles there is a gradual differentiation of the epithelium lining the duct of the sebaceous follicles from keratin-forming epithelium near the opening of the duct to the surface of the skin, to lipid-forming epithelium deeper in the follicle, adjacent to the opening of the sebaceous duct. A failure of the epithelium to differentiate between these two functions was always demonstrated in skin from adolescents with acne vulgaris. In such cases, epithelial cells in the pilary canal near the opening of the sebaceous ducts produced keratin as well as lipid, and the lipid globules, instead of being released into the lumen of the duct by spontaneous rupture of the epithelial cells, became enmeshed and trapped in strands of keratin, comparable to the hyperkeratinizing lesion seen in the light microscope. Most interesting was the observation that this initial defect in sebaceous follicles in acne skin was always associated with the presence of large masses of *P. acnes* in the lumen of the sebaceous duct immediately adjacent to the keratin buildup.

It remains to be elucidated what the exact role of *P. acnes* is in the pathogenesis of acne. But clearly, *P. acnes* represents an example of an anaerobic organism which, although part of the normal bacterial flora, has the capacity in genetically susceptible individuals to produce disease. It is interesting to speculate that this might serve as a model for investigation of other disease states of obscure etiology in which the host's microflora have been considered purely incidental contaminants.

REFERENCES

Altemeier, W.A., and Fuller, W.D.: Prevention and treatment of gas gangrene. *JAMA, 217:*806, 1971.

Ballantine, H.T., Jr., and Shealy, C.N.: The role of radical surgery in the treatment of abscess of the brain. *Surg Gynecol Obstet, 109:*370, 1959.

Blair, C., and Lewis, C.A.: The pigment of comedones. *Br J Dermatol, 82:*572, 1970.

Bornstein, D.L., Weinberg, A.N., Swartz, M.N., and Kunz, L.J.: Anaerobic infections—review of current experience. *Medicine (Baltimore), 43:*207, 1964.

Boughton, B., MacKenna, R.M.B., Wheatley, V.R., and Wormall, A.: The fatty acid composition of the surface skin fats (sebum) in acne vulgaris and seborrheic dermatitis. *J Invest Dermatol, 33:*57, 1959.

Brewer, G.E., and Meleney, F.L.: Progressive gangrenous infection of the skin and subcutaneous tissues following operation for acute perforative appendicitis. A study in symbiosis. *Ann Surg, 84:*438, 1926.

Cullen, S.I., and Crounse, R.G.: Cutaneous pharmacology of the tetracyclines. *J Invest Dermatol, 45:*263, 1965.

Finegold, S.M., Miller, A.B., and Sutter, V.L.: Anaerobic cocci in human infection. *Bacteriol Proc, 68:*94, 1968.

Freinkel, R.K., Strauss, J.S., Yip, S.Y., and Pochi, P.E.: Effect of tetracycline on the composition of sebum in acne vulgaris. *N Engl J Med, 273:*850, 1965.

Hamilton, J.B.: Male hormone substance; a prime factor in acne. *J Endocrinol, 1:*570, 1941.

Imamura, S., Pochi, P.E., Strauss, J.S., and McCabe, W.R.: The localization and distribution of *Corynebacterium acnes* and its antigens in normal skin and in lesions of acne vulgaris. *J Invest Dermatol, 53:*143, 1969.

Kellum, R.E.: Human sebaceous gland lipids: Analysis by thin-layer chromatography. *Arch Dermatol, 95:*218, 1967.

Kimbrell, D.C., Jr.: *Corynebacterium acnes*—a cause of meningitis. *NC Med J, 25:*516, 1964.

Knutson, D.D.: Electron microscopic observations of acne. Paper presented at the 22nd Ann. Symposium on the Biology of Skin, Sebaceous Glands and Acne Vulgaris. Geneden Beach, Oregon, October 8–12, 1972. To be published.

MacLennan, J.D.: The histotoxic clostridial infections of man. *Bacteriol Rev, 26:*177, 1962.

Marples, R.R., Downing, D.T., and Kligman, A.M.: Control of free fatty acids in human surface lipids by *Corynebacterium acnes. J Invest Dermatol, 56:*127, 1971.

Marples, R.R., and Kligman, A.M.: Ecological effects of oral antibiotics on the microflora of human skin. *Arch Dermatol, 103:*148, 1971.

Meleney, F.L.: Bacterial synergism in disease processes with a confirmation of the synergistic bacterial etiology of a certain type of progressive gangrene of the abdominal wall. *Ann Surg, 94:*961, 1931.

Meleney, F.L., and Johnson, B.A.: Further laboratory and clinical experiences in the treatment of chronic, undermining, burrowing ulcers with zinc peroxide. *Surgery, 1:*169, 1937.

Pillsbury, D.M., Shelley, W.B., and Kligman, A.M.: *Dermatology.* Philadelphia, Saunders, 1956.

Pochi, P.E., and Strauss, J.S.: Sebum production, casual sebum levels, titratable acidity of sebum, and urinary fractional 17-ketosteroid excretion in males with acne. *J Invest Dermatol, 43:*383, 1964.

Pien, F.D., Thompson, R.L., and Martin, W.J.: Clinical and bacteriologic studies of anaerobic gram-positive cocci. *Mayo Clin Proc, 47:*251, 1972.

Puhvel, S.M., and Reisner, R.M.: Effect of antibiotics on the lipases of *Corynebacterium acnes in vitro*. *Arch Dermatol, 106:*45, 1972.

Shalita, A.R., and Wheatley, V.R.: Inhibition of pancreatic lipase by tetracyclines. *J Invest Dermatol, 54:*413, 1970.

Smith, J.G., Jr.: Progressive bacterial synergistic gangrene: Report of a case treated with chloramphenicol. *Ann Intern Med, 44:*1007, 1956.

Stokes, E.J.: Anaerobes in routine diagnostic cultures. *Lancet, 1:*668, 1958.

Stone, H.H., and Martin, J.D., Jr.: Synergistic necrotizing cellulitis. *Ann Surg, 175:*702, 1972.

Strauss, J.S., Kligman, A.M., and Pochi, P.E.: The effect of androgens and estrogens on human sebaceous glands. *J Invest Dermatol, 39:*139, 1962.

Strauss, J.S., and Pochi, P.E.: Intracutaneous injection of sebum and comedones. *Arch Dermatol, 92:*443, 1965.

Thomas, C.G.A., and Hare, R.: The classification of anaerobic cocci and their isolation in normal human beings and pathological processes. *J Clin Pathol, 7:*300, 1954.

Waitzkin, L.: Latent *Corynebacterium acnes* infection of bone marrow. *N Engl J Med, 25:*1404, 1969.

Watson, E.D., Hoffman, N.J., Simmers, R.W., and Rosebury, T.: Aerobic and anaerobic bacterial counts of nasal washings: presence of organisms resembling *Corynebacterium acnes*. *J Bacteriol, 83:*144, 1962.

Willis, A.T.: *Clostridia of Wound Infections*. London, Butterworths, 1969.

Wilson, J.W., and Plunkett, O.A.: *The Fungous Diseases of Man*. Berkeley and Los Angeles, U Calif Pr, 1965.

Yelderman, J.J., and Weaver, R.G.: The urologist and necrotizing dermatological infections. *J Urol, 101:*74, 1969.

CHAPTER XXXV

ANTIBIOTIC SUSCEPTIBILITY TESTING OF ANAEROBIC BACTERIA

TRACY D. WILKINS

ABSTRACT: *Problems with the disc diffusion antibiotic susceptibility testing method developed at the V.P.I. Anaerobe Laboratory are discussed briefly. These include failure of some fresh isolates to show confluent growth; the need to establish interpretative criteria for zone diameters for* Clostridium perfringens *before this method can be recommended for tests with this organism; and the fact that either cloudy zones or paradoxical outer zones of inhibition around a zone of growth reflect susceptibility and must be measured accordingly. Despite these problems, this simple method has allowed successful testing of at least 70 percent of anaerobic clinical isolates. A modified disc-tube method now being investigated may allow successful testing of almost all such isolates.*

The mere fact that a symposium such as this exists is evidence of the awakening interest in anaerobic infections by research workers and clinicians. The recent increase in activity in this area might suggest to the uninitiated that an epidemic of anaerobic infections has occurred; advances in isolation methodology are often accompanied by such "epidemics." Unfortunately, the advances that occurred in techniques suitable for isolation of anaerobic bacteria from clinical specimens were not paralleled by improved methodology for testing the antibiotic susceptibility of these isolates. Only within the last year have standardized methods been published for testing the majority of clinical anaerobic isolates, and these methods are the subject of this roundtable. This presentation will concentrate on the disc diffusion method developed at the V.P.I. Anaerobe Laboratory. The other methods will be covered by other members of this roundtable.

The research discussed in this paper was supported by General Medical Science Grant No. 14604. The isolation and characterization of the clinical isolates was performed by Sue Smith under the direction of Drs. Lillian Holdeman and W.E.C. Moore.

The methodology for the V.P.I. method has been presented in detail (Wilkins *et al.*, 1972) and will be covered only briefly here. A single colony is picked into prereduced chopped meat carbohydrate broth (CMC). After maximal growth occurs, 1.5 ml of this culture is added to a tube containing 10 ml of melted and cooled (50°C) brain heart infusion (BHI) agar supplemented with 0.5% yeast extract, hemin, and vitamin K. The inoculated agar medium is mixed and then poured into a 90-mm Petri plate. As soon as the agar solidifies, antibiotic discs are applied to the surface of the agar, and the plates are immediately placed into a GasPak (BBL) anaerobic jar. The zones of inhibition are measured after overnight incubation and compared to published interpretative zone diameters (Wilkins *et al.*, 1972).

This method was designed to be as simple as possible and still allow the majority of anaerobic clinical isolates to be accurately tested. At the risk of sounding too negative, I want to be as objectively critical of this methodology as I can. The purpose of a roundtable, as I see it, is to discuss the current status of a research area, problems in the area, and possible solutions. I am certain that I know most of the problems that occur with this technique, and I would welcome any helpful suggestions as to solutions.

The most common adverse comment that I have received from laboratories trying this technique is that some clinical isolates do not grow on the susceptibility plates or grow so poorly that zones are difficult to see. We also have this problem with some clinical isolates, and I do not know of a solution short of performing the test completely anaerobically in a glove box. The V.P.I. Anaerobe Laboratory routinely performs clinical anaerobic bacteriology for the local 200-bed Montgomery County Community Hospital. Of 125 anaerobic isolates in 1972, 25 (20%) did not give confluent growth on the susceptibility plates on original isolation (Table XXXV-I). The majority of these 25 organisms did grow in the medium, but only as discrete colonies. This indicates that the number of viable cells in the inoculum was probably considerably reduced due to the lethal effects of oxidation before the plates were placed

TABLE XXXV-I

PROPORTION OF CLINICAL ISOLATES WITH NONCONFLUENT GROWTH ON SUSCEPTIBILITY PLATES

Species	*No. of Strains*	*No. with Nonconfluent Growth*
Bacteroides fragilis	25	0
Other bacteroides	18	7
Peptostreptococcus intermedius	16	2
Peptococcus magnus, P. micros	10	1
Other cocci	14	4
Clostridium perfringens	10	0
Other clostridia	8	4
Fusobacteria	9	6
Eubacterium lentum	4	0
Other eubacteria	2	1
Lactobacillus catenaforme	4	0
Other lactobacilli	2	0
Propionibacterium acnes	3	0
	125	25 = 20%

into the anaerobic jar. With this small an inoculum, zones of inhibition are larger than they would have been with the original inoculum; therefore, the interpretative zone diameters cannot be used to give an accurate indication of susceptibility. Extreme resistance can be detected, however, since such zones will be quite small. On continued laboratory cultivation mutants that are less oxygen-sensitive often are selected, and these cultures then yield confluent growth on susceptibility plates. For instance, almost all of our stock collection of fusobacteria give confluent growth on susceptibility plates, but many of the original hospital isolates did not.

Another problem regarding extremely fresh isolates was encountered with *Clostridium perfringens.* This organism is not very oxygen-sensitive and is the fastest growing anaerobe I am aware of. Some isolates of this species grow so fast that the antibiotics cannot diffuse very far from the discs before the organism has reached maximal growth; therefore, the zones are often smaller than with other organisms. We have not yet had enough experience with fresh clinical isolates to define interpretative zone diameters for this species so I presently cannot recommend the test for this organism.

Fortunately, *C. perfringens* strains are uniformly susceptible to penicillin, and, in most cases, isolation of this organism is not sufficient cause for antibiotic therapy unless clinical symptoms of infection are present.

As with all other disc diffusion methods, there are errors inherent when different people measure zones of inhibition. The large inoculum used with this method sometimes results in zones which are not completely clear. Such cloudy zones must be measured. We also occasionally encounter organisms which grow quite well very close to the antibiotic discs, but are inhibited further away from the discs. With all such strains we have tested, the susceptibility correlates with the diameter of the outer zone of inhibition. The explanation of this phenomenon is not known, but the growth next to the disc is not due to resistant mutants.

The V.P.I. disc diffusion method is now being used in several clinical laboratories, including our own, and these are the major problems I am aware of. Of course, several laboratories have reported difficulties that were due to the use of media different from those used to develop the method. Tryptone agar medium does not give the same results as supplemented BHI agar medium, and 1.5 ml of inoculum from thioglycollate broth is not the same as 1.5 ml of inoculum from chopped meat carbohydrate broth. Difficulties have also occurred due to the use of inactive catalysts in the GasPak anaerobic jars. This results in very poor growth of many anaerobes.

As far as we know, the laboratories that have followed the procedures exactly as published have been successful in testing at least 70 percent of their anaerobic clinical isolates. That is 70 percent that could not have been accurately tested one year ago. If the same amount of progress can be made in the coming year, we may soon be able to test almost all clinical anaerobic isolates by simple procedures.

For several months I have been testing a method that has promise of fulfilling this goal. This is a modification of the disc-tube method originally proposed by Schneierson (1954) for use with facultative bacteria. In this method commercial antibiotic discs are used as carriers of antibiotic into separate tubes of prereduced BHI broth (5 ml). By using the appropriate number of discs the attainable blood level of an antibiotic can be approximated in the medium. The tubes containing different antibiotics and a control tube are inoculated with one drop of a turbid culture of the organism, and results are read after overnight incubation on the basis of growth or no growth. This method has the great advantage that all manipulations are anaerobic, and any organism that will grow well enough in BHI broth to produce turbidity within 24 hours after inoculation can be tested with this simple procedure. Preliminary data show that approximately 95 percent of clinical anaerobic isolates can be tested with this method. In tests on 145 anaerobic isolates tested with each of seven antibiotics the results of the broth-disc tests correlated with minimal inhibitory concentrations (MICs) in 97 percent of the tests (Wilkins and Thiel, 1973).

It is apparent that numerous methods will be published for testing the antibiotic susceptibility of anaerobic bacteria. Some people believe that this can only cause confusion, but I would not want to have research stifled in an area in which we still have so much to learn. What may eventually result is an amalgam of the best points from several methods. Great progress has been made in this field in a very short time, and this is only the beginning.

REFERENCES

Schneierson, S.S.: A simple rapid disc-tube method for determination of bacterial sensitivity to antibiotics. *Antibiot Chemother, 4:*125, 1954.

Wilkins, T.D., Holdeman, L.V., Abramson, I.J., and Moore, W.E.C.: Standardized single-disc method for antibiotic susceptibility testing of anaerobic bacteria. *Antimicrob Agents Chemother, 1:*451, 1972.

Wilkins, T.D., and Thiel, T.: A modified broth-disc method for testing the antibiotic susceptibility of anaerobic bacteria. *Antimicrob Agents Chemother, 3:*350, 1973.

CHAPTER XXXVI

IN VITRO SUSCEPTIBILITY TESTING OF ANAEROBES: STANDARDIZATION OF A SINGLE DISC TEST

VERA L. SUTTER
YUNG-YUAN KWOK
SYDNEY M. FINEGOLD

ABSTRACT: *Agar dilution and agar diffusion tests with seven antibiotics were performed with 100 or more strains of* Bacteroides fragilis, *43 strains of* Clostridium perfringens *and 23 to 25 strains of fusobacteria. Correlation of results by the two methods and predictability of the single disc test were analyzed. We found previously that when data were analyzed separately for each species or closely related groups of bacteria, good correlation and predictability were obtained. However, the present study demonstrated that when data on all groups were pooled, there was a wide range of zone diameters at each minimum inhibitory concentration. If predictions of susceptibility of unidentified strains are to be made by results of agar diffusion tests, it may be necessary to establish separate criteria for organisms with different rates of growth.*

During the last 20 or more years, determination of antibiotic susceptibility of facultative bacteria has been a laboratory procedure of prime importance. Because of their simplicity and efficiency in terms of labor, materials and time, the single disc tests of Bauer and others (1966) and Ericsson (1960) have gained widespread use. They have been the subject of a recent international study on standardization of a reference method, wherein a large number of variables presumed to affect results of the test were investigated (Ericsson and Sherris, 1971). Most recently, the method of Bauer and coworkers (1966) has been recommended by the Food and Drug Administration as the method to be given in directions for use of commercially available discs (Fine, 1972). Unspecified dilution methods have been recommended for use with anaerobes.

The use of a single disc method for testing antibiotic susceptibility of anaerobes would have the same advantages for clinical laboratories as does this type of method for facultative and aerobic bacteria. With careful standardization of such variables as medium, inoculum size and age and atmosphere of incubation, as well as incubation time for the test, a test as reliable as that of Bauer *et al.* (1966) may be achievable. Recently, a number of workers in the field of anaerobic bacteriology have given their attention to correlating the inhibition zone diameters obtained with a single disc test to minimum inhibitory concentrations (MIC's) obtained by either broth or agar dilution (Bodner *et al.*, 1972; Kwok *et al.*, 1972; Sapico *et al.*, 1972; Sutter and Finegold, 1973; Sutter *et al.*, 1972, 1973; Thornton and Kramer, 1971; Wilkins *et al.*, 1972).

There are a number of problems associated with susceptibility testing of anaerobes which are not encountered in the testing of aerobes. Several conditions necessary for growth of anaerobes are known to affect the activity of some antibiotics and undoubtedly affect results of the tests. Anaerobiosis itself decreases zone diameters with aminoglycoside antibiotics, while CO_2 in the atmosphere affects the pH of the medium and consequently affects the activity of several antibiotics. The degree of influence of the various factors appears to vary with different drugs and different organisms (Ericsson and Sherris, 1971; Frostell, 1963; Ingham *et al.*, 1970; Rosenblatt and Schoenknecht, 1972; Thornton and Kramer, 1971). Most of these studies have been done with facultative organisms cultivated in an anaerobic or CO_2 containing atmosphere. Exact definition and significance of these effects, especially as they relate to anaerobes, remains to be determined.

Because of the necessity for using many of these conditions to obtain satisfactory growth of the anaerobes, we have approached the problem of standardizing a single disc test for susceptibility testing of these organisms by controlling conditions of the method as carefully as possible and determining the variability of results under these conditions. Once variability of results is known for each antibiotic being considered, then the effect of atmosphere, medium composition or other factors will be determined.

METHODS

Agar diffusion tests and agar dilution tests have been performed as previously described (Sutter *et al.*, 1972). A few colonies (or a 3-mm loopful) of each strain to be tested are inoculated into a tube containing thioglycollate medium without indicator—135C (BBL)—to which is added 5 μg/ml hemin prior to autoclaving and 1 mg/ml $NaHCO_3$ plus 0.5 μg/ml filter-sterilized menadione after autoclaving. These tubes are incubated for 4 to 6 hours. For the agar dilution studies, two-fold dilutions of the antibiotics are prepared in Brucella agar (Pfizer) containing 0.5 μg/ml menadione and 5% laked sheep blood. The inocula are adjusted to the turbidity of the No. 1 McFarland nephelometer standard in freshly boiled and cooled Brucella broth (Pfizer) and then applied to the surface of the plates by means of a Steers replicator (Steers *et al.*, 1959).

For the disc diffusion tests, the same 4 to 6 hour cultures are diluted to the density of the standard recommended by Bauer *et al.* (1966) and applied with cotton swabs (Acme Cotton Co., Valley Stream, Long Island, New York) to freshly prepared Brucella agar plates containing 5 percent defibrinated sheep blood and 0.5 μg/ml menadione. Ninety × 15 mm petri dishes are used; the agar depth is 5 to 6 mm and not more than 4 antibiotic discs are applied on each plate.

All plates are incubated at 37°C in GasPak jars containing palladium-coated alumina catalyst (reactivated after each use). The results of the agar dilution tests are read after about 48 hours' incubation. The MIC is recorded as the lowest concentration of drug at which there is no growth. The inhibition zone diameters around the antibiotic discs are measured by means of calipers after about 24 hours' incubation.

One randomly selected strain and one strain selected because of its susceptibility to penicillin have been tested 50 times with each kind of antibiotic disc to determine variation in inhibition zone diameters for each antibiotic under the conditions used. These same strains are included as controls with each day's tests.

The MIC's are correlated with inhibition zone diameters and regression lines calculated by least squares analysis (Brownlee, 1965) for most of the antibiotics tested. Breakpoints for susceptibility have been selected as those concentrations of antibiotics representing readily achievable blood levels with ordinary dosage schedules.

We feel that only very conservative estimates of susceptibility based on zone diameter measurements should be made until more zone diameter distribution studies and data regarding clinical correlation are available. Whenever overlapping of zone diameters occurs among strains with susceptible MIC's and intermediate MIC's or among strains with intermediate MIC's and resistant MIC's, these zone diameters are considered equivocal. Only those zone diameters which occur with no overlapping among categories are interpreted as susceptible or resistant.

RESULTS

Data have been accumulated for several antibiotics against 100 or more strains of *Bacteroides fragilis* (Sutter *et al.*, 1972, 1973), 43 strains of *Clostridium perfringens* (Sapico *et al.*, 1972), and 23 to 25 strains of *Fusobacterium* of the species *gonidiaformans, mortiferum, necrophorum* and *varium* (Kwok *et al.*, 1972). Our current criteria for estimating susceptibility of anaerobes to various antibiotics on the basis of inhibition zone diameter measurement and MIC are shown in Table XXXVI-I.

Chloramphenicol

Of the anaerobes tested, all had MIC's of 12.5 μg/ml or less and inhibition zone diameters of 21 to 40 mm. Correlation of diameters of zones of inhibition around 30-μg chloramphenicol discs with MIG's is shown in Fig. XXXVI-1. Although resistant strains have not yet been encountered, the regression line predicts that when resistant strains are found, they will probably have zone diameters of 14 mm or less.

TABLE XXXVI-I

ESTIMATES OF SUSCEPTIBILITY BY INHIBITION ZONE DIAMETER MEASUREMENTS*

Antibiotic	*Disc Potency*	*Resistant*		*Equivocal*	*Susceptible*	
		Zone Diameter	*MIC (μg/ml)*†	*Zone Diameter Range*	*Zone Diameter*	*MIC (μg/ml)*‡
Chloramphenicol	30 μg	14 mm or less	≧25	15–20 mm	21 mm or more	≦12.5
Clindamycin	2 μg	8 mm or less	≧12.5	9–14 mm	15 mm or more	≦3.1
	10 μg	14 mm or less	≧12.5	15–25 mm	26 mm or more	≦3.1
Erythromycin	15 μg	16 mm or less	≧6.2	17–26 mm	27 mm or more	≦1.6
Lincomycin	2 μg	—	≧12.5	6–15 mm	16 mm or more	≦3.1
	10 μg	9 mm or less	≧12.5	10–24 mm	25 mm or more	≦3.1
Penicillin§	10 u	12 mm or less	≧12.5	13–28 mm	29 mm or more	≦0.8
Tetracycline	30 μg	15 mm or less	≧12.5	16–28 mm	29 mm or more	≦1.6
Vancomycin	30 μg	14 mm or less	≧25	15–21 mm	22 mm or more	≦6.2

* Revised from Sutter and Finegold (1973).
† Penicillin G expressed in u/ml.
§ 80% inhibition used as endpoint for *F. varium* and *F. mortiferum*.

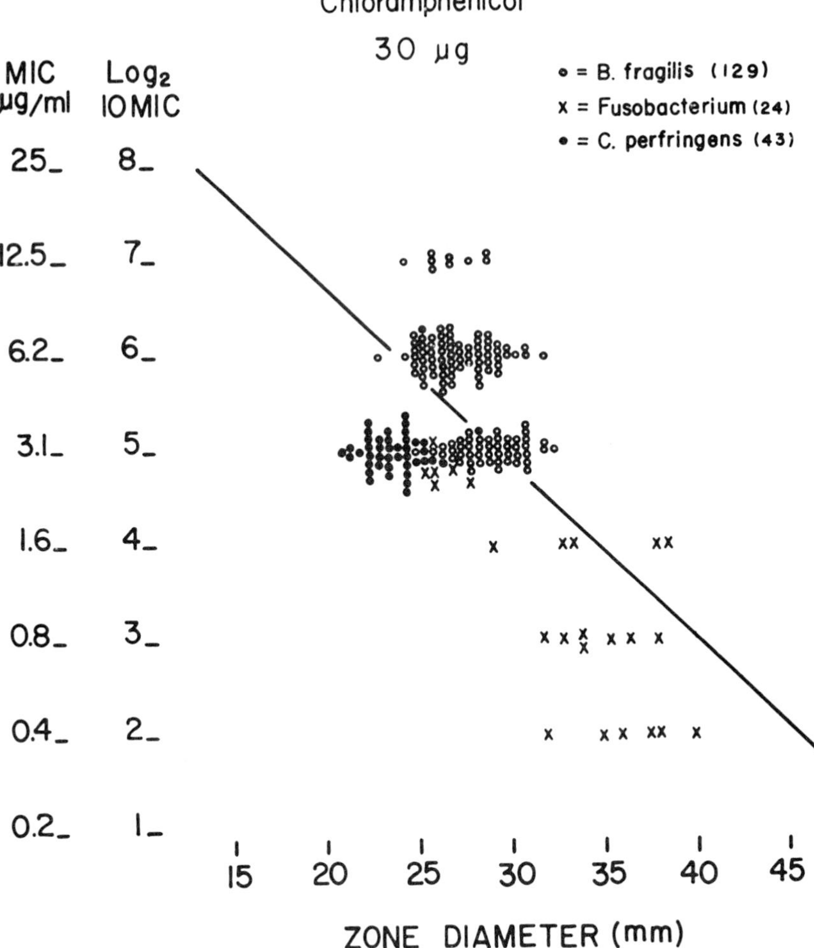

Figure XXXVI-1. Chloramphenicol: relationship of diameters of zones of inhibition around 30-μg discs and MIC values. Least squares line: $y = a + bx$; $a = 10.389$, $b = -0.183$. Correlation coefficient: $r = -0.620$.

Clindamycin

One hundred eighty-three strains of anaerobes have been tested for their susceptibility to clindamycin. The relationship between MIC's and diameters of zones of inhibition around 2-μg and 10-μg clindamycin discs is shown in Figs. XXXVI-2 and XXXVI-3. Strains susceptible to 3.1 μg/ml or less had

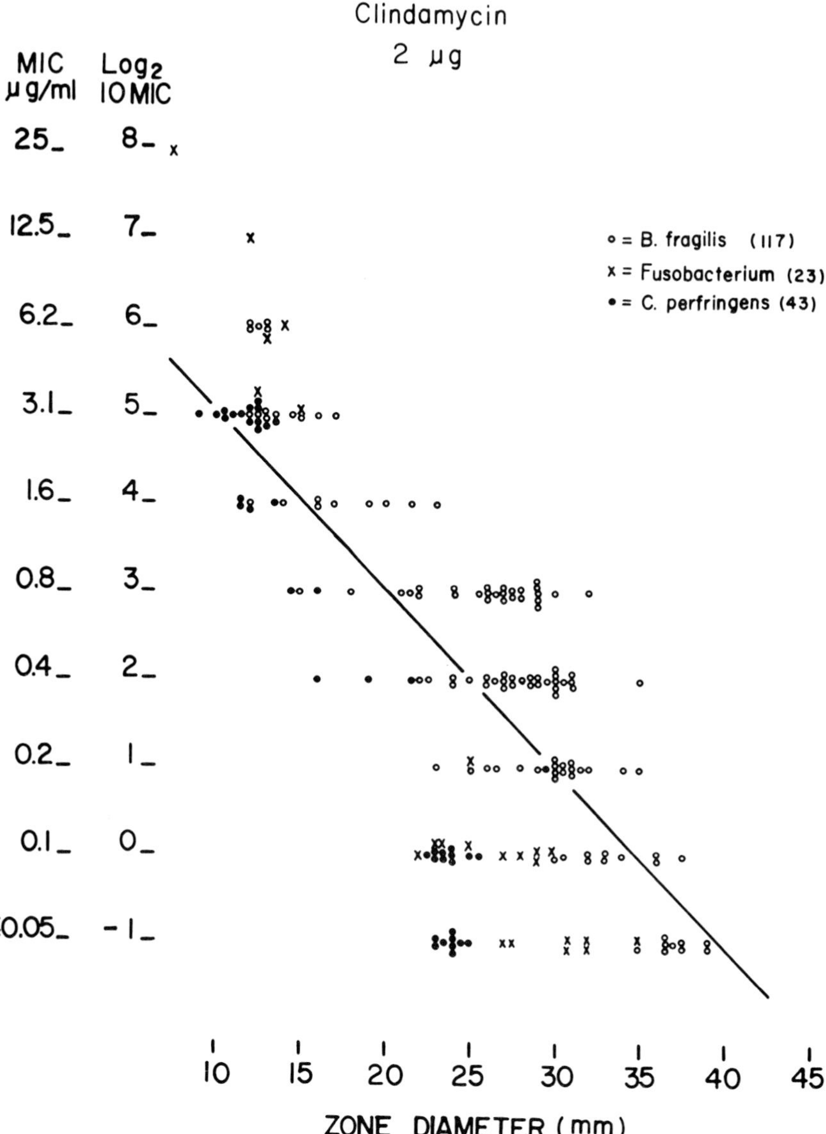

Figure XXXVI-2. Clindamycin: relationship of diameters of zones of inhibition around 2-μg discs and MIC values. Least squares line: $y = a + bx$; $a = 7.087$, $b = -0.204$. Correlation coefficient : $r = -0.733$.

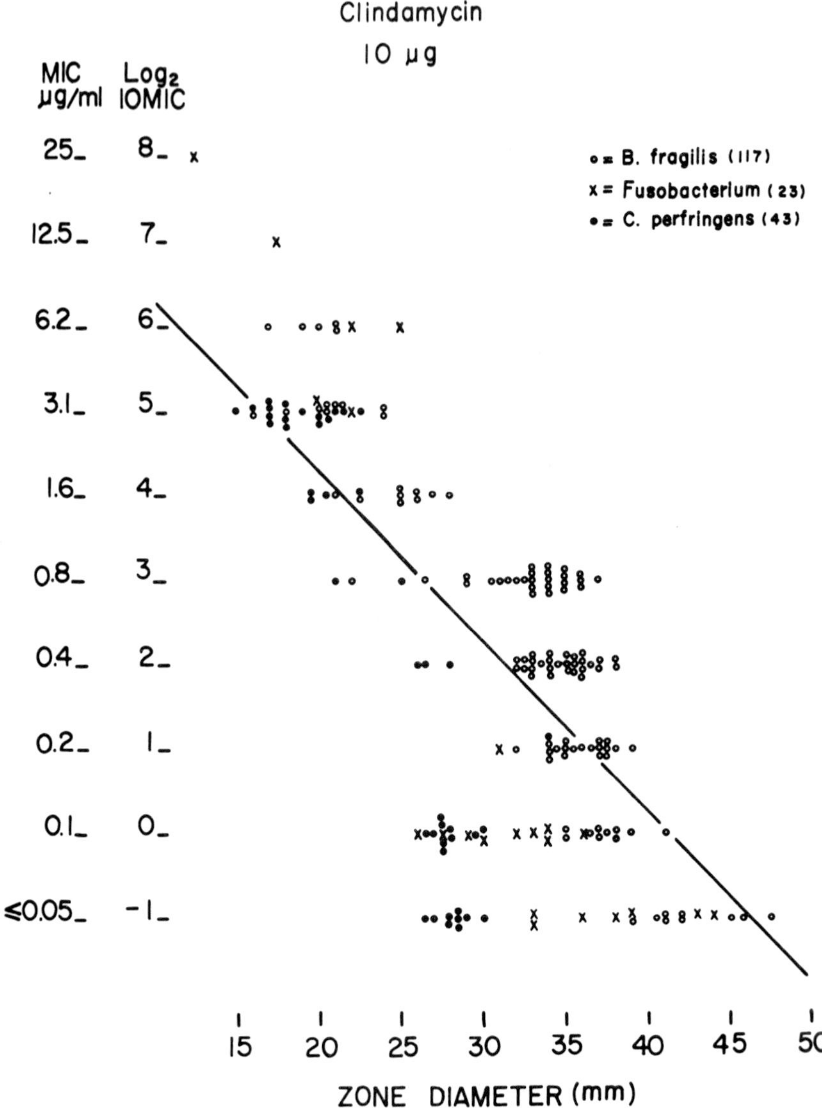

Figure XXXVI-3. Clindamycin: relationship of diameters of zones of inhibition around 10-μg discs and MIC values. Least squares line: $y = a + bx$; $a = 8.226$; $b = -0.199$. Correlation coefficient: $r = -0.664$.

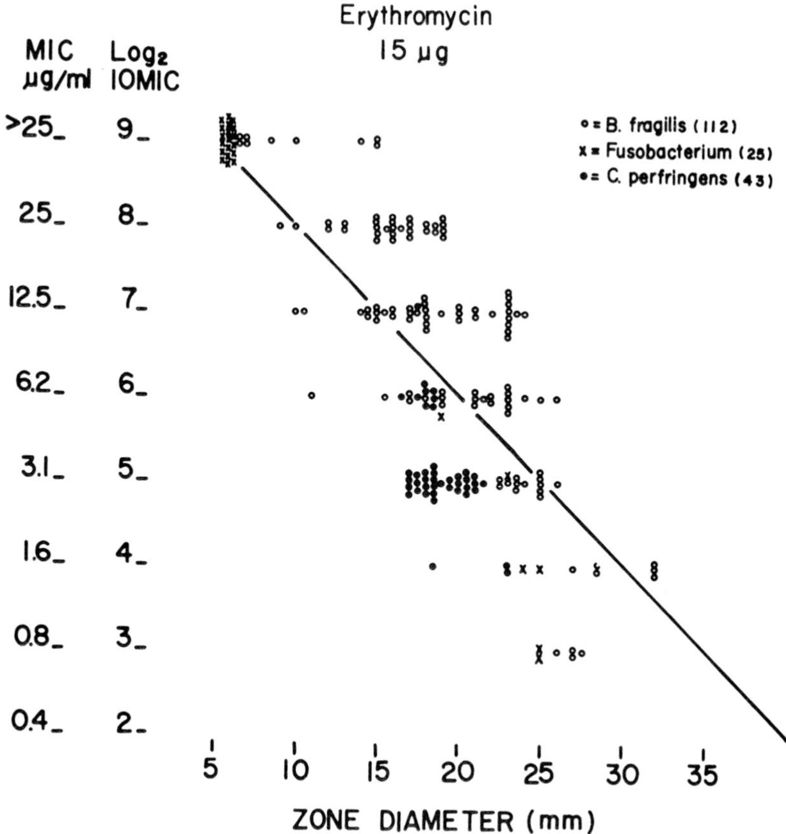

Figure XXXVI-4. Erythromycin: relationship of diameters of zones of inhibition around 15-μg discs and MIC values. Least squares line: $y = a + bx$; $a = 10.004$, $b = -0.198$. Correlation coefficient: $r = -0.584$.

zones of 9 to 39 mm around the 2-μg disc and 15 to 47 mm around the 10-μg disc. One resistant strain and 7 strains with intermediate susceptibility (6.2 μg/ml) had zones of 12 to 14 mm around the 2-μg disc and 17 to 25 mm around the 10-μg disc. These zone diameters are the same as or larger than those exhibited by a number of susceptible strains and we therefore consider zone diameters in the range of 9 to 14 mm with the 2-μg disc and 15 to 25 mm with the 10-μg disc as being equivocal. Fewer strains fell into the equivocal range with the 2-μg disc than with the 10-μg disc.

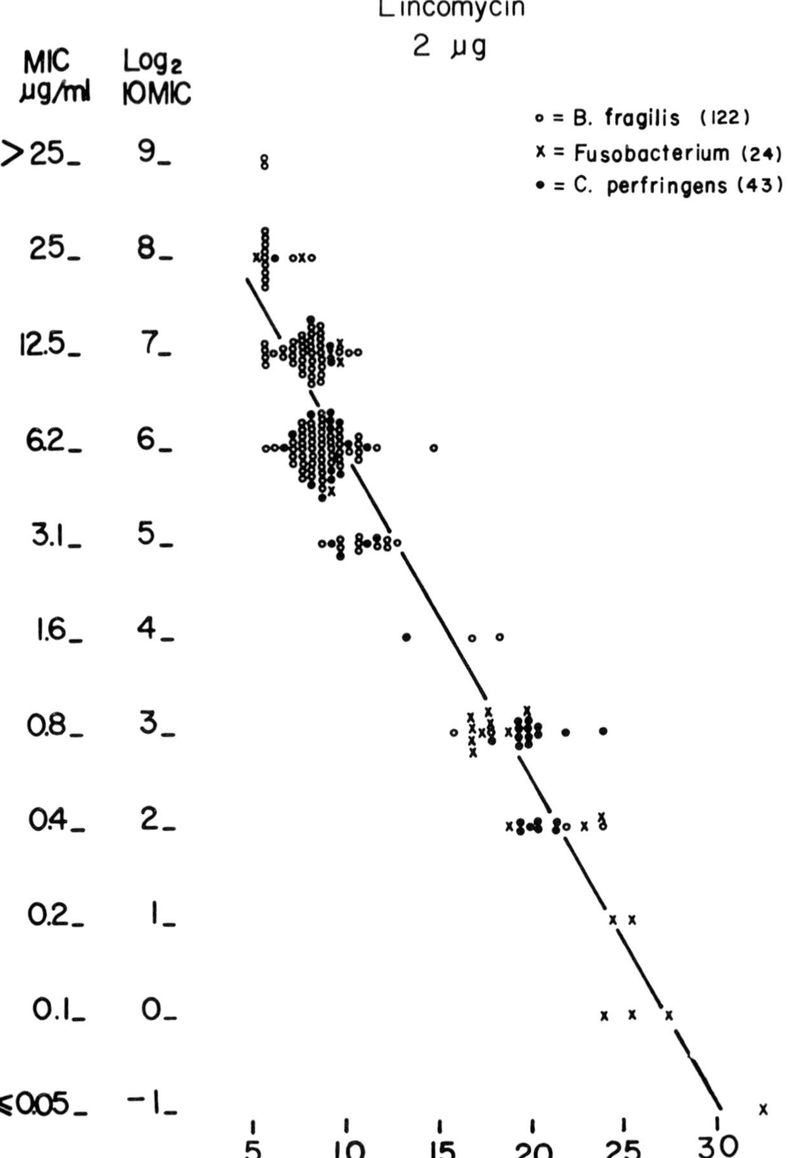

Figure XXXVI-5. Lincomycin: relationship of diameters of zones of inhibition around 2-μg discs and MIC values. Least squares line: $y = a + bx$; $a = 9.404$, $b = -0.339$. Correlation coefficient: $r = -0.941$.

Erythromycin

Strains tested for their susceptibility to erythromycin are depicted in Figure XXXVI-4. Only 10 of 112 strains of *B. fragilis,* 3 strains of *C. perfringens,* 3 strains of *F. necrophorum* and the one strain of *F. gonidiaformans* tested appear susceptible to readily achievable blood levels (1.6 μg/ml or less). All strains of *F. mortiferum* and *F. varium* were resistant to more than 25 μg/ml. There was a wide spread of zone diameters at each MIC with overlapping of zone sizes among susceptible, intermediate and resistant strains. Therefore, our equivocal range of zone diameters encompasses two thirds of the strains tested.

Lincomycin

Results obtained for MIC's related to zone diameters observed with both 2- and 10-μg lincomycin discs are shown in Figs. XXXVI-5 and XXXVI-6. With the 2-μg lincomycin discs some of the strains with intermediate MIC (6.2 μg/ml) exhibited zones of 6 or 7 mm, as did a number of the resistant strains. Some of the susceptible strains also had small zones (9–10 mm). Zones were larger with the 10-μg disc, but a considerable amount of overlapping occurred between susceptible, intermediate and resistant strains. With aerobes and facultatives, it has been recommended that 2-μg clindamycin discs be used for predicting susceptibility of strains to lincomycin (Fine, 1972). In order to see if this were feasible with anaerobes, we plotted the zone diameters obtained with 115 strains of *B. fragilis* and 2-μg clindamycin discs against the lincomycin MIC of each strain (Fig. XXXVI-7). A complete lack of predictability of lincomycin susceptibility based on these zone diameters was observed. A similar lack of predictability was observed when data obtained with 10-μg clindamycin discs was analyzed in this way.

Penicillin G

Results obtained with penicillin G against 185 strains of anaerobes are shown in Fig. XXXVI-8. Endpoints with the fusobacteria, particularly strains of *F. mortiferum* and *F. var-*

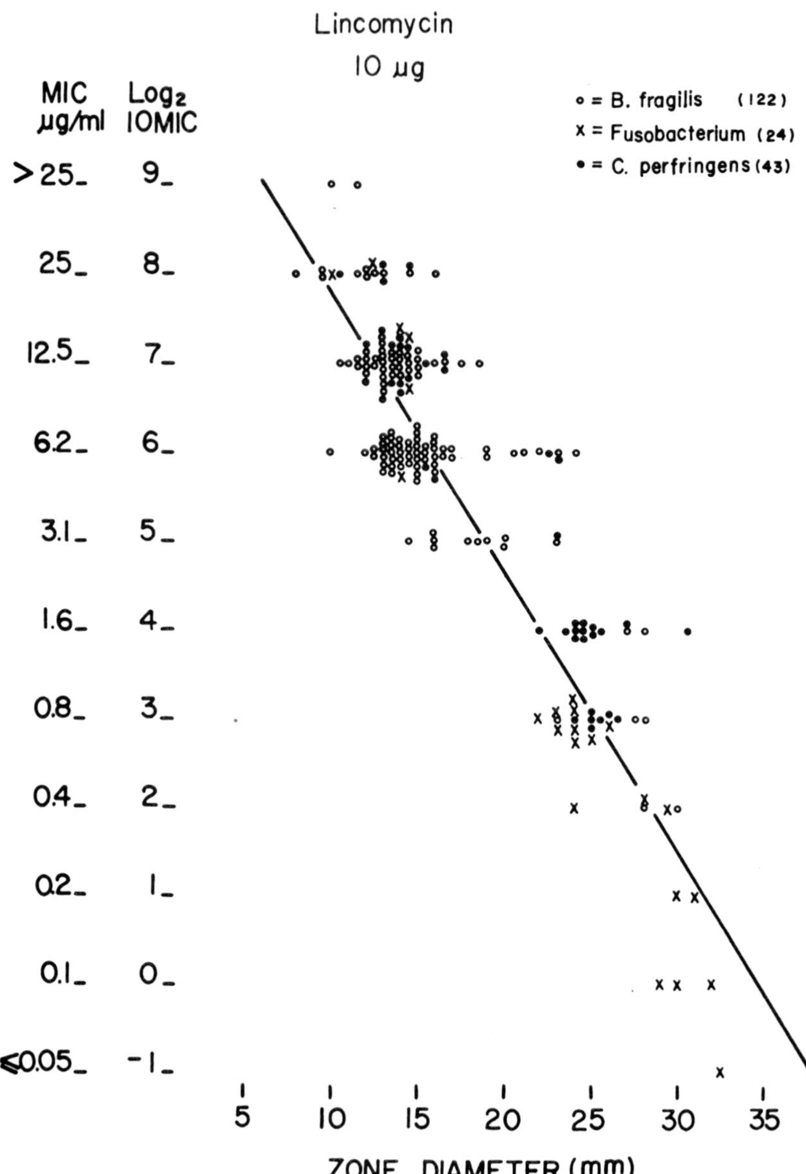

Figure XXXVI-6. Lincomycin: relationship of diameters of zones of inhibition around 10-μg discs and MIC values. Least squares line: $y = a + bx$; $a = 10.912, b = -0.314$. Correlation coefficient: $r = -0.908$.

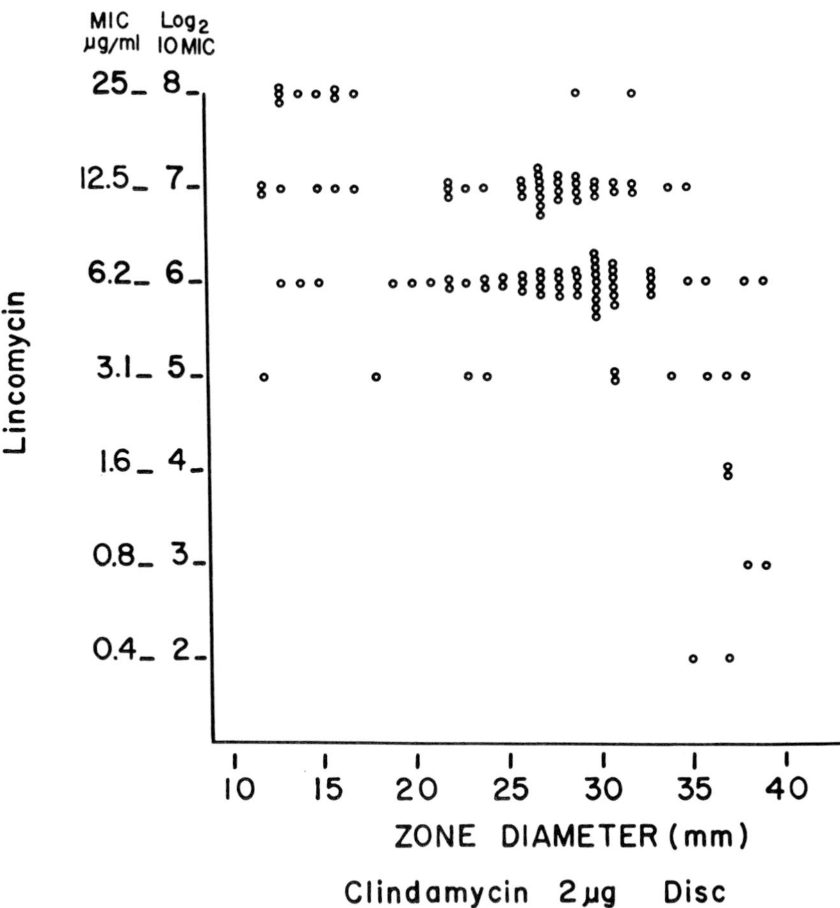

Figure XXXVI-7. Relationship of diameters of zones of inhibition around 2-μg clindamycin discs and lincomycin MIC values for 115 strains of *Bacteroides fragilis*.

ium, are very difficult to define. In the agar dilution tests, at some specific concentration of the drug, there is a definite but not complete inhibition of growth, leaving a thin film at the point of inoculation. In the disc diffusion tests, double zones were common with a thin film extending to the disc. This film is not viable and may represent a phenomenon similar to that observed with the sulfonamides and faculta-

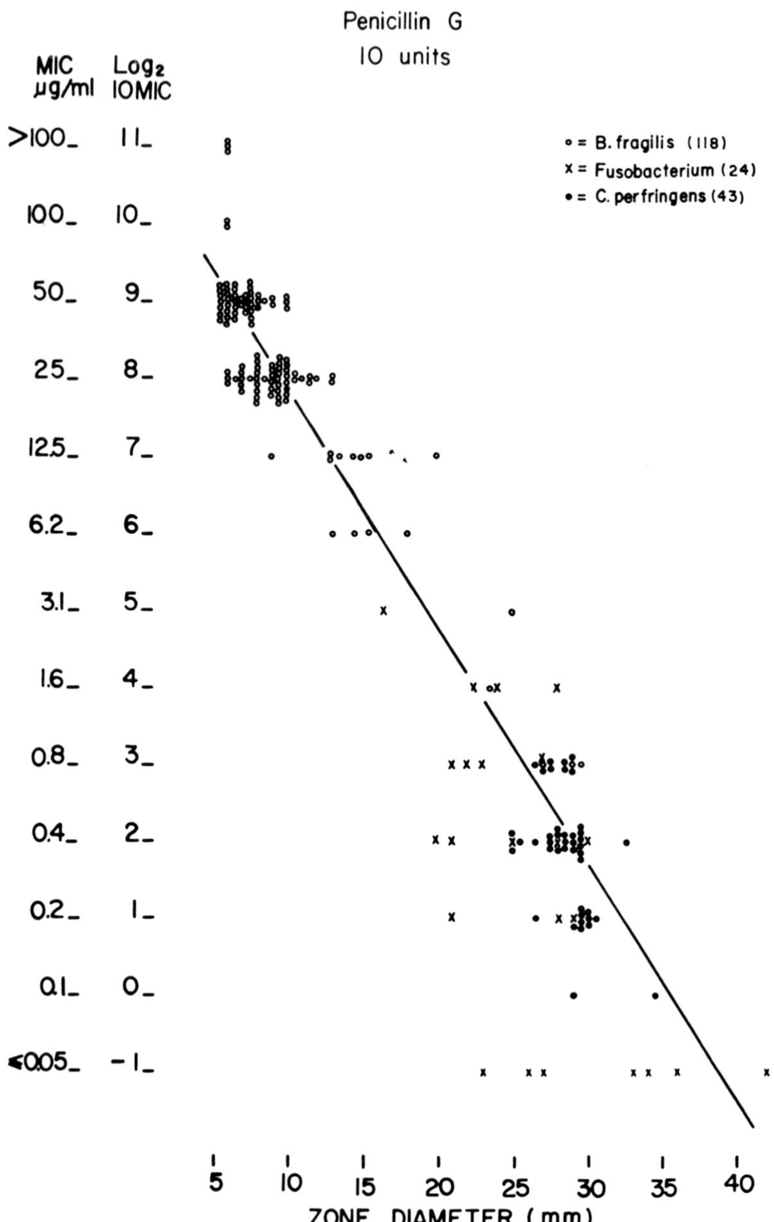

Figure XXXVI-8. Penicillin: relationship of diameters of zones of inhibition around 10-unit discs and MIC values. Least squares line: y = a + bx; a = 10.942, b = −0.309. Correlation coefficient: r = −0.934.

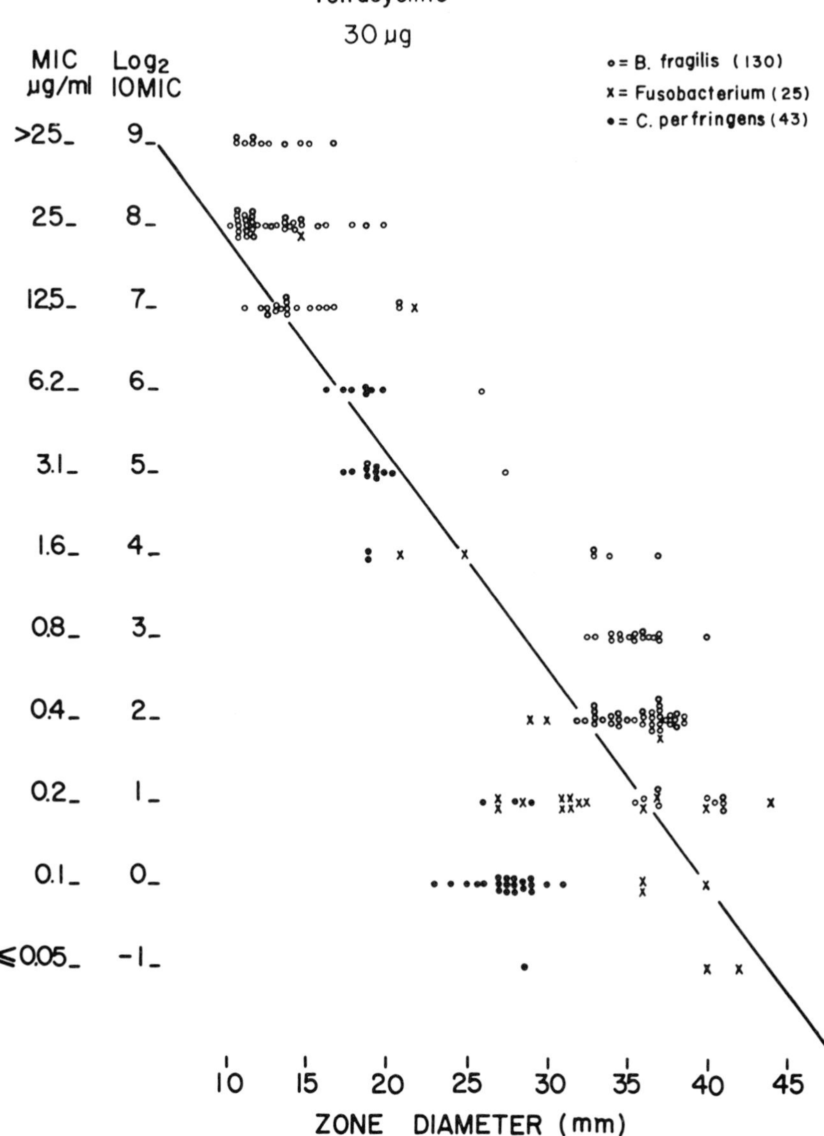

Figure XXXVI-9. Tetracycline: relationship of diameters of zones of inhibition around 30-μg discs and MIC values. Least squares line: y = a + bx; a = 10.492, b = −0.253. Correlation coefficient: r = −0.834.

tives. With these organisms we use 80 percent inhibition as the endpoint.

Tetracycline

Data obtained with tetracycline against 197 strains of anaerobes is shown in Fig. XXXVI-9. Interpretation of susceptibility by MIC and inhibition zone diameters is indicated in Table XXXVI-I.

Vancomycin

Of the 122 strains of *B. fragilis* tested, 4 had an MIC of 12.5 μg/ml with zone diameters of 15 to 21 mm. The remainder were resistant to 25 μg/ml or more and had zones of 7 to 19 mm in diameter. All of the fusobacteria were resistant to more than 25 μg/ml and had zone diameters of 6 to 10 mm. The 43 *C. perfringens* strains were susceptible to 1.6 μg/ml or less; zone diameters were from 19 to 25 mm. The relationship of zone diameters to MIC is shown in Figure XXXVI-10. Interpretation of susceptibility is indicated in Table XXXVI-I.

Variation in Zone Diameters

The variation observed with the two control strains of *B. fragilis* when tested repeatedly by the disc diffusion method is shown in Table XXXVI-II. With most of the discs tested, variation was minimal although strain 1 usually appeared slightly more variable than strain 1887. It was quite variable with erythromycin.

DISCUSSION

The need for a simple, rapid and reliable method for testing susceptibility of clinically significant anaerobic bacteria has become apparent in recent years. Because the single disc agar diffusion method of Bauer *et al.* (1966) has proved satisfactory for use with aerobes and facultatives, a number of workers have been assessing this test or modifications of it for use with anaerobes (Bodner *et al.*, 1972; Kwok *et al.*,

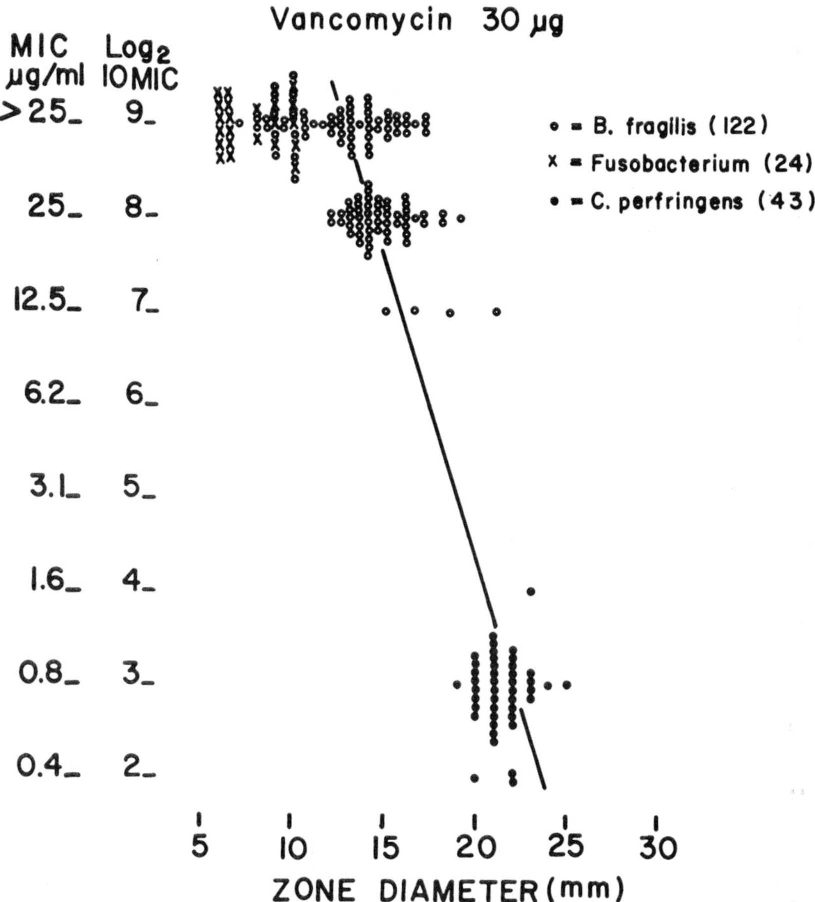

Figure XXXVI-10. Vancomycin: relationship of diameters of zones of inhibition around 30-μg discs and MIC values. Least squares line: y = a + bx; a = 17.083,b = −0.637. Correlation coefficient: r = −0.907.

1972; Sapico *et al.*, 1972; Sutter and Finegold, 1973; Sutter *et al.*, 1972, 1973; Thornton and Kramer, 1971; Wilkins *et al.*, 1972). Some workers have attempted to use the Bauer, Kirby, Sherris and Turck criteria for predicting susceptibility of anaerobes and found them generally unsatisfactory for this purpose (Thornton and Kramer, 1971). Those who have been developing criteria for interpreting susceptibility on the basis of zone diameters related to MIC's obtained anaerobically

TABLE XXXVI-II

VARIATION OF ZONE DIAMETERS*

Antibiotic	*Disc Potency*	*Strain No.*†	*MIC (μg/ml)*§	*Zone Diameters (mm)* *Min-Max*	*Mean*	*S.D.*
Chloramphenicol	30 μg	1	6.2	28–35	32	1.57
		1887	3.1	29–33	31	1.45
Clindamycin	2 μg	1	0.4	24–30	27	1.58
		1887	≤0.05	36–41	39	1.27
Clindamycin	10 μg	1		31–36	34	1.17
		1887		43–47	45	1.36
Erythromycin	15 μg	1	12.5	11–24	16	3.13
		1887	0.8	26–29	28	0.74
Lincomycin	2 μg	1	12.5	9–10	9	0.34
		1887	0.8	17–20	19	1.05
Lincomycin	10 μg	1		13–15	13	0.50
		1887		26–30	28	0.76
Penicillin	10 u	1	25	6–10	7	0.83
		1887	0.8	26–31	28	1.14
Tetracycline	30 μg	1	0.4	27–36	32	2.19
		1887	6.2	17–20	18	0.70
Vancomycin	30 μg	1	50	13–17	14	0.91
		1887	12.5	16–21	18	1.15

* Modified from Sutter, Kwok and Finegold (1973).
† Wadsworth Anaerobic Laboratory (WAL) Culture Collection Number.
§ Penicillin G expressed in units/ml.

have had varying degrees of success with regard to correlation and predictability.

In our own experience, when data is analyzed separately for each species or closely related group of bacteria, statistically good correlation is obtained with most of the antibiotics tested. Satisfactory prediction of susceptibility based on zone diameter measurements can be made (Kwok *et al.*, 1972; Sapico *et al.*, 1972; Sutter *et al.*, 1972, 1973). As demonstrated in this paper, when we compile data for all groups tested thus far, we find a wide range of zone diameters at each MIC. Strains of *C. perfringens* and the fusobacteria usually have smaller zones at a specific MIC than do the *B. fragilis* strains despite the fact that inoculum size has been carefully adjusted. *C. perfringens* and the species of *Fusobacterium* tested grow more rapidly than *B. fragilis* and the differences in zone diameters probably reflect this difference in growth rates. If predictions of susceptibility of unidentified strains are to be made by results of agar diffusion tests, it may be necessary to establish separate criteria for rapid growers, moderately rapid growers and slow growing strains. Having several sets of criteria can pose problems in the clinical microbiology laboratory.

Further studies with other anaerobic bacteria as well as investigations directed toward correlation of *in vitro* results with clinical response remain to be done in order to assess the validity of the several methods now in use for susceptibility testing of anaerobes.

REFERENCES

Bauer, A.W., Kirby, W.M.M., Sherris, J.C., and Turck, M.: Antibiotic susceptibility testing by a single disc method. *Am J Clin Pathol, 45:*493, 1966.

Bodner, S.J., Koenig, M.G., Treanor, L.L., and Goodman, J.S.: Antibiotic susceptibility testing of *Bacteroides. Antimicrob Agents Chemother, 2:*57, 1972.

Brownlee, K.A.: *Statistical Theory and Methodology in Science and Engineering,* 2nd ed. New York, Wiley, 1965.

Ericsson, H.: Rational use of antibiotics in hospitals. *Scand J Clin Lab Invest, 12, Suppl. 50:*1, 1960.

Ericsson, H.M., and Sherris, J.C.: Antibiotic sensitivity testing—report of an international collaborative study. *Acta Pathol Microbiol Scand*[B], *Suppl. 217:*1, 1971.

Fine, S.D.: Antibiotics intended for use in the laboratory diagnosis of disease. *Fed Reg, 37:*20526, 1972.

Frostell, G.: Jämförande resistensbestämningar i aerob och anaerob miljö. *Svensk Lakartidiningen, 60:*1086, 1963.

Ingham, H.R., Selkon, J.B., Codd, A.A., and Hale, J.H.: The effect of carbon dioxide on the sensitivity of *Bacteroides fragilis* to certain antibiotics *in vitro. J Clin Pathol, 23:*254, 1970.

Kwok, Y.Y., Sutter, V.L., and Finegold, S.M.: Standardized antimicrobial disc susceptibility testing of *Fusobacterium.* Abstr. #M90, p. 95. Paper presented at 72nd Ann. Mtg. of American Society for Microbiology. Phila., April 23–28, 1972.

Rosenblatt, J.E., and Schoenknecht, F.: Effect of several components of anaerobic incubation on antibiotic susceptibility test results. *Antimicrob Agents Chemother, 1:*433, 1972.

Sapico, F.L., Kwok, Y.Y., Sutter, V.L., and Finegold, S.M.: Standardized antimicrobial disc susceptibility testing of anaerobic bacteria. II. *In vitro* susceptibility of *Clostridium perfringens* to nine antibiotics. *Antimicrob Agents Chemother, 2:*320, 1972.

Steers, E., Foltz, E.L., and Graves, B.S.: An inocula-replicating apparatus for routine testing of bacterial susceptibility to antibiotics. *Antibiot Chemother, 9:*307, 1959.

Sutter, V.L., and Finegold, S.M.: Antibiotic susceptibility testing of anaerobes. In: Balows, A. (Ed.): *Current Techniques for Antibiotic Susceptibility Testing.* Springfield, Thomas, 1973.

Sutter, V.L., Kwok, Y.Y., and Finegold, S.M.: Standardized antimicrobial disc susceptibility testing of anaerobic bacteria. I. Susceptibility of *Bacteroides fragilis* to tetracycline. *Appl Microbiol, 23:*268, 1972.

Sutter, V.L., Kwok, Y.Y., and Finegold, S.M.: Susceptibility of *Bacteroides fragilis* to six antibiotics determined by standardized antimicrobial disc susceptibility testing. *Antimicrob Agents Chemother, 3:*188, 1973.

Thornton, G.F., and Kramer, J.A.: Antibiotic susceptibility of *Bacteroides* species. In: *Antimicrob Agents Chemother,—1970*. Ann Arbor, American Society for Microbiology, 1971. p. 509.

Wilkins, T.D., Holdeman, L.V., Abramson, I.J., and Moore, W.E.C.: A standardized single-disc method for antibiotic susceptibility testing of anaerobic bacteria. *Antimicrob Agents Chemother, 1:*451, 1972.

CHAPTER XXXVII

FACTORS AFFECTING SUSCEPTIBILITY TESTS AND THE NEED FOR STANDARDIZED PROCEDURES

CLYDE THORNSBERRY

ABSTRACT: *Antimicrobial susceptibility tests on anaerobic bacteria have been performed by agar diffusion and by broth and agar dilution methods. The factors which affect the results of these susceptibility tests have been outlined, and some of the techniques used with anaerobes over the last 25 years have been reviewed in relation to these variables. This review revealed that different media, inoculum densities, and incubation times were used in most studies. Tests for the susceptibility of anaerobes to antimicrobics need to be standardized, so that data from different laboratories can be compared validly. This will provide the physician with more accurate data on which to base his choice of treatment and, consequently, will ultimately benefit the patient.*

DEVELOPMENT OF STANDARDIZED METHODS FOR SUSCEPTIBILITY TESTING

In vitro antimicrobial susceptibility testing began in the 1940's. Antimicrobial agents were incorporated into agar or broth in one or more concentrations, and these media were inoculated with bacteria isolated from clinical infections. After appropriate incubation, the tests were observed for inhibition of growth. Either minimal inhibitory concentrations were determined or an interpretation of sensitive or resistant was made on the basis of inhibition of growth at a fixed concentration of antimicrobic (Erlanson, 1950; Petersdorf and Plorde, 1963).

Then, as now, these techniques were considered too difficult for routine use. Other techniques were soon developed. In one of these methods, tablets containing an antimicrobic were used, and in another, filter paper discs soaked in a solution of the antimicrobic. The tablets or the wet filter paper discs were placed on agar plates that had been seeded with the bacteria to be tested, and the plates were appropriately incubated. If growth was inhibited in a zone around a tablet or disc, the organism was considered to be susceptible to the antimicrobic being tested (Erlanson, 1950; Petersdorf and Plorde, 1963).

Next, filter paper discs that were dried after being soaked in antimicrobic were developed. This was an important step because the discs could be prepared in advance and stored until they were needed. Methods for using the dried filter paper discs were quickly adopted. They included single disc procedures with a variety of concentrations of antimicrobic and multiple disc procedures with varying concentrations of antimicrobic (Erlanson, 1950; Petersdorf and Plorde, 1963).

Few efforts were made to standardize the disc agar diffusion technique until the University of Washington (Seattle) group developed the procedure that was to be called the Kirby-Bauer technique (Bauer *et al.*, 1966). They standardized the procedure by designating use of one medium and a single high-content disc with a standard concentration of antimicrobic; by adjusting the density of the inoculum before seeding the plate; by designating that the diameters of the inhibition zones be measured; and by interpreting the measurements according to standards they developed. These standards were based on the observed relationship between inhibition zone diameters and minimal inhibitory concentrations, consideration of achievable serum levels of the drug after usual dosage, and, finally, the clinical efficacy of the drug. The Seattle group succeeded in getting this technique accepted in many laboratories throughout the country.

Even this standardized method has had problems, most of which result from misuse of the test. Almost every laboratory uses an agar diffusion test, and almost all use the Kirby-Bauer interpretive standards. However, these standards should not be used unless the test is performed as it was described. Neither should the results be based on the mere presence of a zone. These facts have been borne out by proficiency surveys, including one conducted by CDC (Hall, C.T., personal communication).

The limitations of the Kirby-Bauer method should be emphasized because they have a bearing on the testing of anaerobes. The procedure is intended for use with commonly isolated, rapidly growing, aerobic or facultative bacteria. Anaerobes are excluded because the required incubation environment is aerobic and should not contain increased concentrations of CO_2. In general, anaerobes should not be tested by the procedure because the Kirby-Bauer interpretive standards are not based on data obtained with these organisms. Consequently, the published reports on the use of the Kirby-Bauer procedure for testing anaerobes are few; however, I have learned in conversation with laboratorians around the country that the great majority of laboratories that isolate anaerobes also test them by a modified Kirby-Bauer procedure and interpret these susceptibility tests on the basis of the aerobic Kirby-Bauer standards. This practice may be appropriate for a limited number of organisms and drugs, but the test should not be used routinely for anaerobes, unless the laboratorian has accumulated enough data to demonstrate the efficacy of the test for determining the susceptibility of these organisms. If an agar diffusion method is to be used for testing the susceptibility of anaerobes to drugs, interpretive standards for this test and these organisms will have to be developed. Dr. Sutter and Dr. Wilkins have been developing such standards (Sapico *et al.*, 1972; Sutter *et al.*, 1972; Wilkins *et al.*, 1972).

Very little has been done to standardize the dilution tests for antimicrobial susceptibility testing. This work has been limited mostly to the International Collaborative Study on Antibiotic Sensitivity Testing (Ericsson and Sherris, 1971).

Methodology for susceptibility testing of certain bacterial pathogens must be standardized if comparison of results from different laboratories is to be valid. When these standardized methods are developed, all the factors that can affect susceptibility tests should be considered.

FACTORS AFFECTING SUSCEPTIBILITY TESTS

Factors which may affect the results of susceptibility tests include media, density of the inoculum, temperature and composition of the incubation environment, and stability of the antimicrobics.

The kind of medium used influences the size of the zone in a diffusion test, as well as the minimal inhibitory concentration (MIC) in a dilution test. For example, trypticase soy agar will not yield the same results as Mueller-Hinton agar because of differences in their constitution, such as kinds of peptone and agar used, ion content, and ability to promote growth. In the Kirby-Bauer method, only Mueller-Hinton agar may be used. In our laboratory, we have compared different brands of Mueller-Hinton by using them simultaneously in diffusion tests. Differences in zone diameters were generally 2 mm or less, values which are acceptable for this procedure (Thornsberry and Baker, 1973).

The pH of the medium will markedly affect the susceptibility results for some antimicrobics. For example, we have shown in studies with *Staphylococcus aureus* and *Escherichia coli* that the diameters of zones formed with lincomycin and erythromycin approximately double if the pH is raised from 6.5 to 8.0. On the other hand, the diameter of the zone around a tetracycline disc on a medium with a pH of 6.5 will be about half the diameter at pH 8.0 (Thornsberry and Baker, 1973).

The ion content of the medium will affect the susceptibility results with some antibiotics and certain bacteria. For example, if MIC determinations for gentamicin, an aminoglycoside, for *Pseudomonas aeruginosa* are performed in Mueller-Hinton broth and simultaneously in Mueller-Hinton agar, the MIC values will be considerably higher in the agar medium because of the higher Mg^{++} and Ca^{++} concentrations in the agar. This pronounced difference is not seen with *Escherichia coli*.

Adding blood products to Mueller-Hinton agar alters the results of susceptibility tests very little. In studies in our laboratory with diffusion tests, the diameters of zones of inhibition on agar containing blood products were usually 1 to 2 mm smaller than corresponding zones on blood-free agar. We postulate that the smaller zone size is due to the increased growth in the media containing blood. We have had similar results with other supplements, such as yeast extract, and with commercial supplements used in media for growing *Haemophilus influenzae* (Thornsberry and Baker, 1973).

One of the most important variables in susceptibility testing is the number of organisms in the inoculum. It is also, in many laboratories, one of the least controlled. In our laboratory, agar diffusion tests were performed on cultures of *S. aureus* and *E. coli* at two different times in their logarithmic growth phase, so that there was a tenfold difference in the number of bacteria per ml of inoculum. Zone sizes were invariably smaller with the larger inoculum. Tests performed with a 1:10 dilution of the larger inoculum yielded zones the same size as those obtained with smaller inoculum (Thornsberry and Baker, 1973). Likewise, a larger inoculum results in a higher MIC.

The incubation temperature is very important in detecting methicillin-resistant *S. aureus*. We have found that methicillin resistance in *S. aureus* can always be detected at 35°C, but it may not be at 37°C (Thornsberry *et al.*, 1972a, 1972b). Drew *et al.* (1972) have corroborated this finding. Therefore, we recommend that the incubation temperature be no higher than 35°C.

Increased CO_2 concentration in the incubation environment affects susceptibility results by altering the pH on the surface of the agar (Ingham *et al.*, 1970; Rosenblatt and Schoenknecht, 1972). Some reports indicate that certain antibiotics are less active in an anaerobic atmosphere (Rosenblatt and Schoenknecht, 1972; Traub and Raymond, 1971).

Obviously, susceptibility tests can be affected by the deterioration of the antimicrobics being tested. The penicillins and cephalosporins probably deteriorate most quickly. Commercial discs should not be used beyond the stated expiration date. Media containing antimicrobics can be safely stored for at least 1 week if they are stored in closed plastic bags, thereby preventing evaporation. Even though the FDA has approved methicillin discs for use in the Kirby-Bauer procedure (Fine, 1972), Drew and associates (1972) have reported that oxacillin discs are more stable than methicillin discs when used under routine laboratory conditions. We concur in this conclusion.

The potency of all antimicrobics, whether in discs or in media, must be monitored through frequent use of standard control cultures. We recommend that *S. aureus* ATCC 25923, *E. coli* ATCC 25922, and a strain of *P. aeruginosa* be used for this purpose.

METHODS FOR ANAEROBE SUSCEPTIBILITY TESTING

The methods that have been used for susceptibility testing of anaerobic bacteria are spin-offs from the methods used to test aerobic and facultative organisms. Thus, anaerobes have been tested by agar diffusion and by broth and agar dilution methods. Reports on the methods used for anaerobes will not be reviewed here; instead, some of these methods will be discussed in relation to variables that may affect susceptibility results.

Only a few agar diffusion tests with anaerobes have been reported. In the early tests, anaerobic blood agar (Gillespie and Guy, 1956) or thioglycollate agar (Merritt, 1962) was used. More recently (Bodner *et al.*, 1970, 1972), the tests have been modified Kirby-Bauer tests, and Mueller-Hinton agar with 5% blood has been used. In these later studies the inocula were adjusted. In each instance, either a broth culture or a broth suspension of growth from a solid medium was diluted to equal the turbidity of a 0.5 McFarland standard. Times of incubation ranged from 24 to 48 hours. The studies of Sutter and her colleagues (Sapico *et al.*, 1972; Sutter *et al.*, 1972) and Wilkins and his associates (1972) are not included here, since these investigators are participants in this conference.

Broth dilution susceptibility tests were apparently first used for anaerobes in 1947 (Foley, 1947). Other dilution tests for anaerobes were used during this period, but the authors of the reports did not mention the methodology (Beigelman and Rantz, 1949; McVay and Sprunt, 1952). This disregard for outlining methods can also be found in a very recent report.

In it, the susceptibility of anaerobes to antimicrobics is reported without any mention of the methods used (Tracey *et al.*, 1972). In reports where methods are given, many variations in procedure can be found (Fisher and McKusick, 1953; Foley, 1947; Hare *et al.*, 1952; Keusch and O'Connell, 1966; Lewis *et al.*, 1963; Nastro and Finegold, 1972; Werner *et al.*, 1971). Different test broths used were thioglycollate, Brewer's brain-heart infusion, peptone extract broth, and Brucella broth. One group of laboratorians prepared inocula by diluting 2-day broth cultures 1:100 and 1:150; another used about 10^6 bacteria; and another diluted a 2-day broth culture to equal a standard with which I am not familiar. Incubation times were 18 to 24 hrs, 24 to 36 hrs, and 48 hrs.

The agar dilution method has been used more often for testing anaerobes than have agar diffusion and broth dilution. The variations in the agar dilution method have also been extensive. In some reports (Bartlett *et al.*, 1972; Finegold *et al.*, 1966, 1967; Garrod, 1955; Ingham *et al.*, 1968, 1970; Kislak, 1972; Martin *et al.*, 1972; Miller and Finegold, 1967; Pien *et al.*, 1972; Thornton and Cramer, 1971) many different test agars have been used: meat extract peptone agar, blood agar base, eugonagar, nutrient agar, trypticase soy agar, brain-heart infusion agar, heart infusion agar, and Brucella agar. Usually these media contained either sheep or horse blood and, in some cases, hemin and menadione. In most cases, the inoculum was prepared by diluting a broth culture. In some of the early reports, a 48-hr culture was diluted 1:100 and applied to the agar with a Steers replicating device (Steers *et al.*, 1959), which would deliver approximately 0.001 ml of this dilution. In some studies, the culture was taken from an agar plate and diluted in broth. In several of the more recent studies, broth cultures have been diluted to equal the turbidity of $BaSO_4$ standards. Incubation times have ranged from 18 hours to 6 days.

Thus, nearly everyone who has studied the susceptibility of anaerobes to antimicrobics has used a different test medium, a different inoculum, and a different incubation time. Since these variables are known to affect the endpoint readings of susceptibility tests, valid comparison of the results would be very difficult. Standardized methods and collaborative studies are needed so that results from different laboratories can be validly compared.

REFERENCES

Bartlett, J.G., Sutter, V.L., and Finegold, S.M.: Treatment of anaerobic infections with lincomycin and clindamycin. *N Engl J Med, 287:*1006, 1972.

Bauer, A.W., Kirby, W.M.M., Sherris, J.C., and Turck, M.: Antibiotic sensitivity testing by a standardized single-disc method. *Am J Clin Pathol, 45:*493, 1966.

Beigelman, P.M., and Rantz, L.A.: Clinical significance of *Bacteroides. Arch Intern Med, 84:*605, 1949.

Bodner, S.J., Koenig, M.C., and Goodman, J.S.: Bacteremic *Bacteroides* infections. *Ann Intern Med, 73:*537, 1970.

Bodner, S.J., Koenig, M.G., Treanor, L.L., and Goodman, J.S.: Antibiotic susceptibility testing of *Bacteroides. Antimicrob Agents Chemother, 2:*57, 1972.

Drew, W.L., Barry, A.L., O'Toole, R., and Sherris, J.C.: Reliability of the Kirby-Bauer disc diffusion method for detecting methicillin-resistant strains of *Staphylococcus aureus. Appl Microbiol, 24:*240, 1972.

Ericsson, H.M., and Sherris, J.C.: Antibiotic sensitivity testing. Report of an international collaborative study. *Acta Pathol Microbiol Scand* [*B*], *Suppl, 217:*1, 1971.

Erlanson, P.: Determination of the sensitivity *in vitro* of bacteria to chemotherapeutic agents. *Acta Pathol Microbiol Scand, Suppl, 85:*1, 1950.

Fine, S.D.: Antibiotic susceptibility discs. *Fed Reg, 37:*20525, 1972.

Finegold, S.M., Davis, A., and Miller, L.G.: Comparative effect of broad spectrum antibiotics on non-sporeforming anaerobes and normal bowel flora. *Ann NY Acad Sci, 145:*268, 1967.

Finegold, S.M., Harada, N.E., and Miller, L.G.: Lincomycin: activity against anaerobes and effect on normal human fecal flora. In: *Antimicrobial Agents and Chemotherapy—1965.* Ann Arbor, American Society for Microbiology, 1966, pp. 659–667.

Fisher, A.M., and McKusick, V.A.: *Bacteroides* infections. Clinical, bacteriological and therapeutic features of fourteen cases. *Am J Med Sci, 225:*253, 1953.

Foley, G.E.: *In vitro* resistance of the genus *Bacteroides* to streptomycin. *Science, 106:*423, 1947.

Garrod, L.P.: Sensitivity of four species of *Bacteroides* to antibiotics. *Br Med J, 2:*1529, 1955.

Gillespie, W.A., and Guy, J.: *Bacteroides* in intra-abdominal sepsis. Their sensitivity to antibiotics. *Lancet, 1:*1039, 1956.

Hare, R., Wildy, P., Billet, F.S., and Twort, D.N.: The anaerobic cocci: gas formation fermentation reactions, sensitivity to antibiotics and sulfonamides, classification. *J Hyg (Camb), 50:*295, 1952.

Ingham, H.R., Selkon, J.B., Codd, A.A., and Hale, J.H.: A study *in vitro* of the sensitivity to antibiotics of *Bacteroides fragilis. J Clin Pathol, 21:*432, 1968.

Ingham, H.R., Selkon, J.B., Codd, A.A., and Hale, J.H.: The effect of carbon dioxide on the sensitivity of *Bacteroides fragilis* to certain antibiotics *in vitro. J Clin Pathol, 23:*254, 1970.

Keusch, G.T. and O'Connell, C.J.: The susceptibility of *Bacteroides* to the penicillins and cephalothin. *Am J Med Sci, 251:*428, 1966.

Kislak, J.W.: The susceptibility of *Bacteroides fragilis* to 24 antibiotics. *J Infect Dis, 125:*295, 1972.

Lewis, C., Clapp, H.W., and Grady, J.E.: *In vitro* and *in vivo* evaluation of lincomycin, a new antibiotic. In: *Antimicrobial Agents and Chemotherapy—1962.* Ann Arbor, American Society for Microbiology, 1963, pp. 570–582.

Martin, W.J., Gardner, M., and Washington, J.A.: *In vitro* antimicrobial susceptibility of anaerobic bacteria isolated from clinical specimens. *Antimicrob Agents Chemother, 1:*148, 1972.

McVay, L.V., and Sprunt, D.H.: *Bacteroides* infections. *Ann Intern Med, 36:*56, 1952.

Merritt, E.S.: A simple method for determination of the antibiotic sensitivity of anaerobic organisms. *Am J Clin Pathol, 38:*203, 1962.

Miller, L.G., and Finegold, S.M.: Antibacterial sensitivity of *Bifidobacterium. J Bacteriol, 93:*125, 1967.

Nastro, L.J., and Finegold, S.M.: Bactericidal activity of five antimicrobial agents against *Bacteroides fragilis. J Infect Dis, 126:*104, 1972.

Petersdorf, R.G., and Plorde, J.J.: The usefulness of *in vitro* susceptibility tests in antibiotic therapy. *Ann Rev Med, 14:*41, 1963.

Pien, F.D., Thompson, R.L., and Martin, W.J.: Clinical and bacteriologic studies of anaerobic gram-positive cocci. *Mayo Clin Proc, 47:*251, 1972.

Rosenblatt, J.E., and Schoenknecht, F.: Effect of several components of anaerobic incubation on antibiotic susceptibility test results. *Antimicrob Agents Chemother, 1:*433, 1972.

Sapico, F.L., Kwok, Y., Sutter, V.L., and Finegold, S.M.: Standardized antimicrobial disc susceptibility testing of anaerobic bacteria: *In vitro* susceptibility of *Clostridium perfringens* to nine antibiotics. *Antimicrob Agents Chemother, 2:*320, 1972.

Steers, E., Foltz, E.L., and Graves, B.S.: An inocula replicating apparatus for routine testing of bacterial susceptibility to antibiotics. *Antibiot Chemother, 9:*307, 1959.

Sutter, V.L., Kwok, Y., and Finegold, S.M.: Standardized antimicrobial disc susceptibility testing of anaerobic bacteria. I. Susceptibility of *Bacteroides fragilis* to tetracycline. *Appl Microbiol, 23:*268, 1972.

Thornsberry, C. and Baker, C.N.: The agar diffusion antimicrobial susceptibility test. In: Balows, A. (Ed.): *Current Techniques for Antibiotic Susceptibility Testing.* Springfield, Thomas, 1973.

Thornsberry, C., Caruthers, J.Q., and Baker, C.N.: The effect of temperature on *in vitro* susceptibility of *Staphylococcus aureus* to penicillinase-resistant penicillins. Abstr. #M210, p. 115. 72nd Ann. Mtg. American Society for Microbiology, Phila., April 23–28, 1972a.

Thornsberry, C., Caruthers, J.Q., Goodrum, K.J., and Baker, C.N.: Cephalothin resistance in methicillin-resistant staphylococci. Abstr. #73, p. 36. 12th Interscience Conf. Antimicrobial Agents and Chemotherapy. Atlantic City, Sept. 26–29, 1972b.

Thornton, G.F., and Cramer, J.A.: Antibiotic susceptibility of *Bacteroides* species. In: *Antimicrobial Agents and Chemotherapy—1970.* Ann Arbor, American Society for Microbiology, 1971, pp. 509–513.

Tracey, O., Gordan, A.M., Moran, F., Love, W.C., and McKenzie, P.: Lincomycin in the treatment of *Bacteroides* infections. *Br Med J, 1:*280, 1972.

Traub, W.H., and Raymond, E.A.: *In vitro* resistance of *Clostridium perfringens* type A to gentamicin sulfate and reduced activity of the antibiotic under anaerobic atmospheric conditions. *Chemotherapy, 16:*162, 1971.

Werner, H., Pulverer, G., and Reichertz, C.: The biochemical properties and antibiotic susceptibility of *Bacteroides melaninogenicus. Med Microbiol Immunol, 157:*3, 1971.

Wilkins, T.D., Holdeman, L.V., Abramson, I.J., and Moore, W.E.C.: Standardized single-disc method for antibiotic susceptibility testing of anaerobic bacteria. *Antimicrob Agents Chemother, 1:*451, 1972.

CHAPTER XXXVIII

IN VITRO SUSCEPTIBILITY OF ANAEROBIC BACTERIA ISOLATED FROM BLOOD CULTURES

JOHN A. WASHINGTON II,
WILLIAM JEFFERY MARTIN AND PAUL E. HERMANS

ABSTRACT: *The frequency of recognition of anaerobic bacteremias, especially those due to* Bacteroides fragilis, *has increased significantly in the last two decades. In vitro antimicrobial susceptibility tests of blood culture isolates show that most strains of Bacteroidaceae are resistant to tetracycline, that penicillin G has limited activity against* B. fragilis *and moderate activity against other Bacteroidaceae, and that clindamycin and chloramphenicol are highly active against nearly all anaerobic bacteria. Most clinically significant gram-positive anaerobic bacteria have been susceptible to penicillin G and clindamycin. It is therefore recommended that the antibiotic regimen of patients suspected of having gram-negative bacillemia include chloramphenicol or clindamycin in addition to gentamicin.*

The increasing frequency of recognition of anaerobic bacteremia has important therapeutic implications in the initial therapy of bacteremias presumed to be due to gram-negative bacilli. This report reviews the *in vitro* antimicrobial susceptibilities of anaerobic bacteria recovered from blood cultures of patients at the Mayo Clinic and affiliated hospitals suspected of having bacteremia.

MATERIALS AND METHODS

Anaerobic bacteria were identified according to procedures described by Dowell and co-workers (Dowell and Hawkins, 1968; Dowell *et al.*, 1970; Dowell, 1970) and Cato and co-workers (1970). Antimicrobial susceptibility tests were performed with a broth dilution technique described in a previous publication from this laboratory (Martin *et al.*, 1972). Briefly, a 48-hour broth culture of the test organism was diluted 1:100 in brain-heart infusion broth (BHIB) containing 2% sheep blood; 1 ml of this suspension was then added to each one of a row of tubes containing 1 ml of antimicrobial agent serially diluted in BHIB so as to yield final concentrations ranging from 25 to 0.1 μg/ml. The tubes were incubated, along with a control tube without antimicrobial agent, anaerobically (GasPak, BBL) at 37°C for 48 hours. The minimum inhibitory concentration (MIC) was read as the lowest concentration of antimicrobial agent (in μg/ml) resulting in complete inhibition of growth.

At the end of 1970, routine testing of anaerobic bacterial isolates from blood for susceptibility to lincomycin was temporarily discontinued and routine testing with clindamycin was initiated. In July 1972, testing with metronidazole was begun.

In view of the questionable clinical significance of most gram-positive anaerobic bacteria isolated from blood (Wilson *et al.*, 1972) and because of their rather uniform susceptibility, in our experience, to penicillin G, clindamycin, and chloramphenicol (Martin *et al.*, 1972), this group of organisms is no longer routinely tested against antimicrobial agents. Data regarding their antimicrobial susceptibility, published elsewhere (Martin *et al.*, 1972), will be included in this report for purposes of discussion.

RESULTS

The *in vitro* susceptibilities of 159 strains of *Bacteroides fragilis* and 16 strains representing other species of Bacteroidaceae to six antimicrobial agents are listed in Tables XXXVIII-I and XXXVIII-II. Penicillin G had little activity against most isolates of *B. fragilis* except at concentrations of 50 μg/ml or greater; at 50 μg/ml, 75 percent of strains were inhibited. In contrast, 75 percent of isolates representing species other than *B. fragilis* were inhibited by penicillin G at 6.2 μg/ml. Although only a third of the isolates of *B. fragilis* were inhibited by tetracycline at 3.1 μg/ml, slightly more than half of the other species of Bacteroidaceae were inhibited by this concentration of this agent. Greater degrees of activity of lincomycin, chloramphenicol, and metronidazole were noted against species other than *B. fragilis,* while

TABLE XXXVIII-I

IN VITRO ANTIMICROBIAL SUSCEPTIBILITY OF *BACTEROIDES FRAGILIS* ISOLATED FROM BLOOD CULTURES, 1970 TO 1972

Antimicrobial	*No. of Strains Tested*	*Cumulative % Susceptible at Increasing Concentrations (µg/ml)*										
		0.1	*0.2*	*0.4*	*0.8*	*1.6*	*3.1*	*6.2*	*12.5*	*25*	*50*	*100*
Penicillin G	159				1	2	5	13	23	45	62	75
Erythromycin	159	2	4	16	39	60	73	84	92	94	...	...
Tetracycline	159	5	9	21	26	30	33	38	54	72	95	97
Lincomycin	150	5	8	12	19	30	60	81	90	98	99	99
Clindamycin	119	50	62	78	87	95	97	99	100			
Chloramphenicol	159	1	1	2	2	4	20	65	90	96	100	
Metronidazole	21			33	57	76	95	100				

erythromycin and clindamycin were somewhat more active against isolates of *B. fragilis* than against the other species of Bacteroidaceae.

DISCUSSION

Between 1950 and 1970 at the Mayo Clinic, there was a considerable increase in the total number of bacteremias due to the gram-negative bacilli in general (Table XXXVIII-III). Bacteremias due to *Pseudomonas aeruginosa* and to members

TABLE XXXVIII-II

IN VITRO ANTIMICROBIAL SUSCEPTIBILITIES OF BACTEROIDACEAE (OTHER THAN *B. FRAGILIS*)* ISOLATED FROM BLOOD CULTURES, 1970 TO 1972

Antimicrobial	*No. of Strains Tested*	*Cumulative % Susceptible at Increasing Concentrations (µg/ml)*										
		0.1	*0.2*	*0.4*	*0.8*	*1.6*	*3.1*	*6.2*	*12.5*	*25*	*50*	*100*
Penicillin G	16	25	44	44	63	69	69	75	75	94	94	94
Erythromycin	16	6	18	24	30	54	54	62	70	81		
Tetracycline	15	26	53	73	87	100						
Lincomycin	7	71	100									
Clindamycin	14	31	70	75	75	81	81	81	88			
Chloramphenicol	16	6	6	12	18	44	85	75	88		94	
Metronidazole	4	25	50	50	100							

* Includes 8 *Fusobacterium nucleatum*, 1 *F. varium*, 2 *Bacteroides* (CDC group F1), 1 *B. incommunis*, 1*B . oralis*, 2 *B. melaninogenicus*, and 1 *Bacteroides* sp.

TABLE XXXVIII-III

BACTEREMIAS DUE TO GRAM-NEGATIVE ENTERIC BACILLI, BY CAUSATIVE MICROORGANISMS

Organism	*1950–1954*			*1955–1959*			*1960–1964*			*1965*	
	No.	*No./yr*	*%*	*No.*	*No./yr*	*%*	*No.*	*No./yr*	*%*	*No.*	*%*
E. coli	67	13.4	49.0	123	24.6	41.8	219	43.8	40.9	60	37.5
Klebsiella											
Enterobacter											
Klebsiella + Enterobacter	26	5.2	19.0	72	14.4	24.4	145	29.0	27.1	42	26.3
S. marcescens				1	0.2	0.3	1	0.2	0.2		
Proteus	21	4.2	15.3	33	6.6	11.2	49	9.8	9.1	19	11.9
P. mirabilis											
P. morganii											
P. vulgaris											
P. rettgeri											
P. aeruginosa	19	3.8	13.8	59	11.8	20.0	88	17.6	16.4	22	13.7
Bacteroidaceae	4	0.8	2.9	7	1.4	2.3	34	6.8	6.3	17	10.6
Total	137	27.4	100.0	295	59.0	100.0	536	107.2	100.0	160	100.0

TABLE XXXVIII-III

BACTEREMIAS DUE TO GRAM-NEGATIVE ENTERIC BACILLI, BY CAUSATIVE MICROORGANISMS—Continued

Organism	*1966* No.	*1966* %	*1967* No.	*1967* %	*1968* No.	*1968* No.	*1968* %	*1968* %	*1969* No.	*1969* No.	*1969* %	*1969* %	*1970* No.	*1970* No.	*1970* %	*1970* %
E. coli	52	34.7	81	37.0		93		37.5		90		31.1		110		32.6
Klebsiella					43		17.3		54		18.7		46		13.6	
Enterobacter					16		6.5		22		7.6		14		4.1	
Klebsiella + Enterobacter	48	32.0	63	28.8		59		23.8		76		26.3		60		17.7
S. marcescens						3		1.2		3		1.1		16		4.7
Proteus	13	8.7	17	7.7		12		4.8		22		7.6		29		8.6
P. mirabilis					10		4.0		18		6.2		23		6.8	
P. morganii					1		0.4		3		1.1		4		1.2	
P. vulgaris					1		0.4		0		0		1		0.3	
P. rettgeri					0		0		1		0.3		1		0.3	
P. aeruginosa	21	14.0	41	18.8		44		17.7		42		14.5		57		16.9
Bacteroidaceae	16	10.6	17	7.7		37		14.9		56		19.4		66		19.5
Total	150	100.0	219	100.0		248		100.0		289		100.0		338		100.0

of the tribe Klebsielleae appeared to assume a more important role quantitatively, but the difference was not significant by statistical methods. Bacteremias due to members of the family Bacteroidaceae, which previously (1950 through 1954 and 1955 through 1959) constituted less than 3 percent, increased to 19.5 percent ($P<0.01$) in 1970. This increase in recovery of anaerobes was at least partially due to a change in blood culture media in 1968 (Marcoux *et al.*, 1970). In previously published studies of blood culture media, disregarding presumed contaminants, anaerobic bacteria accounted for between 12 and 27 percent of the positive cultures (Washington, 1971, 1972; Washington and Martin, 1973). Of all patients with positive blood cultures between January 1, 1970, and March 31, 1971, at least one anaerobic microorganism was isolated from 20 percent. However, the majority of these organisms other than those in the family Bacteroidaceae were not considered to be clinically significant (Table XXXVIII-IV; see Wilson *et al.*, 1972).

Although the gradual emergence of *P. aeruginosa* as a frequent cause of bacteremia has been generally recognized and has resulted in the current widespread use of gentamicin, with or without carbenicillin, in the initial treatment of patients suspected of having gram-negative bacillemia, the increasing frequency of bacteremias due to the Bacteroidaceae is less well recognized and has important therapeutic implications. It has previously been shown that gentamicin has only limited activity *in vitro* against most strains of anaerobic bacteria (Martin *et al.*, 1972). In our most recent survey (Wilson *et al.*, 1972), 78 percent of clinically significant anaerobic bacteremias were associated with gram-negative nonsporeforming rods, and 90 percent of these bacteremias were due to *B. fragilis*. Therefore, on the basis of the susceptibility data presented in Table XXXVIII-I it is unlikely that the frequently used therapeutic regimen of a penicillin or cephalosporin with gentamicin would result in a favorable clinical response. Despite the fact that between 80 and 90 percent of isolates of *B. fragilis* studied in this laboratory are inhibited by carbenicillin at 100 μg/ml, it is

TABLE XXXVIII-IV

INDIVIDUAL ISOLATES*

Organism	*Patients with Positive Blood Culture (No.)*	*Patients with Clinically Significant Bacteremia (No.)*
Gram-positive nonsporeforming rods		
Propionibacterium acnes	151	1
Eubacterium alactolyticum	1	1
Eubacterium lentum†	2	1
Bifidobacterium sp.†	1	1
Other	10	0
Total	165	3
Gram-negative nonsporeforming rods		
Bacteroides fragilis	57	47
Bacteroides sp.	4	1
Fusobacterium necrophorum	2	1
Fusobacterium nucleatum	1	1
Bacteroides oralis	2	1
Bacteroides terebrans	2	1
Other	3	0
Total	71	52
Gram-positive sporeforming rods		
Clostridium septicum	3	3
Clostridium perfringens	3	2
Clostridium paraputrificum	2	1
Other	4	0
Total	12	6
Gram-positive cocci		
Peptostreptococcus sp.	9	2
Peptococcus sp.	5	2
Peptostrep. + *Peptococcus* sp.	1	1
Total	15	5
Gram-negative cocci		
Veillonella	1	1

* From Wilson *et al.*, 1972.

† *Eubacterium lentum* and *Bifidobacterium* sp. were isolated together from one patient.

unknown at this time whether or not this agent, administered intravenously in large doses, might be helpful in the therapy of "Bacteroides" bacteremia. Therefore, our current recommendations for initial therapy of presumed gram-negative bacillemia would include intravenous administration of chloramphenicol or clindamycin in addition to gentamicin parenterally. Although tetracycline formerly was regarded as the drug of choice for "Bacteroides" bacteremia, it is apparent

TABLE XXXVIII-V

IN VITRO SUSCEPTIBILITY OF GRAM-POSITIVE ANAEROBIC BACTERIA*

		Cumulative % Susceptible at Increasing Concentrations (μg/ml)								
		Penicillin			*Clindamycin*			*Chlor-amphenicol*		
Organism	*Strains Tested (No.)*	*0.2*	*0.8*	*3.1*	*0.2*	*0.8*	*3.1*	*0.4*	*1.6*	*6.2*
Clostridium perfringens	34	97	97	100	44	79	100		15	100
Clostridium spp.	17	88	100		53	76	94	12	53	100
Peptostreptococcus	72	91	98	100	85	98	100	8	67	98
Peptococcus	145	91	96	99	76	94	96	11	63	100

* Adapted from Martin *et al.* (1972).

that resistance of *B. fragilis* to this agent has increased to the extent that only 38 percent of the isolates tested in this study were inhibited by 6.2 μg/ml. Similar findings have been reported by others (Bodner *et al.*, 1970, 1972; Keusch and O'Connell, 1966; Thornton and Cramer, 1970).

From data presented in Table XXXVIII-V, it appears that all strains of gram-positive anaerobic bacteria, exclusive of the anaerobic gram-positive nonsporeforming bacilli, which rarely are of clinical significance (Wilson *et al.*, 1972), are inhibited by concentrations of penicillin which are readily attainable in serum by parenteral therapy. For this and other reasons outlined below, penicillin given intravenously in large doses may be added to the aforementioned antimicrobial regimen of gentamicin with chloramphenicol or clindamycin. The optimal dosage schedule for clindamycin administered parenterally has not yet been established (Bartlett *et al.*, 1972; Medeiros, 1972; Haldane and van Rooyen, 1972). Appropriate adjustments in antimicrobial dosage are made in patients with renal or hepatic impairment. This therapeutic program is based not only on the current distribution of isolates from the blood (Table XXXVIII-III) but also on our previous findings that polymicrobial bacteremia occurred in 6 percent of patients with bacteremia (Hermans and Washington, 1970) and in 32 percent of patients with anaerobic bacteremia (Wilson *et al.*, 1972). Modification of this antimicrobial regimen may be made once microbiologic information regarding the types of organisms causing the bacteremia becomes available and after the clinical response of the patient has been assessed.

Although clostridial species are detectable, on the average, in blood cultures within the first 24 or 48 hours, the Bacteroidaceae require, on the average, between 2½ and 4 days for detection (Washington, 1971, 1972; Washington and Martin, 1973). Furthermore, susceptibility testing of anaerobic bacteria may require an additional 3 to 4 days, so the value of "routine" testing of anaerobes, except to detect significant alterations in patterns of susceptibility, is doubtful.

REFERENCES

Bartlett, J.G., Sutter, V.L., and Finegold, S.M.: Treatment of anaerobic infections with lincomycin and clindamycin. *N Engl J Med, 287:*1006, 1972.

Bodner, S.J., Koenig, M.G., and Goodman, J.S.: Bacteremic *Bacteroides* infection. *Ann Intern Med, 73:*537, 1970.

Bodner, S.J., Koenig, M.G., Treanor, L.L., and Goodman, J.S.: Antibiotic susceptibility testing of *Bacteroides. Antimicrob Agents Chemother, 2:*57, 1972.

Cato, E.P., Cummins, C.S., Holdeman, L.V., Johnson, J.L., Moore, W.E.C., Smibert, R.N., and Smith, L.DS.: *Outline of Clinical Methods in Anaerobic Bacteriology.* Blacksburg, Virginia Polytechnic Inst. and State Univ., 1970.

Dowell, V.R.: Anaerobic infections. In: Bodily, H.L., Updyke, E.L., and Mason, J.O. (Eds.): *Diagnostic Procedures for Bacterial, Mycotic and Parasitic Infections.* New York, Am. Public Health Assoc., 1970, pp. 494–543.

Dowell, V.R., Jr., and Hawkins, T.M.: *Laboratory Methods in Anaerobic Bacteriology* (CDC Laboratory Manual). Atlanta, Center for Disease Control, U.S. Dept. HEW, 1968.

Dowell, V.R., Jr., Thompson, F.S., Whaley, D.N., Alpern, R.J., Felner, J.M., Armfield, A.Y., McCroskey, L.M., and Wiggs, L.S.: *Differential Characteristics of Anaerobic Bacteria.* Atlanta, Center for Disease Control, US Dept. HEW, PHS, Health Services and Mental Health Admin., 1970.

Haldane, E.V., and van Rooyen, C.E.: Treatment of severe bacteroides infections with parenteral clindamycin. *Can Med Assoc J, 107:*1177, 1972.

Hermans, P.E., and Washington, J.A., II: Polymicrobial bacteremia. *Ann Intern Med, 73:*387, 1970.

Keusch, G.T., and O'Connell, C.J.: The susceptibility of *Bacteroides* to the penicillins and cephalothin. *Am J Med Sci, 251:*428, 1966.

Marcoux, J.A., Zabransky, R.J., Washington, J.A., II, Wellman, W.E., and Martin, W.J.: *Bacteroides* bacteremia. *Minn Med, 53:*1169, 1970.

Martin, W.J., Gardner, M., and Washington, J.A., II: *In vitro* antimicrobial susceptibility of anaerobic bacteria isolated from clinical specimens. *Antimicrob Agents Chemother, 1:*148, 1972.

Medeiros, A.A.: "Once, all the world was anaerobic." (Editorial) *N Engl J Med, 287:*1041, 1972.

Thornton, G.F., and Cramer, J.A.: Antibiotic susceptibility of *Bacteroides* species. *Antimicrob Agents Chemother, 10:*509, 1970.

Washington, J.A., II: Comparison of two commercially available media for detection of bacteremia. *Appl Microbiol, 22:*604, 1971.

Washington, J.A., II: Evaluation of two commercially available media for detection of bacteremia. *Appl Microbiol, 23:*956, 1972.

Washington, J.A., II, and Martin, W.J.: Comparison of three blood culture media for recovery of anaerobic bacteria. *Appl Microbiol, 25:*70, 1973.

Wilson, W.R., Martin, W.J., Wilkowske, C.J., and Washington, J.A., II: Anaerobic bacteremia. *Mayo Clin Proc, 47:*639, 1972.

CHAPTER XXXIX

A DISCUSSION OF SUSCEPTIBILITY TESTING

JOHN C. SHERRIS

THE ABSOLUTE RESULTS OF antibiotic susceptibility tests are method dependent and influenced by a considerable number of variables. This is true of both aerobic and anaerobic organisms although the more complex atmospheric and growth requirements of anaerobes add to the problem. Specifically, test results with both groups of bacteria may be influenced by differences in medium components, pH, inoculum concentration, duration of incubation, mean generation time (MGT) and endpoint criteria. The clinically important anaerobes pose somewhat greater difficulties than the most commonly tested aerobes because they have a wider range of MGT and may require CO_2 and more complex media, and because the conditions of incubation for tests on agar media lead to problems of humidity control and to potential atmospheric interaction between different cultures.

The procedures that have been recommended for testing anaerobes are the same in principle as those for aerobes and facultative organisms. They include macro broth dilution methods (Bodner *et al.*, 1972; Williams *et al.*, 1972; Zabransky *et al.*, 1972), micro broth dilution methods (Brier *et al.*, 1972), agar dilution methods (Martin *et al.*, 1972; Sutter *et al.*, 1972a; Thornton and Cramer, 1970; Zabransky *et al.*, 1972) and agar diffusion methods (Bodner *et al.*, 1972; Kwok *et al.*, 1972; Sapico *et al.*, 1972; Sutter *et al.*, 1972a; Sutter *et al.*, 1972b; Wilkins *et al.*, 1972; Zabransky *et al.*, 1972). There are, however, substantial differences between the procedures that have been described among each broad class. For example, many different media with varying supplementation have been used, inocula have varied, and different endpoint criteria have been proposed. Based on comparisons between dilution and diffusion procedures, tentative recommendations for categorizing organisms as "sensitive," "resistant," and into an intermediate category have been made, but with a number of differences in the selected MIC breakpoints by different workers. In some cases different media have been used in comparing agar dilution and diffusion procedures, and this may account for the relatively poor correlation that has sometimes been found with certain antibiotics.

The diversity of procedures recently reported for susceptibility testing of anaerobes reflects a rapid assembling of information about this important group by a number of groups of workers, and mimics to some extent the situation that has existed in the past with aerobes, although at a very much higher level of sophistication and with a greater recognition of the inherent problems of susceptibility testing. Recognition of the problem of method and medium dependencies of sensitivity test results with aerobic and facultative organisms led, belatedly, to moves towards methodological standardization and to recommendations for the acceptance of reference and routine procedures for testing rapidly growing aerobic and facultative organisms (Ericsson and Sherris, 1971). This is exemplified in the recent publication in the Federal Register (1972) of a modification of the Kirby-Bauer (Bauer *et al.*, 1966) procedure that will be used in assessing diffusion test data with new agents and which will be included in disc package inserts together with interpretative recommendations. This should lead to improvements in interconvertibility of results, but, unfortunately, there has been no comparable acceptance of particular dilution test procedures although recommendations from an international collaborative study have made proposals for reference methods. (Ericsson and Sherris, 1971).

With the effort and skill now being put into the susceptibility testing of anerobic organisms, there is a particularly good opportunity to avoid the confusion that bedeviled susceptibility testing of aerobic and facultative organisms in the past. It would be highly desirable to attempt to reach agreement on agar and broth dilution procedures that could serve as both routine methods and as reference procedures against which others, including diffusion procedures, could be compared. These could be interim reference methods which should be described in complete detail. They would certainly differ in media and a number of methodological details from the proposed reference methods for aerobic organisms, but in some respects they could probably follow similar protocols in regard to the specific antibiotic dilutions to be used and so forth. The ICS (Ericsson and Sherris, 1971) recommendation that Log_2 series be used which includes 1 μg (or unit)/ml appears reasonable in this regard. It would be necessary to establish the behavior of two or more "standard" strains which could be made available to control procedures and facilitate interconvertibility of results. Those used for aerobic testing might suffice, but it is more probable that anaerobic organisms should be selected because of the critical nature of suitable anaerobic conditions.

The media and precise technical details to be selected for reference procedures can be based on a number of general principles (Federal Register, 1972; Sherris, 1973; Schoenknecht and Sherris, 1972) to ensure reproducibility and the probability that the results will have clinical relevance. The medium should give good growth support for commonly encountered pathogenic anaerobes. It should have good batch to batch reproducibility and preferably be defined at least as to production details. It should not be subject to wide pH shifts due to bacterial action on fermentable carbohydrates. It should be as free as possible from inhibitors of commonly used antibiotics. The inoculum should be large enough that (1) high mutation rates to resistance are detected and (2) the inoculum is not obviously smaller than the populations anticipated in significant clinical lesions. Temperature and pH should be physiological, although it must be recognized that CO_2 incubation will shift the surface pH on agar plates toward the acid side. Beyond this, decisions on technical factors are arbitrary within quite wide limits, and their selection has to be a best judgment decision which may need to be modified subsequently in the light of clinical experience as to the significance of the results. Special account will have to be taken of the considerable influence of incubation in 5 to 10 percent CO_2 on the activities of a number of antibiotics (Ericsson and Sherris, 1971; Ingham *et al.*, 1970; Rosenblatt *et al.*, 1972). Among potentially clinically important agents for anerobic infections, this is particularly marked with the macrolides and with lincomycin, which show reduced activity under these conditions. The effect appears to be mainly the result of surface pH changes, and also influences other antibiotics whose activity is pH dependent over the range of about pH 6.0 to 7.6. Results in the presence of CO_2 will therefore not be comparable to those obtained with aerobes in the absence of CO_2, and this will have to be taken into account in interpreting MIC data.

Several groups of workers have been developing agar diffusion procedures (Bodner *et al.*, 1972; Kwok *et al.*, 1972; Wilkins *et al.*, 1972; Zabransky *et al.*, 1972) and they differ in a number of technical parameters. Those workers who have compared agar dilution and diffusion methods using the same media have generally found good correlation between log MIC and zone size for particular species. Bodner *et al.* (1972) reported less satisfactory correlations, but were using different media for the two procedures. The wide range of growth rates for clinically important anaerobes results in rather large species differences between the MIC zone size correlative curves (Sapico *et al.*, 1972; Sutter *et al.*, 1972a) and makes it very improbable that a single set of criteria for distinguishing "sensitive," "resistant," and intermediate categories can be achieved for all species. Recommendations for agar diffusion test interpretations will therefore have to specify to which species they apply. Beyond this, agar diffusion methods appear to be reasonably precise and simple, but, once again, it would be highly desirable for agreement to be reached between workers in the field as to a recommended routine procedure. There would be great advantage if they would be compatible as to disc contents at least with the method described in the Federal Register for rapidly growing aerobes and facultative organisms.

Results of diffusion tests have usually been reported in qualitative terms although they can be extrapolated to MIC. This remains a useful approach, certainly pending agreement on MIC procedures. The selection of "breakpoints" involves consideration of the relation of MIC to blood or tissue level data, the distribution of susceptibilities of individual strains within particular species, and the results of clinical experience or trials. Based on these data, best judgment decisions for appropriate breakpoints can be made. The MIC levels of breakpoints so far suggested by different authors vary to some extent and their interconversion is impossible because of different MIC procedures and specific concentrations in the dilution series (Bodner *et al.*, 1966; Kwok *et al.*, 1972; Sutter *et al.*, 1972b; Wilkins *et al.*, 1972). Again, interim agreement with continued review is essential for interconvertibility of results of different laboratories and as a baseline for further work. As with dilution procedures, there is a clear need for the selection of "standard" strains of established performance for control of the test system.

The question can be posed as to the need for susceptibility testing of anaerobes in day to day clinical laboratory practice. Clearly it is essential that adequate data be developed to determine the range of susceptibilities of particular species and to monitor them repeatedly for evidence of the emergence of resistant strains under the selective pressure of chemotherapeutics. This certainly occurred with resistance to tetracycline of *Bacteroides fragilis* (Sutter *et al.*, 1972a) and *Clostridium perfringens* (Sapico *et al.*, 1972). Where such resistance has occurred to clinically important antibiotics, testing in the clinical laboratory may be indicated in many anaerobic infections. Furthermore, because speciation may take longer than determining an antibiogram, testing may be essential in some situations for determining therapy. Thus it seems necessary that clinical laboratories develop the ability to make such tests accurately on anaerobic bacteria. It is, however, unlikely that anaerobic testing will be more than a small proportion of all antibiotic susceptibility testing performed. Although anaerobes can frequently be isolated from clinical material, the proportion of them that play a significant pathogenic role requiring chemotherapy appears to be substantially less than is the case with aerobes and facultative anaerobes. Thus, the choice of methods to be used can be less influenced by considerations of "mass production" and be more selective both in terms of procedures and chemotherapeutics tested. As with aerobic and facultative organisms, unless there is some very specific reason to do so, it is critical to avoid testing anaerobes isolated from the normal flora per se or normal flora contaminating clinical specimens. This is costly in time and personnel and results may be misleading.

The emphasis of this discussion has been on the continuing need for agreement on techniques and their standardization. Susceptibility testing of anaerobes offers a unique opportunity of achieving this, because a relatively small group of workers have been advancing the field, and because their work has been based on sound principles and on understanding of the problems involved. Clearcut procedural recommendations from this group would be of immense value to those responsible for the work in other clinical laboratories.

REFERENCES

Bauer, A.W., Kirby, W.M.M., Sherris, J.C., and Turck, M.: Antibiotic susceptibility testing by a standardized single disc method. *Am J Clin Pathol, 45:*493, 1966.

Bodner, S.J., Koenig, M.G., Treanor, L.L., and Goodman, J.S.: Antibiotic susceptibility testing of *Bacteroides. Antimicrob Agents Chemother, 2:*57, 1972.

Brier, G., Wolny, J., and Griffith, R.S.: A micro-dilution technic for the susceptibility testing of anaerobic bacteria. Abstr. #129, p. 67. Paper presented at 12th Interscience Conf. on Antimicrobial Agents and Chemotherapy. Atlantic City, Sept. 26–29, 1972.

Ericsson, H.M., and Sherris, J.C.: Antibiotic sensitivity testing. Report of an International Collaborative Study. *Acta Pathol Microbiol Scand* [*B*], *Suppl 217:*1, 1971.

Food and Drug Administration, Dept. of HEW: Rules and regulations. Antibiotic susceptibility discs. *Fed Reg, 37(191):*20525, 1972.

Ingham, H.R., Selkon, J.B., Codd, A.A., and Hale, J.H.: A study *in vitro* of the sensitivity to antibiotics of *Bacteroides fragilis. J Clin Pathol, 21:*432, 1968.

Ingham, H.R., Selkon, J.B., Codd, A.A., and Hale, J.H.: The effect of carbon dioxide on the sensitivity of *Bacteroides fragilis* to certain antibiotics *in vitro*. *J Clin Pathol, 23:*254, 1970.

Kwok, Y.Y., Sutter, V.L., and Finegold, S.M.: Standardized antimicrobial disc susceptibility testing of *Fusobacterium*. Abstr. #M91, p. 95. Paper presented at 12th Interscience Conf. on Antimicrobial Agents and Chemotherapy. Atlantic City, Sept. 26–29, 1972.

Martin, W.J., Gardner, M., and Washington, J.A.: *In vitro* antimicrobial susceptibility of anaerobic bacteria isolated from clinical specimens. *Antimicrob Agents Chemother, 1:*148, 1972.

Rosenblatt, J.E., and Schoenknecht, F.: Effect of several components of anaerobic incubation on antibiotic susceptibility test results. *Antimicrob Agents Chemother, 1:*433, 1972.

Sapico, F.L., Kwok, Y.Y., Sutter, V.L., and Finegold, S.M.: Standardized antimicrobial disc susceptibility testing of anaerobic bacteria: *In vitro* susceptibility of *Clostridium perfringens* to nine antibiotics. *Antimicrob Agents Chemother, 2:*320, 1972.

Sherris, J.C.: General considerations of *in vitro* antibiotic susceptibility testing—a summation. In: Balows, A. (Ed.): *Current Techniques for Antibiotic Susceptibility Testing*. Springfield, Thomas, 1973.

Schoenknecht, F.D., and Sherris, J.C.: New perspectives in antibiotic susceptibility testing. In: Dyke, S.C. (Ed.): *Recent Advances in Clinical Pathology*. Series 6. Edinburgh, Churchhill Livingstone, 1973, pp. 275–292.

Sutter, V.L., Kwok, Y.Y., and Finegold, S.M.: Standardized antimicrobial disc susceptibility testing of anaerobic bacteria. I. Susceptibility of *Bacteroides fragilis* to tetracycline. *Appl Microbiol, 23:*268, 1972a.

Sutter, V.L., Kwok, Y.Y., and Finegold, S.M.: Standardized antimicrobial disc susceptibility testing of *Bacteroides fragilis*. Abstr. #M90, p. 95. Paper presented at 12th Interscience Conf. on Antimicrobial Agents and Chemotherapy. Atlantic City, Sept. 26–29, 1972b.

Thornton, G.F., and Cramer, J.A.: Antibiotic susceptibility of *Bacteroides* species. In: *Antimicrobial Agents and Chemotherapy—1970*. Ann Arbor, Amer. Soc. for Microbiol., 1971, pp. 509–513.

Wilkins, T.D., Holdeman, L.V., Abramson, I.J., and Moore, W.E.C.: Standardized single-disc method for antibiotic susceptibility testing of anaerobic bacteria. *Antimicrob Agents Chemother, 1:*451, 1972.

Zabransky, R.J., Johnston, J., and Hauser, K.: A diffusion test for antibiotic susceptibility testing of anaerobic bacteria. Abstr. #130, p. 68. Paper presented at 12th Interscience Conf. on Antimicrobial Agents and Chemotherapy. Atlantic City, Sept. 26–29, 1972.

CHAPTER XL

IN VITRO ANTIBIOTIC SUSCEPTIBILITY TESTING: OPEN DISCUSSION

DR. SONNENWIRTH. I would like to make a plea for testing antibiotic susceptibility of clinical isolates of anaerobes, if not routinely, at least whenever possible. Even in the absence of a standardized procedure valuable information can be obtained if one uses a proper disclaimer to the physician that a standardized procedure was not used. Around 1970, three different laboratories (Dr. Sutter's, Bodner and Goodman's, and ours in St. Louis) recognized that approximately 50 percent of *Bacteroides fragilis* strains had become resistant to tetracycline. These laboratories were using three different methods, only one of which (Dr. Sutter's) was being standardized. If the nonstandardized methods had not been used, it probably would have been 1 to 2 years later before this clinically important information would have been available.

DR. BODNER. I would like to ask members of the panel some questions raised by data on 70 strains of *Bacteroides fragilis*, work done by Dr. Goodman and myself when we were with Dr. Koenig at Vanderbilt University. This is the same slide presented by Dr. Goodman yesterday. The 70 strains were isolated from clinical materials and tested by methods described generally this morning. Both a single high potency disc diffusion method (a modification of the Bauer-Kirby-Sherris method) and an agar dilution method, using a modified Steers replicator, were used. Briefly:

1. As Dr. Goodman pointed out, greater than 90 percent of the strains were sensitive to clindamycin at clinically achievable levels.
2. There was a bimodal distribution of sensitivity to tetracycline, as Dr. Sonnenwirth and others have noted previously. The point I

stress is that the high potency disc method gave good results with tetracycline. Simple modification of the Kirby-Bauer method showed that tetracycline-sensitive strains indeed had large zones of inhibition and tetracycline-resistant strains had practically no zones of inhibition around the high concentration tetracycline disc.

3. More than 90 percent of the strains were inhibited by clinically achievable concentrations of lincomycin; however, the zone of inhibition was practically zero around the disc. I would like to request that Upjohn perhaps make available higher potency discs, at least for lincomycin. These could be helpful in the clinical laboratory for routine susceptibility testing. I think there are studies soon to be published on use of lincomycin for treatment of bacteroides infections.
4. My last comment is in regard to carbenicillin. The cumulative percentages showed that 90 percent of the *B. fragilis* strains were inhibited by carbenicillin at clinically achievable levels. Dr. Finegold asked previously about an effective drug to be used in treatment of patients with bacterial endocarditis. Would the panel members comment on the potential use of carbenicillin in addition to lincomycin and metronidazole for bacteroides infection?

DR. SHERRIS. Perhaps I can clarify one point in regard to lincomycin. The test for susceptibility is quite pH dependent, and the difficulty experienced with lincomycin discs is probably related to the CO_2 atmosphere needed for anaerobes. As I recall, with the aerobic system you can measure up to an MIC in the range of 6 to 10 μg/ml. It may not be possible to perform the diffusion test with a 2-μg lincomycin disc in a CO_2 atmosphere. The reported results are consistent with this view. It seems to me that someone else has shown much larger zones with the 2-μg lincomycin disc.

DR. WILKINS. We did not use lincomycin but a derivative of it, clindamycin, also in a 2-μg disc, and the regression plots were acceptable. You are faced with small differences when the diameter measured is 10 to 12 mm and the zone indicating resistance is about 15 mm. But, with the higher potency disc, if it is too high you may observe extremely large zone diameters. This is a problem with anaerobes because of the small (100-mm diameter) plates we usually use at the present time. Very large zones tend to coalesce and cause what I call "wipe-out."

DR. SUTTER. We used both 2- and 10-μg lincomycin discs

in our studies and saw very little difference between the two in the correlation. Of course, the 10-μg disc gave larger zones and there was some overlapping of the zone diameters in the intermediate range. We have also used 2- and 10-μg clindamycin discs and observed similar phenomena, but the data with clindamycin were somewhat better than with lincomycin discs. There is a recent FDA publication in the Federal Register which Dr. Sherris mentioned regarding package inserts and certification of antibiotic discs for clinical laboratories. In this new rule they are recommending that 2-μg clindamycin discs be used for testing susceptibility to either clindamycin or lincomycin, using 2 sets of standards for the interpretation of results. We have looked at some of our data with *B. fragilis* strains in an attempt to correlate the MIC of lincomycin with the zone diameter of clindamycin and the results are summarized in an extra slide which I brought. The results are self explanatory.

DR. THORNSBERRY. I would like to get away from lincomycin for a minute. Since I was the one that said you shouldn't use the Kirby-Bauer procedure for anaerobes, I would like to respond to Dr. Sonnenwirth and Dr. Bodner. I recognize the excellent correlation that Dr. Bodner obtained in testing bacteroides with tetracycline, also that Dr. Sonnenwirth came up with some similar information. However, I feel that you are being very restrictive, Alex [Dr. Sonnenwirth]. Good results with tetracycline could probably be obtained in most any laboratory but you must be careful with antibiotics other than tetracycline with the Kirby-Bauer procedure. Many people use the Kirby-Bauer interpretative chart incorrectly. It was not designed for use with all bacteria.

DR. KIRBY. Dr. Thornsberry, what is your suggestion?

DR. THORNSBERRY. Someone always has to ask! I feel that if you do use a disc procedure for anaerobes there should be a disclaimer on the report indicating that the tests were performed by a nonstandardized procedure. It would be better to perform a dilution test of some kind if practical.

DR. KIRBY. Dr. Sherris has proposed that we have an interim standardized method for disc testing and for dilution testing. Who would develop these, Dr. Sherris? After 25 years of disc testing the FDA has come out with this proposed recommended method that has been published in the Federal Register. This proposal was based on 25 years of work. Intensive work with the anaerobes has been in progress for only the last two years. It would be fine if we could telescope this, but actually, Dr. Sherris, are we really ready for recommended methods? Who is going to propose them? Also, who is going to adopt them?

DR. SHERRIS. As far as dilution procedures and the reference dilution technique is concerned, yes, I think we are ready. As to who will propose it, it seems to me that the other people at this table who have been working in the field could get together behind a method and use it as a basis for comparing other techniques. It would be very unfortunate if we have to wait 25 years and have the FDA do it.

The other point I would like to make is that except in some specific situation such as with tetracycline and *B. fragilis* it may well be that a disc diffusion technique is just not applicable. Dr. Wilkins indicated that he is working on a tube method, which I imagine uses a disc as a source of antibiotic, that will be convenient for a laboratory to use to perform a dilution procedure. There are automated methods which are being developed which use a disc in essence as a carrier for adding a particular concentration of antibiotic to one or two tubes. For routine testing, some modification of this technique may be better. It just has to be tested and evaluated. I do not feel that it will be necessary to wait a long time because the principle of selecting a reference method is widely accepted.

DR. JACKSON. As a clinician I would like to comment that Dr. Washington's conclusions and statements were music to my ears and I would like to reemphasize them. Also, I would like to emphasize the need for recognizing emerging resistance as Dr. Sonnenwirth indicated. With regard to methodology, until the relevance of *in vitro* results and clinical results are demonstrated, any method that gives reproducible results can be employed. In my opinion, people can use as wide a spectrum as their laboratory and interest permit, as long as the method is reproducible.

I now have some questions for Dr. Sutter and some of the others. Dr. Sutter, you have shown with *B. fragilis* and a single concentration disc that zone sizes of 15 to 30 mm were obtained with an MIC of 1.6. Does that represent strain variability? I ask this because Dr. Washington emphasized the predictability of changes in resistance within the same species. I would like to ask Dr. Washington the same question because he showed a large distribution curve. Has any one of you taken the same strain and repeated it 50 or 100 times so that we can know how much of this is strain to strain variability or whether the limits of the technique influence the reproducibility of the results?

Dr. Sutter. We have tested 2 *B. fragilis* strains 50 times each. We also include these strains in each day's run to serve as controls and determine the variability of the test. We don't get a great deal of variability with these strains and most of the antibiotics. The zones vary from 3 to 7 mm day to day. This does not affect the interpretation of susceptibility or resistance in most cases, as shown on my slides. However, one strain showed a great deal of variability with oleandomycin. This was a resistant strain and we had a zone diameter variation of 13 mm. In most instances we could call the strain resistant from the zone measurement but in some cases the zone was large enough to place it in the equivocal or intermediate range. The other strain we have used is much less variable and the susceptibility pattern does not change from one interpretative zone to another. The MIC of these same strains is checked each day and most of the time a particular MIC is obtained, but it may vary up or down by one dilution. This doesn't happen too often, not even half the time.

Dr. Wilkins. I would like to briefly comment on variations in zone diameters. We also performed 50 tests on the same strain with 2 different organisms and came to the same conclusion as Dr. Sutter. The zone diameter varied from 3 to 7 mm, usually about 5 mm, with the same organism on a day-to-day basis. In respect to *B. fragilis*, we have been looking at variations in the intermediate zone in order to detect problems. These variations do not result in differences in interpretation of susceptibility but do indicate variability in the manner in which the tests are performed and read by different individuals. With very oxygen sensitive organisms, such as some of the cocci, the variation can be increased as the result of oxygen killing the organisms during the inoculation of plates and addition of the antibiotic discs. This is a problem with the diffusion methods we are currently working with and it is one reason I like the tube method which does not have this variable.

DR. WASHINGTON. We have had no experience with the disc diffusion method, so I can't comment on it, but from the standpoint of dilution testing I believe our experience is similar to that of Dr. Sutter's in respect to reproducibility. Dr. Jackson, your point was that if there is strain variation then routine sensitivity testing of isolates is an important consideration to the clinician and I agree. Dr. Wilkins, your data show great strain variations in all of the tests. Because your spectrum of antibiotics was very broad, we need that information if we are to be scientific in our testing. On the other hand the slide you just showed us was of cumulative distribution curves. With the exception of the tetracycline curves, it appears to me that they were very uniform and very steep curves, which suggests that there was *not* much strain variation. I feel that someone should answer the question as to how great the strain-to-strain variation is within a species after the technical artifacts are eliminated.

DR. WILKINS. It depends a lot on which antibiotic and which bacterial species you are talking about. With tetracycline there are two main groups in respect to susceptibility, one which is very susceptible (0.8–1.0 μg/ml) and another up at 12.5 μg and above. This is obviously a large difference in susceptibility. Even with penicillin a few strains of *B. fragilis* are inhibited by fairly low levels, around 2 units. In any of the species we are talking about there will be a spectrum of susceptibility, and whether you consider 90 percent susceptibility or 90 percent resistance there are always at least 5 to 10 percent which approach the line. Susceptibility tests are invaluable to the physician in the treatment of patients with infections due to these exceptional strains. In the case of tetracycline, if 40 percent of the *B. fragilis* strains tested are susceptible, this means that it is possible to use tetracycline for treating 40 percent of patients with *B. fragilis* infections instead of excluding the use of tetracycline entirely. This remains to be discussed by other members of the panel. I believe it was Dr. Washington who said that there is very little strain variability. My point was that at any given concentration which can be achieved with currently recommended dosages of an antibiotic such as clindamycin there is a high level of predictability of susceptibility. The problem is what concentration to use as a break point in determining the difference between susceptible and resistant.

DR. PRESTON. I would like to go back to the effects of CO_2 on susceptibility testing in respect to the point that Dr. Washington made concerning a break point for interpretation of susceptibility on the basis of serum levels achievable clinically. A case in point is the effect of CO_2 on erythromycin, or lincomycin for that matter, *in vitro*. In the absence of CO_2 the MIC of erythromycin for *B. fragilis* may fall in the 2 to 10 μg range but when the strain is tested in the GasPak jar we may see an MIC of 12.5 to 25.0 μg. This effect is very important and can lead to mistakes in interpretation if one simply relates achievable serum levels to an MIC determined by a particular method. We need clinical response data in order to determine what a MIC means when tested by a particular method.

DR. JACKSON. You are simply proposing that we need to correlate the clinical response with the various methods and determine which does work. Not necessarily what is the right MIC, but when an MIC is determined by some method, what does it mean?

DR. CHOW. This was the point I wanted to raise again for the panel. It was pointed out that not only the pharmacological level and population distribution of strains are important in arriving at break points but also the relation of the MIC (obtained by an agreed upon standardized methodology) to the clinical results. We have heard a couple of suggestions today which warrant further discussion on the problems in arriving at break points in relationship to clinical results. For example, this morning Dr. Gorbach mentioned that he had observed strains of bacteroides which were susceptible to chloramphenicol on the basis of *in vitro* results but the patients did not respond to chloramphenicol therapy. However, they did subsequently respond with another drug. If others confirm this experience, it seems that our break point interpretative standards may need to be reevaluated so that this type of data is taken into consideration. On the other hand, it has just been suggested that the cumulative distribution points indicate that penicillin G MICs are very similar, within 1 tube, to those for carbenicillin, and these also are levels which are achievable clinically. So another question that should be faced is whether large doses of penicillin G or carbenicillin are clinically effective in treatment of infections with these organisms.

DR. LAMBE. I would like to make a couple of comments. First, we have been using the Bauer-Kirby disc method to test the susceptibility of anaerobes for 4 years and the dilution method for the past 6 months. We found that the MIC values for tetracycline and some of the other antibiotics are essentially the same as those reported by Dr. Sutter. We also have found that the disc method is useful if the clinicians know its limitations. We always specify that the results are relative and that the method has not been standardized. The results with the disc technique and the MIC determinations do correlate well. Second, I would hate for everyone to leave the meeting with the impression that it takes 4, 5, or 6 weeks to identify anaerobes. Not all of them require this long. If *Bacteroides fragilis* is isolated from blood, pus, fluid or whatever in pure culture, it can be identified and the susceptibility test performed within 48 hours. A tube of prereduced chopped-meat glucose medium (CMG) and an aerobic blood agar plate are inoculated with the clinical specimen. If there is no growth on the aerobic plate at the end of 24 hours and there is growth in the CMG we will do a Gram stain, set up biochemical tests and perform the susceptibility test. At 48 hours we look at the original anaerobic plate which was inoculated with the specimen. If it is a pure culture and the biochemical reactions are compatible we can report the organism as *B. fragilis* along with its susceptibility pattern. Of course, the identification of many other anaerobes is delayed, but *B. fragilis,* if in pure culture is not that hard to identify.

DR. JACKSON. Dr. Lambe, you said that you have been obtaining good results with the disc method and you inform the physician of the limitations of the method. Actually, however, how does the physician know that the organism is sensitive to an antibiotic or not? He wants results he can use.

DR. LAMBE. As I said we use the interpretative criteria, knowing that it is not a standardized method, but the results are correlating well with MIC determinations now. In other words, a large zone size indicates susceptibility and the MIC is low. This is true with *B. fragilis* and tetracycline and also with *B. fragilis* and chloramphenicol. I feel that the results are meaningful. We have found that about 62 percent of our *B. fragilis* strains are resistant to tetracycline and these results agree with those obtained by Dr. Sutter.

DR. JACKSON. So you are finding the results of this testing to be useful clinically?

DR. LAMBE. Well, I hope so.

DR. JACKSON. At least it tells the physicians not to use penicillin for *B. fragilis.* Now on the other hand, Dr. Washington, who works in a large institution, questions whether one at present should charge patients for testing the antibiotic susceptibility of anaerobes for the reasons he raised. Dr. Washington, would you comment? We have a little difference of opinion here.

DR. WASHINGTON. My first point is that we do not have adequate break point information available at present to judge whether or not *in vitro* susceptibility is synonymous with favorable clinical response. The point raised by Dr. Gorbach this morning seems to substantiate this. My second point is that I don't dispute that one can perform susceptibility tests in a hurry, but in situations where we are reasonably certain of an anaerobic infection clinically, it still takes 2½ to 4 days to isolate the organism from the blood. Also, by the time the results of the susceptibility tests are available, the *in vivo* response should be well documented. Therefore, I have reservations about the significance of the results we are now generating. I have reservations about reporting the results of tests at a time when they are no longer useful, and also about passing the charge on to the patient.

Dr. Le Beau. There are two variables not mentioned in regard to the work with tube dilution and agar dilution techniques which could be important in regard to standardization. I refer specifically to shock brought about by contact with oxygen and temperature shock which can influence the number of viable organisms and initiation of growth. Therefore, I would like to ask Dr. Sutter and Dr. Washington, who both use these techniques, whether this work is being done in an anaerobic glove box, in the atmosphere, or is it being performed with dilutions in media at room temperature, ice box temperature, or incubator temperature at the time of inoculation?

Dr. Washington. Neither the agar dilution technique nor the broth dilution technique we use is performed in an anaerobic chamber and the media are at room temperature when inoculated.

Dr. Sutter. Our conditions are probably about the same. We use freshly boiled and cooled broth to dilute the inoculum with and the blood agar plates are inoculated at room temperature, as they have not been refrigerated. Ordinarily, we only perform tests on 10 to 12 strains at a time to minimize exposure to air after inoculation. I know that it can influence the results if the plates are left out in the air too long after inoculation.

Dr. Sonnenwirth. I have a question for the panel or anyone in the audience in regard to the increased interest in clindamycin and its use with anaerobic infections. We have been screening with a modified Kirby-Bauer procedure that has been mentioned several times today. On the basis of the results from testing about 400 strains of *B. fragilis*, our monthly variations for sensitivity range from 95 to 100 percent. In other words, of the 400 strains tested during the last year, 95 to 100 percent seem to be sensitive to clindamycin. My question is—does anyone have further information on the susceptibility or resistance of *B. fragilis* to clindamycin? We have also been testing *C. perfringens* and I believe that Dr. Sutter can tell us more about that. The results with this type of modified Kirby-Bauer technique in testing the sensitivity of *C. perfringens* to clindamycin are absolutely not reliable. Vera, you have observed apparent resistance with the disc technique which was not borne out by the agar dilution results. Would you like to say something about it?

DR. SUTTER. The zone around the clindamycin disc, in general, is much smaller with *C. perfringens* than are the zones with *B. fragilis* or fusobacteria and I believe this was alluded to by someone else today. This probably is due to the fact tht *C. perfringens* grows much more rapidly than the other organisms. It may be necessary to develop two sets of standards.

DR. BORNSTEIN. I was just struck today by the logic of attempting to develop a simplified tube method for performing dilution susceptibility tests. Many of the problems we now have in susceptibility testing are related to the use of discs, particularly in correlating the size of a zone with susceptibility. This is one of the greatest "fudge factors" which can foul up results. The disc procedure is quite good for routine studies of aerobic bacteria, when the laboratory receives a sample one day and by the next day the susceptibility tests can be set up. It also lends itself to convenient processing of huge numbers of samples. If, as Dr. Washington and others have pointed out, the number of clinically important isolates of anaerobes which require antibiotic susceptibility tests is limited in small laboratories, it seems that trying to develop a disc procedure for anaerobes similar to that for aerobes may be the wrong direction to go. The logic of a simplified tube dilution technique which avoids the problem with CO_2 and artifacts caused by the rate of growth is very appealing to me. It is especially appealing if discs can be used for addition of the antibiotics to the tubes, which makes it very simple for a laboratory to perform the test. It seems to me that this technique should be considered in respect to a cooperative study for developing standardized methods. Dr. Wilkins, in essence, isn't this the method that you are proposing?

DR. WILKINS. Yes, I believe so. It basically makes use of

a technique for cultivation of anaerobes in a closed tube system developed by Dr. Hungate and further perfected by Dr. Moore, using prereduced anaerobically sterilized medium. Anaerobes grow very rapidly in this environment and the sensitivity results can be obtained rapidly. Either nitrogen or carbon dioxide can be used for the gas phase, which allows correction of problems with pH, and the procedure can be performed at any desired pH as was previously mentioned. So far we have had very good results with the technique. I feel that it will be necessary to decide exactly what the size of the inoculum should be in order to establish cut-off points, as we do with conventional MIC determinations, in order to correlate the results with the clinical response and determine the efficacy of the technique. This is something we all need to discuss and I am glad that it was brought up in this discussion.

DR. SHERRIS. Dr. Wilkins, are you using a one tube system or a two tube system, or more?

DR. WILKINS. I am using one tube for each antibiotic and we are testing 5 major antibiotics which we consider useful for treatment of anaerobic infections. Of course, the procedure could be modified to use any number of tubes. This is something that could be gained from this discussion. There is also the question of how to interpret the results. Do we want to use such terms as "very resistant," "somewhat resistant," "very susceptible," *etc.* or do we want to just say "susceptible" or "resistant" on the basis of a certain break point? How many tubes to use will depend on this.

I would like to mention just one point in regard to a one tube system. It presents some problems in quality control in respect to knowing what the activity of the antibiotic disc actually is, simply because there is no spectrum of activity to measure against. This is one of the considerations which must also be faced and settled during the development of an automated machine.

May I cover a couple of other points? Dr. Jackson pointed out that the reproducibility of a method, particularly one that has a limited range of susceptibility, is very important. I totally agree with this, because if you know what the population distribution is, it is possible to correlate the results with clinical experience. Then, if you find organisms emerging which have antibiotic susceptibility values outside the predetermined population distribution, you have extremely valuable information. The reason I made the point about standardization is because it came up in another question in regard to the effects of CO_2 on the MIC for erythromycin, tetracycline, *etc.* Dr. Ingham's reports from England have shown that the MIC for erythromycin may vary 10 to 20 fold, depending on whether the procedure is performed in the presence or absence of CO_2. This is the type of consideration where I believe, even though we are developing and evaluating new techniques, it is important to exchange information between different laboratories. The exchange of information would be particularly useful if we can have an agreed upon reference method to allow comparison of results.

DR. GORBACH. Much of the discussion of the antibiotic susceptibility of anaerobes has been based on the concept of disease derived from classical infections with aerobic microorganisms such as the staphylococcus, where there is one microorganism producing one disease. Unfortunately, in the real world of anaerobes, we usually are dealing with mixed infections. Even if you isolate *B. fragilis* in pure culture from the blood of the patient you are deluding yourself if you think the abscess in the belly contains only *B. fragilis.* In fact, it may contain a half dozen of other microorganisms, either anaerobic or facultative. Why only *B. fragilis* is culturable from the blood in this example, I don't know. Thus, we concentrate on the one strain and ignore the very interesting data veterinarians have reported, that factor which has been alluded to several times during the conference, namely synergism. In the case of mixed infections we don't know which of these is the most important clinically. Certainly *B. fragilis* is easy to isolate and common, but is there another organism such as a diphtheroid, as in the animal models, that is somehow feeding the *B. fragilis* or allowing it to create the disease? I am troubled by just having the sensitivity test results on pure cultures and then trying to go back to the patient with the profusion of organisms and figure out which antibiotic(s) to give. Sooner or later we will have to cope with the problem of mixed infections and the clinical response based on improvement of the patient and not necessarily on the basis of the bacteria which can be isolated from the infection. I have the feeling that we are talking in an almost irrelevant manner here about pure cultures, single drugs, and *in vitro* laboratory testing.

DR. WASHINGTON. I would like to collaborate that comment. One thing I did not point out in respect to our blood culture data was that of the patients with anaerobic bacteremia, fully ⅓ of them had more than one organism in their blood. However, is that any reason for a negative attitude toward the development of methods for susceptibility testing? I feel that this is a good perspective but it is off the subject somewhat. We also need some good methods for testing the antibiotic susceptibility of pure cultures as we do get them.

DR. JACKSON. You made the point I was going to. For the record, obviously Dr. Gorbach is correct, as we have heard over and over again. It has been known for a long time in our institution that in the case of lung abscess the results of the antibiotic susceptibility tests on isolates from the abscess are not too useful, relative to the clinical results. In a "helper" type of infection, the elimination of any one of the components may sometimes be beneficial, at least theoretically; one has to start somewhere. I feel that to carry your logic to the end point would be complete anarchy. Therefore, it is my opinion that it is worthwhile to perform susceptibility tests. If you isolate a single strain from the blood and if you direct therapy at this strain and observe improvement, even though the bacteriologist isolated 5 microorganisms from another site, then we are led a little further down the road of meaningful experience. I do believe we must define each species' sensitivities and then observe the response in elimination of these species, knowing full well that there are many organisms there that may be interdependent on one another.

PART VI

CLINICAL RESULTS OF TREATMENT AND SUMMARIES

CHAPTER XLI

TREATMENT OF CLOSTRIDIAL INFECTIONS

W.H. Brummelkamp

ABSTRACT: *In the treatment of clostridial myositis (gas gangrene), the modes of therapy can be divided into primarily surgical therapy and primarily conservative therapy using hyperbaric oxygenation. Surgery to arrest the rapidly advancing progress of myonecrosis must be radical enough to insure that all affected tissue is excised; antitoxin and antibiotic may be used as adjunctive measures. In primarily conservative therapy, oxygen is administered under pressure in a hyperbaric chamber. If required, excisional procedures can be carried out later when the patient's condition has improved. Although the exact mechanism of action of oxygen at high pressure remains to be defined, there is evidence that the rapid clinical improvement and abrupt halt of the progression of the necrotizing phlegmon are due to the inhibition of α-toxin production by the clostridia. Results in 130 cases of gas gangrene treated in the Amsterdam hyperbaric chamber and in other series have confirmed the value of this mode of therapy in this life-threatening infection.*

In tetanus, the mainstays of treatment continue to be nursing care, control of spasms by sedatives and muscle relaxants, administration of antitoxin and surgery of the wound to prevent further absorption of tetanus toxin, and general measures to prevent infection and maintain adequate pulmonary ventilation and electrolyte balance. Hyperbaric oxygenation does not appear to be of value in this disease.

This presentation is confined largely to the treatment of histotoxic gaseous clostridial infections and tetanus.

HISTOTOXIC GASEOUS CLOSTRIDIAL INFECTIONS

As suggested by MacLennan (1962), clostridial gaseous infections can be divided into traumatic infections, uterine infections, nontraumatic infections, and rarely occurring infections such as clostridial brain abscess and infections of the eye. The traumatic infections are subdivided into clostridial cellulitis and clostridial myositis, and the nontraumatic group into idiopathic gas gangrene and infected vascular gangrene. Myositis implies the presence of a progressive necrotizing clostridial phlegmon in the muscular tissues, and cellulitis, the presence of the same process in the subcutaneous tissues. This distinction is not entirely logical. Clostridial myositis without concomitant cellulitis is almost inconceivable, especially considering the damage to skin, subcutaneous tissue and muscle which is inflicted in the type of trauma which often precedes gaseous clostridial infection. Infection starting in muscle is certain to spread via the subcutaneous tissue. MacLennan (1962) justifies the differentiation between myositis and cellulitis by the fact that patients with myositis are often critically ill, while those with clostridial cellulitis are in far less danger. Although this line of reasoning is not entirely without grounds, it is incorrect in its absoluteness. Patients with iatrogenic clostridial infection show a very high mortality rate, although their condition must frequently be classified as cellulitis. I believe it is warrantable on surgical therapeutic grounds to distinguish clostridial myositis from a histotoxic clostridial infection restricted solely to the subcutaneous tissue. Clostridial myositis demands *excisional* therapy whereas *incisional* therapy is sufficient for clostridial cellulitis.

Gas Gangrene (Clostridial Myositis)

GENERAL CONSIDERATIONS. Various *Clostridium* species may be associated with gas gangrene. The most important in respect to pathogenicity and frequency of recovery from gas gangrene are: *Clostridium perfringens* (*Clostridium welchii*), *Clostridium novyi* (*Bacillus oedematiens*), *Clostridium septicum* (*Vibrion septique*), *Clostridium sordellii* (*Bacillus oedematis sporogenes*). When the three last-mentioned species are the causative agents, they produce more edema and induration of the local lesion and less gas is formed than in *C. perfringens* gas gangrene. *C. perfringens*, *C. novyi* and *C. septicum* are more frequently associated with gas gangrene than *C. sordellii*.

The clostridia produce a variety of toxins which play a major role in the initiation of clostridial myositis. Of the numerous toxins produced by *C. perfringens*, the α-toxin, a lecithinase, is the most important. Intramuscular injection of *C. perfrin-*

gens α-toxin produces all of the histological and systemic manifestations of clostridial myositis, including hemolytic anemia and hemoglobinuria. From histological studies on muscle tissue that had been exposed to different toxins of *C. perfringens*, Aikat and Dible (1956) concluded that α-toxin weakens the muscle cell membrane by acting on the lipoprotein complexes, and that it may act on the stromal lipoprotein or directly on the myosin. *C. perfringens* hyaluronidase (μ-antigen) caused dissolution of the endomysial collagenous connective tissue, with wide separation of the endomysium from the sarcolemma. For more extensive information on clostridial toxins, their mode of action and toxicological typing, see Oakley (1954); Prévot (1955); MacLennan (1962) and Willis (1969).

Clostridial gas gangrene is frequently a polymicrobic infection in which several clostridia may play a causative role at the same time (Zeissler *et al.*, 1958; Willis, 1969). However, some clostridia such as *C. tertium* and *C. butyricum* may be present merely as contaminants.

The incubation period for clostridial myositis may vary from 18 hours to 6 days (Table XLI-I). In the majority of cases the diagnosis is established approximately 48 hours after trauma has occurred. The local lesion differs from that in the usual type of pyogenic infection. In an ordinary pyogenic infection, the skin shows an erythematous discoloration in the initial phase and later pus forms in the lesion. In gas gangrene erythema is not the most conspicuous finding; rather, signs of circulatory inadequacy are present, due to the rapid formation of edema and gas. Even in the initial phase the extremity is markedly swollen and the skin is glossy,

TABLE XLI-I

SPECIES-RELATED CLINICAL VARIATIONS IN CLOSTRIDIAL MYOSITIS

Criterion	C. perfringens	C. novyi	C. septicum
Incubation time	18–24 hours	3–6 days	1–3 days
Skin reactions	Bronzing	Pallor	Erythema
Gas	++	++	++
Rapidity of course	++	– or +	++++

taut and pale. The wound does not discharge pus but a brownish watery exudate. Gas in the tissues is so finely distributed that palpation usually fails to detect it, although x-ray may occasionally do so.

During the next phase the skin becomes discolored, the lesion taking on a reddish-brown ("bronze") tint. The discoloration progresses rapidly in a cranial direction; the color is darkest around the lesion. At the same time, complete brownish-black cutaneous necrosis is observed at the wound edges, and this process usually indicates a deeper-seated myonecrosis. The tissues are under high tension, e.g. tension in the lower leg can be so high that the foot, otherwise not affected, presents a pale, edematous appearance. Progression is now extremely rapid; the bronze-colored marginal area of the phlegmon advances at a rate which may be marked by the hour. We observed a patient in whom a phlegmon spread over the upper arm within ¾ hour. Movement of the affected extremity will provoke the rapid spread. However, the time required to make large plaster bandages for immobilization during transport is not warranted by the advantages gained, especially when conservative therapy is anticipated.

During this phase the "gas boundary" may advance ahead of the causative lesion. This gas is already palpable inside the tautly stretched but still normally colored skin areas. Apparently gas is being produced in such massive amounts that part of it is thrust ahead into the adjacent subcutaneous interstices in the form of large palpable bubbles. Obviously this tends to increase the local anoxia, so that the necrotizing phlegmon follows all the more quickly in its wake. On x-ray films, gas can be seen in the muscle tissues as a feather-like shadow, its formation being dependent on the existing tissue structure. At a later stage the skin assumes a darkish brown color which is accompanied by the formation of dark-tinted bullae filled with a fairly clear reddish-brown fluid. The lesion exudes a nauseating sweet, putrescent odor. Black necrosis of the skin spreads and the oldest necrotic portions assume an alarming greenish color. Death supervenes quickly at this stage if no action is taken.

In gas gangrene not caused by *C. perfringens* type A, the

formation of gas may be a less prominent feature. This variation is frequently recognized too late or not at all—which indicates the inadequacy of crepitation due to gas in establishing an early diagnosis.

An incision made into the analgesic swollen area of discolored skin passes through pale yellowish and edematous subcutaneous tissue. This tissue is firmer and more detached from the muscular fascia than usual, forming a "plaque" on top of the fascia. No blood is lost and the veins above the fascia are often thrombosed. The fascia is tautly stretched and incision reveals necrotic muscle tissue. In advanced cases the muscle is dark brown or black and of pasty consistency, so that it can be easily removed with a spoon. Proceeding proximally, the muscular structure again becomes recognizable but the muscle has a dull discoloration and fails to react to mechanical stimuli. More proximally the discolored muscles are again observed to be more glossy and the muscle fibers do react slightly to mechanical stimuli. Inside this transitional area there is so much swelling that when the fascia is incised the muscle at once protrudes far through the opening. This does not occur in the necrotic pasty area.

OTHER SYMPTOMS AND SIGNS. Pain is a prominent symptom. It is usually extremely violent, probably due to the rapid occurrence of swelling in the tissues due to edema and gas formation. We have observed little difference between patients with clostridial myositis and those in whom the process remained restricted to cellulitis.

Hypotension is a poor prognostic sign in very serious degrees of toxicity. Temperature usually shows a rapid initial rise (to 39–40°C in a single day), but may drop spontaneously half a day later, tending to confuse diagnosis. As the lesion progresses rapidly, the temperature rises again rapidly and there is evidence of toxicity.

According to the literature, a very rapid and feeble pulse is characteristic of clostridial myositis. However, this is not a constant sign and not invariably a reliable measure of the severity of the condition. The very low hemoglobin level found so frequently in patients with gas gangrene also tends to influence the pulse rate, as do hypovolemia due to loss of fluids into the markedly edematous tissues, a high temperature, and a low fluid intake. However, after allowance has been made for these factors, a dissociation between temperature and pulse rate is definitely symptomatic of a critical condition. Thus, one may well be seriously concerned about a young man with extensive gas gangrene of an extremity who has a pulse rate of 156 per minute and runs a temperature of 38.6°C, with a hemoglobin content of 6.5 mmol/liter and a hematocrit of 37 percent. While this patient experiences little pain and has an adequate fluid intake and urinary output, which seems reassuring enough, he is gravely ill, and one notes that he keeps looking around without really observing his surroundings. Questions elicit inadequate answers and he appears disoriented in time and place, and may lapse into coma and convulsions.

A psychiatric-neurologic complex of symptoms is one of the toxic manifestations of gas gangrene. One has the impression that the degree of toxicity constitutes a reasonably good measure of the severity of the condition, notably of the activity of the gangrenous process. The degree of toxicity is related to the temperature, which is usually high, but inconstantly related to pulse rate. The toxic effect is generally considered to be directly associated with the presence of large quantities of necrotic tissue (Cooke *et al.*, 1945; MacLennan, 1962), and in fact patients do show rapid improvement following amputation of the affected extremity. However, recent experience with conservative oxygen therapy in a hyperbaric chamber suggests that signs and symptoms of toxicity also clear up rapidly with this therapy, even though all necrotic tissue remains *in situ*. Thus, it is more probable that the toxicity should be ascribed to the presence of bacterial toxins. These considerations do not diminish the importance of recognizing toxic symptoms, for discerning their cause is an incitement to lose no further time in starting therapy.

BACTERIOLOGIC CONSIDERATIONS. When treating a patient with gas gangrene, the physician must consider that the time still available for therapeutic intervention is counted in hours, so there is no time to wait for bacteriological confirmation of the clinical diagnosis. However, there is much value in direct microscopic examination of gram-stained films prepared from wound exudate or fluid obtained by puncturing the phlegmonous area. The clostridia responsible for gas gangrene are large, plump gram-positive rods that rarely form spores in tissue. Strikingly few leukocytes are observed in the film. Numerous workers have developed rapid methods to provide greater bacteriological accuracy (see Willis, 1969, pp. 306–308), and this is a great aid to the surgeon who is wavering between the alternatives of an amputation or such a seriously mutilating operation as an exarticulation. With the adoption of conservative oxygen therapy there is little point in using these emergency diagnostic methods.

PREVENTION. The most important simple measure to prevent clostridial wound infection is debridement, which is frequently carried out indifferently (Kennedy, 1955). The objectives are (1) to remove all devitalized tissue or that which is in jeopardy; (2) to open the wound widely for inspection of all recesses; (3) to remove accessible foreign material; and (4) to control bleeding, since collections of blood and serum act exactly as does avascular tissue (Hoover, 1959). The extent of prophylactic surgical excision must of course be tempered by a regard for the future function of the anatomical region concerned, and this may explain in part the difference in incidence of gas gangrene in different anatomical sites (MacLennan, 1943; Willis, 1969).

The value of prophylactic serum administration is still debated, and the recommended dosages vary (MacLennan, 1962; Parish and Cannon, 1962). Active toxoid immunization has been reviewed by Oakley (1943, 1954) and MacLennan (1962). The results appear promising, but there are still not sufficient data to assess clinical efficiency.

THERAPY OF GAS GANGRENE

Therapy can be divided into two principal modes: primarily surgical and primarily conservative. Some surgical intervention may be required with either mode, and the characteristics of the subcutaneous tissue and necrotic muscle tissue during the rapid progression phase described above should be borne in mind. Minor surgical intervention does not involve loss of blood, but does involve fluid loss. After a very short time the operation sheets become wet with a brownish-yellow exudate which continuously oozes from the wounds. This fluid, which has been drawn from the circulation, may amount to a considerable loss. Decompression restores the blood supply and the lymph circulation to the affected extremity so that very soon renewed fluid losses occur. Therefore, the seemingly simple operation of decompression should be undertaken only when an intravenous infusion is in place and adequate supplies of plasma and blood are available. General anesthesia is not necessary but a well-running intravenous infusion is a primary requirement.

General supporting therapy in all methods of treatment should include blood and plasma transfusions, substitution for electrolyte losses, adequate immobilization of the affected extremity, oxygen inhalation via nasal catheter, etc. If anuria develops, dialysis by means of an artificial kidney has proven valuable.

Primarily Surgical Therapy

This mode consists of surgical, serum and antibiotic therapy. Surgical management, which is still generally considered to be the most important therapeutic measure, involves excision of all affected tissue until obviously healthy tissue, i.e. red, freely bleeding tissue, is exposed. Healthy muscle fibers should contract in reaction to mechanical stimuli. Irrigation of the wound assists in removing blood clot and fragments of necrotic tissue and foreign material; repair of arterial injuries reestablishes an adequate inflow to the part.

Not only must surgery be adequate, it must be undertaken as early as possible. The later it is undertaken in patients with an established infection, the greater the risk that the infection will terminate fatally. Gas gangrene occurs in the arm, thigh and leg in approximately 80 percent of cases, frequently leading to amputation in a guillotine fashion. In limited or localized gas gangrene, complete excision is recommended as a first phase, followed if necessary by higher amputation. Excisional therapy has yielded excellent results in localized gas gangrene, but this procedure is difficult because recognizable anatomical landmarks are absent in and around the necrotic tissue. Nerves and arteries have to be sacrificed, because half-measures have proven useless. Paralysis, poor circulation and instability of the affected extremity are frequent but unavoidable consequences of this intervention. Major excisions on the trunk are often not feasible technically and the prognosis is poor for patients with lesions of the trunk.

Clostridial cellulitis apparently does not necessitate complete excision of the wound area and an attempt may be made to cure this condition by making a number of incisions. These should be carried into the muscular fascia and appropriate drainage insured. In the past attempts were made to keep the wound open and apply oxygen locally. No deeply penetrating effects should be expected from application of local oxidizing agents.

Serum therapy is generally considered valuable. The aims are to prevent the patient succumbing from toxicity and to bridge a period during which the combination of surgical and antibiotic therapy can combat the infection. A suitable dosage recommended by various authors, e.g. Altemeier *et al.* (1957), is intravenous administration every 4 hours of 27,000 IU *C. perfringens*, 27,000 IU *C. novyi* and 13,500 IU *C. septicum* antitoxin. This may be given by continuous drip infusion. The antitoxin should be given before surgery and for 24 to 48 hours after the operation, for a total dosage of 150,000 to 400,000 IU polyvalent antiserum. The British Medical Council recommends an initial IV dose of 45,000 IU of polyvalent antiserum containing 18,000 IU *C. perfringens*, 18,000 IU *C. novyi* and 9,000 IU *C. septicum* antitoxin. Subsequent medication is given by means of an IV drip. A total dose of 150,000 IU has rarely been thought necessary (MacLennan, 1962).

Willis (1969, p. 311–312) has mentioned the extensive literature on chemotherapy and antibiotic sensitivities of clostridia. Sulfonamides, tested during and shortly after World War II, are most active against *C. perfringens*, less active against *C. septicum* and relatively ineffective against *C. novyi*. Dosages of 8×10^6 IU penicillin per day (Altemeier and Furste, 1947) or 500 mg oxytetracycline (Terramycin) IV every 4 to 6 hours are effective against the clostridia, and penicillin is probably the drug of choice (Johnstone *et al.*, 1968). Chloramphenicol is of little value and streptomycin and neomycin are ineffective. Neither chemotherapy nor antibiotic therapy alone is of any value; with this mode, radical excisional therapy is imperative in the treatment of gas gangrene.

Primarily Conservative Therapy (Oxygen Therapy)

It seems logical to treat infections caused by anaerobic bacteria with oxygen because, by definition, these microorganisms cannot grow in the presence of atmospheric oxygen, because oxygen is lethal to them. This statement is merely a play of words, because oxygen therapy involves a phenomenon which is not yet fully understood; no exact explanation can be given for the mode of action.

The partial oxygen pressure compatible with growth differs with different anaerobes (Table XLI-II). The varying oxygen tolerance may be based on the sensitivity of the organisms' enzyme systems to changes in oxidation-reduction potential, and on the degree to which they produce metabolic substances capable of reducing hydrogen peroxide.

LOCAL APPLICATION OF OXYGEN. Hinton (1947) stated that use of oxygen therapy in gas gangrene dates back to ancient Greece, but we have not been able to corroborate this. Numerous ingenious methods of introducing oxygen into the affected

TABLE XLI-II

TOLERANCE OF VARIOUS ANAEROBIC MICROORGANISMS TO OXYGEN*

	mm Hg Pressure of Oxygen Permitting:		
Organism	*Free Growth*	*Scanty Growth*	*No Growth*
C. novyi	0–1	1–2	2–3
C. botulinum	0–1	1–2	2–3
C. tetani	0–1	1–2	2–4
C. sporogenes	0–2	4	8–13
Anaerobic streptococci	0–2	2–10	13–15
C. perfringens	0–30	30–80	90

* According to Gordon *et al.*, 1953.

area have been tried. Locally active oxidizers were discussed by Spiro (1915) and Wesenberg and Hoffmann (1925); the latter warned against subcutaneous injections of hydrogen peroxide solutions because of the risk of fatal gas embolism. Hügel (1938) applied an "oxygen cushion" around a gas phlegmon; this produced a large oxygen emphysema. Löber (1938) stated that good results could be obtained by injecting hydrogen peroxide around the gaseous phlegmon, but other clinicians found that this produced fatal gas embolism (Graf, 1939; Rettig, 1939). Müller (1939) insufflated oxygen via needle around the clostridial phlegmon, allegedly with good results. However, authors such as Dormanns (1942) have commented that these incidental successes in war surgery are difficult to interpret because of lack of bacteriological verification of the diagnosis.

Peters (1939) used a regional oxygen tent and injected a circular serum barrier proximal to the phlegmon. Hinton (1947) advised creating an oxygen barrier proximal to the phlegmon by injection of gaseous oxygen into the intact tissues. Picco *et al.* (1950) used the same principle in two cases. Eckhoff (1930) irrigated excised wounds with hydrogen peroxide and flavine, while Meleney (1936, 1949) advocated application of activated zinc peroxide in aqueous suspension or as an ointment to the excised wounds.

Systemic Application of Oxygen. De Almeida and Pacheco (1941) attempted to circumvent the necessity of local application of oxidizers by saturating the body of small laboratory animals with oxygen by means of oxygen inhalation at increased atmospheric pressure. The entire body is drenched with oxygen in physical solution and the clostridial phlegmon is enclosed by oxygen on all sides, at least at the borderline of healthy and diseased tissue. Conditions are created for an optimal diffusion of oxygen into the phlegmon.

The rationale for this application of oxygen is based on the fact that oxygen is present in the body in two forms: (1) combined with hemoglobin as oxyhemoglobin and (2) in physical solution in plasma and tissue fluids. The former can only be increased to a small extent, because hemoglobin is already virtually saturated with oxygen under normal atmospheric conditions. However, the amount of physically dissolved oxygen can be increased considerably, because this amount is exclusively dependent on and directly related to the partial pressure of the oxygen in contact with the fluid (plasma). A rise in alveolar oxygen tension results in a rise of the oxygen tension of the blood.

Under atmospheric conditions, i.e. at sea level where the weight of the air is 14.7 lb per sq in (760 mm Hg) or one atmosphere absolute (1 ATA), the alveolar oxygen tension is 100 mm Hg, corresponding to 0.3 vol percent of oxygen in physical solution. Arterial saturation is 95 percent and mean venous oxygen saturation usually 65 percent. On inhalation of 100 percent oxygen at 1 ATA the alveolar oxygen tension (pO_2) following complete denitrogenation can theoretically rise to 673 mm Hg, i.e. 760 mm Hg minus 40 mm Hg, the partial pressure of carbon dioxide (pCO_2), and minus 47 mm Hg, the partial pressure of water vapor. At a pressure of 760 mm Hg, 2.3 vols oxygen dissolve in 100 ml plasma, so during inhalation of 100 percent oxygen at 1 ATA, 2 vol percent oxygen can dissolve in blood, i.e. $(673/760) \times 2.3 = 2.06$. Theoretically, 6.6 vol percent oxygen can dissolve in blood at 3 ATA, but such high values are not attained in practice. Possibly arteriovenous shunts and/or obstacles in the diffusion through the alveolar membrane play a role in this respect. Polarographic measurements in the Amsterdam hyperbaric chamber indicated that the arterial oxygen tension in man and dogs during oxygen inhalation at 3 ATA is approximately 1600 mm Hg, corresponding to 4.5 to 5 vol percent oxygen in solution (Schoemaker, 1964). When pulmonary factors are circumvented by use of extracorporeal circulation, the pO_2 can rise to approximately 2200 mm Hg. When oxygen is inhalated at 3 ATA, the oxygen tension in the lymph (thoracic duct) in dogs is 250 to 300 mm Hg (Schoemaker, personal communication). This may be an expression of the oxygen tension in the fluid surrounding tissue elements under these conditions.

ANIMAL EXPERIMENTS. Pacheco and Costa (1941) found that exposure to compressed oxygen accelerated the mortality of *C. perfringens* cultures in Tarozzi broth. De Almeida and

Pacheco (1941) were not able to demonstrate that exposure of guinea pigs to compressed oxygen after inoculation with clostridia protected the animals from experimental gas gangrene. A protective effêct was demonstrated by Kelley and Page (1963) in mice. Brummelkamp *et al.* (1961) found that hyperbaric oxygen drenching could prevent severe clostridial infection after *C. perfringens* inoculation in guinea pigs. Klopper *et al.* (1962) studied the influence of hyperbaric oxygenation on occurrence of clostridial infection of the liver following ligation of the hepatic branches of the hepatic artery in rabbits. Mortality among the treated animals was much lower than among the controls: 4 of 17 treated animals died, compared with 13 of 20 controls.

These experiments with hyperbaric oxygenation immediately following inoculation or ligation of the hepatic artery were of a preventive rather than a therapeutic nature. Therapeutic experiments in small animals have been hampered by signs of oxygen intoxication. The animals soon become dyspneic. Microscopic examination of their lungs shows congestion, edema, atelectasis, interstitial hemorrhages with rupture of alveolar walls, and sometimes hyaline membranes. These pulmonary changes after hyperbaric oxygenation are more frequent and more severe in animals inoculated with clostridia than in controls, perhaps because the metabolism of the inoculated animals is higher. Also, the pulmonary effects of hyperbaric oxygen in small animals seriously disturb the gas exchange so that the animals become hypoxic and the anaerobic infection is encouraged. These factors may account for some of the conflicting results in small animals.

In dogs, Karasewich *et al.* (1964) could not prevent clostridial infection after hepatic artery ligation by treatment with oxygen at 2 ATA, perhaps because of the low pressure. Van Unnik (1965) found that oxygen did not suppress α-toxin production at 2 ATA but did so at 3 ATA.

Clinical Series. Boerema and Brummelkamp (1960) used oxygen at 3 ATA in a last desperate effort to save the life of a patient in whom gas gangrene (myositis) in a lower extremity had progressed past the groin into the pelvic area and up to the thoracic wall. The infection was cured; the patient died two months later from uremia due to arteriosclerotic aortic occlusion. Between 1960 and 1972, 130 patients with gas gangrene were treated in the Amsterdam hyperbaric chamber (Roding *et al.*, 1972).

The Amsterdam unit was described by Boerema (1961) and Brummelkamp (1965a, pp. 53–61). The standard procedure for a patient with gas gangrene is as follows: (1) Sutures if present are removed but surgical manipulation is avoided. (2) Material is obtained from the wound or phlegmon for Gram staining, using an 18-gauge needle and closely fitting Luer-Lock syringe or disposable syringe. Anaerobic and aerobic cultures are made from wound fluid and blood. (3) Boundaries of skin discoloration and gas crepitation are marked. (4) A one-gram dose of chloral hydrate is given rectally to prevent cerebral effects of oxygen intoxication. (5) IV Rheomacrodex (dextran 40) is infused to combat possible shock and to prevent thrombosis. (6) The otorhinolaryngologist performs a myringotomy if: (a) the eardrum is red, retracted or bulging; (b) the patient is under general anesthesia; (c) the patient is too ill, too old or too young to swallow adequately; or (d) the patient develops an earache during compression. (7) The patient and the doctor accompanying him enter the hyperbaric chamber. Standard supplies include surgical gloves; two 1-g vials of chloral hydrate with a syringe and rectal canula; and a Levine tube (which may be used to prevent oxygen swallowing). The atmospheric pressure is raised within 10 minutes to 3 ATA, the chamber is ventilated with a flow of 4000 liter/minute and the air temperature is stabilized at 24°C. Oxygen is administered to the patient through a mask at a flow rate of 10 to 15 liters/minute for 90 minutes, after which decompression is carried out over a 35-minute period (Brummelkamp *et al.*, 1963). Total therapy consists of seven sessions over three days, separated by intervals of at least four hours.

The clinical results obtained by our group with hyperbaric oxygen therapy in clostridial infections have been reported extensively (Boerema and Brummelkamp, 1960, 1962; Brummelkamp, 1961, 1962, 1964a, 1965a,b,c,d, 1966; Brummelkamp and Heins, 1963; Brummelkamp *et al.*, 1961, 1963) and summarized by Roding *et al.* (1972). Of 130 patients with gas gangrene, 70 percent developed the infection after traffic or industrial accidents, 64 after compound fractures. In 31 cases the infection developed after operation for gangrenous lesions of the legs, bowel surgery, *etc*, while in 9, various other diseases were involved. The most common cause was the traffic accident, especially in young male drivers of bicycles and motorbikes. Twenty-nine (22.3%) of the 130 patients died: 12.2 percent of 90 in the trauma group, 45.2 percent of 31 in the postoperative group, and 44.4 percent of 9 in the miscellaneous group. The main reason for the lower mortality rate in the trauma group was that the patients tended to be younger and more vigorous with a disease localized to the affected extremity, while in the postoperative group the patients tended to be older and have complex underlying factors, and the final cause of death might be secondary factors. Of the 29 deaths, 15 were due to causes other than gas gangrene. Fourteen died during manifest gas gangrene; their cultures were still positive at the time of death. We doubt, however, that a positive culture is necessarily indicative of the amount of activity of gas gangrene in patients treated with hyperbaric oxygenation. We have seen many patients clinically cured with live *C. perfringens* bacilli in their wounds (Brummelkamp, 1966). Only continuous active α-toxin production is indicative of manifest gas gangrene. The 14 patients died within 48 hours after treatment had been started or before the first five hyperbaric oxygen sessions had been finished. Some may have actually died from causes other than the gas gangrene, such as shock or consequences of underlying disease.

As stressed by Roding *et al.* (1972), no emergency amputations were performed in this series. This approach almost always results in a more economic operation when operative intervention is undertaken. Brummelkamp (1965a,b) has described in some detail several illustrative cases.

Although 32 of the 130 patients in the series had already received antiserum at the time of admission, no serum therapy was included in the treatment regimen when the patient was submitted to hyperbaric oxygenation. Patients with a mixed flora of clostridia and aerobic microorganisms received penicillin, 1 × 10^6 IU, and streptomycin, 1 g, daily. In cases of pure infection with clostridium, it has been shown that hyperbaric oxygenation alone is sufficient, e.g. see Brummelkamp (1965a, p. 73).

The most impressive feature of hyperbaric oxygen treatment of clostridial infection is the immediate response of the patient. Several comatose patients awakened within 24 hours and started to eat spontaneously. This seems paradoxical in view of the findings of others that oxygen under pressure exerted only a bacteriostatic effect on *C. perfringens* (Fredette, 1965) and did not neutralize cell-free exotoxin, either *in vitro* or *in vivo* (Nora *et al.*, 1969). Van Unnik (1965) found that oxygen at 3 ATA for 90 minutes had no immediate effect on the growth rate of an actively growing culture of *C. perfringens*, but caused an immediate cessation of α-toxin production if the oxygen tension in the broth culture was 250 mm Hg or higher. In patients inhaling oxygen at 3 ATA, we have measured an oxygen tension of approximately 250 to 300 mm Hg with a Beckman polarographic oxygen electrode inserted into an area of clostridial infection (Brummelkamp, 1965a,b). Since it is the α-toxin which determines the clinical fate of the patient, and hyperbaric oxygen does not affect the quantity of this toxin already present before the treatment is started, how can the dramatic improvement of these patients be explained?

Several investigators have maintained that α-toxin is rapidly "fixed" *in vivo*. MacLennan *et al.* (1945) reported that free toxin injected into muscle tissue of rabbits could no longer be demonstrated after 30 minutes. This suggests that within a short time after inhibition of α-toxin production by hyperbaric oxygenation, free toxin has virtually disappeared from the tissues and the hemolytic and necrotizing activity has ended. This could explain the rapid improvement in the general condition of these patients.

Van Unnik (1965) found that clostridia remain capable of producing α-toxin *in vitro* when returned to suitable anaerobic conditions. Our preliminary working hypothesis is that temporary arrest of α-toxin production during hyperbaric oxygenation at 3 ATA and simultaneous elimination of that toxin already present create a situation in which the clostridia cannot surround themselves with α-toxin in sufficient concentration to overcome local tissue defenses. When a second session of hyperbaric oxygenation is instituted several hours after the first, the suppressive influence on the required minimal α-toxin concentration becomes greater. Repeated suppression of toxin production leads to a change in the environment of the clostridia, so that the requirements for optimal clostridial activity are no longer met. The transiently increased oxidation-reduction potential, the arrest of the activity of proteolytic enzymes in the tissue and the consequent arrest of the release of amino acids in the lesion during three days of oxygen treatment may create a condition of the surrounding tissues unsuitable for functioning and multiplication of *C. perfringens*. This change in environment may persist even after hyperbaric oxygenation is discontinued.

This hypothesis would explain the rapid subjective improvement in the general condition of patients with clostridial infection treated with hyperbaric oxygenation. Perhaps the strongest argument for this hypothesis is the fact that no clostridial infection has been seen to progress at all once hyperbaric oxygen therapy at 3 ATA has been started. One must bear in mind that the environmental change in human infections is of an entirely different character from the readily reversible artificial change in broth media *in vitro*.

Clinical results by workers outside the Amsterdam group have confirmed the value of hyperbaric oxygen therapy in clostridial infections (Smith *et al.*, 1962; Van Zijl, 1964, 1966; Van Zijl and Maartens, 1963; Van Zijl *et al.*, 1966; Arnor *et al.*, 1966; Hitchcock *et al.*, 1967; Hanson *et al.*, 1966; Hanson and Slack, 1967; Maudsley, 1967; Maudsley and Colwill, 1966; Trippel *et al.*, 1967; Duff *et al.*, 1967; Bayliss and Cass, 1967; Bertoye *et al.*, 1968; McNally *et al.*, 1968; Pascale and Wallyn, 1968; Wolcott, 1969; Aufranc *et al.*, 1969; Eraklis *et al.*, 1969; Howell, 1969; Kiranow, 1969; Kucher and Riedel, 1969; Slack *et al.*, 1969; Beavis and Watt, 1970; Duff and McLean, 1970; Lulu and Rivera, 1970; Meijne, 1970; Tarbiat,

1970). Although several points regarding the mode of action remain to be elucidated, we can thankfully agree with Pascale and Wallyn (1968) who stated: "The use of OHP in the treatment of anaerobic infections has now become commonplace. The treatment of gas gangrene due to clostridial infection has been a giant step forward in the treatment of this disease as was penicillin in the treatment of pneumonia."

OXYGEN PHYSIOLOGY AND TOXICITY. The quantity of oxygen dissolved in arterial blood when oxygen is inhaled at 3 ATA can virtually cover oxygen requirements of the tissues (Boerema *et al.*, 1956, 1957a,b, 1959, 1960a,b). The increase in the quantity of oxygen combined with hemoglobin is small, for under normal conditions the hemoglobin is already virtually saturated with oxygen. Oxygen does not behave like an inert gas in the organism; the increased quantity of physically dissolved oxygen influences the physicobiochemical equilibrium in the blood and the equilibrium between blood and tissues. Oxygen combined with hemoglobin is not used until the oxygen tension in the blood has fallen below 100 mm Hg, so the venous blood is hardly desaturated, if at all. This has consequences on CO_2 transport, for reduced hemoglobin is important for its buffer action on the venous blood. Only a small part of the CO_2 formed by tissues is transported to the lungs in solution, for the greater part is buffered by the reduced hemoglobin. (The pCO_2 in venous blood exceeds that in arterial blood by only 5–6 mm Hg under normal conditions.) When oxygen is given under increased pressure, more CO_2 must be transported in solution because less reduced hemoglobin is available as a buffer in venous blood. Consequently, the pCO_2 in venous blood (and therefore in tissues) increases to levels of 50 to 55 mm Hg. This increase causes stimulation of the respiratory center and results in hyperventilation, which in turn counteracts part of the increase in pCO_2 in the tissues by lowering the arterial pCO_2. The respiratory quotient does not change under hyperbaric conditions. Therefore, the amount of CO_2 retained in tissues corresponds with the mild increase in pCO_2 in venous blood.

The importance of this slight disturbance in CO_2 transport is not so much in CO_2 retention in tissues as in the cerebral circulation, which is largely regulated by arterial pCO_2. The arterial pCO_2 is decreased to 30 to 40 mm Hg due to hyperventilation, and this decrease causes constriction of the cerebral vascular bed. Lambertson *et al.* (1953) found that the human cerebral circulation is reduced during oxygen inhalation by 5 percent at 1 ATA and by 25 percent at 3.5 ATA. Jacobson *et al.* (1963) reported that the cerebral blood flow in dogs was reduced by 12 percent upon the change from air to oxygen at 1 ATA, and 21 percent at 2 ATA.

It remains to be established whether constriction of the cerebral vessels is explained exclusively by hyperventilation or whether there is also a direct constrictive influence from the increased oxygen tension in the blood. Reduction in cerebral circulation has been considered a possible factor in neurological manifestation of oxygen toxicity. It seems more likely that the cerebral vasoconstriction observed during hyperbaric oxygenation is a homeostatic mechanism which maintains oxygen levels in the cerebral tissue within fairly close limits, thus mitigating the possible deleterious effects of hyperbaric oxygen on the central nervous system (Jacobson *et al.*, 1963).

In a systematic study on human volunteers, Behnke *et al.* (1935) found that the following periods of exposure to oxygen were safe: 4 hours at 1 ATA; 3 hours at 2 ATA; and 2 hours at 3 ATA. Extreme individual variability in sensitivity to oxygen intoxication was confirmed by Donald (1947). Among 36 normal subjects exposed to oxygen at 3.7 ATA, the maximum tolerated exposure times varied from 6 to 96 minutes. Toxic symptoms usually consisted of myoclonic jerks of the lips and very severe dysphoria. Other symptoms included facial pallor, nausea, vertigo, palpitations, depression, euphoria or lack of interest in the environment, anxiety, uncoordinated behavior and disturbed judgment. Auditory and visual hallucinations can occur at later stages, and concentric limitation of the visual field is considered fairly typical. Five of Donald's 36 patients ultimately had a convulsion, but no residual symptoms. Convulsions average 2 minutes in length and disappear almost immediately after inhalation of oxygen is stopped, but if the increased oxygen is maintained convulsions rapidly succeed each other and death may result.

At our brief exposure times of 90 minutes at 3 ATA we have observed oxygen toxicity only in patients with high fever. (In general, hyperthermia increases the activity of peripheral nerves and the CNS.) Cater *et al.* (1962) reported similar experience with test animals. The initial symptoms of oxygen intoxication observed in our patients have been pallor and hyperhidrosis, hyperpnea and salivation. They tossed their heads restlessly from one side to the other and plucked at the blankets. Subsequently they developed a staring gaze with dilated pupils while the head was usually turned to one side. These manifestations disappeared within a few minutes after oxygen administration was stopped, and subsequent oxygen administration was often tolerated without further difficulty. When the initial symptoms were not recognized in time, convulsions occurred.

DECOMPRESSION SICKNESS. According to our routine decompression schedule (Brummelkamp *et al.*, 1963), pressure is reduced in steps from 3 ATA (equivalent to 20 meters depth) to normal over a 35-minute period. The patient continues to inhale oxygen during decompression to minimize the risk of decompression sickness. This condition results from rapid reduction from hyperbaric pressures to 1 ATA or from normal pressure to levels below 1 ATA (high-altitude flight). Gas bubbles form directly in the tissues or in the blood stream and then are transported to various parts of the organism. Carbon dioxide and nitrogen are particularly significant in bubble formation. Decompression sickness can be prevented by oxygen inhalation before or during decompression; preoxygenation washes out nitrogen from the body.

In anaerobic infection, the main effect of body temperature on bubble formation seems to be related to the rate of CO_2 production in anaerobic glycolysis, for CO_2 facilitates nitrogen bubble formation. However, in a patient breathing oxygen, the nitrogen is washed out and this effect of CO_2 does not occur. We have not observed signs of decompression sickness in patients with anaerobic infections treated in the hyperbaric chamber.

TREATMENT OF TETANUS

In general, tetanus in man, which is caused by *Clostridium tetani*, can be divided into the mild type and the severe type. The incubation period is generally assumed to be an indication of the severity of the disease, and it bodes ill for the patient if the interval between inoculation and the first symptoms is less than 8 to 10 days. However, not infrequently, severe tetanus is observed after an apparently longer incubation period, for instance in cases in which the spread of toxin is delayed after the onset of infection (Eckman, 1960). The incubation period is shortened in actively immunized patients who develop tetanus, probably because they are victims of overwhelming infections (Boyd, 1946). Clinically, the incubation period has been distinguished from the period of onset. The latter refers to the time interval between the first symptoms, usually trismus, and the onset of spasms and opisthotonus. Both these intervals are of prognostic significance, for short incubation periods and short periods of onset are associated with a high mortality. The clinical features of tetanus have been reviewed by Willis (1969; pp. 413–417).

General Treatment Measures. The patient with tetanus has to be kept under very quiet conditions (noise, light) with careful observation day and night. Continuous registration of pulse rate, respiration rate, blood pressure, EKG and body temperature is imperative, and the intensity of tonic muscle spasms and/or frequency and intensity of clonic muscle spasms has to be noted. It might be unnecessary to stress the need for daily control of hemoglobin content, electrolytes and x-ray of the thorax.

Prevention of Absorption of Additional Toxin. Administration of antitoxin and surgery of the wound are aimed at prevention of further absorption of toxin. The value of therapeutic administration of antitoxin to patients who have already developed clinical tetanus is disputed (Vaishnara *et al.*, 1966). The poor results obtained in treating tetanus with antitoxin alone suggest that by the time tetanus is clinically evident, the neurotoxin is already fixed to the nervous system; see also Fulthorpe (1956). Experiments in rats (Kaeser and Saner, 1969) confirm that tetanus toxin has, in addition to its central action causing muscle spasms, a peripheral effect interfering with neuromuscular transmission. Probably the toxin inhibits release of acetylcholine from nerve terminals, but a postsynaptic action cannot be ruled out. In man the peripheral action of the toxin may be responsible for the so-called recruiting response in voluntary contraction, for circumscript muscle weakness in local tetanus, and for tetraplegia observed in a small percentage of patients with severe tetanus. There is little doubt that toxin present in the blood and other tissues is readily neutralized by antitoxin, and that production of toxin by the interfering organism is a continuous process, which can be interrupted only by removal or destruction of the tetanus bacilli, e.g. by excising the wound. In very severe cases huge doses of antitoxin (200,000 IU) have been given; the intrathecal route may also be used. Some authors have even stressed the importance of avoiding the intravenous route, e.g. see Eriksson and Ullberg-Olsson (1965) and Athavale and Pai (1966). From a study of 3,295 cases of tetanus, Patel and Aiyar (1966) concluded that a small dose of 10,000 IU of antitoxin would be sufficient for a case of any severity.

In Holland, only human immune globulin is used. Our experience with tetanus is relatively small, but we think that a single therapeutic dose of 20,000 IU will be sufficient. It is important to realize that clinical tetanus does not result in immunity. Active immunization should therefore be started from the beginning of the disease. When the subcutaneous administration of 1 ml tetanotoxoid is sufficiently far away from the administration of serotherapy, the latter will not be disturbed. (See Willis, 1969, p. 425.)

Proper surgical care of wounds is important (Furste, 1969). Usually this consists merely of decompression of a wound, nail extraction or removal of foreign bodies. The necessity to clean the wounds with H_2O_2 is still generally accepted, as well as the fact that any manipulation in the area of entry of the causative microorganism should be timed after the start of serotherapy.

CONTROL OF SPASMS. Sedatives and muscle relaxants are given, and there is no common opinion on the drug(s) of choice or the optimal dosage. The physician should use those drugs with which he has had experience in the treatment of other patients. A fixed dosage scheme is not justified because the amount of drugs needed is directly correlated to the amount and intensity of stimuli; dosage can usually be diminished during the night. An increase in the frequency and intensity of spasms after a week of treatment is often attributable to an increase of stimuli (thrombophlebitis, decubitus, pneumonia, epididymitis, *etc.*), and dosage should not be increased routinely but only after careful evaluation of the cause. Diazepam (Valium) appears to be a safe anticonvulsant for routine use in the management of tetanus (Gautier *et al.*, 1968; Phatak and Shaw, 1970; Lund *et al.*, 1971).

CONTROL OF ADEQUATE PULMONARY VENTILATION AND PREVENTION OF INTERCURRENT PULMONARY INFECTION. Frequent control of blood gasses is imperative because pulmonary ventilation can be diminished by the muscular spasms, the drugs used, or concomitant atelectasis and pneumonia resulting in defects in the ventilation/perfusion ratio. Degenerative changes in respiratory muscles such as the diaphragm and intercostal muscles play a substantial part in respiratory failure (Ebisawa and Matsukura, 1968). Control of adequate ventilation is a difficult problem because physiotherapy is a severe stimulus to the patient. Complete muscle relaxation and mechanical ventilation are indicated: (a) in case of severe asphyxia due to laryngospasm; (b) if respiration is severely disturbed by tonic and clonic spasms of the thoracic musculature; (c) in patients who have had tracheostomy and react too intensely to the stimuli of physiotherapy; (d) in patients prone to severe injury (fracture of a vertebra) due to spasms; (e) in case of high frequency and painfulness of spasms; (f) in patients with hyperthermia who do not respond sufficiently to sedatives; (g) in patients requiring drug therapy which depresses spontaneous ventilation.

Bacteriological control is important during management of the acute phase of the illness. Infection of the tracheostomy wound and airways may develop and must be controlled by antibiotic therapy. Intercurrent pulmonary infection may cause the death of a patient who might have recovered from tetanus. That patients who survive even the most severe form of this disease usually recover completely physically, mentally and emotionally justifies a maximal therapeutic effort. Modern experience dictates that the only place for effective treatment of tetanus is a fully equipped intensive care unit with experienced personnel (Ellis, 1963; Christensen and Thurber, 1968). Monitoring of the EEG has to be included (Kugler and Manz, 1969; Busch *et al.*, 1969). The treatment demands close teamwork and, perhaps most important of all, the highest standard of nursing care.

PREVENTION AND CONTROL OF OTHER INTERCURRENT INFECTIONS. Special attention should be paid to the risk of urinary infection and septicemia from caval catheters used for long-term parenteral feeding.

CONTROL OF FLUID AND ELECTROLYTE BALANCE. In mild cases the patient can take meals or a fluid diet orally. In more severe cases and certainly in those in which mechanical ventilation is necessary, feeding becomes a problem. High caloric feeding is very important (Alder, 1969). One should not use long-term IV alimentation when the gastrointestinal tract is available, and one should not hesitate to perform a gastrotomy in a patient with severe tetanus. The risk of septicemia from caval catheters is high and this complication is difficult to diagnose in case of a concomitant syndrome of sympathetic nervous overactivity which is sometimes seen in patients with severe tetanus (Kerr *et al.*, 1968; Prys-Roberts *et al.*, 1969; Lazar, 1970).

Holloway (1970) studied urine and blood electrolyte levels of 30 patients at the height of their illness in order to help clarify the cause of the vasomotor instability observed in tetanus. All the patients were retaining fluid and electrolytes. Many had vasomotor instability and a clinical picture of low circulating blood volume.

HYPERBARIC OXYGENATION. Several investigators have reported results with hyperbaric oxygenation in the treatment of tetanus (Brummelkamp, 1964; Pascale *et al.*, 1964; Winkel and Kroon, 1964; Van Zijl, 1964). The basic working hypothesis underlying this approach is as follows: (a) part of the toxin, the so-called tetanolysin, is oxylabile; (b) oxidation of the complete toxin by exposure to daylight or oxidants results in a rapid diminution of toxicity (Halter, 1936; Jude *et al.*, 1949); and (c) according to Lippert (1935), guinea pigs inoculated with a lethal dose of tetanus toxin and injected at the same time with the oxygen carrier methylene blue stayed alive as long as the methylene blue was in its oxygenated blue form, but died as soon as it had been transformed into its reduced leucobase. Methylene blue-treated animals lived much longer than controls given only toxin.

In our studies with mice, oxygenated tetanus toxin showed no decrease in toxicity. In mice inoculated in the hindleg with a lethal dose of tetanus toxin and then subjected to hyperbaric oxygenation at 3 ATA for 2 hours, we could detect some favorable effect after one hyperbaric session when these mice were compared to controls. However, after 24 hours, during which three oxygen sessions were accomplished, all animals had died; in fact, the treated animals died sooner than the controls.

From the results in clinical cases, we conclude that hyperbaric oxygenation is of no value in the treatment of tetanus. Perhaps this is because the essential toxin, tetanospasmin, is not oxylabile, and hyperbaric treatment can only be started when the diagnosis has been made, after this toxin has already caused long-term cerebral damage.

REFERENCES

Aikat, B.K., and Dible, J.H.: The pathology of *Clostridium welchii* infection. *J Pathol Bacteriol, 71:*461, 1956.

Alder, A.: Successful intensive care in five cases of tetanus in children. *Anaesthesist, 18:*112, 1969.

Almeida, A.O. de, and Pacheco, G.: Ensaios de tratamento das gangrenas gazosas experimentais pelo oxigênio em atlas pressões e pelo oxigenio em astado nascente. *Rev Bras Biol, 1:*1, 1941.

Altemeier, W.A., Culbertson, W.R., Vetto, M., and Cole, W.: Problems in the diagnosis and treatment of gas gangrene. *Arch Surg, 74:*839, 1957.

Arnor, O., Bitter, J.E., Haglin, J.J., and Hitchcock, C.R.: Hyperbaric oxygen therapy in *Clostridium perfringens* and *Peptococcus* infection. In: Brown, I.W., Jr., and Cox, B.G. (Eds.): *Hyperbaric Medicine.* Washington D.C., Nat. Acad. of Sci. Nat. Res. Council, Publ. 1404, 1966.

Athavale, V.B., and Pai, P.N.: Role of tetanus antitoxin in the treatment of tetanus in children. *J Pediatr, 68:*289, 1966.

Aufranc, O.E., Jones, W.N., and Bierbaum, B.E.: Gas gangrene complicating fracture of the tibia. *JAMA, 209:*2045, 1969.

Bayliss, G.J.A., and Cass, C.: Hyperbaric oxygen used in the treatment of gas gangrene. *Med J Aust, 2:*991, 1967.

Beavis, J.P., and Watt, J.: Hyperbaric oxygen therapy in the treatment of gas gangrene. *J R Nav Med Serv, 51:*26, 1970.

Behnke, A.R., Johnson, F.S., Popper, J.R., and Motley, E.P.: The effects of oxygen on man at pressures from 1 to 4 atmospheres. *Am J Physiol, 110:*565, 1935.

Bertoye, A., Roche, L., and Vincent, P.: L'oxygenotherapie hyperbare dans les infections à bacteries anaerobies. (Treatment with hyperbaric oxygen in infections by anaerobic bacteria.) (French). *Lyon Med, 220:*1501, 1968.

Boerema, I.: An operating room with high atmospheric pressure. *Surgery, 49:*291, 1961.

Boerema, I., and Brummelkamp, W.H.: Inhalation of oxygen at 2 atmospheres for *Clostridium welchii* infections. *Lancet, 2:*990, 1962.

Boerema, I., and Brummelkamp, W.H.: Behandeling van anaerobe infecties met inademing van zuurstof onder een druk van drie atmosferen. *Nederl Tijdschr Geneeskd, 104:*2049, 1960.

Boerema, I., Kroll, J.A., Meyne, N.G., Lokin, E., Kroon, B., and Huiskes, J.W.: High atmospheric pressure as an aid to cardiac surgery. *Arch Chir Neerl, 8:*193, 1956.

Boerema, I., Kroll, J.A., Meyne, N.G., Kroon, B., and Huiskes, J.W.: Interventions sous hyperpression atmosphérique. Un principe auxilliaire dans le développement de la chirurgie intracardiaque. *Minerva Cardioangiol [Europea], 5:*233, 1957a.

Boerema, I., Kroll, J.A., Meyne, N.G., Lokin, E., Kroon, B., and Huiskes, J.W.: Operen onder atmospherische overdruk als hulpmiddel in de hartchirurgie. *Ned Tijdschr v Geneeskd, 101:*1464, 1957.

Boerema, I., Meyne, N.G., Brummelkamp, W.H., Bouma, S., Mensch, M.H., Kamermans, F., Stern Hanf, M., and Aalderen, v.W.: Life without blood. *Arch Chir Neerl, 11:*70, 1959.

Boerema, I., Meyne, N.G., Brummelkamp, W.H., Bouma, S., Mensch, M.H., Kamermans, F., Stern Hanf, M., and Aalderen, v.W.: Life without blood. *J Cardiovasc Surg, I:*133, 1960a.

Boerema, I., Meyne, N.G., Brummelkamp, W.H., Bouma, S., Mensch, M.H., Kamermans, F., Stern Hanf, M., and Aalderen, v.W.: Vida sin sangre. *Gac Sanitarai, 15:3*, 1960b.

Boyd, J.S.K.: Tetanus in the African and European theatres of war 1939–1945. *Lancet, 250:*113, 1946.

Brummelkamp, W.H.: De betekenis van de toediening van zuurstof onder atmosferische overdruk op de behandling van gasflegmone. *Ned Tijdschr Geneeskd, 105:*2430, 1961.

Brummelkamp, W.H., Hoogendijk, J.L., and Boerema, I.: Treatment of anaerobic infections (clostridial myositis) by drenching the tissues with oxygen under high atmospheric pressure. *Surgery, 49:*299, 1961.

Brummelkamp, W.H.: Anaerobic infections. *Bull Soc Int Chir, 21:*481, 1962.

Brummelkamp, W.H., Boerema, I., and Hoogendyk, L.: Treatment of clostridial infections with hyperbaric oxygen drenching. A report on 26 cases. *Lancet, 1:*235, 1963.

Brummelkamp, W.H., and Heins, H.F.: Een acute vorm van voortschrijdend huidgangraen (type Meleney) en haar conservatieve behandeling door middle van doordrencking van de weefsels met behulp van een hyperpressietank. *Ned Tijdschr Verloskd Gynaecol, 63:*245, 1963.

Brummelkamp, W.H.: Hyperbaric oxygen drenching of the tissues in gynaecological and obstetrical infections. In: *Pregnancy, Chemistry and Management.* Springfield, Thomas, 1964a.

Brummelkamp, W.H.: *Hyperbaric Oxygen Therapy in Clostridial Infections type Welchii.* Haarlem, Erven F. Bohn N.V., 1965a.

Brummelkamp, W.H.: Considerations on hyperbaric oxygen therapy at three atmospheres absolute for *Clostridial* infections type *welchii. Ann NY Acad Sci, 117:*688, 1965b.

Brummelkamp, W.H.: Reflections on hyperbaric oxygen therapy at 3 atmospheres absolute for *Clostridium welchii* infections. In: Ledingham, I.M. (Ed.): *Hyperbaric Oxygenation.* London, Livingstone, 1965c.

Brummelkamp, W.H.: Traitement des infections à germies anaerobies par l'inhalation d'oxygène en hyperpression. *Ann Chir Thorac Cardiov, 5:*607, 1966.

Busch, G., Israng, H., and Oettel, P.: Use of the EEG in differential diagnosis of tetanus. (German) *Anaesthesist, 18:*4, 1969.

Cater, D.B., Schoeniger, E.L., and Watkinson, D.A.: Effect on oxygen tension of tumours of breathing oxygen at high pressures. *Lancet, 2:*381, 1962.

Christensen, N.A., and Thurber, D.L.: Current treatment of clinical tetanus. *Mod Treat, 5:*729, 1968.

Cooke, W.T., Peeney, A.C., Thomas, G., Elkes, J.J., Frazer, A.C., Govan, A.D.T., Barlings, S.G., Leather, J.B., and Masonn, R.P.S.: Clostridial infections in war wounds. *Lancet, 248:*487, 1945.

Donald, K.W.: Oxygen poisoning in man. *Br Med J, 1:*667, 1947.

Dormanns, E.: Über das Gasödem der Kriegswunden. *Zentralbl Chir, 69:*610, 1942.

Duff, J.H., Shibata, H.R., VanSchaik, L., Usher, R., Wigmore, R.A., and MacLean, L.D.: Hyperbaric oxygen: A review of treatment in eighty-three patients. *Can Med Assoc J, 97:*510, 1967.

Duff, J.H., McLean, P.H., and MacLean, L.D.: Treatment of severe anaerobic infections. *Arch Surg, 101:*314, 1970.

Ebisawa, I., and Matsukura, M.: Pulmonary and muscular changes in tetanus. *Jap J Exp Med, 38:*27, 1968.

Eckhoff, N.L.: Gas gangrene in civil surgery. *Br J Surg, 18:*38, 1930.

Eckman, L.: *Tetanus.* Basel, Schwabe Verlag, 1960.

Ellis, M.: Human antitetanus serum in the treatment of tetanus. *Br Med J, 1:*1123, 1963.

Eraklis, A.J., Filler, R.M., Pappas, A.M., and Bernhard, W.F.: Evaluation of hyperbaric oxygen as an adjunct in the treatment of anaerobic infections. *Am J Surg, 117:*485, 1969.

Erikson, E., and Ullberg-Olsson, K.: Comparison between modern intensive therapy and ordinary treatment of tetanus. A review of the Swedish tetanus cases 1950–60. In: *Proceedings of the 1st International Conference on Tetanus.* Bombay, 1965, p. 523.

Fredette, V.: Effect of hyperbaric oxygen on anaerobic bacteria and toxins. *Ann NY Acad Sci, 117:*700, 1965.

Fulthorpe, A.J.: Adsorption of tetanus toxin by brain tissue. *J Hyg (Camb), 54:*315, 1956.

Furste, W.: Current status of tetanus prophylaxis. *Ind Med Surg, 38:*41, 1969.

Gautier, J., Lamisse, F., and Lamagnere, J.P.: Treatment of tetanus by diazepam in subjects over 60 years. Remarks on 26 cases. *Sem Hôp Paris, 44:*2831, 1968.

Gordon, J., Holman, R.A., and McLeod, J.W.: Further observations on the production of hydrogen peroxide by anaerobic bacteria. *J Pathol Bacteriol, 66:*527, 1953.

Graf, A.: Umspritzungen mit H_2O_2 bei Gasbrand? *Münch Med Wochenschr, 86:*28, 1939.

Halter, K.: Untersuchungen über die Auswertung unde die Wirkungsweise des Tetanus-toxins. *Z Hygiene, 118:*245, 1936.

Hanson, G.C., Slack, W.K., Chew, H.E.R., and Thomas, D.A.: Clostridial infection of the uterus—a review of treatment with hyperbaric oxygen. *Postgrad Med J, 42:*499, 1966.

Hanson, G.C., and Slack, W.K.: Hyperbaric oxygenation. *Bio-Med Eng, 2:*210, 1967.

Hinton, D.: A method for the arrest of spreading gas gangrene by oxygen injection. *Am J Surg, 73:*228, 1947.

Hitchcock, C.R., Haglin, J.J., and Arnar, O.: Treatment of clostridial infections with hyperbaric oxygen. *Surgery, 62:*759, 1967.

Holloway, R.: Fluid and electrolyte status in tetanus. *Lancet, 2:*1278, 1970.

Hoover, N.W., and Ivins, J.C.: Wound debridement. *Arch Surg, 79:*701, 1959.

Howell, L.M.: Hyperbaric oxygen in gas gangrene. *Northwest Med, 68:*1016, 1969.

Hugel, K.: Zur Behandlung des Gasbrandes mit Sauerstoff. *Zentralbl Chir, 65:*11, 1938.

Jacobson, I., Harper, A.M., and McDowall, D.G.: The effects of oxygen under pressure on cerebral blood-flow and cerebral venous oxygen tension. *Lancet, 2:*549, 1963.

Johnstone, F.R.C., and Cockcroft, W.H.: *Clostridium welchii* resistance to tetracycline. *Lancet, 1:*660, 1968.

Jude, A., Girard, P., and Carrat, P.: Action destructice *in vivo,* de certains agents chimiques sur les toxines botuliques et tétaniques. *C R Soc Biol, 143:*318, 1949.

Kaeser, H.E., and Saner, A.: Tetanus toxin. A neuromuscular blocking agent. *Nature, 223:*842, 1969.

Karasewich, E.G., Harper, A.M., Sharp, N.C.C., Shields, R.S., Smith, G., and McDowall, D.G.: Hyperbaric oxygen in clostridial infections. In: Boerema, I., *et al.* (Eds.): *Clinical Application of Hyperbaric Oxygen.* Amsterdam, London and New York, Elsevier, 1964.

Kelley, H.G., and Page, W.G.: Treatment of anaerobic infections in mice with hyperpressure oxygen. *Surg Forum, 14:*63, 1963.

Kennedy, R.H.: Present-day early care of traumatic wound. *Surg Clin North Am, 35:*355, 1955.

Kerr, J.H., Corbett, J.L., and Prys-Roberts, C.: Involvement of the sympathetic nervous system in tetanus. Studies on 82 cases. *Lancet, 2:*236, 1968.

Kiranow, I.G.: Ueber die operativen Behandlungsmethoden bei anaerober Gasbrandinfektion (Surgical treatment of anaerobic gas gangrene infections). (German) *Zentralbl Chir, 94:*1020, 1969.

Klopper, P.J., Brummelkamp, W.H., and Hoogendijk, J.L.: Recherches expérimentales sur l'effect de l'oxygénotherapie en hyperpression après ligature de l'artère hépatique. *Presse Méd, 70:*1874, 1962.

Kucher, R., and Riedel, W.: Die Behandlung des Gasbrandes in der Sauerstoffnebendruckkammer (The treatment of gas gangrene in the hyperbaric oxygen chamber). (German) *Wien Klin Wochenschr, 81:*308, 1969.

Kugler, J., and Manz, R.: Monitoring of the effect of tetanus treatment by means of EEG. *Anaesthesist, 18:*126, 1969.

Lambertsen, C.J., Kongh, R.H., Emmel, G.L., Loeschke, H.H., and Schmidt, C.F.: Oxygen toxicity. Effects in man of oxygen inhalation at 1 and 3.5 atmospheres upon blood gas transport, cerebral circulation and cerebral metabolism. *J Appl Physiol, 5:*471, 1953.

Lazar, M.: Pathogenesis and treatment of tetanus. *Schweiz Med Wochenschr, 100:*1486, 1970.

Lippert, K.M.: The photodynamic effect of methylene blue on tetanus toxin. *J Immunol, 28:*193, 1935.

Löber, H.J.A.: Gasbrand nach appendizitis. *Münch Med Wochenschr, 85:*1942, 1938.

Lulu, D.J., and Rivera, F.J.: Gas gangrene. *Am J Surg, 36:*528, 1970.

Lund Edelweiss, E., Severo, V., and Martins, S.M.: Tetanus in the adult treated with diazepam. A comparative study of mephenesine and diazepam. *Folia Med, 62:*597, 1971.

MacLennan, J.D.: Anaerobic infections of war wounds in the Middle East. *Lancet, 245:*63, 94, 123, 1943.

MacLennan, J.D., and MacFarlane, R.G.: Toxin and antitoxin studies of gas gangrene in man. *Lancet, 249:*301, 1945.

MacLennan, J.D.: The histotoxic clostridial infections in man. *Bacteriol Rev, 26:*177, 1962.

McNally, J.B., Price, W.R., and MacDonald, W.: Gas gangrene of the anterior abdominal wall. *Am J Surg, 116:*779, 1968.

Maudsley, R.H., and Colwill, M.R.: Hyperbaric oxygen in the management of gas gangrene. *J Bone Joint Surg [B], 48:*584, 1966.

Maudsley, R.H.: Postoperative gas gangrene. *Br Med J, 2:*352, 1967.

Meleney, F.L.: Zinc perioxide in surgical infections. *Surg Clin North Am, 16:*691, 1936.

Meleney, F.L.: *Clinical Aspects and Treatment of Surgical Infections.* Philadelphia, Saunders, 1949.

Meijne, N.G.: *Hyperbaric Oxygen and its Clinical Value.* Springfield, Thomas, 1970.

Müller, Fr.: Beitrag zur Behandlung der Gasphlegmone. *Münch Med Wochenschr, 86:*1797, 1939.

Oakley, C.L.: The toxins of *Cl. welchii.* A critical review. *Bull Hyg 18:*781, 1943.

Oakley, C.L.: Bacterial toxins: demonstration of antigenic components in bacterial filtrates. *Ann Rev Microbiol, 8:*411, 1954.

Pacheco, G., and Costa, G.A.: Influenca do oxygenio sob pressao sobre o *"Clostridium welchii". Rev Bras Biol, 1:*145, 1941.

Parish, H.J., and Canon, D.A.: *Antisera, Toxoids, Vaccines and Tuberculins in Prophylaxis and Treatment*, 6th ed. Edinburgh and London, Livingstone, 1962.

Pascale, L., Wallyn, R.J., Goldfein, S., and Gumbiner, S.: Observations in response of tetanus to hyperbaric oxygenation. In: Boerema, I., *et al.* (Eds.): *Clinical Application of Hyperbaric Oxygen.* Amsterdam, Elsevier, 1964.

Pascale, L.R., and Wallyn, R.: Surgical applications of the hyperbaric chamber. *Surg Clin North Am, 48:*63, 1968.

Patel, Y.C., and Aiyar, A.A.: Tetanus. *Lancet, 2:*1321, 1966.

Peters, K.O.: Zur Frage der Gasbrandbehandlung. *Zentralbl Chir, 66:*1904, 1939.

Phatak, A.T., and Shah, S.H.: Diazepam as adjuvant therapy in childhood tetanus: 477 patients with tetanus in Baroda. *Clin Pediatr, 9:*573, 1970.

Picco, A., and Nicolello, E.: Cura della gangrena gassosa con inierzioni di ossigeno. *Gazz Int Med Chir, 54:*58, 1950.

Prévot, A.R.: *Biologie des maladies dues aux anaérobies.* Paris, Edit. Méd. Flammarion, 1955.

Prys-Roberts, C., Corbett, J.L., and Kerr, J.H.: Treatment of sympathetic overactivity in tetanus. *Lancet, 1:*542, 1969.

Rettig, W.: Umspritzung mit H_2O_2 bei Gasbrand. *Münch Med Wochenschr, 86:*145, 1939.

Roding, B., Groeneveld, P.H.A., and Boerema, I.: Ten years of experience in the treatment of gas gangrene with hyperbaric oxygen. *Surg Gynecol Obstet, 134:*579, 1972.

Schoemaker, G.: Oxygen tension measurements. In: Boerema, I., *et al.* (Eds): *Clinical Application of Hyperbaric Oxygen.* Amsterdam, Elsevier, 1964, p. 330.

Schoemaker, G.: Personal communication.

Slack, W.K., Hanson, G.C., and Chew, H.E.R.: Hyperbaric oxygen in the treatment of gas gangrene and clostridial infection. *Br J Surg, 56:*505, 1969.

Smith, G., Sillar, W., Norman, J.N., Ledingham, I.M., Bates, E.H., and Scott, A.C.: Inhalation of oxygen at 2 atmospheres for *Clostridium welchii* infections. *Lancet, 2:*756, 1962.

Spiro, K.: Die Wirkung von Wasserstoffsuperoxyd und von Zucker auf die Anaerobier. *Münch Med Wochenschr, 62:*497, 1915.

Tarbiat, S.: Ergebnisse der kombinierten chirurgisch-antibiotischen und hyperbaren Sauerstoffbehandlung beim Gasödem. *Chirurg, 41:*506, 1970.

Trippel, O.H., Ruggie, A.N., Staley, C.J., and Van Elk, J.: Hyperbaric oxygenation in the management of gas gangrene. *Surg Clin North Am, 47:*17, 1967.

Vaishnava, H., Goyal, R.K., Neogy, C.N., and Mathur, G.P.: A controlled trial of antiserum in the treatment of tetanus. *Lancet, 2:*1371, 1966.

Van Unnik, A.J.M.: Inhibition of toxin production in *Clostridium perfringens in vitro* by hyperbaric oxygen. *J Microbiol Serol, 31:*181, 1965.

Wesenberg, G., and Hoffman, A.: Die Beeinflussung des Tetanustoxins durch einige oxydierend wirkende Körper. *Zentralbl Bakteriol, 94:*416, 1925.

Willis, A.T.: *Clostridia of Wound Infection.* London, Butterworths, 1969.

Winkel, C.A., and Kroon, T.A.J.: Experiences with hyperbaric oxygen treatment in tetanus. In: Boerema, I., *et al.* (Eds.): *Clinical Appliction of Hyperbaric Oxygen.* Amsterdam, Elsevier, 1964.

Wolcott, M.W.: Comparison of high-pressure oxygen and oxygen-carbon dioxide mixtures on treatment of *Clostridium perfringens* infection in mice. *J Surg Res, 9:*129, 1969.

Zijl, J.J.W. van, and Maartens, P.R.: High pressure oxygen therapy in South Africa. *S Afr Med J, 1:*799, 1963.

Zijl, J.J.W. van: In discussion. In: Boerema, I., *et al.* (Eds.): *Clinical Application of Hyperbaric Oxygen.* Amsterdam, London and New York, 1964, p. 68.

Zijl, J.J.W. van: *Hyperbaric Oxygenation. A General Review with Special Consideration of the Surgical Applications.* Stellenbosch, S.A., Thesis, 1966, p. 209.

Zijl, J.J.W. van, Maartens, P.R., and du Toit, E.D.: Gas gangrene treated in one-man hyperbaric chamber. In: Brown, I.W., Jr., and Cox, B.G. (Eds.): *Hyperbaric Medicine.* Washington, Nat. Acad. of Sci. Nat. Res. Council, Publ. 1404, 1966, p. 515.

CHAPTER XLII

TREATMENT OF BACTEROIDACEAE BACTEREMIA: CLINICAL EXPERIENCE WITH 112 PATIENTS

Anthony W. Chow and Lucien B. Guze

ABSTRACT: *Results of antimicrobial therapy and surgical drainage in 112 patients with Bacteroidaceae bacteremia are presented. Underlying disease was a major determinant of mortality and appropriate choice of antibiotics was of paramount importance. The correlation between* in vitro *susceptibility to antibiotics and clinical outcome was established. Surgical drainage was an important adjunct to therapy, but its beneficial effect was less obvious in the presence of appropriate antimicrobial therapy.*

In the clinical situation, where treatment must be started before the antibiotic susceptibility of the infecting organism is known, knowledge of the antibiotic susceptibility patterns of Bacteroidaceae is of prime importance. In in vitro *studies 98 percent of isolates were susceptible to chloramphenicol and 96 percent to clindamycin, and these antibiotics were also highly effective clinically.*

Anaerobes, particularly of the nonsporesforming, gram-negative bacillary variety, have assumed an increasingly important role in sepsis. Between July 1969 and June 1972, 129 patients with Bacteroidaceae bacteremia were seen at Harbor General Hospital. Bacteroidaceae were isolated from 216 blood cultures obtained from these patients, an incidence approximating 9.3 percent of all positive blood cultures obtained during the same period. The results of therapy in 112 patients were analyzed retrospectively to elucidate the effects of antimicrobial agents and surgical drainage in these patients.

MATERIALS AND METHODS

Hospital records of 123 patients with Bacteroidaceae bacteremia between July 1969 and June 1972 were reviewed and 112 patients were included in the study. The majority were seen by the authors on consultation. Excluded from the study were five patients with positive postmortem heart blood cultures without clinical evidence of sepsis antemortem and six patients in whom gram-negative bacilli were seen on smear of blood culture but failed to grow on solid media. The charts of six other patients were not available for review.

Blood cultures were obtained with 5 ml of blood in 50 ml each of thioglycollate broth (Thio Broth, Difco) and infusion broth (Brain Heart Infusion Broth, Difco), and incubated at 37°C for 14 days. Cultures were examined by Gram stain periodically. If nonsporeforming gram-negative rods were seen, these were subcultured both aerobically and anaerobically on sheep blood agar plates. All anaerobic nonsporeforming gram-negative rods were considered members of Bacteroidaceae. Isolates from 50 patients were available for definitive identification and speciation by biochemical testing and gas chromatography analysis. Prereduced differential media were prepared according to the methods of Holdeman and Moore (1972) and all biochemical tests recommended in their manual for speciation were performed. A Beckman pH-meter was used for recording carbohydrate fermentation reactions and a Dohrmann AnaBac chromatograph was used for fatty-acid analysis. Identification and nomenclature were based on the scheme of Holdeman and Moore (1972).

Fifty-three isolates were available for antibiotic susceptibility testing by agar dilution. Mueller-Hinton agar (Difco) adjusted to pH 7.2 and enriched with 5 percent defibrinated sheep blood was used. Two-fold serial dilutions of antibiotics were added. Vitamin K-hemin was supplemented at a concentration of 0.01% (v/v) when *Bacteroides melaninogenicus* was tested. A 48-hour subculture in thioglycollate broth, enriched with 25% ascitic fluid, was adjusted to a McFarland nephelometer No. 1 standard and a 1 to 10 dilution was used as the inoculum (approximately 10^7 organisms per ml). Plates were inoculated with a Steers' replicating apparatus (Steers *et al.*, 1959) and incubated at 37°C in anaerobic jars after evacuation of air and replacement with a gas mixture containing 80% nitrogen, 10% hydrogen, and 10% carbon dioxide. Aerobic and anaerobic plates without antibiotics were used for controls, and a reference strain with known minimal inhibitory concentration (MIC) was included in each run for reproducibility. All results were read at 48 hours and again at five days. The MIC recorded was the least antibiotic concentration that yielded no visible growth. Criteria for susceptibility to various antibiotics were according to Gavan *et al.* (1971), and based on blood levels achievable by parenteral administration: clindamycin, 2.5 μg/ml (DeHaan *et al.*, 1973); vancomycin, 5 μg/ml (Garrod and O'Grady, 1968; Lepper and McCabe, 1968); and rifampin, 2.5 μg/ml (Reiss *et al.*, 1970).

Agar-diffusion antibiotic sensitivity testing was performed on all isolates by a modified Kirby-Bauer method. In a separate study, 88 strains of Bacteroidaceae, including 40 blood isolates from these patients and 48 isolates from infected sites of other patients, were studied by both agar dilution and agar diffusion for statistical correlation of these two techniques by linear regression (Ericsson and Sherris, 1971).

For comparing the outcome of sepsis in patients with varying severity of underlying disease, patients were categorized by the method of McCabe and Jackson (1962). Patients with acute leukemia, chronic leukemia in blastic crisis, far advanced cancer, *etc*, were considered to have rapidly fatal disease. Those with severe liver disease, arteriosclerosis, diabetes mellitus, end-stage renal disease, solid tumor, collagen vascular diseases, or other illnesses of intermediate severity were considered to have ultimately fatal disease. The remaining patients were considered to have nonfatal disease. "Appropriate antimicrobial therapy" was defined as the administration of therapeutic doses of one or more agents effective *in vitro* against the infecting organisms. "Inappropriate antimicrobial therapy" included administration of antibiotics to which the infecting organisms were resistant *in vitro* as well as omission of antibiotic therapy.

RESULTS

Antibiotic Susceptibility Testing

Agar Dilution Susceptibility

The cumulative percent of strains of Bacteroidaceae inhibited by each dilution of nine antibiotics are summarized in Figure XLII-1. Using the criteria of susceptibility outlined

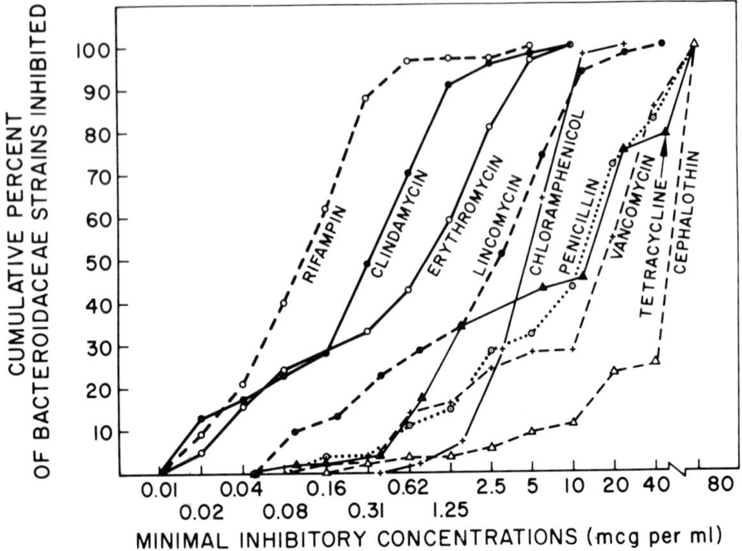

Figure XLII-1. Antibiotic susceptibility of 53 strains of Bacteroidaceae species isolated from patients with bacteremia.

in Table XLII-I, the most effective agents *in vitro* were chloramphenicol, clindamycin, erythromycin, and rifampin (98%, 96%, 81%, and 98% sensitive, respectively). Lincomycin, tetracycline, and penicillin were intermediately effective (51%, 43%, and 32% sensitive, respectively). Vancomycin and cephalothin were least effective (29% and 11% sensitive, respectively). These findings are generally similar to those reported by Martin *et al.* (1972), Kislak (1972), Bodner *et al.* (1972), and others (Finegold *et al.*, 1967; Finegold and Hewitt, 1956; Ingham *et al.*, 1970; Wilkins *et al.*, 1972).

*Correlation of Agar Diffusion and Agar Dilution Techniques**

The results of linear regression studies for each antibiotic are shown in Table XLII-II. For each antibiotic, an analysis

* We are pleased to acknowledge the computing assistance of Dr. Michael A. Fox of the UCLA Health Sciences Computing Facility, sponsored by NIH Special Research Resources Grant RR-3.

TABLE XLII-I

AGAR DILUTION SUSCEPTIBILITY OF BACTEROIDACEAE SPECIES ISOLATED FROM PATIENTS WITH BACTEREMIA

Organism	*Total Strains Tested*	*Chlor-amphenicol* $\leq$ *12.5**	*Clindamycin* $\leq$*2.5*	*Lincomycin* $\leq$ *3.1*	*Tetracycline* $\leq$ *6.25*	*Penicillin* $\leq$ *5.0*	*Cephalothin* $\leq$ *10.0*	*Total Strains Tested*	*Erythromycin* $\leq$ *2.5*	*Vancomycin* $\leq$ *5.0*	*Rifampin* $\leq$ *2.5*
Bacteroides fragilis and subspecies	37	36	35	18	14	10	2	28	21	6	27
ss. *fragilis*	16	16	15	9	8	6	1	15	12	3	14
ss. *vulgatus*	9	9	9	3	2	3	0	6	3	2	6
ss. *distasonis*	4	4	4	3	2	1	1	1	1	1	1
ss. *ovatus*	4	4	4	2	0	0	0	3	2	0	3
ss. *thetaiotaomicron*	3	2	2	0	1	0	0	3	3	0	3
other	1	1	1	1	1	0	0	—	—	—	—
Other *Bacteroides* and *Fusobacterium* species	11	11	11	6	7	4	2	9	8	4	9
B. capillosus	5	5	5	3	4	0	0	5	4	1	5
B. oralis	2	2	2	0	0	0	0	—	—	—	—
B. melaninogenicus	1	1	1	1	1	1	1	1	1	0	1
B. putridinis	1	1	1	1	1	1	1	1	1	1	1
F. nucleatum	1	1	1	1	1	1	0	1	1	1	1
Fusobacterium spp.	1	1	1	0	0	1	0	1	1	1	1
Unspeciated Bacteroidaceae	5	5	5	3	2	3	2	5	5	2	5
	—	—	—	—	—	—	-	—	—	—	—
Total	53	52	51	27	23	17	6	42	34	12	41
(%)		(98%)	(96%)	(51%)	(43%)	(32%)	(11%)		(81%)	(29%)	(98%)

* Criteria for susceptibility in $\mu g/ml$.

TABLE XLII-II

CORRELATION OF ZONE DIAMETER WITH MINIMAL INHIBITORY CONCENTRATION BY LINEAR REGRESSION*; STUDIES OF 88 STRAINS OF BACTEROIDACEAE ISOLATED FROM CLINICAL SPECIMENS

		Test for Linearity			*Test for H_0: b = 0*			
Antibiotic	*Equation* ($y = bx + a$)	*F* (*calc.*) <	*F* (*table*)	*DF*†	*F′* (*calc.*) >	*F′* (*table*)	*DF*†	*Correlation Coefficient*
Clindamycin	$y = -2.92x + 31.1$	0.43	2.07	(8,78)	34.24	3.96	(1,86)	−0.54
Chloramphenicol	$y = -2.03x + 36.5$	1.46	2.34	(5,81)	7.54	3.96	(1,86)	−0.28
Erythromycin	$y = -2.17x + 29.0$	0.74	2.06	(8,78)	29.88	3.96	(1,86)	−0.51
Lincomycin	$y = -3.41x + 26.6$	0.66	2.06	(8,78)	40.58	3.96	(1,86)	−0.57
Penicillin	$y = -3.39x + 25.2$	1.26	2.06	(8,78)	48.47	3.96	(1,86)	−0.60
Tetracycline	$y = -2.47x + 30.5$	1.12	1.90	(8,74)	32.28	3.96	(1,82)	−0.53

* See text.
† DF = degrees of freedom.

of variance (Dixon and Massey, 1957) was carried out to test the hypothesis of a strictly linear relationship between zone size (y) and $\log_2$ MIC (x). Where this is so, the F value (calculated) will be smaller than the corresponding table value. In such cases, a measure of the degree of this linear trend or the variation due to linear regression was performed. To ensure nondegeneracy, the hypothesis beta 0 must be rejected (i.e. F′ value greater than the corresponding table value). As indicated in Table XLII-II, although there was linear fit for clindamycin, chloramphenicol, erythromycin, tetracycline, lincomycin and penicillin, there was considerable variation due to regression of the two variables x and y as reflected by the correlation coefficient values. Vancomycin and rifampin did not show linear correlation and results were not included in the table.

Effect of Antimicrobial Therapy

As summarized in Table XLII-III, 23 of 39 patients (59%) inappropriately treated died while only 12 of 73 patients (16%) who were appropriately treated did not survive ($P < 0.005$). Further analysis indicated that this difference in mortality was significant only among patients with ultimately fatal underlying disease ($P < 0.005$) but was not significant for patients with nonfatal underlying disease ($P > 0.50$). There were too few patients with rapidly fatal underlying disease for a meaningful analysis. The reason for the apparent lack of significant detrimental effect of inappropriate antimicrobial therapy in patients with nonfatal disease may be due to the fact that eight of these 15 patients had postpartal or perinatal bacteremias which were benign and appeared transient in nature. None of these eight patients died. Only four of the 15 patients with nonfatal disease had received surgical drainage for loculated infection. Thus the relatively low mortality in these inappropriately treated patients could not be attributed to other therapeutic measures received. The highly significant difference in mortality between appropriately and inappropriately treated patients with ultimately fatal underlying disease is similar to that in patients with aerobic gram-negative sepsis (Freid and Vosti, 1968). These findings

TABLE XLII-III

EFFECT OF ANTIMICROBIAL THERAPY AND UNDERLYING DISEASE ON MORTALITY OF BACTEROIDACEAE BACTEREMIA

	Underlying Disease Category							
	Nonfatal		*Ultimately Fatal*		*Rapidly Fatal*			
Antimicrobial Therapy	*Total Patients*	*Deaths*	*Total Patients*	*Deaths*	*Total Patients*	*Deaths*	*Total Patients Treated*	*Deaths*
Appropriate	49	5 (10%)	23	7 (30%)	1	0 (0%)	73	12 (16%)
Inappropriate	15	3 (20%)	20	18 (90%)	4	2 (50%)	39	23 (59%)
P value*	>0.5		<0.005		—		<0.005	

* Significance of difference in mortality in the χ^2 test.

indicate that the choice of appropriate antimicrobial agents is of paramount importance and emphasize the value of antibiotic susceptibility testing as the basis of antimicrobial therapy.

Appropriate Antimicrobial Therapy

Seventy-three patients received appropriate antimicrobial therapy. Chloramphenicol, penicillins (including ampicillin and cephalothin), clindamycin, and any combinations of these were the most common appropriate agents (Table XLII-IV). Tetracycline, lincomycin, and erythromycin were infrequently used in our patients. It is interesting to note that as long as the infective organisms were sensitive to the antibiotic used, there was no significant difference in mortality in those who received clindamycin, penicillins, or combinations of appropriate agents when compared with those receiving chloramphenicol alone ($P > 0.975$, $P > 0.10$, and $P > 0.10$ respectively). Moreover, results of combination therapy did not differ significantly from single antibiotic therapy with agents other than chloramphenicol.

Further insight into the effect of antimicrobial therapy was gained by analysis of 12 patients who died despite "appropriate" antimicrobial therapy. Seven had ultimately fatal underlying disease; ten had antecedent trauma or surgery, and nine of these had gastrointestinal involvement. Shock was present in nine patients. Five patients had persistent bacteremia despite antibiotics; three of these had multiple intraabdominal abscesses. Chloramphenicol was used in nine patients, clindamycin in two, and penicillin in one. Surgical drainage was performed in half of the patients. It appeared that the severity of underlying disease, as well as complications of Bacteroidaceae bacteremia, were more important determinants of ultimate outcome than the choice of specific antibiotics.

Inappropriate Antimicrobial Therapy

Fourteen patients received no antibiotics and only four did not survive (Table XLII-V). This was not significantly different from those who received appropriate antimicrobial ther-

TABLE XLII-IV

EFFECT OF APPROPRIATE ANTIMICROBIAL REGIMENS AND UNDERLYING DISEASE ON MORTALITY OF BACTEROIDACEAE BACTEREMIA

	Underlying Disease Category								
	Nonfatal		*Ultimately Fatal*		*Rapidly Fatal*				
*Antimicrobial Regimen**	*Total Patients*	*Deaths*	*Total Patients*	*Deaths*	*Total Patients*	*Deaths*	*Total Patients Treated*	*Deaths*	*P Value*†
Chloramphenicol	12	3	6	3	1	0	19	6 (32%)	
Clindamycin	4	0	5	2			9	2 (22%)	>0.995
Erythromycin	1	0					1	0 (0%)	
Lincomycin			1	0			1	0 (0%)	
Penicillins§	10	0	5	1			15	1 (7%)	>0.1
Tetracycline			1	0			1	0 (0%)	
Any combination of above	22	2	5	1			27	3 (11%)	>0.1
Total	49	5	23	7	1	0	73	12 (16%)	

* With or without aminoglycosides.
† Significance of difference in mortality by χ^2 compared with chloramphenicol-treated patients.
§ Penicillin, ampicillin, or cephalothin.

TABLE XLII-V

EFFECT OF INAPPROPRIATE OR NO ANTIMICROBIAL THERAPY AND UNDERLYING DISEASE ON MORTALITY OF BACTEROIDACEAE BACTEREMIA

	Underlying Disease Category							
	Nonfatal		*Ultimately Fatal*		*Rapidly Fatal*			
*Regimen**	*Total Patients*	*Deaths*	*Total Patients*	*Deaths*	*Total Patients*	*Deaths*	*Total Patients Treated*	*Deaths*
None	8	0	4	4	2	0	14	4 (29%)
Penicillins	4	1	13	12	2	2	19	15 (79%)
Chloramphenicol (alone or in combination)	2	2	2	2			4	4 (100%)
Lincomycin & penicillin	1	0					1	0 (0%)
Sulfisoxazole			1	0			1	0 (0%)
Total	15	3 (20%)	20	18 (90%)	4	2 (50%)	39	23 (59%)

* With or without aminoglycosides.

apy (see Table XLII-III; 12 deaths in 73 patients, $P > 0.25$). However, eight of these 14 patients were in the nonfatal group and all survived, while all four patients in the ultimately fatal group died. In 25 patients who received antibiotics to which the infective organisms were resistant *in vitro*, there were 19 deaths. Nineteen of the 25 patients received penicillin or cephalothin, generally with kanamycin or gentamicin, and 15 of these died. This inappropriate regimen was commonly used for the coverage of aerobic gram-negative sepsis in our hospital. Four patients received chloramphenicol to which the infecting organisms were sensitive, but less than 2 g per day was administered. The intramuscular route was used in three patients, and this would not be expected to achieve adequate blood levels because of poor absorption (DuPont *et al.*, 1970).

Effect of Surgical Drainage

The effect of surgical drainage in patients with or without appropriate antimicrobial therapy is summarized in Table XLII-VI. In either group, addition of surgical drainage did not significantly improve mortality ($P > 0.50$ and $P > 0.10$ respectively). However, the number of patients who received surgical drainage with no or inappropriate antimicrobial therapy was small in our series. In the presence of surgical drainage, appropriate antimicrobial therapy did not significantly improve mortality ($P > 0.950$). Surgical drainage alone (1 of 5, 20%) or appropriate antimicrobial therapy alone (6 of 30, 20%) did not differ significantly in mortality ($P > 0.50$). As would be expected, mortality in those who received neither surgical drainage nor appropriate antibiotics (22 of 34, 65%) was significantly higher ($P < 0.005$) than those who received both (6 of 43, 14%). These data may be interpreted and summarized thus: surgical drainage and appropriate antimicrobial therapy appeared equally effective when employed alone. If either measure was applied, mortality was significantly improved and effect of the other therapeutic measure was less obvious. Conversely, if neither measure was used, mortality was significantly higher than if antimicrobial therapy was employed alone or in conbination with surgical drainage.

TABLE XLII-VI

EFFECT OF SURGICAL DRAINAGE AND ANTIMICROBIAL THERAPY ON MORTALITY OF BACTEROIDACEAE BACTEREMIA

	Inappropriate or No Antibiotics		*Appropriate Antibiotics*		
	Total Patients	*Deaths*	*Total Patients*	*Deaths*	*P Value§*
Surgical Drainage	5	1*	43	6†	>0.950
No Surgical Drainage	34	22†	30	6*	<0.005
P Value§	>0.1		>0.5		

* P > 0.5.
† P < 0.005.
§ Significance of difference in mortality by χ^2.

DISCUSSION

Although appropriate antimicrobial therapy is important in reducing mortality of Bacteroidaceae bacteremia, evaluation of a particular antibiotic regimen has been difficult. The experience reported in the literature has been based on relatively few cases from individual centers collected over extended periods of time. Antimicrobial susceptibility patterns may have altered in the meantime and the general care of patients in sepsis surely has. Moreover, the role of additional surgical drainage is often not separately assessed. All these render comparison between varying treatment schedules extremely difficult and hazardous. Our study is unique in that experience with relatively large number of patients has been collected within one institution over a brief span of three years. Our data suggest that antimicrobial therapy significantly reduces mortality of Bacteroidaceae bacteremia only among patients with ultimately fatal underlying disease, and reinforce the importance of comparing therapeutic regimens only among patients with similar underlying disease. Our data further support the notion that the effect of antimicrobial therapy cannot be meaningfully assessed unless the role of surgical drainage is evaluated. In addition, the value of *in vitro* antibiotic susceptibility testing is clearly demonstrated. Provided that antimicrobial agents used were effective *in vitro* against the infecting organisms, there was no difference in mortality between various antibiotic regimens administered.

In the choice of antibiotics for initial therapy of Bacteroidaceae bacteremia, however, when susceptibility data for the infecting organism are not available, prior knowledge of antibiotic susceptibility patterns of Bacteroidaceae becomes of prime importance. The significant resistance to tetracycline in the present study confirms the observations of others (Bodner *et al.*, 1972; Martin *et al.*, 1972; Thornton and Cramer, 1971). Sutter *et al.* (1972) noted that while 14 of 15 strains isolated prior to 1960 were sensitive to tetracycline, only 24 of 63 isolated after 1970 were susceptible. Thus, tetracycline should no longer be the antibiotic of choice for initial therapy of Bacteroidaceae sepsis as was formerly recommended by Bornstein *et al.* (1964), Marcoux *et al.* (1970), and others (Gelb and Seligman, 1970; Sinkovics and Smith, 1970; Tynes and Frommeyer, 1962). Although rifampin shows promise as a potentially effective antibiotic both in our studies and those of Martin *et al.* (1972), resistance to this antibiotic may develop rapidly as is the case with other microorganisms (McCabe and Lorian, 1968). Our data on erythromycin agree with those of Ingham *et al.* (1968), Hoogendijk (1965), and Kislak (1972), while Finegold *et al.* (1967) and Bodner *et al.* (1972) found it less effective. This discrepancy may be explained by differences in species tested and the techniques of susceptibility testing. Bacterial inhibition by erythromycin may be critically affected by several components of anaerobic incubation including changes in agar pH and CO_2 concentration (Ingham *et al.*, 1970; Rosenblatt and Schoenknecht, 1972).

Chloramphenicol has been recommended by Bodner *et al.* (1970) as the antibiotic of choice for initial therapy of Bacteroidaceae bacteremia and other investigators support this view (Ellner and Wasilauskas, 1971; Thornton and Cramer, 1971). Although our own data confirm the effectiveness of chloramphenicol *in vitro*, it is interesting to note that 21 percent of strains tested required 12.5 μg/ml for inhibition, a concentration close to the limit of attainable therapeutic

TABLE XLII-VII

RESULTS OF PARENTERAL CLINDAMYCIN THERAPY IN 15 PATIENTS WITH BACTEROIDACEAE BACTEREMIA*

	Site of Infection									
	Gastrointestinal		*Female Genital*		*Decubiti & Gangrene*		*Miscellaneous*		*Total*	
Organism	*Total Patients*	*Deaths*	*Total Patients*	*Deaths*	*Total Patients*	*Deaths*	*Total Patients*	*Deaths*	*Patients*	*Deaths*
Bacteroides fragilis	5	2	1	0	3	1	1	0	10	3
ss. *fragilis*	2	1			1	1	1	0	4	2
ss. *vulgatus*	1	1	1	0					2	1
ss. *distasonis*					1	0			1	0
ss. *ovatus*	1	0							1	0
other	1	0			1	0			2	0
Bacteroides capillosus							1	0	1	0
Bacteroidaceae, unspeciated	2	0	2	0					4	0
Total	7	2	3	0	3	1	2	0	15	3 (20%)

* Six patients received other "appropriate" antibiotics without clinical response prior to clindamycin therapy. Three patients had persistent bacteremia while on chloramphenicol or tetracycline even though the infecting organisms were sensitive to these agents.

levels. Moreover, the dose-related marrow suppressive effect as well as the more serious idiosyncratic toxicity of aplastic anemia are real disadvantages. Our *in vitro* data with clindamycin, an analog of lincomycin, suggest that it is highly effective against Bacteroidaceae and this is supported by the findings of Martin *et al.* (1972), Wilkins *et al.* (1972), and others (Bodner, 1972; Kislak, 1972). In a clinical trial with parenteral clindamycin therapy in 15 patients with Bacteroidaceae bacteremia reported here (Table XLII-VII) as well as 25 patients with other types of Bacteroidaceae infection, the results were comparable to those treated with chloramphenicol. The antibiotic was well tolerated and side effects, which included urticarial rash, diarrhea, and phlebitis, were rare. If confirmed by other clinical trials, this antibiotic should replace chloramphenicol as the antibiotic of choice in the initial therapy of Bacteroidaceae bacteremia.

Apart from prior knowledge of antibiotic susceptibility patterns, the other major determinant in appropriate choice of antibiotics might rely on prior knowledge of the prevalence, location and distribution of specific Bacteroidaceae species in the human flora. This in turn would rely on accurate identification and speciation of all Bacteroidaceae isolated from clinical infections. Thus, if *Bacteroides fragilis* and its subspecies were isolated, one might surmise that the gastrointestinal tract was the most likely source of infection and that it was highly likely these organisms would be resistant to penicillin. On the other hand, if *B. oralis* or *B. melaninogenicus* were isolated, one might surmise that the oropharynx was the most likely source of infection and that it was highly likely these organisms would be sensitive to penicillin. Obviously, to support this degree of reliance, much more intensive study is needed in the characterization of both the normal and pathologic occurrence of these very important microorganisms.

REFERENCES

Bodner, S.J., Koenig, M.G., and Goodman, J.S.: Bacteremic *Bacteroides* infections. *Ann Intern Med, 73*:537, 1970.

Bodner, S.J., Koenig, M.G., Treanor, L.L., and Goodman, J.S.: Antibiotic susceptibility testing of *Bacteroides*. *Antimicrob Agents Chemother, 2:*57, 1972.

Bornstein, D.L., Weinberg, A.N., Swartz, M.N., and Kunz, L.J.: Anaerobic infections—review of current experience. *Medicine, 43:*207, 1964.

DeHaan, R.M., Metzler, C.M., Schellenberg, D., and VandenBosch, W.D.: Pharmacokinetic studies of clindamycin phosphate. *J Clin Pharmacol, 13:*190, 1973.

Dixon, W.J., and Massey, F.J., Jr.: *Introduction to Statistical Analysis.* New York, McGraw-Hill, 1957.

DuPont, H.L., Hornick, R.B., Weiss, C.F., Snyder, M.J., and Woodward, T.E.: Evaluation of chloramphenicol acid succinate therapy of induced typhoid fever and Rocky Mountain spotted fever. *N Engl J Med, 282:*53, 1970.

Ellner, P.D., and Wasilauskas, B.L.: *Bacteroides* septicemia in older patients. *J Am Geriatr Soc, 19:*296, 1971.

Ericsson, H.M., Sherris, J.C.: Antibiotic sensitivity testing. A report of an international collaborative study. *Acta Pathol Microbiol Scand* [*B*], *Supp 217:*1, 1971.

Finegold, S.M., Harada, N.E., and Miller, L.G.: Antibiotic susceptibility patterns as aids in classification and characterization of gram-negative anaerobic bacilli. *J Bacteriol, 94:*1443, 1967.

Finegold, S.M., and Hewitt, W.L.: Antibiotic sensitivity pattern of *Bacteroides* species. In: Welch, H., and Marti-Ibanez, F. (Eds.): *Antibiotics Annual, 1955–1956.* New York, Medical Encyclopedia, Inc., 1956, p. 794.

Freid, M.A., and Vosti, K.L.: The importance of underlying disease in patients with gram-negative bacteremia. *Arch Intern Med, 121:*418, 1968.

Garrod, L.P., and O'Grady, F.: *Antibiotics and Chemotherapy,* 2nd ed. Edinburgh and London, Livingstone, 1968, pp. 128, 208, 225.

Gavan, T.L., Cheatle, E.L., and McFadden, H.W., Jr.: *Antimicrobial Susceptibility Testing.* Chicago, Am. Soc. Clin. Pathol., Commission on Continuing Education, 1971, p. 179.

Gelb, A.F., and Seligman, S.J.: *Bacteroidaceae* bacteremia. *JAMA, 212:*1038, 1970.

Holdeman, L.V., and Moore, W.E.C., Eds.: *Anaerobe Laboratory Manual.* Blacksburg, Virginia Polytechnic Institute Anaerobe Laboratory, 1972.

Hoogendijk, J.L.: Resistance of some strains of *Bacteroides* to ampicillin, methicillin and cloxacillin. *Antonie van Leeuwenhoek, 31:*383, 1965.

Ingham, H.R., Selkon, J.B., Codd, A.A., and Hale, J.H.: A study *in vitro* of sensitivity to antibiotics of *Bacteroides fragilis*. *J Clin Pathol, 21:*432, 1968.

Ingham, H.R., Selkon, J.B., Codd, A.A., and Hale, J.H.: The effect of carbon dioxide on the sensitivity of *Bacteroides fragilis* to certain antibiotics *in vitro*. *J Clin Pathol, 23:*254, 1970.

Kislak, J.W.: The susceptibility of *Bacteroides fragilis* to 24 antibiotics. *J Infect Dis, 125:*295, 1972.

Lepper, M.H., and McCabe, W.R.: Chemotherapeutic and antibiotic agents. In: Top, F.H., Sr. (Ed.): *Communicable and Infectious Diseases.* St. Louis, Mosby, 1968, p. 35.

Marcoux, J.A., Zabransky, R.J., Washington, J.A., II, Wellman, W.E., and Martin, W.J.: *Bacteroides* bacteremia. *Minn Med, 53:*1169, 1970.

Martin, W.J., Gardner, M., and Washington, J.A., II: *In vitro* antimicrobial susceptibility of anaerobic bacteria isolated from clinical specimens. *Antimicrob Agents Chemother, 1:*148, 1972.

McCabe, W.R., and Jackson, G.G.: Gram-negative bacteremia. I. Etiology and ecology. *Arch Intern Med, 110:*847, 1962.

McCabe, W.R., and Lorian, V.: Comparison of the antibacterial activity of rifampin and other antibiotics. *Am J Med Sci, 256:*255, 1968.

Reiss, W., Schmid, K., Keberle, H., Dettei, L., and Spring, P.: Pharmacokinetic studies in the field of rifamycins. In: *Proceedings of the 6th International Congress of Chemotherapy—Progress in Antimicrobial and Anticancer Chemotherapy.* Baltimore, Univ Park, 1970, p. 905.

Rosenblatt, J.E., and Schoenknecht, F.: Effect of several components of anaerobic incubation on antibiotic susceptibility test results. *Antimicrob Agents Chemother, 1:*433, 1972.

Sinkovics, J.G., and Smith, J.P.: Septicemia with *Bacteroides* in patients with malignant disease. *Cancer, 25:*663, 1970.

Steers, E., Foltz, E.L., and Graves, B.S.: Inocula replicating apparatus for routine testing of bacterial susceptibility to antibiotics. *Antibiotics and Chemotherapy, 9:*307, 1959.

Sutter, V.L., Kwok, Y.Y., and Finegold, S.M.: Standardized antimicrobial disc susceptibility testing of anaerobic bacteria. I. Susceptibility of *Bacteroides fragilis* to tetracycline. *Appl Microbiol, 23:*268, 1972.

Thornton, G.F., and Cramer, J.A.: Antibiotic susceptibility of *Bacteriodes* species. In: *Antimicrobial Agents and Chemotherapy—1970.* Bethesda, Am Soc Microbiol, 1971, p. 509.

Tynes, B.S., and Frommeyer, W.B., Jr.: *Bacteroides* septicemia. Cultural, clinical, and therapeutic features in a series of twenty-five patients. *Ann Intern Med, 56:*12, 1962.

Wilkins, T.D., Holdeman, L.V., Abramson, J.J., and Moore, W.E.C.: Standardized single disc method for antibiotic susceptibility testing of anaerobic bacteria. *Antimicrob Agents Chemother, 1:*451, 1972.

CHAPTER XLIII

SEROLOGIC SCREENING FOR ACTINOMYCOSIS

PHILLIP I. LERNER

ABSTRACT: *The early diagnosis of actinomycosis remains a problem for most physicians. Delays in diagnosis are associated with various factors relating to difficulty in recovering the organism, its presence as part of the normal oral flora, multiple antibiotic susceptibilities and special growth requirements. A survey of patients with presumed or possible actinomycotic infection has uncovered certain cases where the serologic reaction appears to be quite specific and appropriate to the patient in question with disappearance of antibody following successful therapy. Although the results of this limited survey are not overly encouraging, a broader survey to screen patients with selected pulmonary infections seems in order. Necrotizing pneumonias of unknown etiology, lung abscess involving more than a single pulmonary segment, chronic or relapsing pneumonias and pleural empyemas deserve further evaluation.*

Delayed diagnosis is an unfortunate characteristic of actinomycosis. My earliest interest in this infection was kindled by a tragic case that wandered among several hospitals in Boston for approximately ten years before succumbing to intracranial complications of cervicofacial actinomycosis. The correct diagnosis was appreciated only just prior to death.

A number of factors, listed in Table XLIII-I, contribute in varying degree to thwart the early diagnosis of actinomycosis. Shortly after I arrived at the V. A. Hospital in Cleveland, two interesting cases stimulated the present evaluation of serologic methods in the diagnosis of actinomycosis.

1. C.F. was a 39-year-old black male admitted on 9/24/69 with ulcerations of the anterior chest wall, face, hand and right thigh of six weeks duration. One year earlier the patient suffered a traumatic fracture of the second metacarpal bone of the right hand which

---

Supported by a grant from the Eli Lilly Company and V.A. Research Fund No. 4–68

The author is indebted to Dr. Lucille Georg and R. Marie Coleman of the Mycology Laboratory, Center for Disease Control, Atlanta, Georgia, for their cooperation in performing the serologic determinations reported in this study.

TABLE XLIII-I

FACTORS CONTRIBUTING TO ERRORS OR DELAYS IN THE RECOGNITION OF ACTINOMYCOSIS

Organisms part of normal "oral" flora
Mixed infection usually present
Readily confused with diphtheroids
Erroneous classification as "fungus"
Anaerobic or microaerophilic growth requirement
Multiple antibiotic susceptibility:
—interference in culture media
—growth of organisms suppressed in patient on antibiotic
Prolonged incubation usually needed
Non-specificity of "sulfur granules" (?)

became infected but resolved with antibiotic treatment. Six months later, he developed an abscess of the right cheek and right thigh; treatment with tetracycline, dose and duration unknown, led to resolution. He remained well until six weeks before admission when a new ulcer developed in the right thigh, small ulcerations appeared on the anterior chest wall with recurrent ulcerations on the face and hands. Physical examination revealed an afebrile patient with poor dental repair, a scarred healed ulcer on the right cheek, three small ulcers on the sternum, two scarred lesions on the dorsum of the right hand and a large (15 × 16 cm) indurated, ulcerated draining abscess on the medial aspect of the right thigh, biopsy of which revealed a sulfur granule consistent with actinomycosis. Pertinent laboratory workup included a hematocrit of 36 percent, ESR of 31 mm/hr and a markedly elevated serum globulin level. Chest x-ray revealed a right paratracheal infiltrate with decreased volume of the right upper lobe and a moderate shift of the mediastinum (Fig. XLIII-1). Because of the biopsy finding, he was started on intravenous aqueous crystalline penicillin (6,000,000 u/day) prior to recovery of an organism. Serum tested at the Center for Disease Control reacted strongly with *Actinomyces israelii* serogroup II antigen. Subsequently, organisms grown from the thigh lesion were identified as *A. israelii*, serogroup II (Brock and Georg, 1969; Slack *et al.*, 1969). On the fifth hospital day, penicillin was discontinued because of severe phlebitis and intramuscular cephaloridine (4 gm/day) was substituted, with rapid clearing and healing of the thigh lesion. Cephaloridine therapy was continued for one month and then penicillin V, 4 gm/day, was continued until January, 1970, when the patient finally consented to surgical removal of the right upper lobe. The lung lesion had changed not at all during 3½ months of antibiotic therapy, despite complete healing at all other sites. Right upper lobectomy was performed 1/28/70; histology revealed chronic pneumonitis with focal abscesses and marked fibrosis with a dense pleural reaction. One sulfur granule was seen, but cultures were negative. All sera tested preoperatively as late as December, 1969 continued to react with *A. israelii* type II antigen. Serum drawn 2/10/70, approximately two weeks postoperatively, no longer reacted.

In summary, this represents an unusual case of disseminated actinomycosis, probably secondary to pulmonary disease (Butas *et al.*, 1969; Graybill and Silverman, 1969).

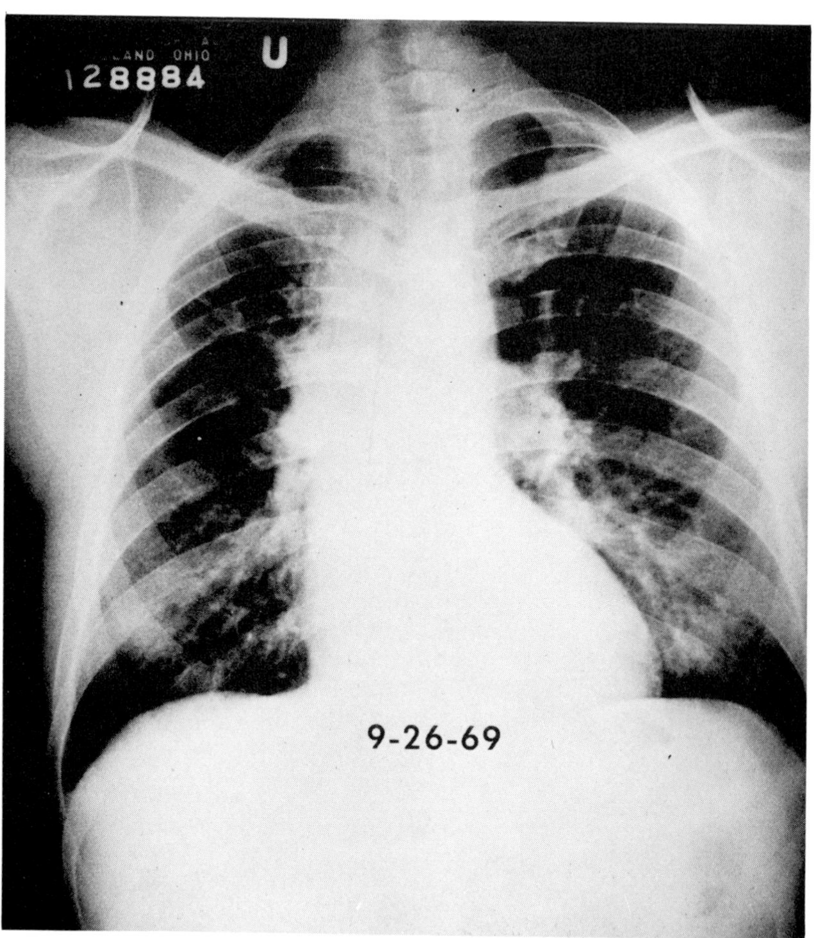

Figure XLIII-1. Chest x-ray of patient C.F., demonstrating a right paratracheal mass and collapse of the right upper lobe.

Serological identification of a specific *Actinomyces* species preceded recovery of that organism from the wound. Antibody persisted until the last obvious focus of actinomycotic infection, the collapsed right upper lung, had been removed surgically. Cephaloridine therapy proved an effective substitute for benzylpenicillin G.

2. C.H. was a 53-year-old white male first admitted in May, 1969 because of a "spot" in the left lung discovered on a routine chest film elsewhere. He was found to have pulmonary and pleural tuberculosis and was treated with isoniazid and ethambutol. Recovery was slow, with resolution of the pulmonary infiltrate and pleural effusion mirroring the gradual disappearance of fever over a period of several months. He was discharged on antituberculous therapy 10/20/69, to be followed as an outpatient. Chest x-ray in January, 1970 showed further reduction in size of the pleural reaction. In May, 1970, he began to complain of fever and chest pain on the left side. Chest x-ray revealed a return of infiltrate in the left upper lung field. Admission was unavoidably delayed until 6/4/70, when a repeat film showed further increase in the infiltrate and the appearance of a well-circumscribed mass in the center of the infiltrate (Fig. XLIII-2). Pleuritic left chest pain had increased and fever and sweating had become persistent. He steadfastly denied any interruption in his antituberculous therapy. Work-up revealed a hematocrit of 36 percent and a white blood count of 16,000/mm^3 with 83 percent polymorphonuclear leukocytes. Gram stains of sputum revealed many polymorphonuclear leucocytes but few bacteria; AFB smears were negative. Sputum cultures yielded normal flora on three occasions, but sputum cytology revealed "pus cells and actinomycotic granules." He was started on intravenous aqueous penicillin (6,000,000 u/day) and defervesced within 48 hours, with marked clinical improvement and rapid roentgenographic clearing. Subsequently all sputum cultures for AFB were negative. Serum tested at the Center for Disease Control demonstrated precipitins against *Arachnia propionica* on 6/24/70 and 7/21/70, but was negative on 12/12/70 when the new pulmonary lesions had completely resolved.

Although actinomyces were not recovered from his sputum, this patient may represent an example of pulmonary *Arachnia propionica* infection in man. Penicillin therapy produced striking clinical and roentgenographic response in a subacute pulmonary infection associated with sulfur granules in a purulent sputum with subsequent disappearance of specific antibody. It should be noted that at this hospital all sputa are routinely cultured for anaerobes, but the plates are held only for 48 hours. Actinomyces may easily be missed by this technique.

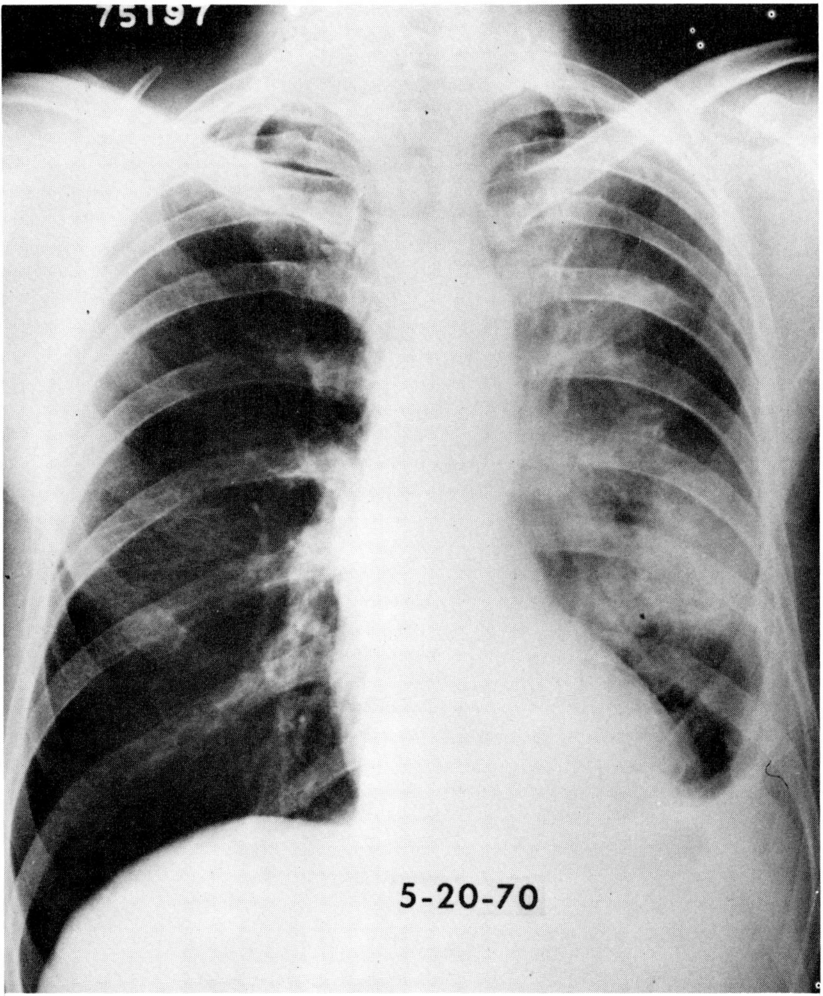

Figure XLIII-2. Chest x-ray of patient C.H., demonstrating infiltrate throughout the entire left upper lobe with a well-circumscribed mass in the left mid-lung field.

SEROLOGIC SURVEY

Encouraged by the serologic aspects of these two interesting cases, we began a survey for serum actinomyces precipitins in patients with known or possible actinomycosis, especially cervicofacial and pleuropulmonary disease, with particular emphasis on cavitary lung diseases such as lung abscess and necrotizing pneumonias. All serologic testing was carried out at the Center for Disease Control in Dr. Lucille Georg's laboratory. Sera from a group of miscellaneous patients were submitted as interesting "controls."

RESULTS

Table XLIII-II lists the pertinent clinical data on 11 patients from whom *Actinomyces* species were recovered. Only one of four patients with cervicofacial actinomycosis had a positive serology. This was an 11-year-old boy with an acute submandibular abscess and serum precipitins for *Actinomyces israelii* types I and II as well as *Arachnia propionica*, whose abscess grew *A. israelii* and an anaerobic streptococcus. A repeat serum drawn seven weeks later, when the lesion had healed, no longer contained precipitin antibodies. Two patients had *Actinomyces naeslundii* isolated repeatedly as the predominant organism in bronchoscopic sputum during the course of a work-up for cavitary pulmonary disease but neither demonstrated antibody to this organism. In two patients with acute empyema, no precipitins were found. Indeed, the only impressive result listed is the patient previously mentioned with disseminated *A. israelii* type II infection.

Nineteen patients had probable primary lung abscesses. This diagnosis was based on the nature and location of the abscess, presence of factors predisposing to aspiration, response to penicillin, characteristic Gram stains and cultures. Sixteen patients with lung abscess had negative serologies, one despite repeated isolation of *A. naeslundii* as the predominant organism of his mixed sputum flora. This chronic lung abscess was ultimately resected and there was no histologic confirmation of actinomycosis. Three patients in the lung

TABLE XLIII-II

SEROLOGIC RESULTS IN PATIENTS WITH POSITIVE ACTINOMYCES CULTURES

Pt.	*Age*	*Sex*	*Lesion*	*Species*	*Source*	*Pathology*	*Serology*	*Duration*† (*Months*)
S.E.	11	M	Cervicofacial	*A. israelii*	Neck abscess	Not done	*A. israelii* I, II *Ar. propionica*	½
R.B.	40	M	Cervicofacial	*A. israelii*	Neck abscess	Sulfur granule	0	½
L.M.	40	M	Cervicofacial	*A. israelii*	Neck abscess	Sulfur granule	0	10
G.E.	72	M	Cervicofacial	*A. israelii*	Neck abscess	Sulfur granule	0	15
C.F.	39	M	Disseminated	*A. israelii* II	Thigh abscess	Sulfur granule	*A. israelii* II	6
D.E.	48	M	Lung nodule	*A. israelii*	Lung biopsy	Chronic infection	0	12
J.H.	40	M	None	*Actinomyces* sp.*	Sputum	Not done	Weak + *A. israelii*	
W.B.	38	M	Lung abscess	*A. naeslundii*	Sputum	Chronic abscess	0	4
C.S.	38	F	Apical cavity	*A. naeslundii*	Sputum	Granuloma	0	5
L.S.	55	M	Empyema	*A. israelii*	Pl. fluid	Not done	0	½
M.O.	78	F	Mixed empyema	*A. israelii*	Pl. fluid	Not done	0	½

* Unclassified aerobic species.
† Duration of illness prior to first serum specimen.

abscess group, however, had positive serologic reactions (Table XLIII-III). G.D. had only a weakly positive reaction for *A. israelii* antigens and appeared to have a typical, uncomplicated lung abscess. The other two, F.B. and C.D., are of interest since their lesions involved more than one pulmonary segment, the only two such patients among the entire lung abscess group. One had two abscesses while the other had an abscess in the superior segment of the right lower lobe and a pneumonia in the posterior segment of the right upper lobe.

Other cavitary pulmonary infections surveyed deserve comment. Three of four patients with necrotizing pneumonias, in whom specific pathogens could not be identified, had positive precipitin reactions (Table XLIII-III). D.S. failed to respond to either penicillin or erythromycin, but had prompt clinical and x-ray improvement following the administration of chloramphenicol. L.W. had a necrotizing left upper lobe pneumonia and responded to penicillin and antituberculous therapy, although AFB smears and cultures were ultimately negative. A repeat serum taken eight months later, when he was asymptomatic and the left upper lobe almost entirely atelectatic, no longer reacted with actinomyces antigens. One of two patients with cavitary pulmonary carcinoma and one of two patients with active cavitary pulmonary tuberculosis had positive serum precipitin reactions (Table XLIII-III). As noted in Table XLIII-II, of four patients with active cervicofacial actinomycosis, proven either by biopsy or culture, only one had serum precipitins which disappeared after seven weeks, following successful treatment. However, two patients with malignant neck lesions also had precipitin antibodies (Table XLIII-III). J.L. ultimately had a diagnosis of Hodgkin's disease established after several biopsies; his serum had a weakly positive reaction with *A. israelii* antigens. W.J. had erythroleukemia and recurrent, bilateral, massive indurated neck swellings which ultimately softened and drained in response to penicillin therapy. Only coagulase-positive staphylococci could be recovered from these lesions. Successive serologies over a four-month period revealed initially a weak, but later definitely positive reaction for *A. israelii*

TABLE XLIII-III

CLINICAL DATA IN PATIENTS WITH SERUM ACTINOMYCES PRECIPITINS

Pt.	*Age*	*Sex*	*Serology*	*Diagnosis*	*Actinomyces Culture*	*Pathology*	*Response to penicillin*
R.B.	63	F	*A. israelii* I, II	Active cavitary TB	0	Not done	No
F.B.	52	M	*A. israelii* I, II *Ar. propionica*	Lung abscess*	0	Not done	Yes
C.F.	39	M	*A. israelii* II	Disseminated actinomycosis	*A. israelii* II	Sulfur granules	Yes
J.C.	78	M	*A. israelii* I	Pleural effusion	Not done	Not done	
C.D.	42	M	*A. israelii* I, II	Lung abscess*	0	Not done	?
G.D.	52	M	*A. israelii* I, II†	Lung abscess	0	Not done	Yes
S.E.	11	M	*A. israelii* I, II *Ar. propionica*	Cervicofacial actinomycosis	*A. israelii* I	Not done	Yes
S.H.	37	M	*A. israelii* I, II	Necrotizing pneumonia	0	Not done	?
D.S.	42	M	*A. israelii* I, II	Necrotizing pneumonia	0	Not done	No
L.W.	52	M	*A. israelii* I, II *Ar. propionica*	Necrotizing pneumonia	0	Not done	Yes
G.R.	58	M	*A. israelii*	Carcinoma, lung, cavity	0	0	No
J.L.	29	M	*A. israelii*†	Hodgkins	0	0	No
W.J.	58	M	*A. israelii*	Erythroleukemia, neck masses	0	0	Yes

* Involvement of 2 pulmonary segments.
† Weakly positive reactions.

I, II (Table XLIII-III). Unfortunately, autopsy was refused. Table XLIII-III summarizes all patients with positive serologic reactions for actinomyces precipitins. Only two of the 13 patients had positive cultures for actinomyces, but preantibiotic culture specimens were usually not available.

Only two patients with empyema were studied (Table XLIII-II). L.S. had an acute purulent empyema and grew a single colony of *A. israelii* as the sole organism from his pleural fluid, but his serum contained no precipitins. M.O., an elderly lady with *A. israelii* and bacteroides empyema, also lacked serum precipitins. Ten patients with other diagnoses, including brain abscess, tuberculosis of the rib, *Hemophilus influenzae* empyema and pulmonary cryptococcosis, had negative serum precipitin reactions, as did four additional men with carcinoma metastatic to cervical and submandibular lymph nodes. Two patients with lymphoproliferative malignancies and nocardiosis had non-specific reactions to actinomyces antigens.

Table XLIII-IV lists the serologic reactions and clinical data in five patients with active or past tuberculous infection. Three of the five demonstrated serum actinomyces precipitins.

DISCUSSION

There are three major clinical forms of actinomycosis: cervicofacial, pulmonary and abdominal. Each type usually

TABLE XLIII-IV

SEROLOGIC REACTIONS AND CLINICAL DATA IN 5 PATIENTS WITH CURRENT OR OLD TUBERCULOSIS*

Pt.	*Age*	*Sex*	*Diagnosis*	*AFB Culture*	*Actinomyces Serology*	*Response to Penicillin*
R.B.	63	F	Apical cavity	+	*A. israelii* I, II	No
R.R.	51	M	Lung abscess	+	0	Yes
C.H.	53	M	Pneumonia	0	*Ar. propionica*	Yes
C.D.	42	M	Lung abscess	0	*A. israelii* I, II	?
C.E.	53	M	Lung abscess†	0	0	Yes

* Actinomyces cultures negative.
† *Mycobacterium kansasii.*

escapes identification for periods of months to years, on occasion. Biopsy of cutaneous or cervicofacial lesions ultimately alerts the physician to the correct diagnosis. However, pulmonary actinomycosis poses special problems for diagnosis, since the organism represents part of the normal oral flora (Kay, 1948).

In recent years, a number of species other than *A. israelii* (Table XLIII-V) have been associated with human disease (Coleman and Georg, 1969; Leers *et al.*, 1969; Gerencser and Slack, 1967, 1969; Coleman *et al.*, 1969; Georg *et al.*, 1965; Slack and Gerencser, 1970; Merline and Mattman, 1971). It thus becomes increasingly important to establish a means of definitive diagnosis other than recovery of the organism from pulmonary secretions in puzzling cases. A diagnostic serologic test would be most welcome, since patients with puzzling pulmonary or pleuropulmonary infections have already received multiple antibiotics before they are identified as problem cases. "Unfortunately, by the time the need for special cultural technics is appreciated, if indeed it is considered at all, the rather ready susceptibility of *A. israelii* to most available antibiotics has so suppressed its growth that cultures, even when properly sown, are likely to be negative" (Peabody and Seabury, 1960).

It is apparent from this report that the serum precipitin reaction may be quite specific in selected instances, as in the two cases summarized above. However, the overall survey is not encouraging and generally supports the findings of Georg, Coleman and Brown (1968), who reported positive

TABLE XLIII-V

INFECTIONS IN MAN WITH ACTINOMYCES OTHER THAN *A. ISRAELII**

A. naeslundii
—empyema of gall bladder, wounds, cervicofacial, suppurative thyroiditis, bacteremia
Ar. propionica
—lacrimal canaliculitis, cervicofacial
A. eriksonii
—lung and wound abscess, empyema
A. viscosus (aerobic; ? catalase + *A. naeslundii*)
—wound, sinus tract

* See Kay (1948); Leers *et al.* (1969); Gerencser and Slack (1967, 1969); Coleman *et al.* (1969); Georg *et al.* (1965); Slack and Gerencser (1970); Merline and Mattman (1971).

precipitin reactions in 21/24 cases with disseminated disease but only 1/10 with a localized form of actinomycosis. They also found reactions in 13/30 patients with tuberculosis and in patients with a variety of other infections and conditions, including hidradenitis suppurativa, nocardiosis, streptococcal infections, Hodgkin's disease and pneumonia. It should be noted that many of their sera reacted only after two- or four-fold concentration.

Despite these limitations, it still appears worthwhile to continue investigating the possible diagnostic value of precipitin antibodies in selected forms of pulmonary infection. These are (1) primary aspiration lung abscess, particularly involving more than one pulmonary segment; (2) penicillin-responsive necrotizing pneumonias where the usual pathogens are not recovered; (3) empyemas; and (4) all chronic or relapsing pneumonias, especially if sulfur granules are seen in sputum cytologies. The specificity of sulfur granules in sputum remains to be determined (Graybill and Silverman, 1969; Speir *et al.*, 1971; Robboy and Vickery, 1970).

The number of cases is few, but there is a suggestion that specific serologic reactions, when present, may aid both diagnosis and management. Specific precipitins were no longer present in follow-up sera from the two patients discussed at the beginning of this paper. C.F. retained *A. israelii* II antibody until his infected atelectatic right upper lobe was resected. C.H. no longer demonstrated serum antibody to *Arachnia propionica* antigens six months after successful penicillin treatment. Albeit his acute serum contained antibody reacting with both *A. israelii* and *Arachnia propionica* antigens, serum taken from S.E. (Table XLIII-II) when cured seven weeks later no longer reacted with any of the antigens tested. The same was true of L.W. (Table XLIII-III), the only patient with necrotizing pneumonia in whom a follow-up serum was obtained. Despite progressive and persistent atelectasis of his left upper lobe, the patient was asymptomatic eight months later and his serum no longer reacted with any actinomyces antigens.

Therapeutic principles for managing actinomycosis remain unchanged since the excellent review by Peabody and Seabury (1960), who emphasized a judicious combination of prolonged antibiotic therapy with either penicillin or tetracycline coupled with appropriate surgical drainage or excision of sinus tracts for optimum results. The introduction of newer antimicrobial categories, particularly the cephalosporins and lincomycins, enlarges the alternative choices to penicillin in patients who are allergic to that drug or in those unable to tolerate the tetracyclines. Clinical experiences are not extensive, but reports already suggest that cephalothin and cephaloridine as well as lincomycin and clindamycin are effective alternative drugs in the management of actinomycosis (Mohr *et al.*, 1970; Caldwell, 1971; Rose and Rytel, 1972). The oral cephalosporin compounds, however, have not yet been tested. *In vitro* susceptibility data suggest that the parenteral cephalosporins and the lincomycin compounds should be active against all of the species thus far associated with human disease (Lerner, 1968, 1969, and unpublished data).

REFERENCES

Brock, D.W., and Georg, L.K.: Characterization of *Actinomyces israelii* serotypes 1 and 2. *J Bacteriol, 97:*589, 1969.

Butas, C.A., Read, S.E., Coleman, R.E., and Abramovitch, H.: Disseminated actinomycosis. *Can Med Assoc J, 103:*1069, 1970.

Caldwell, J.L.: Actinomycosis treated with cephalothin. *South Med J, 64:*987, 1971.

Coleman, R.M., and Georg, L.K.: Comparative pathogenicity of *Actinomyces naeslundii* and *Actinomyces israelii. Appl Microbiol, 18:*427, 1969.

Coleman, R.M., Georg, L.K., and Rozzell, A.R.: *Actinomyces naeslundii* as an agent of human actinomycosis. *Appl Microbiol, 18:*420, 1969.

Georg, L.K., Coleman, R.M., and Brown, J.M.: Evaluation of an agar gel precipitin test for the serologic diagnosis of actinomycosis. *J Immunol, 100:*1288, 1968.

Georg, L.K., Robertstad, G.W., Brinkman, S.A., and Hicklin, M.D.: A new pathogenic anaerobic *Actinomyces* species. *J Infect Dis, 115:*88, 1965.

Gerencser, M.A., and Slack, J.M.: Identification of human strains of *Actinomyces viscosus. Appl Microbiol, 18:*80, 1969.

Gerencser, M.A., and Slack, J.M.: Isolation and characterization of *Actinomyces propionicus. J Bacteriol, 94:*109, 1967.

Graybill, J.R., and Silverman, B.D.: Sulfur granules. *Arch Intern Med, 123:*430, 1969.

Kay, E.B.: *Actinomyces* in chronic bronchopulmonary infections. *Am Rev Tuberc, 57:*322, 1948.

Leers, W.D., Dussault, J., Mullens, J.E., and Volpe, R.: Suppurative thyroiditis: an unusual case caused by *Actinomyces naeslundii. Can Med Assoc J, 101:*714, 1969.

Lerner, P.I.: Susceptibility of *Actinomyces* to cephalosporins and lincomycin. In: *Antimicrob Agents Chemother—1967*. Ann Arbor, Am Soc Microbiology, 1968, p. 730.

Lerner, P.I.: Susceptibility of *Actinomyces* to lincomycin and its halogenated analogues. In: *Antimicrob Agents and Chemother—1968.* Ann Arbor, Am Soc Microbiology, 1969, p. 461.

Lerner, P.I.: Unpublished data.

Merline, J.R., and Mattman, L.H.: Cell wall deficient forms of *Actinomyces* complicating two cases of leukemia. *The Michigan Academician, 3:*113, 1971.

Mohr, J.A., Rhoades, E.R., and Muchmore, H.G.: Actinomycosis treated with lincomycin. *JAMA, 212:*2260, 1970.

Peabody, J.W., and Seabury, J.H.: Actinomycosis and nocardiosis. *Am J Med, 28:*99, 1960.

Robboy, S.J., and Vickery, A.L.: Tinctorial and morphologic properties distinguishing actinomycosis and nocardiosis. *N Engl J Med, 282:*593, 1970.

Rose, H.D., and Rytel, M.W.: Actinomycosis treated with clindamycin. *JAMA, 221:*1052, 1972.

Slack, J.M., and Gerencser, M.A.: Two new serological groups of *Actinomyces. J Bacteriol, 103:*266, 1970.

Slack, J.M., Landfried, S., and Gerencser, M.A.: Morphological, biochemical, and serological studies on 64 strains of *Actinomyces israelii. J Bacteriol, 97:*873, 1969.

Speir, W.A., Mitchener, J.W., and Galloway, R.F.: Primary pulmonary botryomycosis. *Chest, 60:*92, 1971.

CHAPTER XLIV

INFECTIONS DUE TO ANAEROBIC COCCI

DWIGHT W. LAMBE, JR., DAVID H. VROON
AND CHARLES W. RIETZ

ABSTRACT: *Introduction of improved culture and identification procedures and other measures increased the yield of clinical isolates of gram-positive anaerobic cocci from 3 in 39 months to 368 in 41 months. Identification of the species of gram-positive anaerobic cocci isolated in pure culture from patients and evaluation of clinical findings demonstrated the capability of these organisms to produce clinically significant infection in the human.* Peptococcus *species were isolated in pure culture in 15 infections, ranging from simple, benign skin infections responsive to local measures to more severe diseases such as deep cavity abscesses, osteomyelitis and septicemias. Only one* Peptostreptococcus *isolate was found in pure culture; this was from thoracentesis fluid. In several cases in which a single anaerobic coccus was the predominant organism while aerobic bacterial growth (usually* Staphylococcus epidermidis*) was very sparse, it appeared that the anaerobic coccus was probably the etiologic agent. When a single anaerobic coccal species was mixed with aerobes of known pathogenicity, or when mixed anaerobes and aerobes were present, the precise etiologic role of each could not be determined. Further studies are needed to delineate the potential pathogenicity of individual anaerobes.*

Relatively little has been reported in the literature regarding the pathogenicity of various species of the anaerobic cocci and so this problem has been of particular interest to us for the past two years. Since the early 1900s, the anaerobic streptococci have been implicated in certain disease processes including septic abortion, pelvic abscesses, and puerperal sepsis (Bornstein *et al.*, 1964; Sandusky *et al.*, 1942; Smith and Holdeman, 1968; Thomas and Hare, 1954). Peptostreptococci have been implicated in a number of other diseases such as anaerobic myositis (MacLennon, 1943); various types of abscesses (Finegold *et al.*, 1968; Heineman and Braude, 1963; Moore *et al.*, 1969; Patterson *et al.*, 1967; Sandusky *et al.*, 1942; Thomas and Hare, 1954); subacute bacterial endocarditis (Vogler *et al.*, 1962); peritonitis (Sandusky *et al.*, 1942); crepitant anaerobic gangrene (Altemeier and Culbertson, 1948); and progressive synergistic gangrene (Brewer and Meleney, 1926; Meleney, 1931).

Stokes (1958) reported pure cultures of anaerobic cocci,

which were not further identified, from sebaceous cysts, finger infections, breast abscesses, and paronychial infection. Pien, Thompson and Martin (1954) also reported anaerobic gram-positive cocci from ischemic ulcers, skin infections, and surgical wounds. However, in no study was speciation of the anaerobic cocci accomplished.

It is not uncommon to find anaerobic gram-positive cocci mixed with multiple anaerobes and aerobes; we ourselves have seen more than a hundred such infections. The role of a particular organism is extremely hard to define in such infections. In the study described here, we have definitively speciated anaerobic cocci by present-day diagnostic methods and evaluated clinical findings in cases in which the isolates were obtained in pure culture in an effort to determine more about the disease potential of each anaerobic coccus. Examples are presented also of infections in which anaerobic cocci were isolated with other organisms.

MATERIALS AND METHODS

Bacteriological Isolation

The system for isolation of anaerobic bacteria from clinical specimens used in our laboratory until four years ago consisted of culturing specimens into thioglycollate broth, and then subculturing the broth onto blood agar plates that were incubated in GasPak anaerobic jars. The new procedure for cultivation of anaerobic bacteria from clinical specimens consisted of streaking pus or fluid directly onto two plated media and into a tube of prereduced anaerobically sterilized (PRAS) chopped meat glucose (CMG) broth (Scott Laboratories). A specimen submitted on a swab was swirled in 0.5 ml of PRAS CMG broth and then the inoculum was transferred to two plates of media plus a tube of CMG broth. The first plated medium was blood agar prepared from brain heart infusion agar (Difco), 0.5% yeast extract, and 5% sheep blood. The second medium was 4% agar with laked blood prepared with brain heart infusion agar (Difco), 0.5% yeast extract, 4% agar, 5% laked sheep blood, and hemin-menadione solution (Holdeman and Moore, 1972). The two plated media were used within two to three hours after preparation unless stored in an anaerobe jar.

Bacteriological Identification

Identification of the anaerobic cocci in our laboratory was carried out according to the methods described by Holdeman and Moore (1972). The biochemicals inoculated to characterize these anaerobes included: cellobiose, fructose, glucose, lactose, maltose, mannitol, melizitose, sucrose, gelatin, indol, nitrate, aesculin for hydrolysis and a brain-heart infusion agar slant for catalase production. All biochemical media were prereduced media from Scott Laboratories. Gas chromatographic analyses were carried out on ether extracts of glucose cultures of all isolates, and analyses were also performed on pyruvate and threonine extracts if needed for speciation or subpeciation.

Selection of Cases

The cases of infection associated with anaerobic cocci described here are those we have seen in our hospital within the past nine months. These serve as examples of the types of infections associated with these organisms. Some other examples mentioned in this paper were described previously by Lambe and Del Bene (1972).

After evaluation of each case, the judgment was made whether or not there were signs of significant infection which were directly related to the specific culture in question. Criteria for infection were based upon the responses of the host: systemic signs of fever, leukocytosis or toxicity indicative of septicemia, and/or local signs of induration and inflammation or pus accumulation. In no case was the judgment based on fever or leukocytosis alone. Unless noted otherwise, the cases described here met the clinical criteria for presence of infection.

The cases are divided into groups based on the identification of the isolates:

Infections yielding a single gram-positive anaerobic coccus in pure culture.

Infections yielding gram-positive anaerobic cocci mixed with anaerobic gram-negative rods but no aerobes.

Infections involving a single gram-positive anaerobic coccus mixed with one or more aerobes.

Infections yielding gram-negative anaerobic cocci.

RESULTS AND DISCUSSION

Bacteriological Isolates

The isolation of anaerobic cocci in the Microbiology Section of the Clinical Pathology Laboratory at Emory University Hospital was rare until 4 years ago when we: (1) began working closely with the hospital clinician and nurses to improve specimen sampling; (2) achieved at least some progress in more rapid culture of clinical specimens; and (3) established an anaerobic bacteriology laboratory to culture and to definitively identify anaerobic organisms. Table XLIV-I compares the numbers of anaerobic cocci from wounds, abscesses, tissues and fluids that were isolated in our laboratory before and after these changes.

During the thirty-nine months preceding improved isolation methods, when thioglycollate broth was used, only three strains of unspeciated peptococci were found. During the subsequent forty-one months, using improved culture methods, we have isolated a total of 368 strains of gram-positive anaerobic cocci and speciated most of them. During the first two years of operation of our anaerobe laboratory we did not include gelatin liquefaction in the biochemical group of tests to differentiate *Peptococcus magnus* from *Peptococcus variabilis*. Since incorporating this differential test into our identification scheme, we have noted that *P. variabilis* is much more common isolate in our laboratory than *P. magnus*. Therefore, a number of strains that were originally classified as *P. magnus* may have been *P. variabilis* if, indeed, the separation of these two species is valid.

In the gram-positive group, the peptococci far outnumbered the peptostreptococci. Of the 302 strains of *Peptococcus*, the most common species were *Peptococcus prevotii*, *Peptococcus asaccharolyticus*, *Peptococcus variabilis*, and *Peptococcus magnus*. Of the 66 strains of peptostreptococci isolated, the most common species was *Peptostreptococcus anaero-*

bius. Of 57 isolates in the gram-negative group, the single most common species was *Veillonella parvula*, followed closely by *Veillonella alcalescens*.

Selected Cases of Infection with Gram-positive Anaerobic Cocci

Pure Cultures of Peptococcus *or* Peptostreptococcus *species*

PEPTOCOCCUS VARIABILIS. Lambe and Del Bene (1972) described several significant infections from which *P. variabilis* was isolated in pure culture. These included osteomyelitis of the femur, chronic abscess of the neck, pneumonitis, and secondarily infected stasis ulcer.

One strain of *P. variabilis* was cultured from a urinary catheter tip, but there was no evidence of infection; thus, the organism was possibly part of the normal flora of the urethra.

P. variabilis fits the description of *Micrococcus variabilis* by Prevot, who reported isolating this organism in cases of metritis, tonsillitis, sinusitis and septicemia (Prevot, 1966).

TABLE XLIV-I

ANAEROBIC COCCI ISOLATED BEFORE AND AFTER IMPROVED CULTURE TECHNIQUES

Organisms	*No. Isolated During 39 Months Before*	*No. Isolated During 41 Months After*
Gram-positive cocci		
Peptococcus asaccharolyticus	0	83
Peptococcus constellatus	0	1
Peptococcus micros	0	1
Peptococcus magnus	0	83
Peptococcus morbillorum	0	1
Peptococcus prevotii	0	92
Peptococcus variabilis	0	33
Peptococcus species	3	8
Peptostreptococcus anaerobius	0	56
Peptostreptococcus intermedius	0	6
Peptostreptococcus species	0	4
	3	368
Gram-negative cocci		
Veillonella alcalescens	0	24
Veillonella parvula	0	30
Veillonella species	0	3
	0	57

Peptococcus magnus. Seven significant infections associated with pure isolates of *P. magnus* included one scrotal cyst; three postoperative wound infections; one subcutaneous thigh abscess; and one case of septicemia (Lambe and Del Bene, 1972). We have seen another case recently:

> Patient H.T., a 64-year-old male with an atherosclerotic aneurysm of the abdominal aorta, had an elective resection. Postoperatively he developed a wound infection, and pus from the midline wound grew out a pure culture of *P. magnus*. The patient responded readily to cephalothin therapy and incision and drainage.

P. magnus corresponds to Prevot's description of *Diplococcus magnus*, which he reported being isolated from puerperal infection, pyelonephritis, pulmonary gangrene, an osteomyelitic fistula and certain other infections (Prevot, 1966).

Peptococcus asaccharolyticus. We have isolated *P. asaccharolyticus* in pure culture from two significant infections (Lambe and Del Bene, 1972). One came from the uterine cavity in a case of septic abortion. The infection responded readily to treatment with penicillin and streptomycin and dilatation and curettage. The other isolate was from a case of bacteremia associated with a postoperative subphrenic abscess. Successful treatment included parenteral cephalosporin and incision and drainage of the abscess.

A third pure culture of *P. asaccharolyticus* was obtained from axillary lymph node tissue in the absence of any clinical evidence of infection.

Peptococcus prevotii. Only one of our four pure cultures of *P. prevotii* was associated with significant infection: pus from a secondarily infected epidermal inclusion cyst yielded a pure culture. In three other cases there was no evidence of infection; one isolate came from a resected aneurysm at surgery and the others from blood from patients with cardiac diseases. We can only speculate regarding the origin of these latter three isolates, but the skin seems to be a likely source.

Prevot (1966) mentioned four strains of *Micrococcus prevotii* which caused lung abscesses, purulent cystitis, suppuration of the lips and recurring subcutaneous abscesses. The two saccharolytic strains are now called "*Gaffkya*" *anaerobia*, while the two nonsaccharolytic strains are classified as *P. prevotii* (Holdeman, L.V., personal communication).

TABLE XLIV-II

BACTERIOLOGIC FINDINGS IN A PATIENT WHO DEVELOPED SEPTICEMIA DUE TO *PEPTOCOCCUS CONSTELLATUS*

Date of Culture	*Source*	*Anaerobes*	*Aerobes*
5/15/72	Gastrostomy tube	*Lactobacillus acidophilus**	Enterococcus group
		*Peptococcus constellatus**	Yeast
		Veillonella parvula	*Neisseria* species
		Peptostreptococcus intermedius	*Eickonella corrodens*
		Bacteroides oralis	α-hemolytic streptococcus (not group D)
			Staphylococcus epidermidis
5/18/72	Gastrostomy tube	*L. acidophilus*	α-hemolytic streptococcus
		P. constellatus	Enterococcus group
		Bacteroides ruminicola ss. *brevis*	*Staphylococcus aureus*
			Acinetobacter calcoaceticus
			S. epidermidis
5/18/72	Blood (5 bottles; 2 venipunctures)	*L. acidophilus*	None
5/19/72	Blood (6 bottles; 2 venipunctures)	*L. acidophilus* (6/6)	None
		L. casei var. *casei** (1/6)	
6/7/72	Blood (2 bottles; 1 venipuncture)	*L. casei* var. *rhamnosus**	None
	1 Pleural fluid	*L. casei* var. *rhamnosus*	
6/17/72	Blood (2 bottles; 1 venipuncture)	*Peptococcus constellatus*	None

* VPI identification.

PEPTOCOCCUS CONSTELLATUS. In 1924 Prevot reported isolation of *Diplococcus constellatus* from a case of tonsillitis and also found the organism associated with appendicitis and purulent pleurisy (Prevot, 1966). However, we have found no previous report of *Peptococcus constellatus* as the causative agent of septicemia. In the case described below and summarized in Table XLIV-II, *P. constellatus* was isolated in pure culture from the blood during an episode of septicemia in a debilitated patient.

A 52-year-old male had inoperable carcinoma of the head of the pancreas complicated by the occurrence of two pancreatic pseudocysts requiring drainage. He was being greated with 5-fluorouracil. Approximately four days after the second drainage and insertion of a gastrostomy tube, fever of 103.2° F and leukocytosis (25,000 cells/mm^3) were noted. Multiple blood cultures taken at this time were negative. Clinically, the patient had evidence of a left subdiaphragmatic infection. He did not respond to treatment with ampicillin, tetracycline and oral cephalosporin. Two weeks postoperatively (5/15/72) a culture of purulent drainage from around the gastrostomy tube yielded multiple aerobes and anaerobes (Table XLIV-II), including *Lactobacillus acidophilus* and *P. constellatus*. A second culture three days later (5/18/72) yielded a somewhat different mixture of aerobes and anaerobes, still including *L. acidophilus* and *P. constellatus*. That same day blood cultures were positive for *L. acidophilus;* these were the first positive blood cultures during the period of sepsis. The patient had persistent fever and leukocytosis. On 5/19/72 his blood cultures were positive for *Lactobacillus casei* var. *casei* as well as *L. acidophilus*. He was treated with intravenous cephalothin, which resulted in a gradual decrease in temperature. However, during this therapy, blood cultures (and pleural fluid) were positive for *Lactobacillus casei* var. *rhamnosus*.

Recurrent fever after cephalothin therapy was discontinued prompted additional blood cultures, which yielded *P. constellatus* alone. The patient responded to another course of intravenous cephalothin and had no subsequent septicemias.

Besides the isolation of *P. constellatus* in pure culture in the presence of signs of septicemia, there are two other noteworthy points: (1) the local infection around the gastrostomy tube was apparently the origin of the etiologic agents of septicemia; and (2) certain aerobes in the local infection, such as *Staphylococcus aureus* and *Acinetobacter calcoaceticus*, to which we often attribute potential pathogenicity, did not cause systemic involvement, whereas certain anaerobic bacteria to which we have previously attributed little pathogenicity caused septicemia. This patient had septicemias due to four anaerobic organisms.

PEPTOSTREPTOCOCCUS INTERMEDIUS (*Microaerophilic*). In the case summarized below, a microaerophilic strain of *P. intermedius* was isolated in pure culture from thoracentesis fluid.

A 54-year-old male presented with a cerebral infarct and aspiration pneumonia. *Pseudomonas aeruginosa*, *Klebsiella* species and *Serratia marcescens* were isolated from the sputum. The patient developed a pleural effusion. The thoracentesis fluid was cloudy, yellow-brown and purulent, and grew out a pure culture of a microaerophilic strain or *P. intermedius*. Blood cultures were negative. The pneumonia and pleural effusion cleared with ampicillin, cephalothin and gentamicin. The role of anaerobes in the pneumonia was unknown, since a proper anaerobic specimen was not obtained.

Other investigators have attributed infections such as septic abortion, abscesses and stasis ulcers to "anaerobic peptostreptococci" (Bornstein *et al.*, 1964; Finegold *et al.*, 1968; Heineman and Braude, 1963; Moore *et al.*, 1969; Patterson *et al.*, 1967; Pien *et al.*, 1954; Prevot, 1966; Sandusky *et al.*, 1942; Thomas and Hare, 1954).

COMMENTS ON ANAEROBIC COCCI ISOLATED IN PURE CULTURE. Very little information regarding the pathogenicity of specific species of *Peptococcus* has been available. The cases cited above illustrate the capability of specific anaerobic cocci to produce infections ranging from simple, benign skin lesions responding to local treatment to more severe diseases such as deep cavity abscesses, osteomyelitis and septicemias. In this series *Peptococcus* species were isolated in pure culture in 15 cases of significant infection. Only one significant infection due to *Peptostreptococcus* species, a microaerophilic strain of *Peptostreptococcus intermedius,* was noted. However, others have reported "anaerobic peptostreptococci" in a variety of infections. Close attention to complete identification of anaerobic cocci isolated in pure culture from clinically significant infections will help clarify the pathogenic potential of each species. Eventually we should be able to speak as knowledgeably about each anaerobe as we now do about aerobic organisms such as *Staphylococcus aureus* and *Pseudomonas aeruginosa.* Animal pathogenicity studies, further investigation of the immunological response in man, and additional clinical studies should help achieve this goal.

Gram-Positive Anaerobic Cocci Mixed with Anaerobic Gram-Negative Rods

We have seen nine cases of infection due to anaerobic cocci mixed with anaerobic gram-negative rods. Seven previously reported by Lambe and Del Bene (1972) were caused by one of five different cocci mixed with one or more anaerobic gram-negative rods. These cases demonstrate the spectrum of infections in which anaerobic bacteria can be involved as well as the sources of such infections. They ranged from simple, low-grade infection of a sebaceous cyst to anaerobic septicemia secondary to a decubitus ulcer. One case each of pelvic abscess and Bartholin gland abscess were secondary to organisms from the vagina, while an intraperitoneal abscess and two periappendiceal abscesses were secondary to contamination by intestinal contents. We have seen two additional cases recently:

Patient A.L.G., a 57-year-old female with ovarian adenocarcinoma and pelvic carcinomatosis treated with irradiation and chemotherapy, developed recurrent pelvic infection and was treated with tetracycline over a long period of time. Two cultures of cul-de-sac fluid and vaginal drainage yielded anaerobic gram-negative and gram-positive rods, as well as *P. prevotii, P. variabilis, P. asaccharolyticus, Peptostreptococcus anaerobius, "Gaffkya" anaerobia* and three aerobes. She also developed septicemia due to *Bacteroides fragilis* ss. *fragilis*, and we demonstrated an immunological response similar to those reported at this conference by Danielsson *et al.* A serum sample collected five weeks post-septicemia produced no precipitin band by agar gel diffusion, but gave a 1:80 titer in the agglutination test. A second sample 13 weeks post-septicemia produced one precipitin band and a 1:40 agglutination titer.

On a subsequent admission the patient had recurrent fever and pelvic pain, and treatment with ampicillin, kanamycin and procaine penicillin G failed to resolve the infection. One week after admission a pelvic abscess was drained surgically, and 300 ml of cul-de-sac fluid yielded *P. prevotii, P. magnus* and *B. fragilis* ss. *fragilis*. Her subsequent clinical course was not complicated by evidence of infection.

Patient J.C.M., a 41-year-old male with a herniated lumbar disc, underwent posterior fusion and developed a postoperative wound infection due to *Bacteroides melaninogenicus* ss. *intermedius, Eubacterium lentum, S. aureus*, diphtheroids, *Staphylococcus epidermidis*, and α-hemolytic streptococcus, not group D. He was treated with cephalothin. Three weeks postoperatively he developed a lumbar abscess and was readmitted for treatment. The abscess was drained and yielded mixed anaerobic flora only: *P. variabilis, Peptococcus micros, P. anaerobius, Bacteroides* species (group 1)* and *B. melaninogenicus* ss. *intermedius*. The patient was treated with incision and drainage and cephalothin and had a good clinical response, but five weeks later recurrent fever and purulent drainage were noted. Repeat culture yielded *S. aureus* and diphtheroids but no anaerobes. Cephalothin therapy was continued with resolution of the infection.

These two cases illustrate persistent or recurrent infections

* VPI identfication.

with changing bacterial flora, which at some point yielded mixtures of anaerobic cocci and anaerobic gram-negative rods without aerobes. Patient A.L.G. had recurrent pelvic infection, first yielding mixtures of anaerobes and aerobes and finally anaerobes only. Patient J.C.M. had persistent postoperative wound infection, yielding first anaerobes and aerobes, then anaerobes only, and finally aerobes only.

Gram-Positive Anaerobic Cocci Mixed with Aerobic Bacteria

In these cases a single anaerobic coccus was isolated with aerobic bacteria. Since the role of certain aerobes in infection has been well documented in the past, we wanted to determine if possible what role the anaerobic cocci played in combination with aerobes.

PEPTOCOCCUS VARIABILIS MIXED WITH AEROBIC BACTERIA. Nine clinically significant infections with *P. variabilis* in combination with one or more aerobes included five postoperative infections: one followed a Thompson procedure in a patient with Milroy's disease; one infection developed in a skin flap following disarticulation of the hip for osteogenic sarcoma of the femur; one wound infection of the thigh developed after saphenous vein bypass in a patient with diabetes mellitus; and two skin grafts became infected, one on an obese patient with stasis ulcer and the other in a paraplegic patient with a decubitus ulcer. In both cases infection resulted in loss of the graft. Spontaneous infections developed in four patients: acute nasopharyngitis and a superficial skin infection in one patient each with acute myelogenous leukemia, and a skin infection and infected sebaceous cyst in one each who had no predisposing disease.

None of these infections followed bowel surgery. Most of the patients had predisposing disease other than a surgical procedure which probably contributed to the occurrence of the infections. The infections were localized and not unusually severe, and responded to the usual modes of therapy.

Table XLIV-III lists the sources of the bacterial isolates and the aerobes isolated. In the first five cases listed, aerobic bacterial growth was very sparse: the aerobes either grew in broth only or only one or two colonies were present on

TABLE XLIV-III

SOURCE OF BACTERIAL ISOLATES IN NINE CASES OF INFECTION WITH *PEPTOCOCCUS VARIABILIS* MIXED WITH AEROBIC BACTERIA

Aerobic Organisms	*Amount of Growth*	*Number of Cases*	*Source*
Staphylococcus epidermidis	Broth only; 2 colonies	2	Skin pustule; skin flap abscess
S. epidermidis Enterococcus group	2 colonies; broth only	1	Purulent drainage from thigh wound
S. epidermidis *Enterobacter aerofaciens* *Proteus* species	Broth only	1	Purulent drainage from naso-pharynx
S. epidermidis Diphtheroids Streptococcus (presumptive group D)	Light	1	Superficial skin infection of face
Escherichia coli *S. epidermidis* Enterococcus group	Heavy	1	Fetid pus from thigh wound
Pseudomonas aeruginosa	Heavy	1	Purulent drainage from skin graft
P. aeruginosa *Staphylococcus aureus*	Heavy	1	Infected skin graft left leg
Proteus mirabilis *S. epidermidis*	Heavy	1	Thick purulent material from breast abscess

the original isolation plate. *P. variabilis* growth varied from light to heavy, and this was the predominant organism. Thus, it would appear that *P. variabilis* was the etiologic agent of disease in these cases. In the last four cases listed, however, heavy growth of well-known aerobic bacterial pathogens occurred and in these cases we can only say that *P. variabilis* was mixed with a heavy growth of aerobes.

PEPTOCOCCUS MAGNUS MIXED WITH AEROBIC BACTERIA. We had two cases of clinically significant infection with *P. magnus* mixed with aerobes.

Patient A.O., a 63-year-old male with rheumatoid arthritis who was being maintained on steroids, developed a secondarily infected rheumatoid nodule of the fifth toe. Pus from the superficial skin infection yielded *P. magnus*, *S. epidermidis* and α-hemolytic streptococcus, not group D. The infection responded readily to local treatment with warm compresses and Neosporin ointment.

Patient L.B.L. was a 64-year-old female with diabetes mellitus and occlusive vascular disease who had chronic infection of the right foot. She presented with chills and fever, cellulitis of the foot and abscess of the right great toe. The bacterial flora of the infected toe changed over a six-month period from *Fusobacterium* species, β-hemolytic streptococcus and *S. aureus* to mixed multiple anaerobes and aerobes, and finally to *P. magnus*, *Enterobacter cloacae*, *Proteus vulgaris* and *Klebsiella* species. In spite of incision and drainage and nafcillin and chloramphenicol therapy, the lesion progressed to osteomyelitis of the foot, necessitating a below-knee amputation. The postoperative course was not complicated.

PEPTOCOCCUS ASACCHAROLYTICUS MIXED WITH AEROBIC BACTERIA. Two clinically significant infections yielding *P. asaccharolyticus* mixed with aerobes are summarized below.

Patient S.P.H., a 73-year-old female with chronic venous hypertension and stasis dermatitis, developed a surgical wound infection four days after venous stripping. Pus from the ankle wound yielded *P. asaccharolyticus* and *S. aureus*. The infection slowly responded to treatment with nafcillin, cephalothin and debridement.

Patient S.W. was a 77-year-old female with severe generalized atherosclerosis and hypothyroidism who presented with mesenteric vascular insufficiency. Eight days after exploratory laparotomy she suffered a deep abdominal surgical wound infection which extended to the fascial plane; she had a leukocytosis of 13,500 cells/mm^3 and a temperature of 100.6° F. *P. asaccharolyticus*, *S. epidermidis* and *E. cloacae* were cultured from abdominal wound pus. Treatment included debridement (X3), topical gentamicin, and cephalothin therapy. The patient responded over a three-week period.

PEPTOCOCCUS PREVOTII MIXED WITH AEROBIC BACTERIA. In the cases summarized below, infections with *P. prevotii* mixed with aerobes occurred in devitalized tissue; all were localized infections responding to the usual surgical and antibiotic modes of therapy.

Patient M.H. was a 14-year-old male who suffered multiple injuries in an automobile accident. Purulent drainage from subcutaneous tissue from a suprapubic laceration yielded *P. prevotii* plus multiple aerobes: *Serratia marcescens*, *E. cloacae*, *S. epidermidis*, *Citrobacter freundii* and streptococcus (presumptive group D). Although the patient was placed on cephalothin therapy, he died of shock and renal failure before clinical improvement of the infection was noted.

Patient M.S.B., a 33-year-old female with stage 2A carcinoma of the cervix, developed a vesicovaginal fistula subsequent to irradiation therapy. She had a chronic urinary tract infection. Pelvic exenteration with an ileo-conduit was performed. Three cultures obtained at surgery contained predominantly anaerobic cocci of various species. One of these, urine from the left ureter, yielded *P. prevotii* and (from broth only) *Proteus mirabilis*. The right ureter yielded *P. asaccharolyticus*, *P. anaerobius*, *B. melaninogenicus* ss. *asaccharolyticus* and *P. mirabilis*. The vaginal culture yielded *P. anaerobius* and one colony of *P. mirabilis*. The patient was treated with cephalothin, ampicillin and kanamycin, with resolution of the infection.

Patient J.P., a 24-year-old female developed cellulitis and abscess subsequent to traumatic injury of the right great toe. Purulent material obtained at the time of incision and drainage yielded *P. prevotii*, *S. epidermidis*, *a*-hemolytic streptococcus (not group D) and diphtheroids. The lesion responded to incision and drainage and penicillin G (given IM).

PEPTOSTREPTOCOCCUS INTERMEDIUS (MICROAEROPHILIC) MIXED WITH AN AEROBE. The only case yielding a *Peptostreptococcus* species mixed with aerobic bacteria was the following:

Patient J.W.W., an 85-year-old female, had a diagnosis of biliary abscess with underlying diseases of chronic cholecystitis, cholelithiasis and choledocholithiasis. Culture of the biliary abscess at surgery yielded predominantly a microaerophilic strain of *P. intermedius* plus a few colonies of α-hemolytic streptococcus, not group D. Following cholecystectomy the patient was treated with ampicillin and cephalothin.

Cases of Infection with Gram-Negative Anaerobic Cocci

VEILLONELLA SPECIES. We had only one pure culture of *Veillonella* (Lambe and Del Bene, 1972), and that was from a 60-year-old male with a liver abscess found at laparotomy for resection of adenocarcinoma of the colon. Culture of the necrotic debris at surgery yielded *Veillonella parvula*. The patient was afebrile three days later and had no subsequent signs of infection.

Twenty-nine additional strains of *V. parvula* were isolated from other patients, but each isolate was mixed with aerobes or anaerobes plus aerobes.

We have isolated 24 strains of *V. alcalescens* from various types of clinical material, but all were mixed with other anaerobes as well as multiple aerobes.

REFERENCES

Altemeier, W.A., and Culbertson, W.R.: Acute nonclostridial crepitant cellulitis. *Surg Gynecol Obstet, 87:*206, 1948.

Bornstein, D.L., Weinberg, A.N., and Swartz, M.N.: Anaerobic infections: review of current experience. *Medicine (Baltimore), 43:*207, 1964.

Brewer, G.E., and Meleney, F.L.: Progressive gangrenous infection of the skin and subcutaneous tissues, following operation for acute perforative appendicitis: a study in symbiosis. *Ann Surg, 84:* 438, 1926.

Finegold, S.M., Miller, A.B., and Sutter, V.L.: Anaerobic cocci in human infection. In *Bacteriologic Proceedings—1968. Abstr #* M166, Ann Arbor, Am Soc Microbiol, 1968, P. 94.

Heineman, H.S., and Braude, A.L.: Anaerobic infection of the brain: observations on eighteen consecutive cases of brain abscess. *Am J Med, 35:*682, 1963.

Holdeman, L.V., and Moore, W.E.C. (eds.): *Anaerobe Laboratory Manual.* Blacksburg, Virginia Polytechnic Institute and State Univ., 1972.

Lambe, D.W., and Del Bene, V.E.: The incidence and clinical significance of anaerobic cocci in certain infections. Oral presentation at the 72nd Ann. Mtg. American Soc. for Microbiology. Philadelphia, April 23–28, 1972.

MacLennon, J.D.: Streptococcal infection of muscle. *Lancet, 1:* 582, 1943.

Meleney, F.L.: Bacterial synergism in disease processes: with a confirmation of the synergistic bacterial etiology of a certain type of progressive gangrene of the adominal wall. *Ann Surg, 94:*961, 1931.

Moore, W.E., Cato, E.P., and Holdeman, L.V.: Anerobic bacteria of the gastrointestinal flora and their occurrence in clinical infections. *J Infect Dis, 119:*641, 1969.

Patterson, D.K., Ozeran, R.S., and Glantz, G.J.: Pyogenic liver abscess due to microaerophilic streptococci. *Ann Surg, 165:*362, 1967.

Pien, F.D., Thompson, R.L., and Martin, W.J.: Clinical and bacteriologic studies of anaerobic gram-positive cocci. *Mayo Clin Proc, 47:*251, 1954.

Prevot, A.R.: *Manual for the Classification and Determination of the Anaerobic Bacteria.* Philadelphia, Lea and Febiger, 1966.

Sandusky, W.R., Pulaski, E.J., and Johnson, B.A.: The anaerobic nonhemolytic streptococci in surgical infections on a general surgical service. *Surg Gynecol Obstet, 75:*145, 1942.

Smith, L.DS., and Holdeman, L.V.: *The Pathogenic Anaerobic Bacteria.* Springfield, Thomas, 1968.

Stokes, E.J.: Anaerobes in routine diagnostic cultures. *Lancet, 1:*668, 1958.

Thomas, C.G.A., and Hare, R.: The classification of anaerobic cocci and their isolation in normal human beings and pathological processes. *J Clin Pathol, 7:*300, 1954.

Vogler, W.R., Dorney, E.R., and Bridges, H.A.: Bacterial endocarditis: a review of 148 cases. *Am J Med, 32:*910, 1962.

CHAPTER XLV

INVITED DISCUSSION: CLINICAL ANAEROBIC INFECTIONS

H. BEERENS

FIRST IN THIS DISCUSSION I would like to comment on the clinical results of treatment, and secondly, I shall discuss some ideas concerning future work on anaerobes.

TREATMENT

In regard to Dr. Brummelkamp's lecture on gas gangrene, often one has no time to apply treatment after the diagnosis is made before the patient dies. In two cases of gas gangrene due to *Clostridium perfringens* this year in Lille, France, death occurred less than ten hours after the first symptoms appeared. No treatment could be applied quickly enough. In any case of gas gangrene, remember that it is important to treat very early.

As to Dr. Chow's results in treating bacteremias, we agree that clindamycin may be the best treatment for *Bacteroides fragilis* infection. Yesterday Dr. Sonnenwirth told us that the proportion of blood cultures positive for *B. fragilis* has grown from 3 to 7 percent. In our personal experience the increase is about the same, to approximately 10 percent. In contrast, the number of blood cultures due to *Fusobacterium necrophorum* or *F. funduliformis* has decreased. We are very far from the 30 cases observed in two years by Lemierre in 1936. Why such a difference?

B. fragilis bacteremias are probably increasing because this species is becoming more resistant to antibiotics. Our experience shows that 70 percent of *B. fragilis* strains isolated from blood cultures are now resistant to tetracycline. If we consider that this antibiotic is the one most commonly administered as a preventive treatment, we may conclude that the selection of resistant strains is an obvious consequence. Conversely, the decrease in *F. necrophorum* blood stream infections is due to the fact that this organism is still sensitive to antibiotics.

Concerning infections with anaerobic cocci, I mention the efficacy of preventive treatments, especially in puerperal infections. In 1910, Schottmuller described puerperal fever due to *Streptococcus putridus* and presented 600 cases. Neither this infection nor this important species has been mentioned during this symposium. Here again, it seems that there is a modification of the flora concordant with preventive treatment with antibiotics.

I wish to mention an old treatment of certain gram-negative anaerobic infections which may be used when modern treatments have no effect: chrysotherapy—treatment with gold thiosulfate. We published ten years ago a case of septicopyemia due to a *Fusiformis fusiformis* (*Fusobacterium nucleatum*) strain which was resistant to all antibiotics. (Clindamycin did not exist at that time.) Recovery was obtained when gold thiosulfate was given at a dose of 0.1 g per day for 30 days.

Treatment of cervicofacial actinomycosis has been a problem, in my opinion. Although all of the species concerned are very sensitive to penicillin, treatment with this antibiotic is not quite satisfactory, because of the type of lesion. Antibiotics do not appear in a sufficient concentration in the "wooden lesion" characteristic of cervicofacial actinomycosis.

FUTURE WORK

Dr. Gibbons said that future work will involve *Bacteroides melaninogenicus*. The same approach may be applied to other anaerobic genera, e.g. *Dialister*, *Vibrio* and *Spirella*, as well as some spirochetes.

Dialister species are widespread as normal indigenous organisms. The species *Dialister pneumosintes* (*Bacteroides pneumosintes*) has been mentioned only once these three days (in Dr. Bartlett's lecture). This species is very difficult to grow and purify because it is the smallest anaerobic bacterium. It gives very small transparent colonies which are very difficult to observe. Often this species is included in a symbiotic system, which is common in anaerobiosis. Nobody mentioned the importance of symbiosis among anaerobes.

The genus *Vibrio* is also widely distributed in the mouth and intestine. Cultures are difficult to obtain and to maintain. We personally isolate pure cultures many times but lose them rapidly.

The genus *Diplococcus*—suppressed in the new edition of Bergey's Manual—might be studied. Such species as *Diplococcus paleopneumoniae (Peptostreptococcus paleopneumoniae)* and *D. morbillorum (Peptococcus morbillorum)* are widely distributed as endogenous flora but their role in infections is unknown.

For all of these difficult problems, we hope that new culture technics will be available.

Two interesting topics which have not been mentioned are (1) corynebacteria (propionibacteria) and especially *Corynebacterium parvum* as an adjuvant to immunity (Prevot, 1968; Stiffel *et al.*, 1971) and (2) the role of these organisms in Whipple's disease.

In another field, although the problem of *C. perfringens* enterotoxin was discussed by Dr. Hauschild, no one spoke about a particular technic used to look for *C. perfringens* spores in food or in feces. This technic consists of heating a suspension of the sample and inoculating sulfite medium with ferrous sulfate to obtain black colonies after 18 to 20 hours of incubation at 44°C. Dr. Sebald has shown that many spores of this species become lysozyme dependent after heating. We confirmed her work, and we now use a medium containing lysozyme, so the number of colonies keeps increasing. This lysozyme dependence can also be obtained in the presence of thioglycollate. Can lysozyme dependence of spores be demonstrated for other species? We know that Dr. Sebald has just published the same results with *C. botulinum* type E spores (Sebald and Ionesco, 1972).

Finally, in regard to *B. fragilis*, each of the subspecies of this important species possesses a major specific antigen: E for ss. *fragilis*, A for ss. *thetaiotaomicron*, D for ss. *distasonis*, B for ss. *ovatus*, and C for ss. *vulgatus*. These and other common antigens permit serological typing. In the near future we hope to determine a common antigen for all of the pathogenic strains. At present, 93 percent of the strains isolated in pure culture as a single species from pathological material possess E antigen, 80 percent as a major antigen and 13 percent as a minor antigen (Shinjo *et al.*, 1971).

All of the comments in discussions show that our knowledge concerning the anaerobic bacteria are far from satisfactory and that much more work is needed.

REFERENCES

Lemierre, A.: On certain septicaemias due to anaerobic organisms. *Lancet, I:*701, 1936.

Prevot, A.R.: Corynébactériose anaérobie. *Laval Med, 39:*308, 1968.

Schottmuller, H.: Die puerperale Sepsis. *Munch Med Wochenschr, 75:*1580, 1634, 1928.

Sebald, M., and Ionesco, H.: Germination lzP-dépendente des spores de *Clostridium botulinum* type E. *C R Acad Aci* [*D*] (*Paris*), *275:*2175, 1972.

Shinjo, T., Beerens, H., Wattre, P., and Romond, C.: Classification sérologique de 131 souches de *Bacteroides* du groupe *fragilis* (*Eggerthella*). *Ann Inst Pasteur* (*Lille*), *22:*85, 1971.

Stiffel, C., Mouton, D., and Biozzi, G.: Rôle des macrophages dans l'immunité non spécifique. *Ann Inst Pasteur* (*Paris*), *120:*412, 1971.

SUMMARY OF THE CONFERENCE

Louis DS. Smith

A CONFERENCE OF THIS TYPE is particularly worthwhile at two times in any area of research. One is when research has slowed down to a steady state, at which time it is useful for the various people engaged in it to get together, compare opinions, and see what new concepts have evolved that can be used as basis for further research. The other time when it is useful to have a conference such as this is when things are just starting and when the amount of active research is increasing every year. Then it is especially valuable for the investigators to get together to help one another with advice and encouragement and to stimulate one another with criticism and argument.

In summarizing some of what has been said here, Dr. Hook, who will take care of the clinical side, and I will perhaps overlap somewhat in our comments. If we do, please forgive us, and also please forgive us for any omissions. It is not easy to distinguish between the clinical and the purely microbiological in a conference such as this, nor do I think we should attempt to do so.

With regard to the microbiological aspect, this conference started out with a few comments about the care and treatment of specimens. There was general agreement that proper collection and transport are important and that specimens should be protected against oxygen and against drying on the way between the patient and the isolation media. Whether this is done by transporting them in syringes, in small anaerobic containers or in gassed out tubes is a matter of individual preference. There was difference of opinion, however, on the value of transport media and this disagreement I think must remain. Not much evidence was given either way, some people considering transport media potentially damaging, others thinking they are very convenient. After a specimen gets to the laboratory, the next step, and a most valuable one, is a direct gram stain from the specimen. This provides a quality control check on isolation procedures and is about the only one that can conveniently be used.

On primary isolation of anaerobes from clinical material we again heard disagreement among the various workers. We have the enthusiasts for roll tubes, for anaerobic glove boxes, and for anaerobe jars. Deciding among these is not easy. Decision is complicated by the need for speed to get a report to the waiting clinician as quickly as possible and is also complicated by the fact that the longer the primary plates or tubes are incubated, the more isolates will be obtained. Consequently, the answer is a compromise, but what that compromise should be is not certain. It was pointed out that about 3 times as many isolates will be obtained using the best techniques. In the discussion of isolation nothing was said about the preparation of media for use in the anaerobe jars. Some workers feel that all significant anaerobes can be cultivated in anaerobe jars and this probably is true if exceeding care is paid to every detail of procedure and preparation of media. The only comment that can be made on the isolation methods is that here, above all, we must mistrust our methods and must continually attempt to better them. The need for relatively good methods was illustrated by several speakers who pointed out that strict methods enabled the easier isolation of some of the species or subspecies particularly associated with clinical conditions.

In blood cultures we find the same organisms that we find most commonly in stool specimens: bacteroides, fusobacteria, and anaerobic cocci. However, we also heard that not all species that are part of the predominant intestinal flora are found in infections, although most of the anaerobes which are found in infections are part of the normal flora. Two discussants pointed out that positive blood cultures tend to run high in young women and in old men, but no one chose to elaborate further on this correlation.

On the identification of the anaerobes there was no real disagreement. Identification to species can be carried out in about 80 percent of the isolates from clinical material and about half of these fall into only 8 species. It might be more convenient if we had new species set up for *B. fragilis* ss. *fragilis* and *B. melaninogenicus* ss. *asaccharolyticus* but this remains in the hands of the taxonomists. One suggestion on which I heard no comment whatsoever, but which I thought was interesting, and worthy of consideration, was that there was a need for Regional Anaerobic Reference Laboratories to aid in identification.

We had considerable information on the normal microbial flora at two sites. One was the intestine, where there is considerable variation with the diet. There was an interesting comparison of the incidence of fusobacteria in stool specimens in persons on a Japanese diet, in which the fusobacteria tend to run high, and those on a western diet, in which the fusobacteria are fewer in number. Moreover, on a western diet a large gram-negative pleomorphic bacteroid is present that is not found in stool specimens of people on a Japanese diet.

In the intestinal disorders due to the normal flora, we have the blind loop syndrome in which the microbial flora of the large intestine grows in the small intestine. In the small intestine two changes are brought about, one being the production of volatile fatty acids (acetic, propionic, isobutyric, butyric, isovaleric, valeric, etc.) and more significantly, the competitive uptake of vitamin B_{12} by the bacteria. This prevents absorption of B_{12} by the intestinal mucosa. This uptake does not depend on vitamin B_{12} being combined with the intrinsic factor.

Cancer of the colon and its relation to the normal flora were also discussed. Here again there is an association with diet. Apparently the higher concentration of *Bacteroides fragilis,* of whatever subspecies, in the stools of persons on diets high in proteins and high in fats seems to be involved. Not only is the incidence of cancer of the colon increased in persons on high fat diets, but the incidence of cancer of the breast is also increased. The high incidence of cancer of the colon is possibly associated with the presence of degraded steroids and that of cancer of the breast with the possible production of estrogen. There is a possible production of carcinogenic compounds from the breakdown of protein. Nitrosamine has been much in the news, but as was pointed out, it is doubtful if nitrosamine is formed in the intestinal tract, though it might be formed in other portions of the body. The carcinogenic compound resulting from the breakdown of tryptophane by the normal intestinal flora does seem another likely possibility.

Clostridium perfringens food poisoning, another intestinal disorder, is caused by one of the normal flora although it is not endogenous. *C. perfringens* is, of course, carried in the intestine of men and all animals. Perfringens food poisoning has been definitely known since about 1945. Until recently, its cause has not been very well understood, although we knew that an infective dose of about 10^9 living cells was required. We now know that the abdominal cramping, the diarrhea, and the nausea without vomiting or fever is caused by an enterotoxin that is produced by *Clostridium perfringens* when it forms spores. Anyone who has worked with *C. perfringens* in the laboratory knows that this organism rarely forms spores, at least not on the usual laboratory media. However, it has been known since about 1915 that it does form spores readily in the intestine. We heard at this meeting that the enterotoxin formed at the time the ingested organisms sporulate in the intestinal tract was responsible for the symptoms associated with *C. perfringens* food poisoning.

At this conference also was presented one of the first reports of immune response to some of the nonsporing anaerobes. A certain amount of strain specificity in bacteroides and fusobacteria appeared to be present. This immune response was demonstrated by hemagglutination and by indirect immunofluorescence methods.

In experimental animals the bacteria of the normal flora are necessary for the development of normal intestines or at least of the normal cecum. In germ-free animals the cecum grows to an exaggerated size. It does not do so when bacteria are present, apparently because the enzymes that cause the intestine to increase in size in the germ-free animal are broken down by the bacteria of the normal gut. Nutrition of the host may also be involved here because germ-free animals must have a better diet to survive than animals with a normal intestinal flora. The importance of the endotoxins produced by fusobacteria and bacteroides in cases of infection by these organisms is not known. They may possibly contribute to lysosomal breakdown.

The flora of the mouth is considerably different from that of the intestine. The mouth seems to have several microbial habitats: the cheek, tongue, the surface of the teeth, and the gingival sulcus. The distribution of the bacteria in these areas of the mouth very largely is affected by one property—the adherence of bacteria to surfaces. The specificity of the bacterial flora in different parts of the mouth probably depends on the adherence of the bacterial cell wall to the teeth or to the cell membrane of the epithelial cells.

The last word on actinomyces was also given to us. *Actinomyces israelii* has two serotypes which can be neatly demonstrated with the fluorescent antibody technique. *A. israelii*, of course, causes a number of infections ranging from gum boils to disseminated lesions. *Arachnia propionica* is very much like *A. israelii* and again it has two serotypes, highly specific serologically. It can readily be told from *A. israelii* if one happens to have a gas chromatograph at hand because unlike *A. israelii* it produces propionic acid in almost all of the media in which it grows. Fluorescent antibody again can serve the same purpose. *Actinomyces naeslundii* differs from the other two in that it is facultative but needs a fair concentration of carbon dioxide to grow. This is a relatively fast growing organism, much faster than *Actinomyces israelii*, and also has two serotypes. *A. naeslundii* seems to be a pathogen for humans; this organism has been reported in man from the gall bladder to the brain.

The clostridia were covered thoroughly in the clinical section. One was to be surprised at the amazing versatility of an organism such as *Clostridium perfringens* that can cause not only food poisoning but also wound contamination, cellulitis, meningitis, septicemia, and gas gangrene. Fortunately for the microbiologists it is easily grown and readily recognized. Tetanus, as was pointed out, is quite another situation. *C. tetani* grows relatively slowly, does not invade any tissue and is often a problem for the microbiologist because by the time the patient shows tetanus the organism may be dead. The following series of incidents is involved. The organism has penetrated beneath the skin, sometimes with a foreign body, sometimes with a small amount of clothing. It grows, dies, lyses, and when it lyses, tetanus toxin is released. The toxin is absorbed, carried up the nerves to the central nervous system, and the patient starts showing symptoms of tetanus. By this time almost all the organisms may have died and lysed and recovery from the site of infection may be impossible. *Clostridium septicum* was mentioned several times in association with malignancies. The reason for this is not known and has been only suspected for a few years. The first report of the association of *C. septicum* with malignancies in humans came, I think, about 2 years ago from the Center for Disease Control.

The infection of wounds by *Clostridium botulinum* is also mentioned and described. There are 12 known cases of botulism in man from wound infection. All these cases were caused by type A and all occurred in the United States and most largely in the western part of the United States. So wound botulism probably will continue to be an American specialty occuring rarely, if ever, in other countries. In each of these cases of wound botulism the disease was definitely diagnosed only when the clinician suspected he had a case of botulism and alerted the laboratory to look specifically for *Clostridium botulinum* or its toxin.

The roundtable on antibiotic susceptibility was most heartening and stimulating. As the situation now stands, we have no one standard method for the determination of the antibiotic susceptibility of anaerobic bacteria, but with the amount of wholehearted cooperation and free give and take that was shown this morning, I am sure that problem will be solved within the next few years.

In conclusion I would like to speak, so far as I may, for my fellow microbiologists and thank the Center for Disease Control, Emory University and the Upjohn Company for this very stimulating conference.

SUMMARY OF THE CONFERENCE

Edward W. Hook

During the past 3 days we have spent 22 hours talking about anaerobes. To summarize the conference in a few minutes is an impossible task. Certainly all of us have a better understanding of anaerobes than before we came here. In fact, for me this has been one of the most stimulating and rewarding meetings that I have attended in some time.

After this meeting, no one could doubt even for a moment that there has been a resurgence and reawakening of interest in anaerobic bacteria. Whatever the reasons accounting for this, it is clear that we now have the desire and the capability to look carefully at the role of anaerobes in health and disease.

It is amazing to me that organisms that form the predominant flora from mouth to anus, that cover the skin and important organs such as the female genitalia, and outnumber aerobes in feces 100- to 200-fold could have received so little attention in the past.

Despite the steep upsurge in knowledge about anaerobes in recent years, we must have only scratched the surface in understanding the implications for man of this giant component of his ecologic system—irrespective of whether man is functioning optimally with health or suboptimally with disease. Dr. Tabaqchali presented a single example in showing that bacterial overgrowth consequent to intestinal stasis could contribute to B12 deficiency by competitive uptake by bacteria of bound B12. Other mechanisms of this type are recognized, and there must be dozens, perhaps hundreds, of other unrecognized biologic events related to intestinal flora that influence man's nutritional state, susceptibility to infection, reactivity to exogenous chemicals, and so forth. This is an exciting area for future studies.

Persons interested in the pathogenesis of infection have long been fascinated by the capacity of certain bacteria to localize and grow at specific sites or in specific organs. Our understanding of these phenomena is not at all complete but is being augmented by studies such as those described by Dr. Gibbons. The implications of his exciting observations are that there are receptors for specific bacteria on the surfaces of certain epithelial cells and that adherence of bacteria at these sites, an event that can be modified by specific antibody, determines localization and growth of bacteria in certain body niches or on certain mucous membrane surfaces. I am sure that Dr. Gibbons will find in the months to come that he has stimulated a number of other investigators to join him in this area of research. Perhaps we can look forward to major extensions of these important observations in the future.

There have been many comments regarding the appropriate posture of the clinical laboratory in the diagnosis of anaerobic infections. Not being so familiar with the techniques of anaerobic bacteriology I find that I am confused! I have gotten the message that thioglycollate is not the answer, but I don't know whether to recommend that our medical-center type laboratory or a near-by community hospital laboratory use a streak tube, roll tube, glove box, GasPak, prereduced media or whatever other refinements that are available. As far as I can determine, and it seems that Dr. Smith and I agree on this point, the best isolation techniques for the clinical laboratory have not yet been defined. In most situations of this type, we need more data. I wonder if this is the case in this instance! At any rate there seems to be a need for a meeting of the minds of our experts so as to develop a series of guidelines on the collection and processing of specimens for anaerobic culture in the clinical microbiology laboratory.

I have somewhat the same reaction regarding the detailed identification of anaerobic isolates. Specific detailed identification is always desirable but unless it has implications for the management of the patient, there is no need to have this capability in every clinical laboratory. Witness the situation that exists at the present time with salmonella or *Escherichia coli* serotyping or even in the identification of group A streptococci. It seems to me that at this moment in time the function of detailed identification of anaerobes could be a function for reference laboratories. I certainly would rather see emphasis placed on implementation of modern efficient techniques for isolation of anaerobes rather than on techniques for detailed identification. I sense also from Dr. Hofstad's comments that other means of identification, specifically immunologic means, may be forthcoming in the future.

Many fascinating bacteriologic and clinical aspects of anaerobic infections have been mentioned at this meeting—many which require additional study and explanation! Many people have mentioned the high frequency of mixed cultures—usually mixtures of two or more anaerobes but also mixtures of anaerobes and facultatives. Dr. Weinberg mentioned an incidence of mixed cultures of 50 to 60 percent. Dr. Goodman said in relation to his bacteremic patients that "most cultures were mixed", and Dr. Rotheram mentioned a 50 percent incidence. The explanation for this phenomenon requires study. Dr. Gibbons pointed out that complex mixtures are not necessarily infectious, emphasizing in his experimental model that *B. melaninogenicus* seemed to play a key role in the disease-producing capacity of the mixtures. What is required, as has been pointed out several times, is to define the nature of this synergistic interaction, to determine differences in pathogenicity for man of the various anaerobes, to study the factors accounting for pathogenicity, and so forth. Another fascinating aspect of these infections is the frequent occurrence of thrombophlebitis. We need to know more about this complication and extend the studies on the influence of these organisms and their products on the blood coagulation mechanism. It is interesting to note that Dr. Weinberg viewed thrombophlebitis as possibly enhancing anaerobiosis and thereby perpetuating infection, whereas Dr. Rotheram viewed the process as a mechanism for localizing infection. Obviously we don't know the answer but all of us can think of several experiments that might resolve these different concepts.

The implications for clinical medicine of the observations presented at this conference are impressive. The wide spectrum of disease caused by anaerobes has been emphasized, as well as the importance of impaired host resistance in making possible invasion by these relatively noninvasive organisms. Dr. Bartlett, in elegant studies, has demonstrated better than anyone the important role of anaerobes as etiologic agents of certain pleuropulmonary diseases, especially lung abscess, empyema, and necrotizing pneumonia. He has suggested that these organisms are second only to the pneumococcus as a cause of bacterial infections of the lung and pleura, a remarkable observation if correct.

I always have been concerned about the dominance of pathologists over our clinical microbiology laboratories. Considerable justification for this view has come from Dr. Rotheram in his account of what happened when an internist was placed in charge of the clinical microbiology laboratories of his predominantly Ob-Gyn hospital. In a few words, Rotheram's studies indicate that the textbooks and other descriptions of septic abortion must be rewritten to show what he clearly established—that is, that nonclostridial anaerobes are the major invasive pathogens in patients with septic abortion. Dr. Swenson has established an equally important role for these organisms in patients with endometritis and pelvic abscesses.

Conference participants have documented in various papers the importance of anaerobes in many other situations, such as brain abscess, peritoneal infections, liver abscesses, certain types of dental erosions, and in the manifestations of acne. It was rather refreshing to find that anaerobes did not seem to be important as etiologic agents in at least two types of infections—urinary tract infections and bacterial endocarditis. Dr. Martin's observations at the Mayo Clinic show that the incidence of anaerobes in urinary infections is so low that routine culture of urine for anaerobes is not warranted except in patients with major neurologic problems. Dr. Felner has shown that the occasional case of endocarditis caused by anaerobes occurs most often after insertion of a prosthetic heart valve or in the setting of extensive intraabdominal suppuration.

I would like to mention a few major points that were made during the discussions on therapy. One of the key points is the role of surgery—a role which must not be relegated to a secondary position. Dr. Altemeier and Professor Brummelkamp appropriately emphasized the importance of surgery in the prophylaxis and treatment of infections caused by anaerobes. The drainage of abscesses, closure of perforations, removal of devitalized tissues and foreign bodies, *etc*, are of primary therapeutic importance and may be curative in these infections. In evaluating results of antimicrobial therapy, the role of surgery must also be considered. The role of surgery was not considered in evaluating therapeutic results in all of the studies that have been presented at this meeting; this makes evaluation of some of the results difficult or impossible.

The question of choice of the most appropriate antimicrobial agent is not a simple matter, although I will try to make it as simple as possible in summarizing the attitudes of the participants at this conference. First, I would like to say that *in vitro* activity cannot be equated with *in vivo* effectiveness. Many more well designed clinical studies are required to define the relative roles of antibiotics in the management of patients with anaerobic infections.

Quite clearly clindamycin and chloramphenicol are in favor today as antimicrobials of choice for the treatment of the commonly observed anaerobic infections. Chloramphenicol has the disadvantage of serious toxicity which occurs on a very low order of magnitude; however, many patients with anaerobic infections require prolonged therapy, which is inadvisable with chloramphenicol. The relative disadvantage of clindamycin is that we have had less clinical experience with it, and no long term usage, and thus less testing for toxicity in man as we have done with chloramphenicol and many other antibiotics. Both chloramphenicol and clindamycin are quite active *in vitro;* however, it would appear that on a weight basis that clindamycin has a slight advantage over chloramphenicol *in vitro.* Everything considered, it appears to me that today the agent of choice for the initial treatment of patients suspected or proved to have clinically significant systemic infections with the common anaerobes—primarily the Bacteroidaceae—is clindamycin. Perhaps it is a bit too early to make such a commitment to clindamycin but I sense from conversations with many people here that this probably is the consensus. There are exceptions to the use of clindamycin; in patients with central nervous system infections caused by anaerobes, chloramphenicol or some other agent that crosses the blood-brain barrier is indicated because clindamycin apparently does poorly in gaining access to the brain and spinal fluid.

Regarding susceptibility testing, I agree with Dr. Chow that susceptibility testing can be helpful, not only by establishing the antibiotic susceptibility pattern of organisms in a specific geographic area, but also in the management of certain patients with anaerobic infections, for example, the patient who is not responding to antibiotics. It is important to point out that occasional isolates of clinically important anaerobes are resistant even now to chloramphenicol or clindamycin, although the number of such strains is quite low at present. It is at least historically accurate to anticipate an increase in the proportion of resistant strains in the future. I agree with Dr. Washington that we should continue to work to improve methods for susceptibility testing of clinically significant isolates; during the present developmental phase, the experts in susceptibility testing could perform a valuable service by developing a series of recommendations on methods of susceptibility testing of anaerobes for use by small clinical laboratories.

Dr. Chow, in his series of over 100 cases of bacteremia caused by Bacteroidaceae has emphasized the importance of associated or underlying diseases in determining the outcome of therapy. In fact, in his study significant reduction in mortality as a result of appropriate therapy could be documented only in patients with ultimately fatal underlying disease. These studies show once again, as previously demonstrated by Jackson, McCabe and others with aerobic gram-negative bacilli, the importance of comparing therapeutic programs only among patients with similar underlying diseases or diseases of essentially similar severity.

It is important to note that the mortality rate in severe systemic "nonobstetric" anaerobic infections continues to be quite high despite the utilization of effective antibiotics and surgery. The mortality rate in Goodman's series of cases of bacteremia was 35 percent and in Bartlett's series of pulmonary infections, where one would anticipate the process to be a little less severe, 14 percent. Dr. Brummelkamp reported a 12 percent mortality in gas gangrene despite "the best" management and Chow and Guze in their series of patients with bacteremia have described a mortality rate between 16 and 59 percent, depending on whether antimicrobial therapy was appropriate or not. Thus, the results of therapy, even with clindamycin, chloramphenicol, surgery, hyperbaric oxygen, and whatever else you would like to add, are far from satisfactory. These observations point to areas that require additional study.

From the standpoint of prevention, it may be worth noting that many of the problems that we have discussed—the alcoholic with lung abscess, the patient with septic abortion—are somehow products of our society. This view at least points to another approach to prevention if one cares to tackle it.

In closing, I would like to apologize to those contributors to the symposium whose papers I failed to include in this brief and incomplete review.

Like Dr. Smith, and for all of the people here, I would like to thank the sponsors of the symposium for a most delightful conference.

AUTHOR INDEX

C

D

E

F

G

H

I

J

K

Mc

M

N

O

P

Q

R

S

T

Y

Z

SUBJECT INDEX

B

D

E

F

G

H

I

J

K

L

N

O

P

Q

R

S

T

U

V

W

X

Y

Z